# Modern Control Systems and Engineering

# Modern Control Systems and Engineering

Edited by
Ashley Potter

Willford Press
www.willfordpress.com

Published by Willford Press,
118-35 Queens Blvd., Suite 400,
Forest Hills, NY 11375, USA



ISBN: 978-1-68285-384-9

**Cataloging-in-Publication Data**

Modern control systems and engineering / edited by Ashley Potter.
p. cm.
Includes bibliographical references and index.
ISBN 978-1-68285-384-9
1. Automatic control. 2. Control theory. 3. Systems engineering. I. Potter, Ashley.
TJ213 .M63 2017
629.8--dc23

For information on all Willford Press publications
visit our website at www.willfordpress.com

WILLFORD PRESS

Printed in the United States of America.

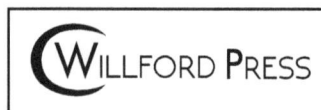

# Contents

# Preface

The fundamentals as well as modern approaches of control systems have been discussed in this book. Application of control theory to systems to control their behavior is known as control systems engineering. In this engineering discipline, input actuators collect the feedback generated by the output sensors to control behavior of the system under observation. The ever growing need of advanced technology is the reason that has fueled the research in the field of control systems in recent times. This book is ideal for the readers who wish to develop a better understanding of the modern applications of control systems. Coherent flow of topics, student-friendly language and extensive use of examples make this book an invaluable source of knowledge.

All of the data presented henceforth, was collaborated in the wake of recent advancements in the field. The aim of this book is to present the diversified developments from across the globe in a comprehensible manner. The opinions expressed in each chapter belong solely to the contributing authors. Their interpretations of the topics are the integral part of this book, which I have carefully compiled for a better understanding of the readers.

At the end, I would like to thank all those who dedicated their time and efforts for the successful completion of this book. I also wish to convey my gratitude towards my friends and family who supported me at every step.

**Editor**

# 1

# A comparative study of neuro fuzzy and recurrent neuro fuzzy model-based controllers for real-time industrial processes

B. Subathra[a]*, S. Seshadhri[b] and T.K. Radhakrishnan[c]

[a]*ICE Department, Kalasalingam University, Madurai, Tamil Nadu, India;* [b]*International Research Centre, Kalasalingam University, Madurai, Tamil Nadu, India;* [c]*National Institute of Technology, Tiruchirappalli, Tamil Nadu, India*

Nonlinearities in system dynamics and the multivariable nature of processes offer a stiff challenge in designing predictive controllers that improve process performance in industries. This investigation presents a recurrent neuro fuzzy network (RNFN) model for a nonlinear multivariable system in process industries and a methodology to design model-predictive controllers (MPCs) using the proposed model. The RNFN model combines the learning features of artificial neural networks with human cognition capabilities of fuzzy systems. Therefore, RNFN leads to a modelling framework that has the ability not only to learn the model parameters, but also makes decision on operating region of the nonlinear model depending on the input–output data. Furthermore, the recurrent structure and the introduction of a memory unit between the fuzzy inference and fuzzification layer enhance the prediction capability due to the use of past input–output data, making the model more suitable for designing predictive controllers. Next, the MPC design methodology that exploits the advantages of the RNFN model to optimize the control moves is presented. The proposed MPC uses the gradient descent algorithm to minimize the control moves as against the traditional state-space approaches that require complex computations and solvers. Therefore, implementing the proposed MPC in embedded hardware becomes easier. The proposed modelling framework and the MPC design methodology are illustrated using experiments on a laboratory-scale quadruple tank. Our experiments show that the proposed RNFN-based MPC performs better than the neuro fuzzy network-based MPC for both servo and regulatory responses.

**Keywords:** modelling; intelligent techniques; hybrid soft computing; recurrent neuro fuzzy network; MPC; gradient descent algorithm

## 1. Introduction

Competition stemming from globalization and increasing operating cost necessitates new technologies that can optimize in process performance without significant investments industries. In this backdrop, a model predictive controller (MPC) that use process models, estimation on disturbance and an optimization routine to improve the process performance has emerged as a promising solution. However, model complexity due to system nonlinearities and the multivariable nature of the process, time-varying disturbances and the need for solvers and complex optimization routines offer stiff challenges in adapting MPCs within process industries. On the other hand, optimization has become a necessity to derive technological and market leadership. Therefore, a new modelling framework that can handle model complexities such as non-linearity and the multivariable nature of the process, and controller design methodologies using such models are required in process industries. In particular, models capturing the interactions among the manipulated and controlled variables are required for multivariable systems. Our objective in this investigation is to propose a new modelling framework for interacting multivariable and nonlinear systems that inherently estimates possible future disturbances, and then uses them for designing an MPC that optimizes process performance while simultaneously reducing the fluctuations in control moves.

Modelling and designing controllers for multivariable and nonlinear systems have attracted significant research. Zhang and Morris (1999) classified process models into two categories: first principle and empirical. Furthermore, the authors point out that the first principle models are computationally cumbersome and time consuming to develop, especially for complex systems. To overcome this difficulty, process models based on input–output data have been developed.

Empirical models using neural networks have shown better accuracy and simplicity over other methods such as system identification, and have been used successfully for modelling complex processes by many researchers (Bhat & McAvoy, 1990; Bulsari, 1995; Morris, Montague, & Willis, 1994). However, neural network models

*Corresponding author. Email: clk0602@gmail.com

are difficult to interpret and lack robustness when applied to unseen data. Model accuracy can be improved by using process knowledge along with process input–output data. For instance, the process knowledge can be used to derive local models for a given set of operating points using fuzzy approach as in Yager and Filev (1994). In the fuzzy modelling approach developed by Takagi and Sugeno (1985), the model input space is portioned into several fuzzy regions. A local linear model is used within each region and a global model is obtained using centre of gravity defuzzification. The modelling approach illustrated that process knowledge can be used to significantly reduce model complexity. A similar approach has been used to construct Nonlinear Auto Regressive Moving Average with eXgenous inputs models by Johansen and Foss (1993). The fuzzy modelling approach used process knowledge, but the potential of input–output data was not completely exploited in fuzzy models. Consequently, researchers combined fuzzy and neural approaches to derive process models that exploited the learning capability of neural networks and decision-making of fuzzy systems. This led to the development of adaptive-network-based fuzzy inference system (ANFIS) architecture to represent fuzzy models (Jang, 1992; Jang & Sun, 1995; Jang, Sun, & Mizutani, 1997); two types of feed-forward neuro fuzzy networks were proposed by the authors for nonlinear process modelling (Zhang & Morris, 1995b). The use of dynamic neural network for long range prediction models has been studied in globally recurrent neural networks (RNN, e.g. Su, McAvoy, & Werbos, 1992; Werbos, 1990), Elman networks (Elman, 1990; Scott & Ray, 1993), dynamic feed-forward network with filters (Montague, Tham, Willis, & Morris, 1992; Morris et al., 1994) and locally recurrent networks (Frasconi, Gori, & Soda, 1992; Tsoi & Back, 1994; Zhang & Morris, 1995a). Zhang and Morris (1995a) presented a sequential orthogonal training strategy which allows for hidden neurons to be added gradually, avoiding an unnecessarily large network structure. However, Su et al. (1992) proved that these models are not suitable for long-term predictions. In Zhang and Morris (1999), a recurrent neuro fuzzy network (RNFN) was proposed that allowed construction of a global nonlinear multi-step ahead prediction model from the fuzzy conjunction of a number of local dynamic models for the pH neutralization process. This approach used both process data and knowledge to build multi-step ahead models. However, the role of the MPC design has not been explored, which is required for improving process performance. The RNFN model and the structure obtained in Zhang and Morris (2000) have been used for long-term prediction of outputs in level control of a conical tank. Juang and Chen (2003) proposed a six-layer, TSK-type, RNFN structure-based controller for a thermal process. Lia and Cheng (2007) proposed a six-layer RNF system, in which recurrence is introduced in the membership layer (self) and the results are illustrated for a simulated system. In the six-layer structure proposed, the network consists of two external inputs and a single output. Hence, only one output can be obtained at the end of training of the network and it is therefore not suitable for modelling multivariable processes. Review of the literature reveals that the use of RNFN for modelling real-time multivariable processes with interactions and design of predictive controllers has not been explored (Subathra & Radhakrishnan, 2011b). Motivated by this research gap, this investigation aims to model a nonlinear multivariable process with interactions using RNFN and then propose an MPC design methodology for the model.

To reach the objectives, this investigation first presents a new seven-layer RNFN structure for modelling and identification of a real-time nonlinear multivariable process, and then uses it to design the predictive controller. Recurrent structure is obtained by introducing delay units between the fuzzy inference and fuzzy membership layer. Therefore, weights in the inference layer determine the local operating regions, while the weights in the output of the membership layer represent the singleton values in the consequent part of the rules. Neural back propagation (BP) algorithm in fuzzy inference layer is used to tune the antecedent part of the rule and this can be used to control the input variable. In Juang and Chen (2003), firing strength of the fuzzy rules has been varied based on internal parameters and internal inputs and the role of varying the consequent part of the fuzzy rule has not been explored. In this study, the firing strength is varied by changing the consequent part of the fuzzy rule in addition to the conventional RBFN structure. This small modification leads to significant improvements in model accuracy. Furthermore, the long-range prediction capability of the nonlinear multivariable system is enhanced due to the recurrent structure and this can be observed from our experimental studies on a quadruple process. The model thus obtained is used to design the MPC for the nonlinear multivariable system. Multi-step ahead prediction based on the RNFN model and the set point calculations are used to predict the future control moves that optimize the process performance. To design the MPC using the RNFN model, gradient descent algorithm is used for optimizing the control moves. This eliminates the need for dedicated solvers and facilitates implementation in dedicated embedded hardware.

The main contributions of this investigation are: (i) a new seven-layer RNFN structure for modelling and identification of a nonlinear multivariable process (ii) an MPC design methodology employing the RNFN model and (iii) illustration of the model accuracy and controller performance using experiments on a laboratory-scale quadruple tank prototype. The paper is organized in seven sections. Section II presents the problem and discusses the model and working of the quadruple process. The RNFN model is presented in Section 3, and the empirical model for the quadruple process is discussed in Section 4. The MPC controller design is discussed in Section 5. Results are presented in Section 6 and conclusions are drawn from the obtained results in Section 7.

## 2. Problem description

The problem is to propose a new modelling framework and MPC design methodology for multivariable and nonlinear processes with interactions among manipulated variables. The idea of the modelling technique is to use input–output data and knowledge on process operating points to obtain a process model that can be used for designing MPC. The proposed predictive controller uses the model and computes the future control moves that optimize the performance.

### *2.1. Quadruple tank process*

The input–output data and process knowledge from a laboratory-scale quadruple process first studied by Johansson and Nunes (1998) are used. The authors have derived the mathematical model of the process and have studied the working of the process in detail. Furthermore, the study established that linearized dynamics of the system has a multivariable zero that can move along the real-axis by changing the valve position. Because of right half plane zero performance limitations are present in the process. The tank process has received significant attention recently in both modelling and control due to its dynamic characteristics (Doyle, Gatzke, Vadigepalli, & Meadows, 1999). By modifying the valve position and location of zero the dynamic characteristics of the process can be modified and exhibits non-minimum phase behaviour. The quadruple tank system can be used to demonstrate not only the conventional controllers, but also sophisticated controllers such as MPC using linearized models.

In the quadruple process, two pumps are used to transfer water from a storage tank to the process tanks. The two tanks at the upper level drain into the two tanks at the bottom level as shown in Figure 1. The liquid levels in all the four tanks are measured and bottom two tanks controlled. The piping system is such that each pump affects the liquid levels of both the tanks. A part of the flow from

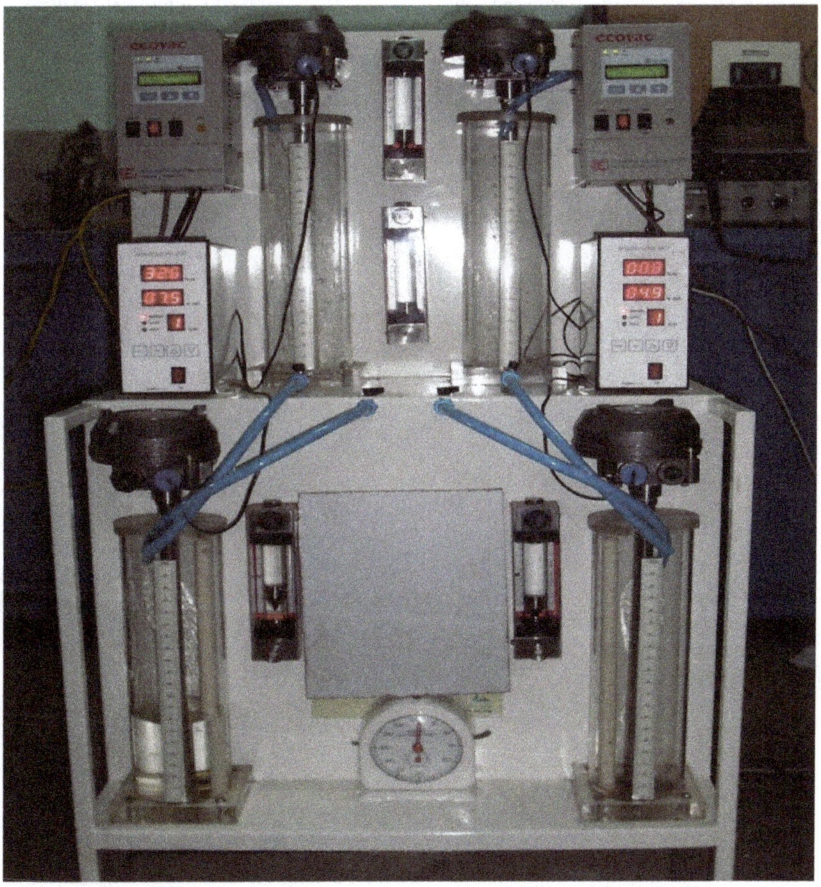

Figure 1. Photograph of experimental set-up.

Table 1. Nominal Operating conditions and Parameter values.

| Symbol | State/parameters | Value |
|---|---|---|
| $h^0$ | Nominal levels | [11.65; 9.6; 8.6; 5.6] cm |
| $v^0$ | Nominal pump settings | [60; 60] |
| $a_i$ | Area of the drain in Tank $i$ | [0.236; 0.1985; 0.159; 0.173] $cm^2$ |
| $A_i$ | Areas of the tanks | 65.755 $cm^2$ |
| $\gamma_1$ | Ratio of flow in Tank 1 to flow in Tank 4 | 0.453 |
| $\gamma_2$ | Ratio of flow in Tank 2 to flow in Tank 3 | 0.307 |
| $k_j$ | Pump proportionality constants | [5.5143; 4.9728] $cm^3$/(V-sec) |
| $G$ | Gravitation constant | 980 cm/$sec^2$ |

one pump is fed to one of the lower level tanks (where the level is monitored). A photograph of the experimental prototype is shown in Figure 1. The amount of interaction between inputs and outputs can be adjusted by varying the bypass valves of the process. External flow disturbances can be introduced into the upper level tanks. The process dynamics changes with respect to the ratio of the valve openings. The nomenclature and the nominal operating conditions used in this study are given in Table 1. The mass balance equation for the quadruple process is a nonlinear multivariable equation (Johansson & Nunes, 1998):

$$\begin{aligned}
\frac{dh_1}{dt} &= -\frac{a_1}{A_1}\sqrt{2gh_1} + \frac{a_3}{A_1}\sqrt{2gh_3} + \frac{\gamma_1 k_1}{A_1}v_1,\\
\frac{dh_2}{dt} &= -\frac{a_2}{A_2}\sqrt{2gh_2} + \frac{a_4}{A_2}\sqrt{2gh_4} + \frac{\gamma_2 k_2}{A_2}v_2,\\
\frac{dh_3}{dt} &= -\frac{a_3}{A_3}\sqrt{2gh_3} + \frac{(1-\gamma_2)k_2}{A_3}v_2,\\
\frac{dh_4}{dt} &= -\frac{a_4}{A_4}\sqrt{2gh_4} + \frac{(1-\gamma_1)k_1}{A_4}v_1.
\end{aligned} \tag{1}$$

Johansson and Nunes (1998) analysed the performance of the quadruple tank process and reported that the process yields an inverse response during $(\gamma_1 + \gamma_2) < 1$. Prior to the control analysis the manipulated inputs are scaled by 25%. This scaling corresponds to actual input levels in the range of 25–75%. Because of the nonlinearities introduced by the water head in the piping and limited pump capacities, the pump cannot operate satisfactorily below the 25% level. Hence, the pump settings are constrained to remain within 25–75% in our analysis.

## 3. Hybrid RNFN model

### 3.1. *Neuro fuzzy network*

Neuro fuzzy technique is widely used in modelling of complex processes (Brown & Harris, 1994) and the hybridization combines the unique advantages of the fuzzy and neural networks. The model can be developed in such a way as to mimic the process exactly with the help of empirical data and linguistic fuzzy rules can be used to represent the model. The advantages of hybrid modelling techniques are:

(a) Modelling and identification of a nonlinear process can be obtained by using both input–output data and knowledge about the process.
(b) The resulting models are simple.

Knowledge on process operations is used to define a suitable model structure; this model is then synthesized during the modelling process such that it can successfully represent and reproduce the available empirical data used for training. This adaptation step is often called learning. The main objective of neuro fuzzy modelling is to construct a model that accurately predicts the value(s) of the output variable(s) even when new values of the input variables are presented. The major drawback of this modelling technique is that it cannot be used for dynamic processes and yields a good one-step ahead prediction model because of its feed-forward structure (Brdys & Kulawski, 1999). While developing a long-range predictive controller, the history of input and output data is needed for dynamic processes. For this purpose, a feedback loop is introduced in this structure.

### 3.2. *Recurrent neuro fuzzy modelling*

Most of the industrial processes are complex and dynamic in nature. Therefore, modelling such processes requires the history of input–output data. The feedback structure offers a memory unit to store the past data, which enhances the long-range prediction capability of the models required for designing predictive controllers (Zhang, 2003). In literature, five- and six-layer RNFN structures have been proposed for modelling and control of the SISO process (Lia & Cheng, 2007). The role of RNFN to model a nonlinear process has been studied in Zhang and Morris (1999), where the authors proposed a five-layer Jordon network to model the pH neutralization (SISO) process. In the six-layer RNFN structure proposed in Juang Huang, and Duh (2006) to control the temperature, an Elman-type recurrent structure has been used. The role of RNFN in modelling a nonlinear and multivariable process has not been investigated.

### 3.3. *Proposed seven layer with RNFN structure*

In this investigation, a Takagi–Sugeno (TS)-type fuzzy inference system is implemented with an rRNN.

#### 3.3.1. *Basic structure of RNFN*

The proposed approach uses the Elman-type recurrent network structure in which delay units act as memory elements that are located in the feedback path between the fuzzy inference layer and the fuzzy membership layer so that consequent parts of the fuzzy rules are tuned considering the dynamics of the process. The RNFN approaches in literature modify the fuzzy rule and firing strengths are fixed using the rule layer output and external input (for, e.g. Zhang & Morris, 1999; Juang et al., 2006). This investigation varies the firing strength by changing the consequent part of the fuzzy rule or in other words the internal parameters in addition to the conventional RBFN structure. Thus, proper tuning of fuzzy rules can be obtained by modifying the firing strength. The comprehensive rule base for this proposed structure is

$$\begin{aligned} &\text{RULE } i: \text{IF } x_1(k) \text{ is } A_{i1} \ldots x_n(k) \text{ is } A_{\text{in}} \text{ AND } h_i(k) \text{ is } G \\ &\text{THEN } \hat{y}_i(k+1) = a_{i0} + a_{i1}x_1(k) + a_{i2}x_2(k) + a_{i3}x_3(k) \\ &\quad + e(k) \text{ AND } h_i(k+1) \text{ is} W_i, \end{aligned} \tag{2}$$

where the fuzzified inputs are $x_1(k), \ldots, x_n(k)$, the fuzzy sets are '$A$' and '$G$', consequent parameters are '$W$' and '$a$' for the inference output '$h$' and '$y$', respectively, number of external input variables is denoted as '$n$', error $e(k) = y_{\exp}(k) - \hat{y}(k)$, where $y_{\exp}(k)$ is the experimental output and $\hat{y}(k)$ is the predicted output. The TS-type fuzzy method is used to model the process and the linear combination of the input variables $x$ and internal variables $h$ plus a constant. Process variables are defined with fuzzy operating regions using Gaussian membership functions and are the inputs to the fuzzification (input) layer. In the feedback path, sigmoidal membership functions are used for fuzzification to the rule layer as shown in Figure 2.

The rules and operating regions of the fuzzy controller are framed based on the knowledge about the process. The fuzzy operating regions of the process are defined with neurons and they are trained with neural network training algorithms. The fuzzified values of the process variables and delayed feedback from the fuzzy inference layer are the inputs to the rule layer which determine the firing strength. Output of the fuzzy rule layer is the result of AND operation applied to inputs and its membership function. Each process variable is represented as a fuzzy set

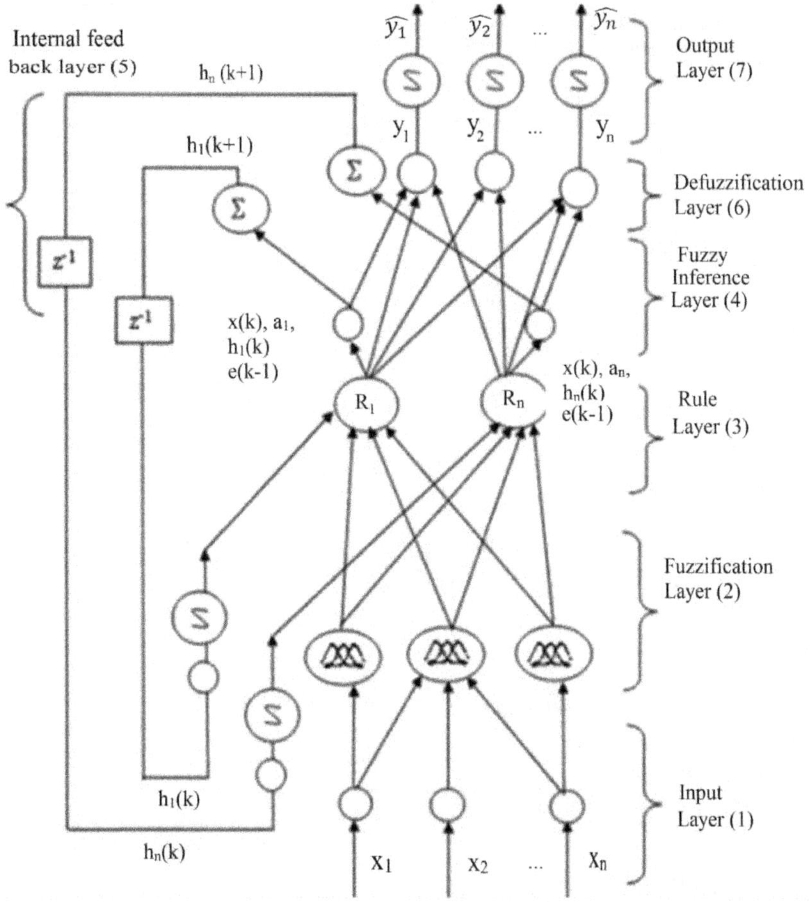


Figure 2. A new seven-layer RNFN model.

with several fuzzy operating regions (Subathra, Raja Rao, & Radhakrishnan, 2009; Subathra & Radhakrishnan, 2010, 2011a).

### *3.3.2. Description of RNFN*

(i) Input Layer (1)

In this structure, inputs to the fuzzification layer are process variables and they are used to define fuzzy operating regions. Each process variable can be defined as fuzzy sets in the fuzzification layer. Every fuzzy set can be represented by neurons; fuzzy set output gives the membership function. Two different types of neurons are employed; the nodes in the first layer represent an input variable. This layer does not perform any computation and the values are transmitted directly to the next layer.

(ii) Membership Layer (2)

A process variable can be represented in terms of the fuzzy membership function. Each node in the second layer corresponds to a linguistic variable and is represented using the fuzzy membership function. Two fuzzy membership functions are used in the proposed structure. For external input $x_j$, Gaussian membership function is used,

$$O_i^{(2)}(k) = \exp\left\{-\frac{(x_j^{(2)}(k) - m_{ij})^2}{\sigma_{ij}^2}\right\} \quad \text{and}$$

$$x_j^{(2)}(k) = O_j^{(1)}(k), \tag{3}$$

where $m_{ij}$ and $\sigma_{ij}$ are the centre and the width of the Gaussian membership function of the $i$th term of the $j$th input variable $x_j$. All the weights in membership layer are set to unity. The sigmoid membership function is used for internal feedback variables $h_i$,

$$O_i^{(2)}(k) = \frac{1}{1 + \exp\{-x_i^{(2)}(k)\}} \quad \text{and} \quad x_i^{(2)}(k) = O_i^{(5)}(k), \tag{4}$$

where $O_i^{(5)}(k) = h_i(k)$ and $O_i^{(2)}(k) = p_i(k)$.

Membership value is calculated in each node in order to specify the degree to which an input variable belongs to a linguistic label. Other types of membership functions like trapezoidal and triangular could be used.

(iii) Rule Layer (3)

The output of each node in this layer is determined by fuzzy AND operation.

$$O_i^{(3)}(k) = \prod_{j=1}^{n+r} O_j^{(2)}(k) = \prod_{R=1}^{r} \frac{1}{1 + \exp\{-O_r^{(5)}(k)\}} * \exp\left\{-\left(\sum_{j=1}^{n} \frac{(O_j^1(k) - m_{ij})^2}{\sigma_{ij}^2}\right)\right\}, \tag{5}$$

where '$n$' is the number of external inputs and $r$ is the number of rules. The link weights are all set to unity. In this layer, the product operation is utilized to determine the firing strength of each rule.

(iv) TS Fuzzy Inference Layer (4)

The nodes in this layer perform a linear summation. Linear function in each RNFN rule consequent is replaced by a constant value. The mathematical function of each node $i$ is

$$O_i^{(4)}(k) = \sum_{j=1}^{3} a_{4i-4+j}(k)x_j(k) + a_{4i}p_i(k-1) + \mathrm{err}(k), \tag{6}$$

(v) Internal Feedback Layer (5)

The outputs of inference layer are fed back to the fuzzification layer, thus acting as internal feedback. The output of the nodes of the previous layer is normalized in this layer by the following operation.

$$h_i(k) = \frac{O_i^{(4)}(k)}{\sum_{i=1}^{r} O_i^{(4)}(k)}, \tag{7}$$

The link weights represent the singleton values in the consequent part of the internal rules. The simple weighted sum is calculated in each node. As shown in Figure 1, the delayed value of $h_i(k)$ is fed back to the fuzzification layer and acts as an input variable to the precondition part of a rule. Each rule has a corresponding internal variable $p_i(k)$ and is used to decide the current rule.

(vi) Output Layer (6)

The node in this layer computes the output linguistic variable $y$ of the RNFN. The output node along with links will be proceeding like the defuzzifier. The mathematical function is

$$y(k) = O_i^{(6)} = \frac{\sum_{j=1}^{r} O_j^{(3)} O_j^{(4)}}{\sum_{j=1}^{r} O_j^{(3)}}, \tag{8}$$

(vii) Normalization Layer (7)

The output from the layer (6) has to be formatted such that it is suitable for real-world application and is called normalization. The final output of the network has to be suitable for the real-world process; this node yields the normalized output of the RNFN model.

$$\hat{y}(k) = \frac{1}{1 + \exp\{-y(k)\}}. \tag{9}$$

This new seven-layer RNFN can be used to develop a model for any nonlinear chemical process. The recurrent structure enhances the prediction capability. In this study a quadruple tank model has been developed by using the proposed RNFN structure. The NFN differs from RNFN by the internal feedback layer.

#### 3.3.3. BP learning

The network structure assumes the initial values of the parameters. The problem is to adjust these parameters in such a way that it acquires the required knowledge and produces solutions to the required classification problems. The types of classification problems of general interest are too complex to solve a priori by analytical techniques. Hence, it is necessary to develop an adaptive training algorithm that is driven by example data. If there are adequate features, number of processing elements and sufficient representative training data samples, the parameters will slowly adjust correctly through training.

They adjust in such a way as to end up with a set of network parameters that will give a satisfactory classification performance for other inputs that the network has not seen during training. This optimization can be achieved most effectively by adjusting the parameters to minimize the mean square error (MSE) of the network outputs compared with desired responses. This can be very time consuming if it is necessary to compute the MSE of all the training pairs before the parameters can be incrementally adjusted once. The main idea behind BP -of-error learning is to adjust the parameters a little each time as a new random training input–output vector pair is presented to the network. This is done repeatedly until a satisfactory convergence occurs.

- Step-by -step BP algorithm

Initially, parameters of the network are chosen randomly, and then a BP algorithm computing the parameters and necessary modifications was implemented in the network. The algorithm consists of the following few steps:

- Feed-forward computation
- BP to the output layer
- BP to the hidden layer
- Parameter updates

The algorithm is repeated until the error value is sufficiently small.

#### 3.3.4. NFN and RNFN model estimation

In the model estimation phase, the model parameters are estimated by minimizing the square of sum errors. The optimization problem is solved using parameter estimation, and new parameter values are found for the new model. The reason behind parameter learning is to tune the free parameters of the constructed network in an optimal manner. There are a number of training methods, such as the BP method (Morris et al., 1994), the conjugate gradient method (Frasconi et al., 1992), Levenberg–Marquardt optimization (Tsoi & Back, 1994) or methods based on genetic algorithms (Zhang & Morris, 1995b), available in literature. Among them, BP algorithm opts for training the neural network. BP algorithm is a common method of training networks. As the algorithm name implies the errors (and therefore the learning) propagate backwards from the output nodes to the inner nodes. BP usually allows quick convergence on satisfactory local minima for error in the kind of networks to which it is suited. To train the coefficients of the fuzzy inference layer, the BP algorithm Equation (10) is used.

$$a_i(k+1) = a_i(k) + \Delta a_i(k) \Rightarrow a_i(k) + \lambda\left(-\frac{\partial(e^2(k))}{\partial a_i(k)}\right), \quad (10)$$

where '$a$' denotes the tuning parameter, $e(k)$ is the error difference between the measured output and the target that is the desired output and learning rate is denoted by '$\lambda$'. Several values of '$\lambda$' are considered and the one resulting in the smallest error on the testing data is adopted.

In the proposed model, output of the inference layer is fed back to the consequent part of the newly generated rule via delay units to the fuzzification layer; hence all the rules have memory elements separately to memorize the firing strength of past instants. One important point to note here is that parameter learning occurs concurrently with structure learning. In neural computation, this stage is usually called training of network. The internal parameters of the recurrent neuro fuzzy structure are trained using this BP algorithm. Often, the aim of the estimation is to minimize the sum of the error squares between the model output and the desired output. The cost function 'e' for the multi output case is defined as

$$e(k) = \frac{1}{p}\sum_{k=1}^{p}(y_{sp}(k) - \hat{y}(k))^2, \quad (11)$$

where $\hat{y}(k)$ is the model output, $y_{sp}(k)$ is the desired output and '$p$' denotes the number of output nodes. The weighting vector of the NFN/RNFN model is adjusted such that the error defined in Equation (11) is less than the desired threshold value following much iteration. The BP algorithm is used to train the network parameters recurrently.

#### 3.3.5. Model validation

Validation ensures that the model meets its intended requirements in terms of the methods employed. The newly developed model must be validated. Normalized error (NE) is calculated.

$$\text{Normalized error} = \frac{1}{N}\sum_{k=1}^{N}\frac{|y_{exp}(k) - \hat{y}(k)|}{y_{exp}(k)} \times 100, \quad (12)$$

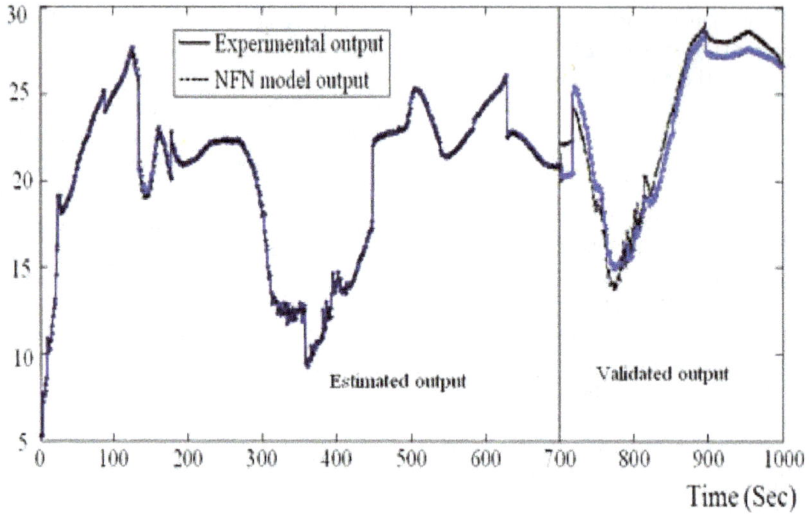


Figure 3. Observed $h_1$.

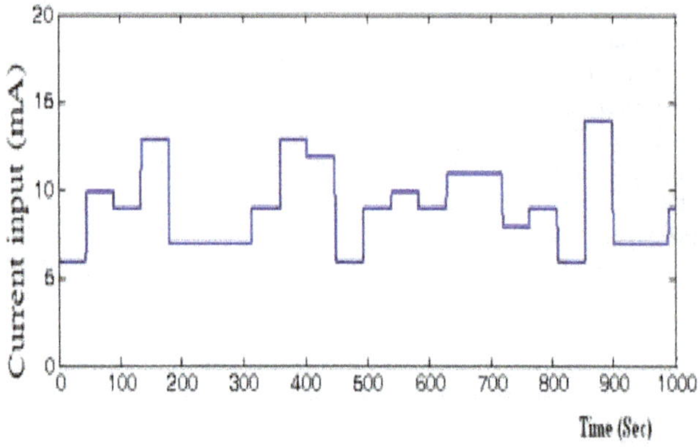


Figure 4. Excitation input $u_1$.

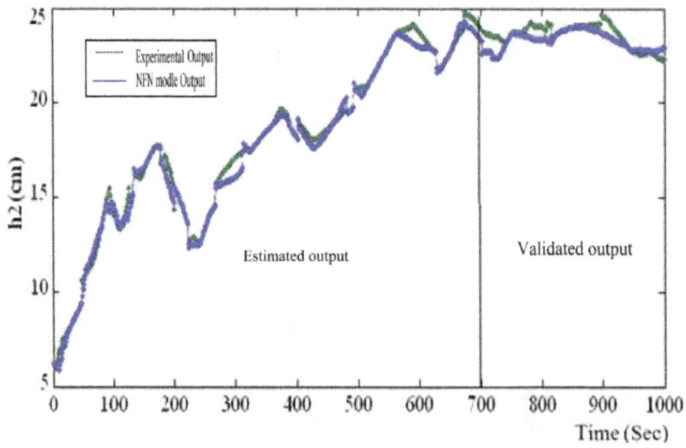


Figure 5. Observed $h_2$.

where $y_{exp}$ is the experimental (process) output, $\hat{y}$ is the NFN/RNFN model output and $N$ is the total data used for validation. The purpose of model validation is to make the model useful in the sense that the model addresses the problem exactly, yields accurate results about the system being modelled.

### *3.3.6. Multistep prediction of outputs*

Identical inputs are given to the experimental process and the outputs from the NFN and RNFN models are stored. The outputs predicted from the RNFN model and obtained from the experiments are compared. The percent prediction error (PPE) is calculated (Henson & Seborg, 1997; Subathra et al., 2009).

$$\mathrm{PPE} = \frac{\sum_{k=1}^{N} (y(k) - \hat{y}(k))^2}{\sum_{k=1}^{N} (y(k) - \bar{y})^2}, \qquad (13)$$

where $\bar{y}$ is the mean value of the sequence $y(k)$. In this study, the model used is recurrent in nature; it can be used for multistep prediction of process outputs with great accuracy.

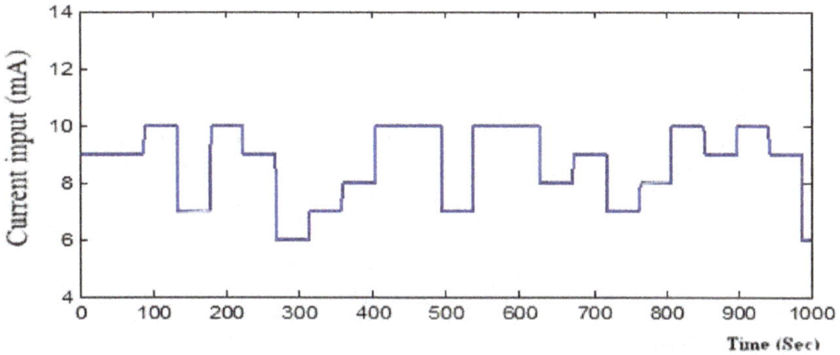


Figure 6. Excitation input $u_2$.

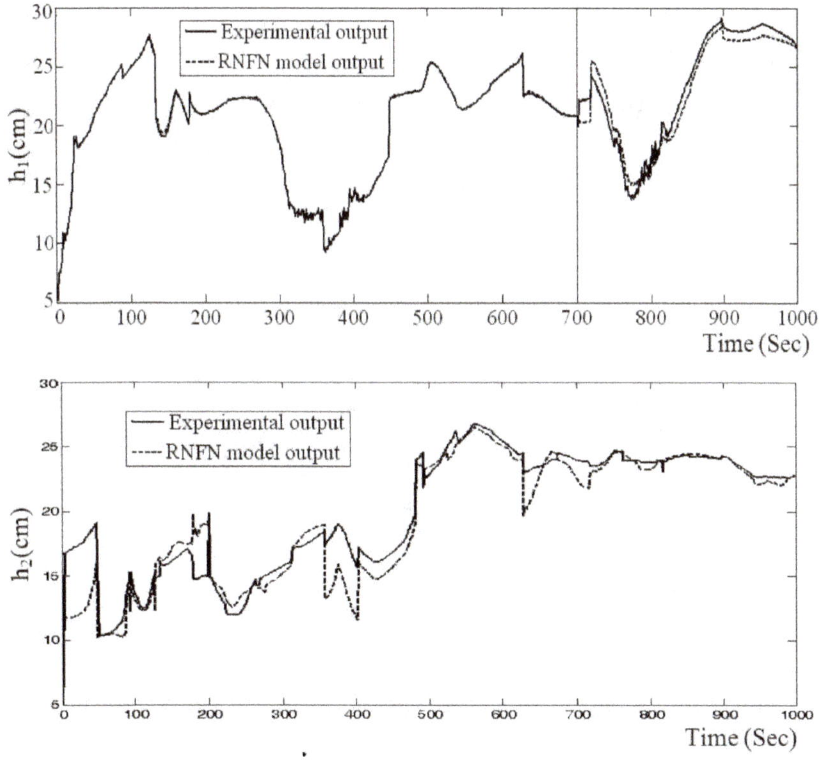


Figure 7. Comparison of outputs from experiments and the RNFN model.

Table 2. Comparison of NE values for RNFN and NFN.

| | NE | | | |
|---|---|---|---|---|
| | RNFN model | | NFN model | |
| Variable | Estimation | Validation | Estimation | Validation |
| $h_1$ | 3.7756 | 1.9867 | 4.0246 | 7.3489 |
| $h_2$ | 2.6228 | 1.6468 | 4.9823 | 6.0386 |

## 4. Modelling of a quadruple tank using a recurrent network

### *4.1. Neuro fuzzy network modelling*

Intelligent techniques also have some limitations such as poor decision-making capability of NN and lack of learning algorithms of FL. Such kind of problems can be overcome using hybrid soft computing techniques like neuro fuzzy networks (NFNs). Identification of the model is performed in a similar fashion to that performed for a two-tank heating system.

#### *4.1.1. Model estimation and validation using the NFN model*

Random excitation signals are given to both the manipulated variables (current inputs to drive 1 and drive 2) for both experimental and NFN model. The experimental and NFN model outputs are recorded. The data set is divided into training (estimation) and test (validation) sets. Among the 1000 data set, 700 are used for training and the remaining 300 are used for testing. In order to examine the fitness of the model most of the tests require a set of data that is not used in training. It is called validation set. It is necessary that the test set must satisfy the same demands as the training set regarding representation of the entire operating range. The NE is calculated using Equation (12) (Subathra et al., 2009; Subathra & Radhakrishnan, 2010). The estimation and validation results of NFN are shown in Figures 3–6 along with excitation inputs.

### *4.2. RNFN modelling*

The same sets of random inputs in the manipulated variables used in the NFN modelling are given to the RNFN model. The estimation and validation plots are shown in Figure 7. The RNFN model output is compared with experimental outputs and the NE is calculated. From Table 2, it can be concluded that the recurrent model is endowed with lesser error. Among the two models studied in this investigation, the RNFN model gives good model accuracy.

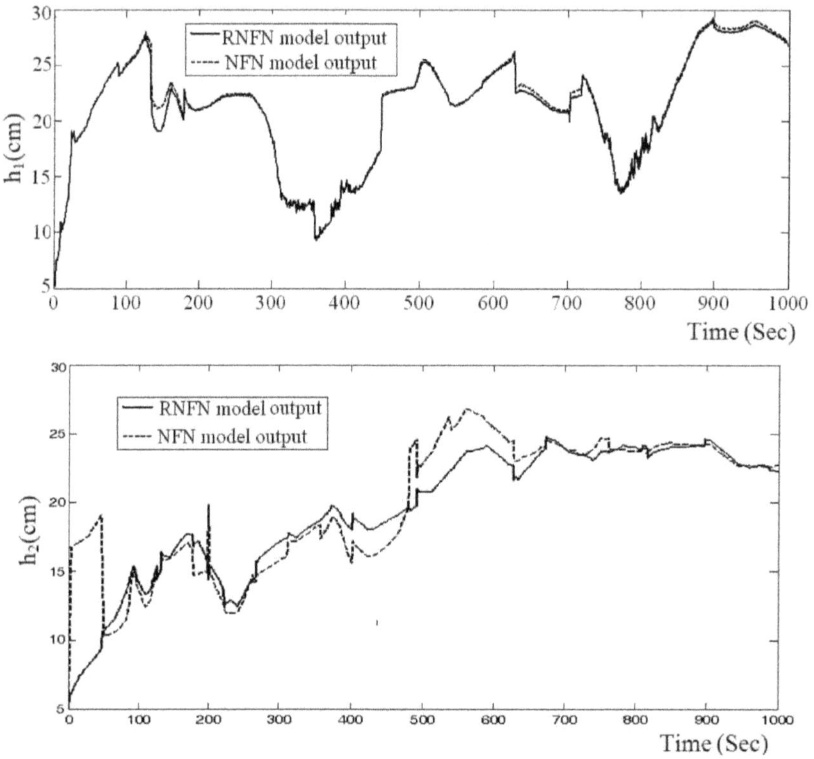


Figure 8. Comparison of one-step ahead RNFN and NFN prediction models.

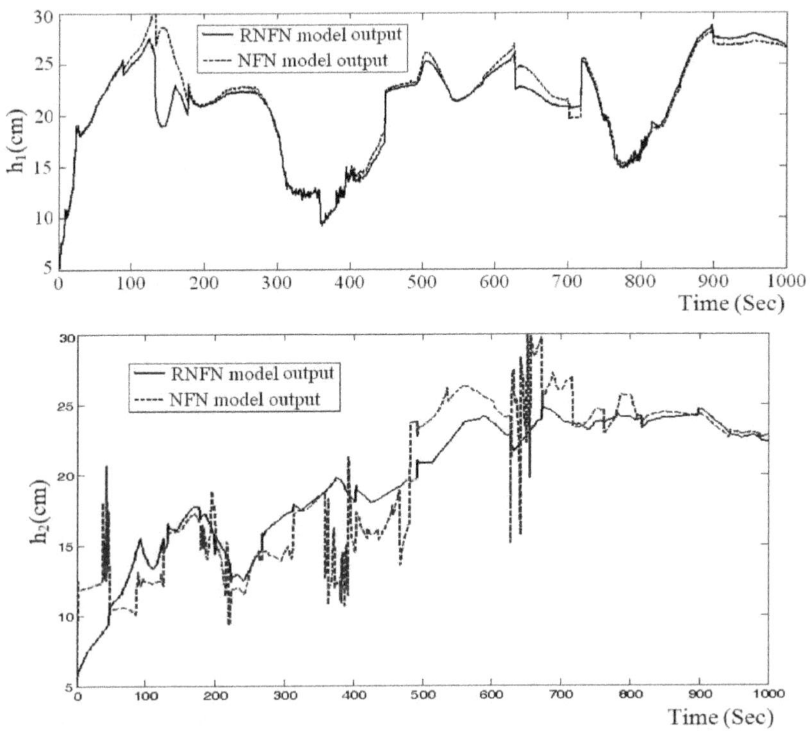


Figure 9. Comparison of five-step ahead RNFN and NFN prediction models.

Table 3. Comparison of PPE for RNFN and NFN models.

| | PPE (%) | | | |
|---|---|---|---|---|
| | $h_1$ | | $h_2$ | |
| Model | One step | Five step | One step | Five step |
| RNFN | 0.03 | 0.14 | 0.02 | 0.07 |
| NFN | 0.10 | 0.22 | 0.12 | 0.12 |

### 4.3. *Prediction using the RNFN model*

Long-range prediction models are obtained to design the multistep ahead predictive controllers. The heights of tanks 1 and 2 are predicted for one and five steps ahead using the NFN and the proposed RNFN model and the plots are shown in Figures 8 and 9. The model outputs are predicted for a random input sequence given to the process.

In this study, for the process variables $h_1$ and $h_2$, one-step and five-step prediction models are developed. The percentage prediction error is calculated using Equation (13) and the results are tabulated in Table 3. From the results given in Table 3, it can be seen that the proposed RNFN model can be used to develop multistep predictive controller outputs with good accuracy.

## 5. MPC design

The MPC approach uses a receding horizon principle and handles disturbances, constraints and model complexities inherently in its design. Therefore, it is more suited for process industries that are trying to optimize process performance in the presence of disturbances and constraints. The MPC design requires prediction of output using an approximate process model. Traditional MPCs use step response, impulse response or state-space models to predict the future changes based on current estimate of the disturbance. However, long range predictions with empirical models and a controller design that uses an empirical model are challenging.

A block diagram of an MPC that uses the proposed RNFN model is shown in Figure 10. The recurrent structure makes it possible to perform multi-step ahead predictions using the prior knowledge of the process operating conditions and process input–output data. Furthermore, model complexities such as non-linearity, inverse response and non-minimum phase behaviour can also be handled in the controller design inherently. This model is used for multi-step ahead prediction of the process outputs with more accuracy (Subathra & Radhakrishnan, 2010, 2011a). The prediction methodology used is illustrated in Figure 11.

The prediction obtained from the RNFN model is used as input to the control calculations where the set-points

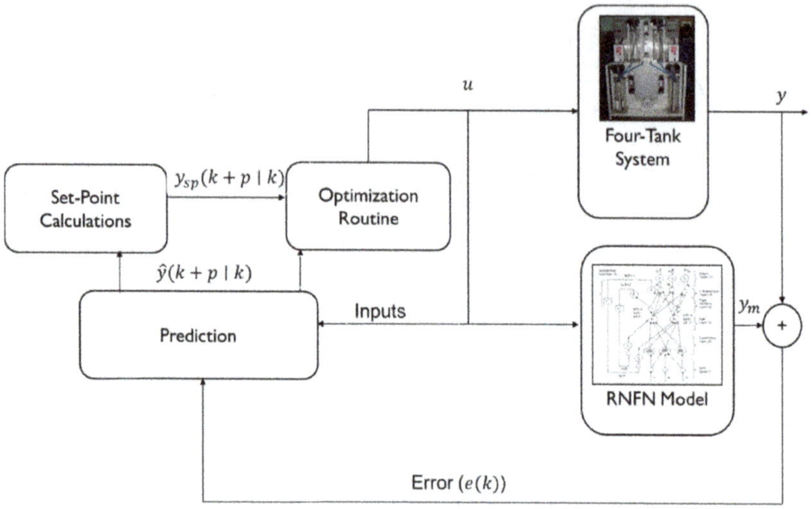


Figure 10. Block diagram of model predictive control.

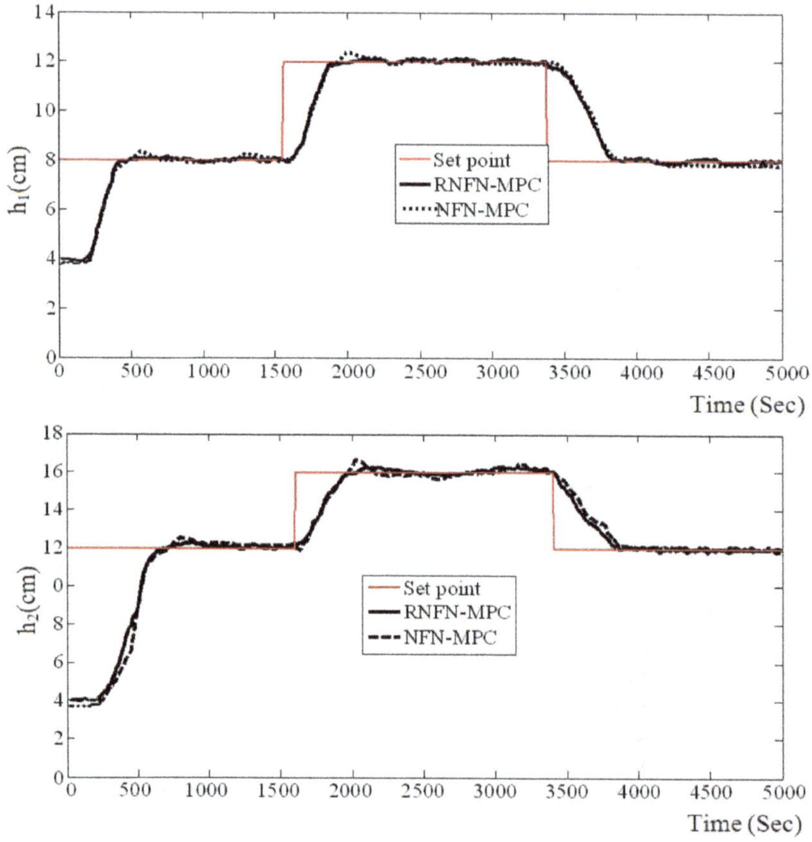


Figure 11. Comparison of servo response of RNFN-MPC with NFN-MPC.

are generated by the set-point calculations block depending on the operating mode. The predicted output and set-point changes are the input to the optimization routine (control calculations) block. Typically, the optimization objectives include maximizing a profit function or production rate or minimizing a cost function. The optimum values of set-points are changed frequently owing to varying process conditions. The MPC accounts for this and the set-point trajectory is calculated each time when the control computations are performed. The control calculations are based on current measurements and predictions of future values of the outputs. The objective of the MPC control calculations is to determine a sequence of control moves so that the predicted response moves to the set-point in an optimal manner. The actual output is $y$, predicted output is $\hat{y}$ and the manipulated input is $u$. An optimization problem is formed that minimizes the magnitude of the future control moves. Therefore, the objective function is given by

$$J(k) = \sum_{p=N_1}^{N_2} \|y_{sp}(k+p|k) - \hat{y}(k+p|k)\|^2 + \lambda \sum_{p=0}^{N_u} \|\Delta u(k+p|k)\|^2, \quad (14)$$

where $y_{sp}$ is the set-point value, $\hat{y}$ is the predicted output and $\lambda$ is a scalar. The $\lambda \geq 0$ defines in fact a ratio of the weight attributed to damping of the input moves versus the (unity) weight attributed to a reduction of the control errors. In this study, $N_1 = 1$; $N_2 = 5$; $N_u = 3$ and $\lambda = 1$. The control inputs that optimize the performance of the multivariable nonlinear system is the minimizer of the unconstrained optimization problem:

$$J(k) = \min_u \sum_{p=N_1}^{N_2} ||y_{sp}(k+p|k) - \hat{y}(k+p|k)||^2 + \lambda \sum_{p=0}^{N_u} ||\Delta u(k+p|k)||^2. \quad (15)$$

The objective function in Equation (14) is a nonlinear optimization problem and solving it using traditional optimization methods, for example, sequential quadratic programming, is difficult due to the absence of a mathematical model. Furthermore, implementing MPCs based on state-space models requires complex computations to be performed on embedded hardware and this makes the design complex. To overcome these computation difficulties and to facilitate implementation on dedicated hardware, this investigation uses the gradient descent method to find the optimal control input of the objective function in Equation (15).

## 6. Results and discussions

### *6.1. RNFN model validation*

In order to validate the RNFN and NFN models, manipulated variables (current inputs to drive 1 and drive 2) are varied randomly to both the process and the models. The output thus obtained are recorded. The data set is divided into training (estimation) and test (validation) sets. Among the 1000 data, 700 are used for training and the remaining 300 data for testing. The experimental outputs and RNFN model outputs and the excitation inputs are shown in Figures 6–9. The fitness function of the model is examined using a set of data that is not used when training. That data set is known as validation data. It is very important that the test set must satisfy the same demands as the training set regarding representation of the entire operating range. The results of both estimation and validation for both NFN and RNFN models are compared with the experimental results as shown in Figures 6 and 8. The NE is calculated using Equation (12). It can be concluded from Table 2 that the recurrent model is endowed with lesser error compared with the feed-forward model. Among the two models proposed in this investigation, it is found that RNFN gives good model accuracy than the NFN models.

Table 4. MISE values for NFN-MPC and RNFN-MPC.

| Response | Controller | $h_1$ | $h_2$ |
|---|---|---|---|
| Servo | NFN-MPC | 0.32 | 0.81 |
| | RNFN-MPC | 0.26 | 0.73 |
| Regulatory | NFN-MPC | 1.70 | 1.12 |
| | RNFN-MPC | 1.03 | 0.87 |

### *6.2. Comparison of NFN and RNFN model-based controllers*

#### *6.2.1. Servo response*

The RNFN-based model predictive control is designed and implemented in the quadruple tank process. Figure 11 gives the comparison between NFN and RNFN, with various reference trajectories. The step input changes with a positive magnitude of 4 cm at 1600 and the negative step change is given at 3400 s for $h_1$. The positive step changes with a magnitude of 4 cm at 1600 and the negative step change is given at 3400 s for $h_2$. The MPC will decide the future control action based on the error and set-point. The set-points for $h_1$ are 8-12-8 cm and for $h_2$ are 12-16-12 cm. Performance measure (MISE) of the developed controller is given in Table 4. From the comparative results, it can be seen that the time taken by the NFN and RNFN model-based controller to reach the new step change set-point is lesser.

#### *6.2.2. Regulatory performance*

The process is subjected to an external disturbance after the level has reached a steady state.

A dozer pump, which has a high precision ejection, is used to inject an external step disturbance, which can be calibrated in terms of quantity of the nominal diaphragm pump discharge. The change in the level in tank 1 and tank 2 and also the response of the controllers are stored.

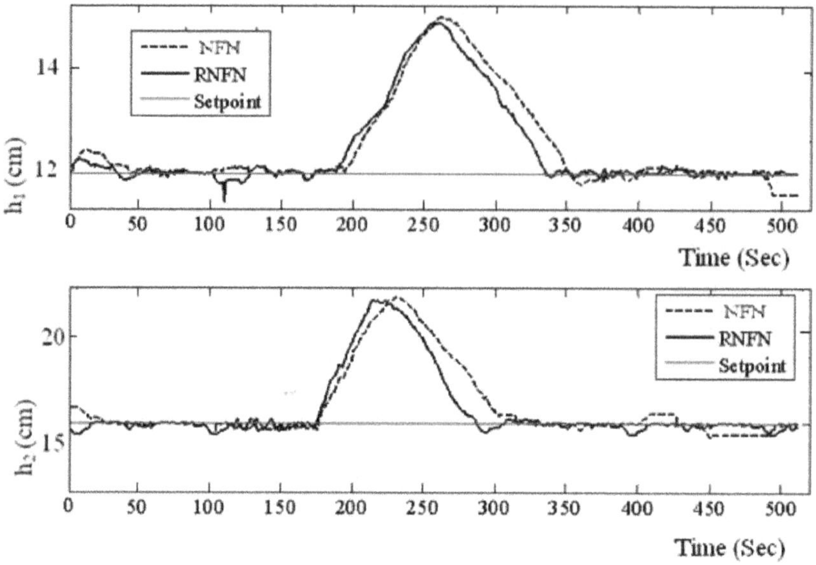


Figure 12. Comparison of regulatory response of RNFN-MPC with NFN-MPC.

The disturbance in tank 3 and tank 4 is about 10% of the nominal flow.

The process is subject to an external disturbance after the level reaches steady state ($h_1$ is 12 cm and $h_2$ is 16 cm) and for the respective input, the changes in the process variables and also the response of the controller and its action to reject the disturbances are obtained and are shown in Figure 12. From MISE (Mean Integral Square Error) values shown in Table 4, one can infer that the developed RNFN-MPC has a better servo tracking performance than the NFN-MPC.

## 7. Conclusions

This investigation presented a new seven-layer RNFN structure for modelling a multivariable nonlinear process and an MPC design methodology using the proposed model. The modelling approach and controller design methodology were illustrated on a quadruple process.

In this study, the RNFN structure is used to model the process. Based on the open loop input–output data sets fuzzy technique is used to divide the entire operating regions and the NN is used for learning and training. The proposed model used the RNFN that combined the learning feature of ANN with the cognitive features of fuzzy systems. This led to a modelling framework that can be used to develop empirical models and it can use both input–output data and prior process knowledge. The introduction of memory elements between the fuzzy inference and fuzzification layer makes it possible to perform multi-step ahead predictions required for designing an MPC. In the RNFN model, the firing strength has been varied by changing the consequent part of the fuzzy rule in addition to the conventional RNFN structure, resulting in significant improvements in model accuracy. The obtained model was used to design an MPC and the use of empirical models necessitated new methods to solve the optimization problem. Gradient descent algorithm was used to obtain the control inputs that minimize the objective function or optimize the plant performance. As a result, the proposed MPC solution method does not require solvers for solving the quadratic optimization problem and therefore is more suitable for implementation in embedded platforms. Experimental studies on the quadruple process indicate that the proposed RNFN-based MPC performs better than the NFN-based MPC for both servo and regulatory performance. The effectiveness of the designed RNFN-MPC for the control of the quadruple tank process demonstrates that it can be used effectively for the control of multivariable interacting nonlinear processes. Implementation of the proposed MPC in dedicated hardware, studying the role of RNFN models in industries and studying the possibility of hardware-in-loop (HIL) simulation are the future prospects of this investigation.

## Disclosure statement

No potential conflict of interest was reported by the authors.

## References

Bhat, N. V., & McAvoy, T. J. (1990). Use of neural nets for dynamical modeling and control of chemical process systems. *Computers & Chemical Engineering*, *14*, 573–583.

Brdys, M. A., & Kulawski, G. J. (1999). Dynamic neural controllers for induction motor. *IEEE Transaction of Neural Networks*, *10*, 340–355.

Brown, M., & Harris, C. J. (1994). *Neuro fuzzy adaptive modeling and control*. Englewood Cliffs, NJ: Prentice-Hall.

Bulsari, A. B. (Ed.). (1995). *Computer-aided chemical engineering, Vol. 6, neural networks for chemical engineers*. Amsterdam: Elsevier.

Doyle, F. J., Gatzke, E. P., Vadigepalli, R., & Meadows, E. S. (1999). Experiences with an experimental project in a graduate control course. *Chemical Engineering Education*, *4*(33), 270–275.

Elman, J. L. (1990). Finding structures in time. *Cognitive Science*, *14*, 179–211.

Frasconi, P., Gori, M., & Soda, G. (1992). Local feedback multilayered networks. *Neural Computing*, *4*, 120–130.

Henson, M. A., & Seborg, D. E. (1997). *Nonlinear process control*. Upper Saddle River, NJ: Prentice-Hall PTR.

Jang, J. S. R. (1992). Self-learning fuzzy controllers based on temporal back propagation. *IEEE Transaction on Neural Networks*, *3*, 714–723.

Jang, J. S. R., & Sun, C. T. (1995). Neuro-fuzzy modeling and control. *Proceedings of the IEEE*, *83*, 378–406.

Jang, J. S. R., Sun, C. T., & Mizutani, E. (1997). *Neuro-fuzzy and soft computing: A computational approach to learning and machine intelligence*. Englewood Cliffs, NJ: Prentice-Hall.

Johansen, T. A., & Foss, B. A. (1993). Constructing NARMAX models using ARMAX models. *International Journal of Control*, *58*(5), 1125–1153.

Johansson, K. H., & Nunes, J. L. R. (1998). *A multivariable laboratory process with an adjustable zero*. Paper presented at the 17th American Control Conference, Philadelphia, PA.

Juang, C. F., & Chen, J. S. (2003). A recurrent neural fuzzy network controller for a temperature control system. *Proceedings of the 12th IEEE International Conference on Fuzzy Systems*, *1*, 408–413.

Juang, C. F., Huang, S. T., & Duh, F. B. (2006). Mold temperature control of a rubber injection-molding machine by TSK-type recurrent neural fuzzy network. *Neuro Computing*, *70*, 559–567.

Lia, C., & Cheng, K. H. (2007). Recurrent neuro-fuzzy hybrid-learning approach to accurate system modeling. *Fuzzy Sets and Systems*, *158*, 194–212.

Montague, G. A., Tham, M. T., Willis, M. J., & Morris, A. J. (1992). *Predictive control of distillation columns using dynamic neural networks*. Proceeding of the 3rd IFAC Symposium on Dynamics Control of Chemical Reactors, Distillation Columns, and Batch Processes, Maryland, USA, pp. 231–236.

Morris, A. J., Montague, G. A., & Willis, M. J. (1994). Artificial neural networks: Studies in process modeling and control. *Trans. IChemE*, *72*(pt. A), 3–19.

Scott, G. M., & Ray, W. H. (1993). Creating efficient nonlinear network process models that allow model interpretation. *Journal of Process Control*, *3*(3), 163–178.

Su, H. T., McAvoy, T. J., & Werbos, P. (1992). Long-term prediction of chemical processes using recurrent neural networks: A parallel training approach. *Industrial & Engineering Chemical Research*, *31*, 1338–1352.

Subathra, B., & Radhakrishnan, T. K. (2010). Modeling and control of multivariable process using intelligent techniques. *Sensors and Transducers*, *121*(10), 68–79.

Subathra, B., & Radhakrishnan, T. K. (2011a). Modeling and control of MIMO nonlinear process using recurrent neuro fuzzy networks instrumentation science and technology. *Instrumentation Science and Technology*, *39*(2), 211–230.

Subathra, B., & Radhakrishnan, T. K. (2011b). Recurrent neuro fuzzy and fuzzy neural hybrid networks: A review. *Instrumentation Science and Technology*, *40*(1), 29–50.

Subathra, B., Raja Rao, N., & Radhakrishnan, T. K. (2009). A comparative study of conventional and recurrent fuzzy technique for chemical processes. *Instrumentation Science and Technology*, *37*(6), 660–675.

Takagi, T., & Sugeno, M. (1985). Fuzzy identification of systems and its application to modeling and control. *IEEE Transaction on System, Man, Cybernetics, SMC*, *15*, 116–132.

Tsoi, A. C., & Back, A. D. (1994). Locally recurrent globally feedforward networks: A critical review of architectures. *IEEE Transaction on Neural Networks*, *5*, 229–239.

Werbos, P. J. (1990). Backpropagation through time: What it does and how to do it. *Proceedings of the IEEE*, *78*, 1550–1560.

Yager, R. R., & Filev, D. P. (1994). *Essentials of fuzzy modeling and control*. New York, NY: Wiley.

Zhang, J. (2003). Multi-objective optimal control of batch processes using recurrent neuro-fuzzy networks. *Neural Networks. Proceedings of the International Joint Conference*, *1*, 304–309.

Zhang, J., & Morris, A. J. (1995a). *Dynamic process modeling using locally recurrent neural networks. Proceedings of the American Control Conference (Vol. 4), Seattle, WA*, pp. 2767–2771.

Zhang, J., & Morris, A. J. (1995b). Fuzzy neural networks for nonlinear systems modeling. *Proceeding of the Institution of Electrical Engineers, Control Theory Applications*, *142*, 551–561.

Zhang, J., & Morris, A. J. (1999). Recurrent neuro-fuzzy networks for nonlinear process modeling, *IEEE Transaction on Neural Networks*, *10*(2), 313–326.

Zhang, J., & Morris, A. J. (2000). Long range predictive control of nonlinear processes based on recurrent neuro-fuzzy network models. *Neural Computing & Applications*, *9*, 50–59.

# 2

## Simultaneous state and input estimation with partial information on the inputs

Jinya Su[a], Baibing Li[b]* and Wen-Hua Chen[a]

[a]*Department of Aeronautical and Automotive Engineering, Loughborough University, Loughborough LE11 3TU, UK;*
[b]*School of Business and Economics, Loughborough University, Loughborough LE11 3TU, UK*

This paper investigates the problem of simultaneous state and input estimation (SSIE) for discrete-time linear stochastic systems when the information on the inputs is partially available. To incorporate the partial information on the inputs, matrix manipulation is used to obtain an equivalent system with reduced-order inputs. Then Bayesian inference is drawn to obtain a recursive filter for both state and input variables. The proposed filter is an extension of the recently developed state filter with partially observed inputs to the case where the input filter is also of interest, and an extension of the SSIE to the case where the information on the inputs is partially available. A numerical example is given to illustrate the proposed method. It is shown that, due to the additional information on the inputs being incorporated in the filter design, the performances of both state and input estimation are substantially improved in comparison with the conventional SSIE without partial input information.

**Keywords:** Bayesian inference; partial information; state filter; unknown input filter

### 1. Introduction

State estimation for discrete-time linear stochastic systems with unknown inputs has been receiving increasing attention (see, e.g. Cheng, Ye, Wang, & Zhou, 2009; Darouach & Zasadzinski, 1997; Darouach, Zasadzinski, & Boutayeb, 2003; Hsieh, 2000; Kitanidis, 1987, among many others) due to its widespread applications in the fields of weather forecasting (Kitanidis, 1987), fault diagnosis (Mann & Hwang, 2013), etc.

In some applications such as population estimation, traffic management (Li, 2013), and chemical engineering (Mann & Hwang, 2013), however, information on the input variables is not completely unknown; rather, it is available at an aggregate level. Li (2013) has recently proposed a unified filtering approach to incorporate this kind of information. It is shown that this approach includes two extreme scenarios as its special cases, that is, the filter where all the inputs are unknown (i.e. the scenario investigated in Kitanidis, 1987; Gillins & De Moor, 2007, etc.) and the filter where the inputs are completely available (i.e. the classical Kalman filter can be applied). Later, Su, Li, and Chen (2015a) further investigated some properties of the aforementioned unified filter such as existence, optimality and asymptotic stability. However, Li (2013) and Su et al. (2015a) only considered the problem of sole state estimation; the problem of simultaneous state and input estimation (SSIE) with partial information on the inputs has not been investigated.

Gillins and De Moor (2007) developed a SSIE method using the approach of minimum-variance unbiased estimation (MVUE), then Fang and Callafon (2012) further investigated its asymptotic stability. Potentially, the SSIE can be applied to a wide range of problems such as fault diagnosis (see, e.g. Gao & Ding, 2007; Patton, Clark, & Frank, 1989), fault-tolerant control (Jiang & Fahmida, 2005), disturbance rejection control (Profeta, Joseph, William, & Marin, 1990). In the field of fault detection and fault-tolerant control, for example, actuator, sensor and/or structure faults are usually modelled as inputs to the system with unknown dynamics. One can monitor system status by estimating the inputs for fault diagnosis purposes where the estimated inputs can provide valuable information for the fault-tolerant control system; see, for example, Jiang and Fahmida (2005) and Su, Chen, and Li (2014). In the field of disturbance rejection control, the uncertainties in system model are usually modelled as lumped system inputs (which may include system mismatches, parameter uncertainties, external disturbances); see Chen, Ballance, Gawthrop, and O'Reilly (2000), Yang, Li, Su, and Yu (2013), and Yang, Su, Li, and Yu (2014) for a detailed discussion. When inputs are approximately obtained based on disturbance estimation algorithms, one can attenuate their effects on dynamic systems by directly feedthrough of the estimated value.

In this paper, we investigate the problem of SSIE. Unlike Gillins and De Moor (2007) where the inputs are

*Corresponding author. Email: b.li2@lboro.ac.uk

assumed to be completely unknown, we consider the scenario where the information on the inputs is partially available. To incorporate the partial information on the inputs, the original inputs are decoupled into two parts, where the first part is completely known based on the available information on the original inputs, whereas the second part is completely unknown which will serve as the unknown inputs of the new system. On this basis, we draw Bayesian inference (see, e.g. Li, 2009, 2013) and obtain simultaneous estimates of the state and new unknown inputs. According to the Bayesian theory, the obtained estimates are optimal in the sense of minimum mean square estimation under the assumption of Gaussian noise terms (Li, 2013). Finally, the estimates of the original inputs can be worked out by pooling together all the available information on the inputs.

Compared with the filter in Li (2013) where only state estimate is of interest, the proposed method obtains SSIE, and hence the estimated inputs can be used in fault detection and other applications. Compared to the results in Gillins and De Moor (2007), this paper takes into account the additional information on the inputs, and hence it results in a better estimate of state and input vectors. In addition, we show that Bayesian inference can provide an alternative derivation of the filter in Gillins and De Moor (2007) for the SSIE problem. We further investigate the relationships of the proposed filter with some existing approaches. In particular, we show in this paper that (a) when the inputs are completely available, the proposed filter reduces to the classical Kalman filter (Simon, 2006); (b) when no information on the unknown inputs is available, it reduces to the results of Gillins and De Moor (2007) where both state and input estimation are concerned; and (c) if only state estimation is of interest, it is equivalent to the filter for partially available inputs developed in Li (2013).

The rest of paper is structured as follows. Section 2 formulates the considered problem. The main results of the paper are provided in Section 3. In Section 4, a simulation study is carried out to illustrate the proposed filter. Finally, Section 5 concludes the paper.

## 2. Problem formulation

Consider a discrete-time linear stochastic time-varying system with unknown inputs:

$$\begin{aligned} x_{k+1} &= A_k x_k + G_k d_k + \omega_k, \\ y_k &= C_k x_k + \upsilon_k, \end{aligned} \tag{1}$$

where $x_k \in R^n, d_k \in R^m, y_k \in R^p$ are the state vector, input vector, and measurement vector at each time step $k$ with $p \geq m$ and $n \geq m$. Following Li (2013), the process noise $\omega_k \in R^n$ and the measurement noise $\upsilon_k \in R^p$ are assumed to be mutually independent, and each follows a Gaussian distribution with zero mean and a known covariance matrix, $Q_k = E[\omega_k \omega_k^{\mathrm{T}}] > 0$ and $R_k = E[\upsilon_k \upsilon_k^{\mathrm{T}}] > 0$, respectively. $A_k$, $G_k$, $C_k$ are known matrices. Following the existing researches (e.g. Gillins & De Moor, 2007; Kitanidis, 1987; Li, 2013; Su et al., 2015a), $G_k$ is assumed to have a full column-rank; otherwise, the redundant input variables can be removed.

We consider the scenario where the input vector $d_k$ is not fully observed at the level of interest but rather it is available only at an aggregate level. Specifically, let $D_k$ be a $q_k \times m$ known matrix with $0 \leq q_k \leq m$ and $F_{0k}$ an orthogonal complement of $D_k^{\mathrm{T}}$ such that $D_k F_{0k} = O_{q_k \times (m-q_k)}$. It is assumed that the input data are available only on some linear combinations:

$$r_k = D_k d_k, \tag{2}$$

where $r_k$ is available at each time step $k$, whereas no information on $\delta_k = F_{0k}^{\mathrm{T}} d_k$ is available. Hence, $\delta_k$ is assumed to have a non-informative probability density function $f(\delta_k)$ such that all possible values of $\delta_k$ are equally likely to occur:

$$f(\delta_k) \propto 1. \tag{3}$$

Without loss of generality, we assume that $D_k$ has a full row-rank; otherwise, the redundant rows of $D_k$ can be removed from the analysis (see Su et al., 2015a).

As pointed out in Li (2013), the matrix $D_k$ characterizes the availability of input information at each time step $k$. It includes two extreme scenarios that are usually considered: (a) $q_k = 0$, that is, no information on the input variables is available; this is the problem investigated in Kitanidis (1987) and Gillins and De Moor (2007); (b) $q_k = m$ and $D_k$ is an identity matrix, that is, the complete input information is available. This is the case that the classical Kalman filter can be applied (Simon, 2006).

The objective of this paper is to simultaneously estimate the state and input vectors based on Equations (1) and (2).

## 3. Main results

In this section, the main results of the paper will be given. To incorporate the partial information on the inputs $d_k$, $G_k d_k$ is decoupled into two parts based on a decoupling matrix, that is, the known part given by the prior information (2) and unknown part $\delta_k$. Note that $\delta_k$ has a lower dimension than the original input vector $d_k$ and it plays the role of unknown inputs in the new system. Next, Bayesian inference is drawn to obtain recursive estimates of both state variables $x_k$ and unknown inputs $\delta_k$, upon which the estimate of the original input vector $d_k$ can be worked out. Finally, the relationships between the proposed method and the relevant existing filters are discussed. The diagram of the system and the proposed filter structure is shown in Figure 1.

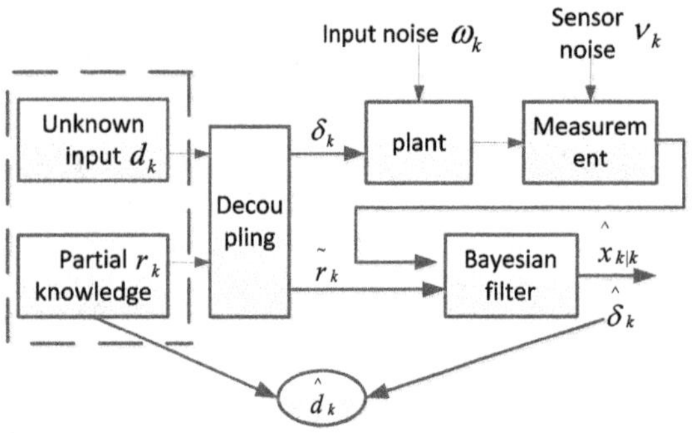


Figure 1. Diagram of the system and filter structure.

### 3.1. *Transformation*

To incorporate the information $r_k = D_k d_k$, a decoupling method is used here (see Su et al., 2015a). Define a nonsingular decoupling matrix $M_k$ of appropriate dimension as follows:

$$M_k = \begin{bmatrix} D_k & O \\ O & I \\ F_{0k}^{\mathrm{T}} & O \end{bmatrix} [G_k, \ G_k^{\perp}]^{-1},$$

where $G_k^{\perp}$ denotes an orthogonal complement of $G_k$, $O$ and $I$ represent the zero matrix and identity matrix of appropriate dimensions, respectively. $F_{0k}$ is the orthogonal complement of $D_k^{\mathrm{T}}$ such that $D_k F_{0k} = O$ and $F_{0k}^{\mathrm{T}} F_{0k} = I$.

Then, $M_k G_k d_k$ can be expressed as follows:

$$\begin{aligned} M_k G_k d_k &= [D_k^{\mathrm{T}}, O, F_{0k}]^{\mathrm{T}} d_k \\ &= [(D_k d_k)^{\mathrm{T}}, O, O]^{\mathrm{T}} + [O, O, I]^{\mathrm{T}} F_{0k}^{\mathrm{T}} d_k \\ &= \tilde{r}_k + \tilde{G}_k \delta_k, \end{aligned} \tag{4}$$

where $\tilde{r}_k := [r_k^{\mathrm{T}} \ O \ O]^{\mathrm{T}}$ is completely available due to the available information on the inputs, $\tilde{G}_k := [O \ O \ I]^{\mathrm{T}}$ and $\delta_k := F_{0k}^{\mathrm{T}} d_k$.

Multiplying $M_k^{-1}$ (the explicit form of $M_k^{-1}$ is given in the appendix) on both sides of Equation (4), $G_k d_k$ can be decoupled into two parts:

$$G_k d_k = M_k^{-1} \tilde{r}_k + M_k^{-1} \tilde{G}_k \delta_k. \tag{5}$$

Consequently, the dynamics of $x_{k+1}$ can be written as

$$\begin{aligned} x_{k+1} &= A_k x_k + M_k^{-1} \tilde{r}_k + M_k^{-1} \tilde{G}_k \delta_k + \omega_k \\ &= A_k x_k + M_k^{-1} \tilde{r}_k + F_k \delta_k + \omega_k, \end{aligned}$$

where $F_k := M_k^{-1} \tilde{G}_k = [G_k, \ G_k^{\perp}] \left[ \begin{smallmatrix} F_{0k} \\ O \end{smallmatrix} \right] = G_k F_{0k}$.

Hence, the linear system (1) with the additional information on the inputs, $r_k = D_k d_k$, can equivalently be represented by the following system:

$$\begin{aligned} x_{k+1} &= A_k x_k + M_k^{-1} \tilde{r}_k + F_k \delta_k + \omega_k, \\ y_k &= C_k x_k + v_k. \end{aligned} \tag{6}$$

*Remark 1* An alternative approach to incorporating the unknown input information is to use pseudo-inverse theory. From Equation (2), one can obtain the general solution of $d_k$

$$d_k = D_k^{+} r_k + F_{0k} \bar{\delta}_k, \tag{7}$$

where $D_k^{+} = D_k^{\mathrm{T}} (D_k D_k^{\mathrm{T}})^{-1}$ and $\bar{\delta}_k$ is completely unknown. If we select $\bar{\delta}_k := \delta_k = F_{0k}^{\mathrm{T}} d_k$, we can show that this approach is equivalent to the decoupling matrix based method.

*Remark 2* It should be noted that the partial information on the inputs $r_k = D_k d_k$ has been fully incorporated into the system (6). We also note that the dimension of the inputs has been reduced from $m$ to $m - q_k$.

### 3.2. *Filter design*

It can be seen from Equation (6) that $y_k$ is a function of $x_k$, and $x_k$ is related to the inputs $\delta_{k-1}$. Hence, the input estimate of $\delta_k$ is delayed by one time unit (Gillins and De Moor, 2007). The objective of filter design is to obtain the estimate of $x_k$ and $\delta_{k-1}$ based on the available measurement sequence $Y_k = \{y_1, y_2, \ldots, y_k\}$. For the new system (6), we can either solve the filtering problem based on the approach of MVUE (e.g. Gillins & De Moor, 2007) or Bayesian inference (e.g. Li, 2009, 2013). In the paper, we use the Bayesian method that can be seen as an alternative approach to that of Gillins and De Moor (2007).

In the context of Bayesian inference, the first step is to predict the dynamics of $x_k$ and $\delta_{k-1}$ based on the available measurement sequence $Y_{k-1} = \{y_1, y_2, \ldots, y_{k-1}\}$. Since we do not assume that the unknown input vector $\delta_k$ satisfies any transition dynamics, prediction is only performed to determine the dynamics of $x_k$, that is, $p(x_k|Y_{k-1})$. The likelihood function can be determined based on the observation equation of system (6). The second step is to obtain the posterior distribution of the concerned variables after the measurement vector $y_k$ is received based on Bayes' chain rule:

$$p(x_k, \delta_{k-1}|Y_k) \propto p(y_k|x_k) p(x_k, \delta_{k-1}|Y_{k-1}). \tag{8}$$

The main results on filtering design are summarized in Theorem 1.

THEOREM 1 *For state space model* (6), *suppose the matrix $C_k F_{k-1}$ has a full column-rank, then the prior and posterior distributions for $x_k$ and $\delta_{k-1}$ at any time step $k$ can be obtained sequentially as follows:*

(i) *Posterior of $x_{k-1}$ for given $Y_{k-1}$ :*

$$x_{k-1} \sim N(\hat{x}_{k-1|k-1}, P_{k-1|k-1}^{x}).$$

(ii) *Prediction for* $x_k$ :

$$N(\hat{x}_{k|k-1}, P^x_{k|k-1}),$$

*with* $\hat{x}_{k|k-1} = A_{k-1}\hat{x}_{k-1|k-1} + M^{-1}_{k-1}\tilde{r}_{k-1},$

$$P^x_{k|k-1} = A_{k-1}P^x_{k-1|k-1}A^{\mathrm{T}}_{k-1} + Q_{k-1}. \quad (9)$$

(iii) *Posterior of* $\delta_{k-1}$ *for given* $Y_k$ :

$$\delta_{k-1} \sim N(\hat{\delta}_{k-1}, P^{\delta}_{k|k}),$$

*where the posterior mean is given by*

$$\hat{\delta}_{k-1} = P^{\delta}_{k|k}(C_kF_{k-1})^{\mathrm{T}}\tilde{R}^{-1}_k(y_k - C_k\hat{x}_{k|k-1}), \quad (10)$$

*and the posterior covariance matrix is given by*

$$P^{\delta}_{k|k} = (F^{\mathrm{T}}_{k-1}C^{\mathrm{T}}_k\tilde{R}^{-1}_kC_kF_{k-1})^{-1}, \quad (11)$$

*while the posterior of* $x_k$ *for given* $Y_k$ *is*

$$x_k \sim N(\hat{x}_{k|k}, P^x_{k|k}),$$

*where the posterior mean is given by*

$$\hat{x}_{k|k} = \hat{x}_{k|k-1} + P^x_{k|k-1}C^{\mathrm{T}}_k\tilde{R}^{-1}_k(y_k - C_k\hat{x}_{k|k-1}) + (F_k - P^x_{k|k-1}C^{\mathrm{T}}_k\tilde{R}^{-1}_kC_kF_k)\hat{\delta}_{k-1}, \quad (12)$$

*and the posterior covariance matrix is given by*

$$P^x_{k|k} = P^x_{k|k-1} - P^x_{k|k-1}C^{\mathrm{T}}_k\tilde{R}^{-1}_kC_kP^x_{k|k-1} + (F_{k-1} - P^x_{k|k-1}C^{\mathrm{T}}_k\tilde{R}^{-1}_kC_kF_{k-1})(P^{\delta}_{k|k})^{-1}()^{\mathrm{T}}, \quad (13)$$

*where* $\tilde{R}_k = C_kP^x_{k|k-1}C^{\mathrm{T}}_k + R_k$, $()^{\mathrm{T}}$ *in* $(*)A()^{\mathrm{T}}$ *stands for the transpose of* $*$.

*Proof* From Equation (8), the posterior distribution $p(x_k, \delta_{k-1}|Y_k)$ is governed by

$$p(x_k, \delta_{k-1}|Y_k) \propto \exp\{-(y_k - C_kx_k)^{\mathrm{T}}R^{-1}_k() - (x_k - \hat{x}_{k|k-1} - F_{k-1}\delta_{k-1})^{\mathrm{T}}(P^x_{k|k-1})^{-1}()\}.$$

By completing the square on $[x^{\mathrm{T}}_k, \delta^{\mathrm{T}}_{k-1}]^{\mathrm{T}}$, the exponent can be rewritten as $-([x^{\mathrm{T}}_k, \delta^{\mathrm{T}}_{k-1}] - [\hat{x}^{\mathrm{T}}_{k|k}, \hat{\delta}^{\mathrm{T}}_{k-1}])P^{-1}_{k|k}()^{\mathrm{T}}$, where

$$\begin{bmatrix} \hat{x}_{k|k} \\ \hat{\delta}_{k-1} \end{bmatrix} = P_{k|k}\begin{bmatrix} C^{\mathrm{T}}_kR^{-1}_ky_k + (P^x_{k|k-1})^{-1}\hat{x}_{k|k-1} \\ -F^{\mathrm{T}}_{k-1}(P^x_{k|k-1})^{-1}x_{k|k-1} \end{bmatrix}$$

and

$$P_{k|k} = \begin{bmatrix} C^{\mathrm{T}}_kR^{-1}_kC_k + (P^x_{k|k-1})^{-1} & -(P^x_{k|k-1})^{-1}F_{k-1} \\ -F^{\mathrm{T}}_{k-1}(P^x_{k|k-1})^{-1} & F^{\mathrm{T}}_{k-1}(P^x_{k|k-1})^{-1}F_{k-1} \end{bmatrix}^{-1}.$$

This indicates that the posterior distribution is a Gaussian distribution with mean $[\hat{x}^{\mathrm{T}}_{k|k}, \hat{\delta}^{\mathrm{T}}_{k-1}]^{\mathrm{T}}$ and covariance matrix $P_{k|k}$. When $C_kF_{k-1}$ is of full row-rank, based on the inverse of partitioned matrix, we can obtain the recursive estimation of both $x_k$ and $\delta_{k-1}$ as shown in Equations (9)–(13).

So far, we have obtained the state estimate $\hat{x}_{k|k}$ and estimate $\hat{\delta}_{k-1}$ for the transformed system. When $F^{\mathrm{T}}_{0k-1}d_{k-1} = \hat{\delta}_{k-1}$ is obtained, based on Equation (5), we can further obtain the estimate of the original inputs $d_{k-1}$ as follows:

$$\hat{d}_k = (G^{\mathrm{T}}_kG_k)^{-1}G^{\mathrm{T}}_k(M^{-1}_k\tilde{r}_k + M^{-1}_k\tilde{G}_k\hat{\delta}_k).$$

It can be verified that the obtained unknown input estimate satisfies the unknown input information (Equation (2)), that is,

$$D_k\hat{d}_k = r_k. \quad (14)$$

The proof is given in the appendix. ■

### 3.3. Relationships with the existing results

In this section, we investigate the relationships between the proposed approach and the relevant results in the existing literature. This is summarized in the following theorem.

THEOREM 2 *The set of recursive formulas* (9)–(13) *reduces to*

(1) *the classical Kalman filter when all entries of the input vector* $d_k$ *are available;*
(2) *the filter in* Gillins and De Moor (2007) *when no information on the unknown inputs* $d_k$ *is available;*
(3) *the filter in* Li (2013) *when only state estimation is concerned.*

*Proof* For the case where all the input variables are available at the level of interest, $D_k$ becomes an $m \times m$ identity matrix, and $F^{\mathrm{T}}_{0k}$ becomes an zero-by-zero empty matrix. Consequently, the last term on the right-hand side of Equations (12) and (13) vanishes, and Equations (12) and (13) reduce to

$$P^x_{k|k} = P^x_{k|k-1} - P^x_{k|k-1}C^{\mathrm{T}}_kH^{-1}_kC_kP^x_{k|k-1}.$$

Since $M^{-1}_{k-1}\hat{r}_{k-1} = G_{k-1}d_{k-1}$, Equation (12) becomes

$$\hat{x}_{k|k} = A_{k-1}\hat{x}_{k-1|k-1} + G_{k-1}d_{k-1} + P^x_{k|k-1}C^{\mathrm{T}}_kH^{-1}_k(y_k - C_k(A_{k-1}\hat{x}_{k-1|k-1} + G_{k-1}d_{k-1})).$$

Clearly, these recursive formulas are identical to the classical Kalman filter equations (see Simon, 2006).

Next, we consider the case where no input information is available. Clearly $\tilde{r}_k$ in Equation (4) is an empty vector, $F_k$ becomes $G_k$, and $\delta_k = d_k$. Hence, Equation (13) reduces to

$$P^x_{k|k} = P^x_{k|k-1} - P^x_{k|k-1}C_k^{\mathrm{T}}\tilde{R}_k^{-1}C_kP^x_{k|k-1} + [G_k - P^x_{k|k-1}C_k^{\mathrm{T}}H_k^{-1}C_kG_{k-1}]P^{\delta}_{k|k}[]^{\mathrm{T}}$$

and the unknown input covariance matrix (11) becomes

$$P^{\delta}_{k|k} = (G_{k-1}^{\mathrm{T}}C_k^{\mathrm{T}}\tilde{R}_k^{-1}C_kG_{k-1})^{-1}.$$

In addition, Equation (12) becomes

$$\hat{x}_{k|k} = \hat{x}_{k|k-1} + P^x_{k|k-1}C_k^{\mathrm{T}}\tilde{R}_k^{-1}(y_k - C_k\hat{x}_{k|k-1}) + (G_k - P^x_{k|k-1}C_k^{\mathrm{T}}\tilde{R}_k^{-1}C_kG_k)\hat{\delta}_{k-1}$$

and the unknown input estimation Equation (10) becomes

$$\hat{\delta}_{k-1} = P^{\delta}_{k|k}(C_kG_{k-1})^{\mathrm{T}}\tilde{R}_k^{-1}(y_k - C_k\hat{x}_{k|k-1}).$$

These recursive formulas are identical to (a) the results in Kitanidis (1987) when only state filtering is of interest; and (b) the results in Gillins and De Moor (2007) for both unknown input and state estimations obtained using the approach of MVUE.

Finally, if only state estimation is concerned, the proposed method leads to the same results as those in Li (2013). To show this, we note that the state estimation error covariance matrix Equation (13) is the same as the one in Li (2013). In addition, inserting Equations (10) and (11) into Equation (12), Equation (12) can be rewritten in the following form:

$$\hat{x}_{k|k} = \hat{x}_{k|k-1} + K_k(y_k - C_k\hat{x}_{k|k-1}),$$

where the gain matrix $K_k$ is defined as

$$K_k = P^x_{k|k-1}C_k^{\mathrm{T}}\tilde{R}_k^{-1} + [F_{k-1} - P^x_{k|k-1}C_k^{\mathrm{T}}\tilde{R}_k^{-1}C_kF_{k-1}](P^{\delta}_{k|k})^{-1}F_{k-1}^{\mathrm{T}}C_k^{\mathrm{T}}\tilde{R}_k^{-1}.$$

We can further show that (see the appendix for details)

$$M_{k-1}^{-1}\tilde{r}_{k-1} - K_kC_kM_{k-1}^{-1}\tilde{r}_{k-1} = P_{k|k}\bar{M}_{k-1}^{\mathrm{T}}(\bar{M}_{k-1}P_{k|k-1}\bar{M}_{k-1}^{\mathrm{T}})^{-1}\bar{r}_{k-1}, \tag{15}$$

where the left-hand side of Equation (15) is the term associated with the prior information of the proposed filter, whereas the right-hand side of Equation (15) is the term associated with the prior information of the filter in Li (2013). This completes the proof. ■

## 4. Simulation study

In this section, we use a simple numerical example to illustrate the developed filter. First, we will show that, when only state estimation is of interest, the proposed filter can obtain the same result as that of Li (2013). Next we further demonstrate that incorporating the partially available information on the inputs can effectively improve both state estimation and unknown input estimation in comparison with the one without using the input information (Gillins & De Moor, 2007).

The system for the simulation is chosen the same as that of Su, Li, and Chen (2015b) that has been widely used in many previous studies (see, e.g. Cheng et al., 2009). However, to better assess the performance of the proposed filter under uncertainties, we considered a system subject to larger random variation: the covariance matrices $Q_k$ and $R_k$ of the system and measurement noises were taken 10 times as those of Cheng et al. (2009). The initial values of system model is chosen as $x_0 = [3,\ 1,\ 2,\ 2,\ 1]^{\mathrm{T}}$, the initial state and covariance matrix of filter are chosen as $\hat{x}_0 = 0_{5\times1}$ and $P^x_{0|0} = 0.2 \times I_5$.

We applied the recursive formulas in this paper to estimate the state and unknown input vectors at each time step. To evaluate the quality of the state estimate and unknown input estimate obtained using the developed filter, we calculated the trace of the error covariance matrix $P^x_{k|k}$ and the trace of the error covariance matrix $P^{\delta}_{k|k}$ at each time step, as displayed in Figure 2(a) and Figure 2(b) (real line), respectively. For comparison, we also considered the state estimation using the filter in Li (2013) (only state estimation is concerned) and Gillins and De Moor (2007) (assuming that the inputs were completely unknown). The traces of $P^x_{k|k}$ are superimposed in Figure 2(a) (dotted line for Li, 2013 and dashed line for Gillins & De Moor, 2007), and the trace of $P^{\delta}_{k|k}$ is superimposed in Figure 2(b) (dashed line for Gillins & De Moor, 2007).

It can be seen from Figure 2(a) that the trace of state estimation error covariance using the proposed filter is the same as that of Li (2013). Both the method in Li (2013) and the proposed method have a smaller trace of the covariance matrix than that of Gillins and De Moor (2007).

In addition, Figure 2(b) shows that the trace of the error covariance matrix of the unknown input estimate using the proposed filter is smaller in comparison with that of Gillins and De Moor (2007). This is because more information on the unknown inputs was used by the filter developed in this paper. This demonstrates that when the unknown inputs are of practical interest, the proposed method in this paper will have a better performance than Gillins and De Moor (2007) if there is additional information available on the unknown inputs for filtering.

We also compared the state estimates obtained using the three filters, that is, the filter in Li (2013) (Figure 3), the proposed filter (Figure 4) and the filter in Gillins and De Moor (2007) (Figure 5). The upper graphs of Figures 3–5

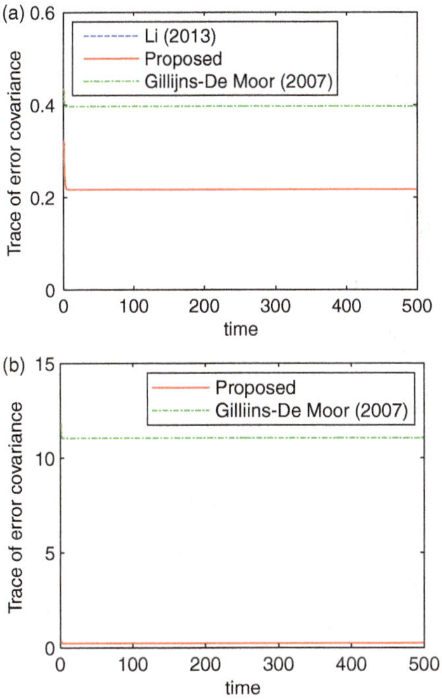


Figure 2. (a) Traces of the covariance matrix $P^{x}_{k|k}$ for three different filters; (b) traces of the covariance matrix $P^{\delta}_{k|k}$ for the proposed approach and the filter in Gillins and De Moor (2007).

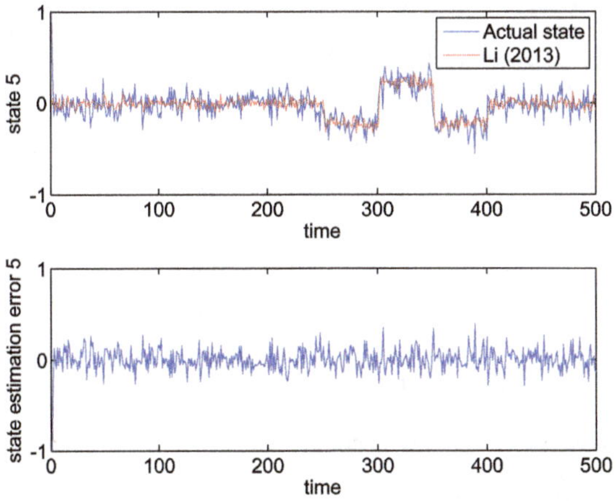


Figure 3. State estimation of the filter in Li (2013) and its estimation error.

display the simulated true values of the fifth state variable (real line) and the estimated state using the filters (dotted line), while the lower graphs plot the corresponding state estimation error for each filter.

It can be seen from Figures 3–5 that the three methods can provide a reasonably good estimate of the state vector. However, overall the state estimation errors using the

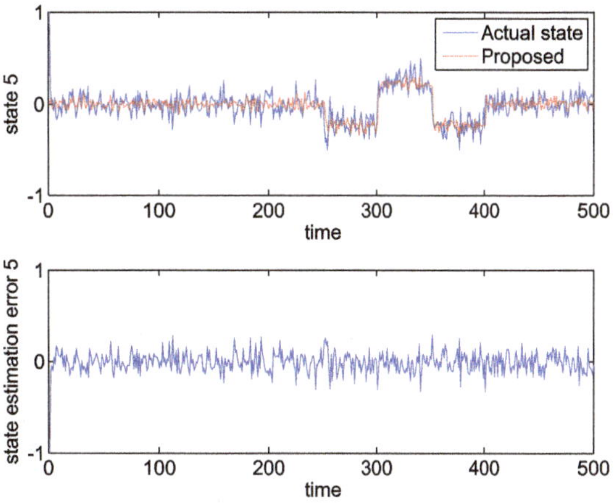


Figure 4. State estimation of the proposed filter and its estimation error.

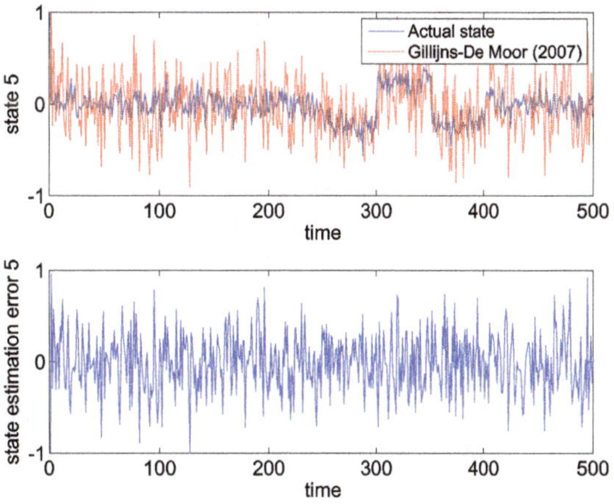


Figure 5. State estimation of the filter in Gillins and De Moor (2007) and its estimation error.

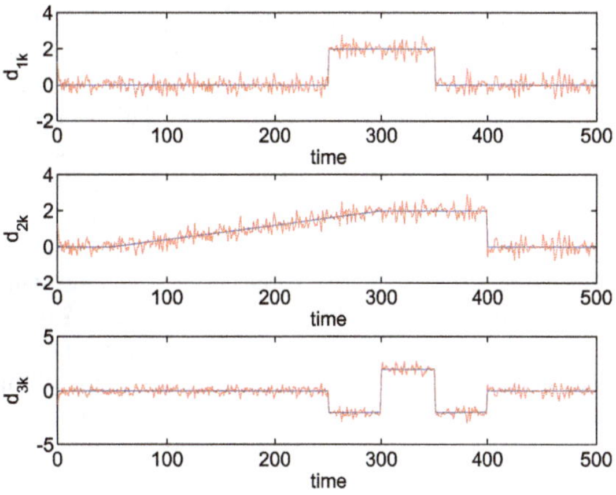


Figure 6. Unknown input estimation based on the proposed filter.

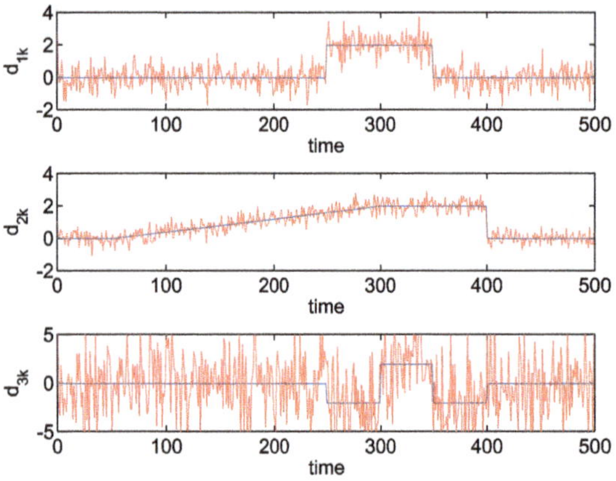


Figure 7. Unknown input estimation based on the filter in Gillins and De Moor (2007).

proposed filter and the filter in Li (2013) are smaller compared with that of Gillins and De Moor (2007) because the additional unknown input information was incorporated into the proposed filter and that of Li (2013).

Finally, we further compared our proposed method with the results in Gillins and De Moor (2007) for the purpose of unknown inputs estimation. The comparison results are shown in Figure 6 (the proposed method) and Figure 7 (the method in Gillins and De Moor (2007)), where real unknown inputs are depicted by real lines, and the unknown input estimations are depicted by the dotted lines.

We can see from Figures 6 and 7 that, by incorporating the information on the unknown inputs, the proposed method can obtain a much better performance for the unknown input estimation.

## 5. Conclusions

In this paper, the problem of SSIE has been investigated when partial information on the unknown inputs is available at an aggregate level. A decoupling approach is used to incorporate the unknown input information into the system dynamics. Then Bayesian inference is drawn to obtain the recursive state and input filter. The relationships of the proposed approach with the existing results are also discussed. Finally, the numerical example shows that, in comparison with the filter without using any input information, the proposed filter that makes use of the input information available at an aggregate level can substantially improve on the quality of both the state and input estimations. Future research can be done to extend the result to the case where there exists direct feedthrough of the partially observed inputs.

## Disclosure statement

No potential conflict of interest was reported by the author(s).


## Funding

This work was jointly funded by UK Engineering and Physical Sciences Research Council (EPSRC) and BAE System (EP/H501401/1).


## References


Chen, W., Ballance, D., Gawthrop, P., & O'Reilly, J. (2000). A nonlinear disturbance observer for robotic manipulators. *IEEE Transaction on Industrial Electronics*, *47*(4), 932–938.

Cheng, Y., Ye, H., Wang, Y. Q., & Zhou, D. H. (2009). Unbiased minimum-variance state estimation for linear systems with unknown input. *Automatica*, *45*(2), 485–491.

Darouach, M., & Zasadzinski, M. (1997). Unbiased minimum variance estimation for systems with unknown exogenous inputs. *Automatica*, *33*(4), 717–719.

Darouach, M., Zasadzinski, M., & Boutayeb, M. (2003). Extension of minimum variance estimation for systems with unknown inputs. *Automatica*, *39*(6), 867–876.

Fang, H.-Z., & Callafon, R. A. D. (2012). On the asymptotic stability of minimum-variance unbiased input and state estimation. *Automatica*, *48*(12), 3183–3186.

Gao, Z. W., & Ding, S. X. (2007). State and disturbance estimator for time-delay systems with application to fault estimation and signal compensation. *IEEE Transaction on Signal Processing*, *55*(12), 5541–5551.

Gillijns, S., & De Moor, B. (2007). Unbiased minimum-variance input and state estimation for linear discrete-time systems. *Automatica*, *43*(1), 111–116.

Hsieh, C. S. (2000). Robust two-stage Kalman filters for systems with unknown inputs. *IEEE Transaction on Automatic Control*, *45*(2), 2374–2378.

Jiang, B., & Fahmida, N. C. (2005). Fault estimation and accommodation for linear MIMO discrete-time systems. *IEEE Transaction on Control System Technology*, *13*(3), 493–499.

Kitanidis, P. K. (1987). Unbiased minimum-variance linear state estimation. *Automatica*, *23*(6), 775–778.

Li, B. (2009). A non-Gaussian Kalman filter with application to the estimation of vehicular speed. *Technometrics*, *51*(2), 162–172.

Li, B. (2013). State estimation with partially observed inputs: A unified Kalman filtering approach. *Automatica*, *49*(3), 816–820.

Mann, G., & Hwang, I. (2013). State estimation and fault detection and identification for constrained stochastic linear hybrid systems. *IET Control Theory and Application*, *7*(1), 1–15.

Patton, P., Clark, T., & Frank, P. M. (1989). *Fault diagnosis in dynamic systems: Theory and applications*. Upper Saddle River, NJ: Prentice-Hall International Series in Systems and Control Engineering, Prentice-Hall.

Profeta, III., Joseph, A., William, G. Vogt, & Marin, H. Mickle (1990). Disturbance estimation and compensation in linear systems. *IEEE Transaction on Aerospace and Electronics Systems*, *26*(2), 225–231.

Simon, D. (2006). *Optimal state estimation: Kalman, $H_\infty$, and non-linear approaches*. New York: Wiley.

Su, J., Chen, W., & Li, B. (2014). Disturbance observer based fault diagnosis. *Proceedings of the 33rd Chinese control conference*, Nanjing, China, pp. 3024–3029.

Su, J., Li, B., & W.-H. Chen (2015a). On existence, optimality and asymptotic stability of the Kalman filter with partially observed inputs. *Automatica*, *53*, 149–154.

Su, J., Li, B., & W.-H. Chen (2015b). Recursive filter with partial knowledge on inputs and outputs. *International Journal of Automation and Computing*, *12*(1), 35–42.
Yang, J., Li, S., Su, J., & Yu, X. (2013). Continuous nonsingular terminal sliding mode control for systems with mismatched disturbances. *Automatica*, *49*(7), 2287–2291.
Yang, J., Su, J., Li, S., & Yu, X. (2014). High-order mismatched disturbance compensation for motion control systems via a continuous dynamic sliding-mode approach. *IEEE Transactions on Industrial Informatics*, *10*(1), 604–614.

## Appendix

### *A.1. Proof of Equation (14)*

First, we can obtain the inverse of $M_k$ as follows:

$$M_k^{-1} = [G_k,\ G_k^{\perp}]\begin{bmatrix} (I - F_{0k}F_{0k}^{\mathrm{T}})D_k^{\mathrm{T}}(D_kD_k^{\mathrm{T}})^{-1} & O & F_{0k} \\ O & I & O \end{bmatrix}.$$

Then, Equation (14) can be obtained as follows:

$$\begin{aligned} D_k\hat{d}_k &= D_k(G_k^{\mathrm{T}}G_k)^{-1}G_k^{\mathrm{T}}M_k^{-1}[r_k\ O\ O]^{\mathrm{T}} \\ &\quad + D_k(G_k^{\mathrm{T}}G_k)^{-1}G_k^{\mathrm{T}}M_k^{-1}[O\ O\ I]^{\mathrm{T}}\hat{\delta}_k \\ &= D_k(G_k^{\mathrm{T}}G_k)^{-1}G_k^{\mathrm{T}}G_k(I - F_{0k}F_{0k}^{\mathrm{T}})D_k^{\mathrm{T}}(D_kD_k^{\mathrm{T}})^{-1}r_k \\ &\quad + D_k(G_k^{\mathrm{T}}G_k)^{-1}G_k^{\mathrm{T}}G_kF_{0k}\hat{\delta}_k \\ &= r_k. \end{aligned}$$

### *A.2. Proof of Equation (15)*

Define $M_{k-1}^P = P_{k|k}\bar{M}_{k-1}^{\mathrm{T}}(\bar{M}_{k-1}P_{k|k-1}\bar{M}_{k-1}^{\mathrm{T}})^{-1}$. Then, we have

$$\begin{aligned} &M_{k-1}^{-1}\tilde{r}_{k-1} - K_kC_kM_{k-1}^{-1}\tilde{r}_{k-1} \\ &= (I - K_kC_k)M_{k-1}^{-1}\tilde{r}_{k-1} \\ &= M_{k-1}^P\bar{M}_{k-1}M_{k-1}^{-1}\tilde{r}_{k-1} \\ &= M_{k-1}^P\bar{M}_{k-1}G_{k-1}D_{k-1}^{\mathrm{T}}(D_{k-1}D_{k-1}^{\mathrm{T}})^{-1}D_{k-1}d_{k-1} \\ &= M_{k-1}^P\begin{bmatrix} D_{k-1} & O \\ O & I \end{bmatrix}[G_{k-1},\ G_{k-1}^{\perp}]^{-1} \\ &\quad \times G_{k-1}D_{k-1}^{\mathrm{T}}(D_{k-1}D_{k-1}^{\mathrm{T}})^{-1}D_{k-1}d_{k-1} \\ &= M_{k-1}^P\begin{bmatrix} r_{k-1} \\ O \end{bmatrix} = M_{k-1}^P\bar{M}_{k-1}G_{k-1}d_{k-1} \\ &= P_{k|k}\bar{M}_{k-1}^{\mathrm{T}}(\bar{M}_{k-1}P_{k|k-1}\bar{M}_{k-1}^{\mathrm{T}})^{-1}\tilde{r}_{k-1}, \end{aligned}$$

where in the above derivation, we have used the following identities:

$$M_{k-1}^{-1}\tilde{r}_{k-1} = G_{k-1}D_{k-1}^{\mathrm{T}}(D_{k-1}D_{k-1}^{\mathrm{T}})^{-1}D_{k-1}d_{k-1}, \tag{A1}$$

$$I - K_kC_k = M_{k-1}^P\bar{M}_{k-1}. \tag{A2}$$

Now we show Equation (A2):

$$\begin{aligned} &I - K_kC_k - M_{k-1}^P\bar{M}_{k-1} \\ &= I - K_kC_k - P_{k|k}\bar{M}_{k-1}^{\mathrm{T}}(\bar{M}_{k-1}P_{k|k-1}\bar{M}_{k-1}^{\mathrm{T}})^{-1}\bar{M}_{k-1} \\ &= I - P_{k|k}C_k^{\mathrm{T}}R_k^{-1}C_k - P_{k|k}[\bar{M}_{k-1}^{\mathrm{T}}(\bar{M}_{k-1}P_{k|k-1}\bar{M}_{k-1}^{\mathrm{T}})^{-1} \\ &\quad \times \bar{M}_{k-1} + C_k^{\mathrm{T}}R_k^{-1}C_k - C_k^{\mathrm{T}}R_k^{-1}C_k] \\ &= I - P_{k|k}C_k^{\mathrm{T}}R_k^{-1}C_k - [I - P_{k|k}C_k^{\mathrm{T}}R_k^{-1}C_k] \\ &= O, \end{aligned}$$

where $\bar{M}_k = \begin{bmatrix} D_k & O \\ O & I \end{bmatrix}[G_k,\ G_k^{\perp}]^{-1}$.

# 3

# Modeling and model predictive control of dividing wall column for separation of Benzene–Toluene-*o*-Xylene

Rajeev Kumar Dohare[a]*, Kailash Singh[a] and Rajesh Kumar[b]

[a]*Department of Chemical Engineering, Malaviya National Institute of Technology, Jaipur 302017, Rajasthan, India;* [b]*Department of Electrical Engineering, Malaviya National Institute of Technology, Jaipur 302017, Rajasthan, India*

In this paper, dividing wall column (DWC) has been chosen for a BTX (Benzene–Toluene–*o*-Xylene) system. A MATLAB® program has been written for nonlinear unsteady-state DWC, which is used in Simulink environment for control of the system by Model Predictive Control (MPC). Compositions of the three products (benzene, toluene, and *o*-xylene) are indirectly controlled by controlling the corresponding temperatures of the respective tray due to requirement of online analyzer. The temperature of uppermost tray in the rectifying section, stage temperature in the main column corresponding to the side stream withdrawn, and the bottom stage temperature in the stripping section have been chosen in order to maintain the compositions of the three products. The manipulated variables are reflux rate ($L_0$), side-stream flow rate (SSRF), and reboiler heat duty ($Q_B$). It has been observed that MPC shows good performance even in the presence of $\pm$ 10% change in the feed flow rate, feed composition, and liquid split factor in comparison with conventional controllers. The MPC has less settling time (almost 1.5 h) compared with the PI controller (approximately 3–4 h).

**Keywords:** BTX separation; dividing wall column; multivariable control; model predictive control

## 1. Introduction

Distillation is the most desirable process for the separation of multi-component liquid mixture. However, when separation of high-purity products is required, a single distillation column is not sufficient; for the separation of $n$-component mixture, $n - 1$ columns are required (Sotudeh & Shahraki, 2007). An important effort has been focused on the development of a new design, optimization, and control method for thermally coupled distillation columns, which provide saving up to 30–40% of the total annual cost for the separation of some multi-component liquid mixtures as compared with classical distillation sequences (Triantafyllou & Smith, 1992). The use of fully thermally coupled distillation arrangements, such as the dividing wall column (DWC), leads to significant reductions in both energy and capital costs when compared with conventional two-column arrangements in the separation of ternary mixtures.

Although some authors have studied control of DWC to improve the process operation, yet model-based control finds its good applicability on this type of the distillation column. DWC is a nonlinear system due to its inherent complex dynamics because of the middle wall. Therefore, this system is challenging to control at its optimum operating parameters. Rewagad and Kiss (2012) showed the model predictive control (MPC) results for composition control; however, in this study, we have focused the control of temperatures. Besides disturbance in feed flow rate and feed composition, the disturbance in liquid split factor has also been considered. The parameter-based controllers, like proportional-integral-derivative controllers, are characterized by a short development time and smaller development efforts. The development of MPC for a DWC is a promising approach as the advantages of this control methodology can be intensified in view of the operational difficulties of the column. For maintaining the product purity at its desirable conditions, proper control is required. Therefore, model predictive control was selected to analyze its suitability.

The huge impact of distillation processes in both operation and investment costs has motivated the development of various types of fully thermally coupled distillation columns that can be used in saving energy and capital cost (Halvorsen & Skogestad, 2004). Conventionally, a ternary mixture can be separated via direct sequence (most volatile component is separated first), indirect sequence (heaviest component is separated first), or distributed sequence (mid-split) consisting of two to three distillation columns (Van Diggelen, Kiss, & Heemink, 2010). Eventually, this led to the concept known today as DWC that integrates in fact the two columns of a Petlyuk system into one column shell (Kaibel, 1987; Kolbe & Wenzel, 2004; Schultz et al.,

*Corresponding author. Email: rajeevdohare@gmail.com

2002). Mostly, the position of the wall in the DWC is in the middle, but off-center positions of the dividing wall may also be possible. This might be useful in situations, when the concentration of the medium boiling component is small as compared with the overhead and bottom (Asprion & Kaibel, 2010). Woinaroschy and Isopescu (2010) investigated the ability of computational dynamic programming to solve time optimal control of DWC. They focused on start-up control of the DWC.

In DWC, a single column is divided into four sections by inserting a wall in it. DWC is especially advantageous for separating ternary mixtures. The schematic diagram of the DWC is shown in Figure 1. It is divided into four sections: top section of the column is known as a rectifying section, left section as prefractionator, right section as main column, and the bottom section as stripping section. Feed consisting of 30 mol% benzene, 30 mol% toluene, and 40 mol% *o*-xylene is introduced on the 12th stage in the prefractionator. Almost pure toluene is withdrawn from the 11th stage of the main column. Benzene and *o*-xylene are obtained as top and bottom products, respectively.

Van Diggelen et al. (2010) proposed a model of DWC with the assumptions of constant pressure, no vapor flow dynamics, linearized liquid dynamics, and neglecting energy balance and changes in enthalpy. The researchers used this model to compare various control strategies. Hiller, Buck, Ehlers, and Fieg (2010) developed a non-equilibrium stage model by assuming heat and mass transfer between the liquid and vapor phases for the ideal component system. Ignat and Woinaroschy (2011) used a dynamic model for minimizing the distillation start-up time for the separation of an ideal benzene–toluene–ethylbenzene ternary mixture and the separation of a non-ideal methanol–ethanol–1-propanol mixture. The effect of

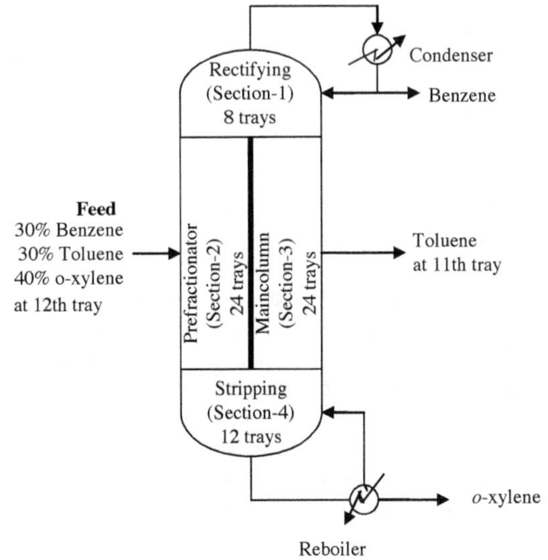


Figure 1. Schematic diagram of the DWC.

the liquid split ratio and vapor split ratio is responsible for internal disturbance and the dynamics of the DWC. Ignat and Woinaroschy (2011) also assumed the hold-up volume in the column. The vapor split ratio is assumed to be constant and liquid split ratio as a load change in this study.

Recently, Buck, Hiller, and Fieg (2011) also reported experimentally, temperature profile of the column by implementing MPC. Their study proves the real-life practicability of MPC; they did not provide an analysis of the transient behavior of DWC under desired disturbances. Kvernland, Halvorsen, and Skogestad (2010) applied MPC on the Kaibel column that separates a feed stream into four product streams using only a single column. The main objective for optimal operation was to minimize the total impurity flow. Finally, they concluded that MPC obtained typically less total impurity flow as compared with conventional decentralized control.

In this work, a mathematical model of DWC has been developed assuming non-constant volatility. Several simulation runs of the model have been used to investigate the effects of several parameters and dynamics of the system. BTX (Benzene–Toluene–Xylene) has been chosen as a component system. A MATLAB® program was written for unsteady-state DWC for investigating the MPC methodology. The controlled variables selected are: temperature of uppermost tray in the rectifying section, stage temperature corresponding to the side stream withdrawn, and the bottom stage temperature in the stripping section in order to maintain the compositions of the three products (Ling & Luyben, 2009). The manipulated variables are reflux rate, side-stream flow rate (SSRF), and reboiler heat duty. The performance of MPC controller has been verified by giving load changes in the feed flow rate and feed composition. The performance of MPC in the presence of disturbances variables has been compared with that of PI controller.

## 2. Mathematical model

The following assumptions have been taken into account for model development:

(1) Constant volume holdup of condenser/reflux
(2) Fast energy dynamics on trays (molar enthalpy change on trays with respect to time is negligible)
(3) Vapor split ratio is constant as it is fixed by column design.
(4) Raoult's law is assumed as BTX is close to ideal solution.

For the rectifying section, prefractionator, main column, and stripping section:

Mass balance for component $i$ at tray $j$,

$$\frac{\mathrm{d}(M_j x_{j,i})}{\mathrm{d}t} = V_{j+1}, y_{j+1,i} + L_{j-1}x_{j-1,i} - V_j y_{j,i} - L_j x_{j,i} + F_j z_{j,i} - S_j x_{j,i}, \quad (1)$$

where $n_c$ denotes the number of components and $M_j$ denotes the total liquid holdup on the $j$th stage.

Summation Equations:

$$\sum_i x_{j,i} = 1; \quad \sum_i y_{j,i} = 1, \tag{2}$$

where $x_{j,i}$ and $y_{j,i}$ represent the mole fraction of the $i$th component in liquid and vapor phase at the $j$th stage, respectively.

Energy Balance at tray $j$:

$$\frac{d(H_{L,j} M_j)}{dt} = V_{j+1}H_{V_{j+1}} + L_{j-1}H_{L_{j-1}} - V_i H_{V_j} - L_j H_{L_j} + F_j H_{F_j} - S_j H_{L_j}. \tag{3}$$

In energy balance equation, $H_{L_j}$ and $H_{V_j}$ represent the liquid and vapor enthalpy on the $j$th stage, respectively.

Equilibrium relationship:

$$y_{j,i} = K_{j,i}x_{j,i} \text{ where } K_{j,i} = \frac{\gamma_{j,i}P_{j,i}^{\text{sat}}}{P_j}, \tag{4}$$

where $\gamma_{j,i}$ denotes the activity coefficient of the $i$th component in liquid phase at the $j$th stage.

For the condenser,

Material Balance:

$$\frac{d(M_0 x_{D,i})}{dt} = V_1 y_{1,i} + L_0 x_{D,i} - D x_{D,i}. \tag{5}$$

Energy Balance:

$$\frac{d(M_0 H_D)}{dt} = V_1 H_{V,1} - L_0 H_{L,0} - D H_D - q_C. \tag{6}$$

Summation equation:

$$\sum_i x_{D,i} = 1. \tag{7}$$

For the reboiler,

material Balance:

$$\frac{d(M_R x_{w,i})}{dt} = L_{n_4}x_{n_4,i} - V_{n_4+1}y_{n_4+1,i} - w x_{n_4+1,i}, \tag{8}$$

where $M_R$ denotes liquid holdup in reboiler

Energy Balance:

$$\frac{d(M_{n_4+1}H_{L,n_4+1})}{dt} = L_{n_4}H_{L,n_4} - H_{V,n_4+1}V_{n_4+1} - wH_{L,n_4+1} + q_R. \tag{9}$$

Equilibrium Relationship:

$$y_{n_4+1,i} = K_{n_4+1,i}x_{w,i}. \tag{10}$$

Summation Equations:

$$\sum_i x_{n_4+1,i} = 1 \quad \text{and} \quad \sum_i y_{n_4+1,i} = 1. \tag{11}$$

At the intersection of rectifying section (Section 1) with prefractionator (Section 2) and main column (Section 3):

Vapor Mixing:

$$V_{n_1+1}^{(1)} = V_1^{(2)} + V_1^{(3)}, \tag{12}$$

$$V_{n_1+1}^{(1)}y_{n_1+1,i}^{(1)} = V_1^{(2)}y_{1,i}^{(2)} + V_1^{(3)}y_{1,i}^{(3)}. \tag{13}$$

Liquid Splitting:

$$\begin{aligned} L_0^{(2)} &= \alpha L_{n_1}^{(1)}, \\ L_0^{(3)} &= (1-\alpha)L_{n_1}^{(1)}, \end{aligned} \tag{14}$$

where $\alpha$ is liquid split factor.

$$x_{0,i}^{(2)} = x_{n_1,i}^{(1)}, \quad x_{0,i}^{(3)} = x_{n_1,i}^{(1)}.$$

At the intersection of Sections 2 and 3 with Section 4 (Stripping section):

Vapor splitting:

$$\begin{aligned} V_{n_2+1}^{(2)} &= \beta V_1^{(4)}; \\ V_{n_2+1}^{(3)} &= (1-\beta)V_1^{(4)}; \end{aligned} \tag{15}$$

where $\beta$ is a vapor splitting factor.

$$y_{n_2+1,i}^{(2)} = y_{1,i}^{(4)};$$

$$y_{n_2+1,i}^{(3)} = y_{1,i}^{(4)}.$$

Liquid mixing:

$$L_0^{(4)} = L_{n_2}^{(2)} + L_{n_2}^{(3)}, \tag{16}$$

$$L_{n_2}^{(2)}x_{n_2,i}^{(2)} + L_{n_2}^{(3)}x_{n_2,i}^{(3)} = L_0^{(4)}x_{0,i}^{(4)}. \tag{17}$$

These model equations are set of ordinary differential equations-initial value problems, which were solved by ode15s solver (an inbuilt function in MATLAB® to solve ordinary differential equations). For simulation, physical properties and the nominal operating conditions of the investigated system (Benzene, Toluene and *o*-Xylene) are given in Table 1.

## 3. Model predictive control of a DWC

The controllability indicator such as relative gain array (RGA) is a useful method to understand the behavior of the system (Segovia-Hernandez, Hernandez-Vargas, & Marquez-Munoz, 2007; Skogestad & Postlethwaite, 2005). The RGA provides information about the interactions among the controlled and manipulated variables. The RGA element has been calculated as the ratio of open-loop gain to the closed-loop gain for a pair of variables.

For a selected pair of variables, values of the RGA element close to unity are preferred as the best combination. Due to complex interactions among the manipulated and controlled variables, decentralized controller's performance is not very good.

Table 1. Properties of the system and nominal operating parameters.

| Properties of the system | Benzene | Toluene | Xylene |
|---|---|---|---|
| Molecular formula | $C_6H_6$ | $C_7H_8$ | $C_8H_{10}$ |
| Molecular weight | 78.11 | 92.14 | 106.17 |
| Boiling point | 80.2 °C | 110.7 °C | 144.5 °C |
| Density (mol/cm$^3$) | 0.01133 | 0.00941 | 0.00829 |
| Critical temperature | 289.1 °C | 318.7 °C | 357.2 °C |
| Antoine equation and its constants | | | |
| $T_{\text{sat}} = B/(A - \log P) - C$, units of T in K and P in atm | | | |
| Components | A | B | C |
| Benzene | 15.9008 | 2788.51 | 220.79 |
| Toluene | 16.0137 | 3096.52 | 219.48 |
| Xylene | 16.1156 | 3395.57 | 213.69 |
| Nominal operating parameters | Value | | |
| Feed flow rate | 1 kmol/s | | |
| Feed temperature | 85 °C | | |
| Feed composition (mol%) | 30% B, 30% T, 40% X | | |
| Reflux rate | 0.860 kmol/s | | |
| Vapor split factor | 0.627 | | |
| Liquid split factor | 0.353 | | |
| Side stream flow rate | 0.296 kmol/s | | |
| Reboiler heat duty | 40.544 MW | | |
| Bottoms flow rate | 0.401 kmol/s | | |
| Reflux ratio | 2.84 | | |

The DWC is a multivariable process that gives motivation for use of a MPC. The MPC offers a large number of operational advantages for diverse processes like reaction, separation, etc. (Backx, Bosgra, & Marquardt, 2000; De Temmerman, Dufour, Nicolaï, & Ramon, 2009). In view of the multi-input–multi-output system, the MPC is characterized by many manipulated and controlled variables and it may be performed for a complex system (Buck et al., 2011). It is an appropriate controller to control a multivariable process with considerable interactions. The MPC shows a good result in eliminating loop interaction when it is compared with a decentralized control scheme. Model predictive controller is good for creating a linear model at any operating point. Total operating cost and settling time can be reduced by controlling the temperature of the different sections in spite of the compositions of the products (Rodríguez Hernández & Chinea-Herranz, 2012). The manipulated variables, reflux ratio, side stream flow rate, and reboiler heat duty are used to control the temperatures (and therefore product purities). For investigating the controllability of MPCs, MPC Toolbox in MATLAB® was used. The cost function in MPC methodology is

$$\text{Min}\,F = \sum_{j=1}^{N_y}\sum_{i=1}^{P}(y_{\text{sp}_{i,i}} - y^*_{i,j})^2 w_{y,j} + \sum_{j=1}^{N_u}\sum_{i=1}^{M}(\Delta u_{i,j})^2 w_{u,j}, \tag{18}$$

where $N_u$ and $N_y$ denote the number of manipulated and controlled variables, respectively. $p$ denotes the length of prediction horizon and $M$ denotes the length of control horizon with respect to the sequence of input increments $\Delta u$.
subject to:

$$\Delta u_{\text{LB}} \le \Delta u \le \Delta u_{\text{UB}}, \tag{19}$$

$$u_{\text{LB}} \le u \le u_{\text{UB}}, \tag{20}$$

where $u_{\text{LB}}$ and $u_{\text{UB}}$ represent the lower and upper bounds to control the oscillations of the manipulated variables.

Model predictive control works on prediction horizon, control horizon and robustness factor parameters. Prediction horizon is the number of sampling intervals over which the cost function is minimized. Control horizon is the number of sampling intervals over which the control moves are estimated by optimization routine. The best suitable MPC parameters based on several runs by selecting different values are given in Table 2. Robustness factor controls the sluggishness and the speed of the controller. The working scheme of MPC is shown in Figure 2.

These parameters have a high impact on the total controller performance. The lower and upper limits on the manipulated variables (reflux rate, side stream flow rate,

Table 2. Best suitable model predictive control parameters.

| Parameter | Value |
|---|---|
| Control interval (sampling interval) | 10 s |
| Prediction horizon | 20 |
| Control horizon | 6 |
| Robustness factor | 0.25 |

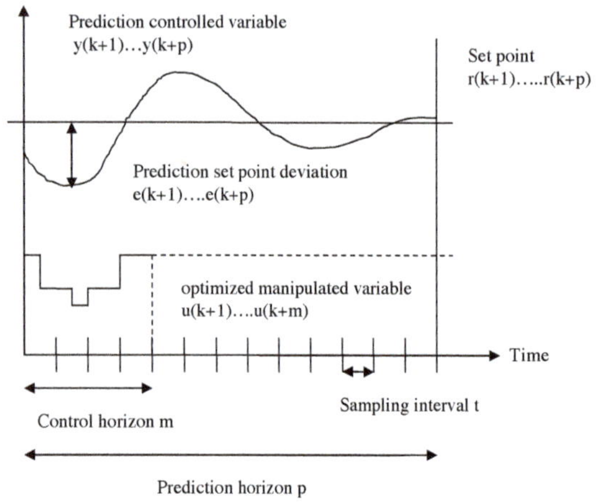


Figure 2. Working scheme of model predictive control.

and reboiler heat duty) have been set as [360.5 96 20.54] and [2000 496 60.54], respectively. The prediction horizon is coupled with the system's time constants and the chosen sampling rate (Agachi, Nagy, Cristea, & Imre-Lucaci, 2006).

In this study, temperature of tray 6 in Section 1 (Tsec1 (6)), temperature of tray 18 in Section 3 (Tsec3 (18)), and temperature of tray 6 in Section 4 (Tsec4 (6)) have been controlled by the reflux rate ($L_0$), side stream flow rate (SSFR) and reboiler heat duty ($Q_B$).

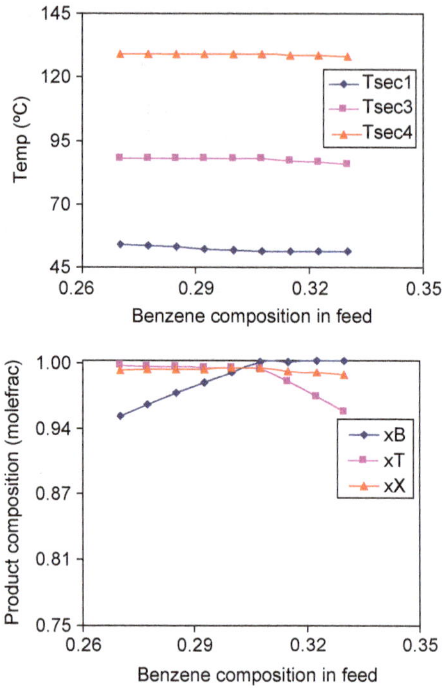


Figure 3. Effect of benzene composition in feed on temperatures and product composition.

## 4. Effect of parameters on temperature and composition

Some important parameters of DWC such as feed composition of benzene, toluene, and xylene, feed flow rate, reflux rate, SSRF, and reboiler heat duty were selected for investigation of the effect on temperatures and the resulting product compositions.

### 4.1. *Effect of benzene composition in the feed*

The effect of benzene composition of the feed is shown in Figure 3. While changing the benzene composition, the other two components were assumed to be in the same proportion. As can be seen, on increasing benzene composition in the feed, there is no significant change in *o*-xylene composition in the bottoms product. Toluene composition in the middle product does not change much until 30 mol% of benzene composition in the feed, after which there is a declining trend because extra benzene is mixed up in the middle product. The benzene composition of the top product increases up to 30 mol% of benzene in the feed, beyond which, there is no significant change. The temperatures also follow the trend accordingly. Therefore, it is evident that 30 mol% composition of benzene in feed is an optimum value giving the maximum purity of all the products.

### 4.2. *Effect of toluene composition in the feed*

The effect of toluene composition of the feed is shown in Figure 4. On increasing the feed composition of toluene, the benzene composition in the top product and *o*-xylene composition of the bottom product do not show significant effect up to 30 mol% of toluene composition in the feed but decreases afterwards. However, toluene composition in the middle product increases and becomes almost constant after 30 mol% of toluene composition in the feed. Therefore, the toluene composition of 30% seems to be an optimum feed composition as it gives the highest purity of all the products. The temperatures also follow the trend accordingly in equilibrium with the product composition.

### 4.3. *Effect of o-xylene composition in the feed*

The effect of *o*-xylene composition in the feed is shown in Figure 5. While changing the *o*-xylene composition, the other two components are supposed to be in the same proportion. Purity of benzene is constant up to 40 mol% *o*-xylene in the feed after which, it suddenly decreases. Moreover, *o*-xylene purity of the product increases up to 40% and then it becomes constant on further increasing xylene composition. Toluene purity decreases beyond 40

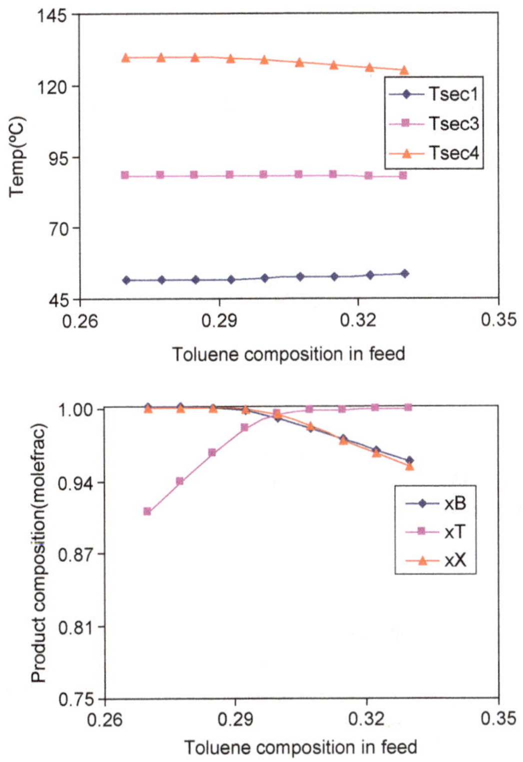


Figure 4. Effect of toluene composition in feed on temperature and product composition.

mol% composition of the *o*-xylene in the feed. Therefore, the temperature in the bottom section (stripping section) increases and the temperatures in the top section (rectifying section) and middle section decrease slightly.

### *4.4. Effect of feed flow rate*

As the feed flow rate increases up to 1000 mol/s, benzene and toluene compositions in the product increase, but the xylene composition remains maximum at almost 100%. On increasing the feed flow rate further, the benzene composition remains constant; however, other two product

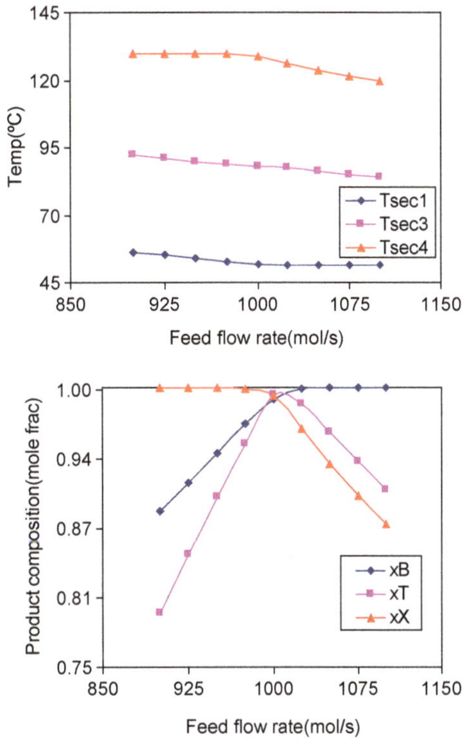


Figure 6. Effect of feed flow rate on temperature and product composition.

compositions decrease as shown in Figure 6. Accordingly, there is a small decrease in the temperature of the rectifying section, while the temperatures of the main column and stripping section decrease significantly.

### *4.5. Effect of reflux rate*

The reflux rate plays an important role in the product purities as shown in Figure 7. Below reflux rate of 860 mol/s, benzene and toluene compositions increase and *o*-xylene composition in the bottom product remains constant at almost 100%. For more than reflux rate of 860 mol/s, benzene composition of the top product remains

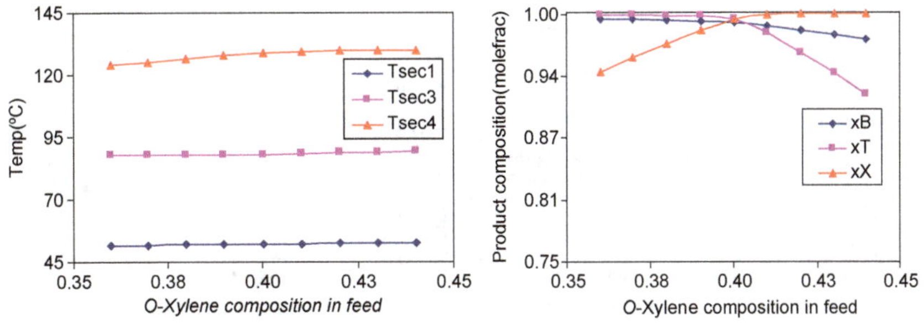


Figure 5. Effect of *o*-xylene composition in feed on temperature and product composition.

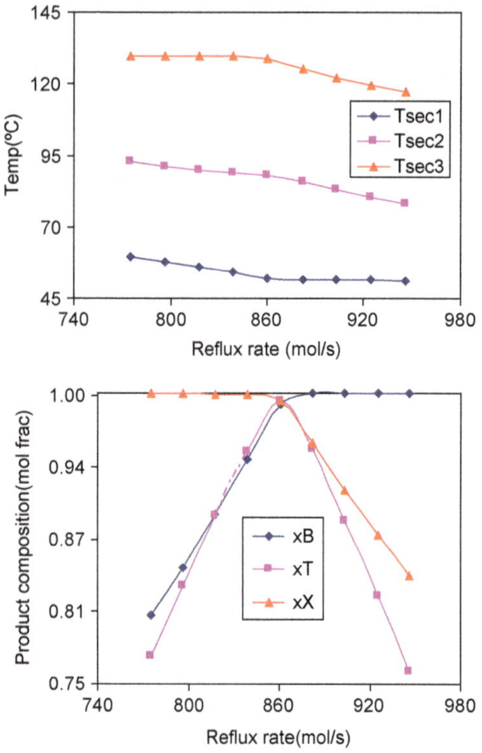


Figure 7. Effect of reflux rate on temperature and product composition.

constant, but *o*-xylene and toluene compositions decrease sharply. Accordingly, the temperature in the rectifying section (Tsec1) initially decreases and then becomes constant after 860 mol/s of reflux rate, while the temperature in the stripping section (Tsec4) decreases after 860 mol/s of reflux rate.

### 4.6. *Effect of side stream flow rate*

For side stream flow rate less than 296 mol/s, benzene composition of the top product does not change much for a change in the side stream flow rate; however, toluene composition in the middle product decreases after 296 mol/s as shown in Figure 8. The *o*-xylene composition in the bottom product increases and becomes constant after 296 mol/s of side stream flow rate. Therefore, side stream flow rate of 296 mol/s is the optimum flow rate as it gives the highest purity of all the products.

### 4.7. *Effect of reboiler heat duty*

The effect of the reboiler heat duty is shown in Figure 9. As can be seen, benzene composition of the top product does not show significant change until 41 MW, beyond which it starts decreasing. Toluene composition first increases and then decreases, and *o*-xylene in the bottom product increases up to 41 MW after which it remains almost constant. Therefore, 41 MW is the optimum reboiler heat duty

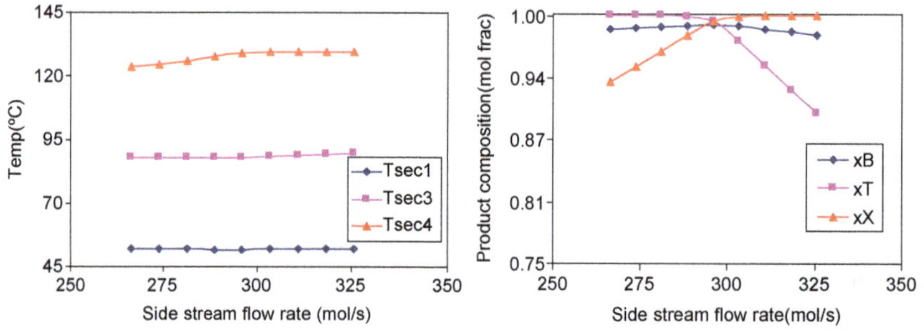


Figure 8. Effect of side stream flow rate on temperature product composition.

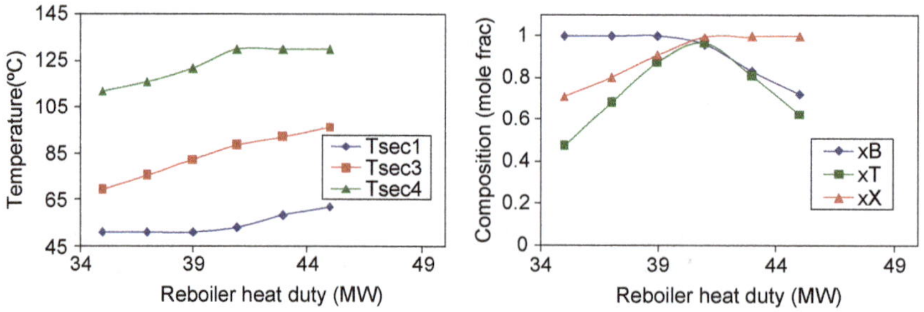


Figure 9. Effect of reboiler duty on temperature and product composition.

at which all the three products have highest purity at 99.9%. The temperatures increase on increasing reboiler heat duty.

## 5. Results for model predictive control of DWC

### *5.1. Load change in benzene composition of feed*

± 10% load changes were given in the feed flow rate to observe the performance of the MPC as shown in Figure 10. Temperature overshoots in Section 1 are approximately 0.3 °C and 0.2 °C for − 10% and + 10% change, respectively. Similarly, there is only very small overshoot (only 0.5%) in the product composition. Moreover, the offset is negligibly very small in both cases.

To compare the performance of MPC with conventional controller, Ling and Luyben's (2010) results have been considered for the study. The PI controller stabilized the controlled variables in about 3–4 h; however, time taken by MPC controller is one-fourth time of that by PI controller. Temperature overshoot in Section 1 is 0.35 and 0.4 °C for − 10% and + 10% change, respectively, in case of PI controller; in case of MPC, this is marginally lower, that is, 0.3 °C and 0.2 °C, respectively. Similarly, in Section 3, temperature overshoot is 0.6 in PI control; however, in case of MPC, there is very small overshoot, that is, 0.13 °C and 0.12 °C for − 10% and + 10% load change, respectively. In Section 4 also, the overshoot is marginally better for MPC performance.

### *5.2. Load change in toluene composition of feed*

The MPC performance for ± 10% load change in toluene feed composition is also studied as shown in Figure 11. The maximum overshoot was 0.1 °C in the temperatures. The controller brought back the temperature to their set points within approximately half an hour. Accordingly, the composition was also controlled with negligible offset.

PI control achieved stabilization in 4–5 h as reported by Ling and Luyben (2010); however, MPC stabilized in 0.6 h upon ± 10% disturbance in toluene feed composition. Temperature overshoot in Section 1 is 0.5 °C in PI and 0.1 °C in MPC Controller. Temperature overshoot in Section 3 is 0.7 in case of PI controller and 0.1 in MPC. Similarly, in Section 4, the overshoot is marginally better.

### *5.3. Load change in o-xylene composition of feed*

The MPC performance for load change of ± 10% in *o*-xylene composition of feed is shown in Figure 12. The maximum overshoot is approximately 0.2 °C in temperature with almost no offset. The product compositions were also controlled according to their initial values. PI control achieved stabilization in 4–5 h as reported by Ling and Luyben (2010); however, MPC stabilized in 1.5 h upon ± 10% disturbance in the *o*-xylene feed composition. Temperature overshoot in Sections 1 and 3 are a little bit better in the case of MPC; however, in Section 4, MPC shows more overshoot than PI control (a difference of 0.5 °C).

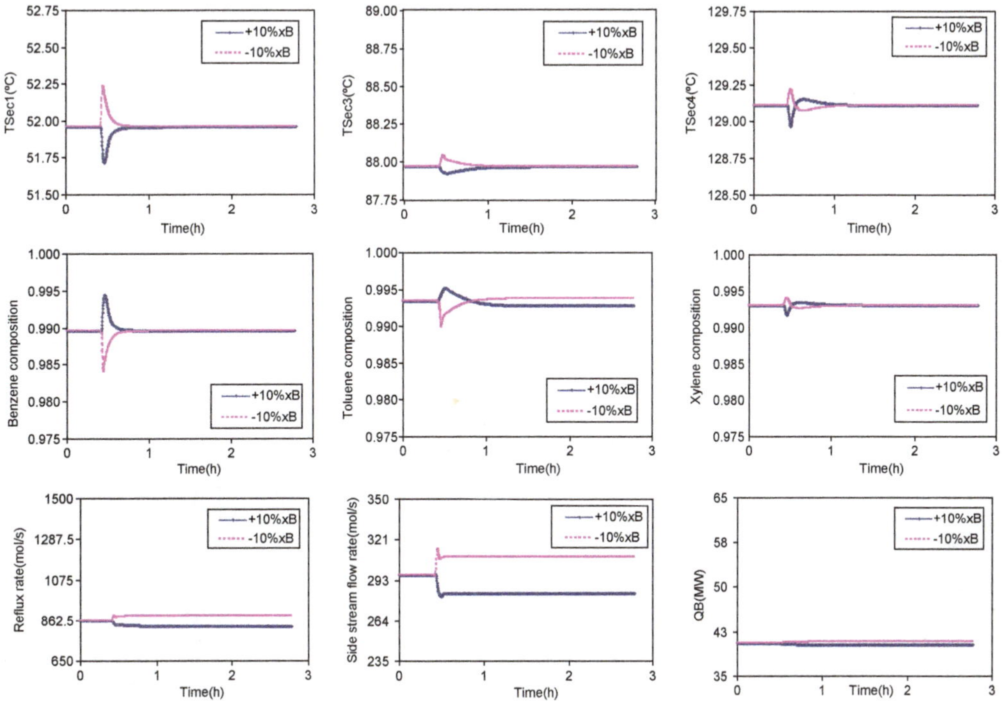


Figure 10. MPC performance for ± 10% disturbance in benzene feed composition.

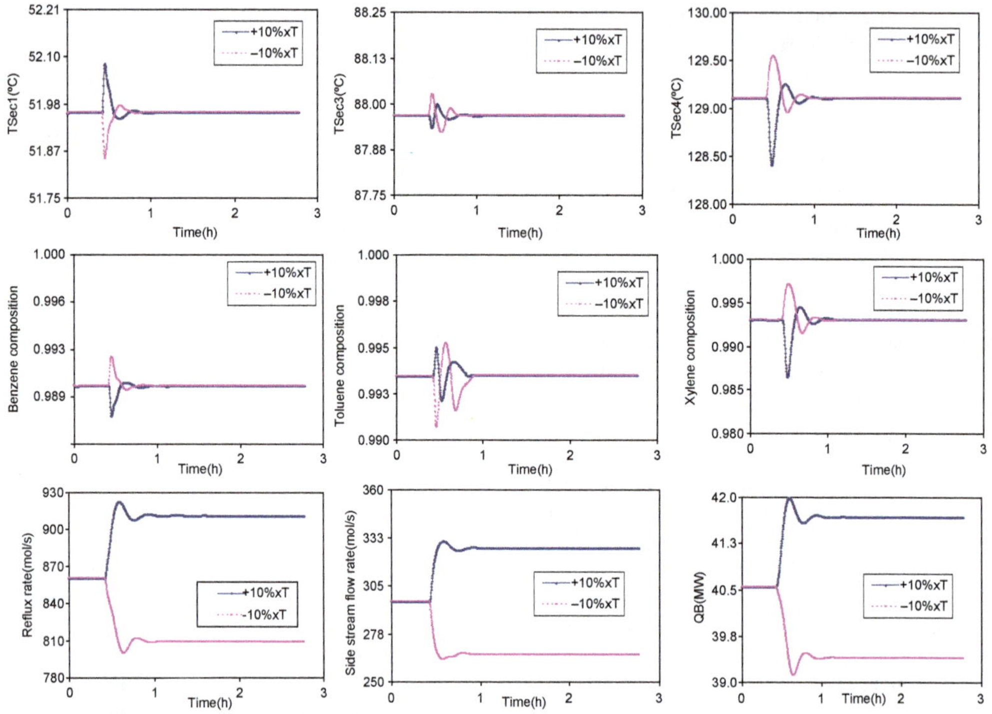

Figure 11. MPC performance for ± 10% disturbance in toluene feed composition.

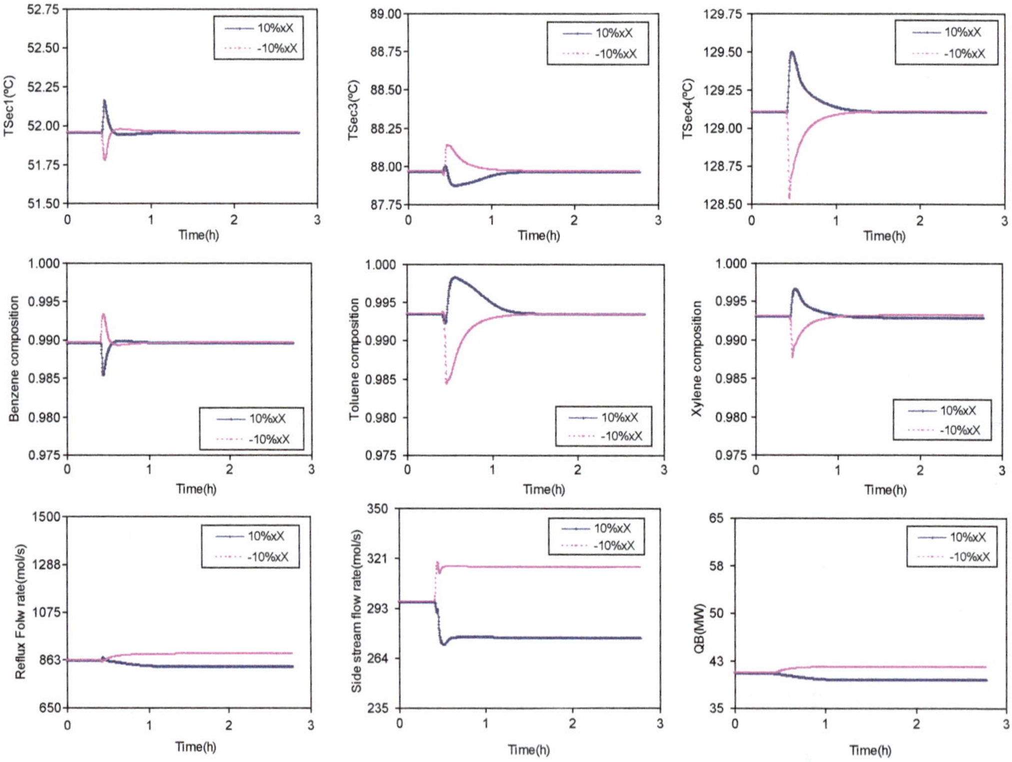

Figure 12. MPC performance for ± 10% disturbance in *o*-xylene feed composition.

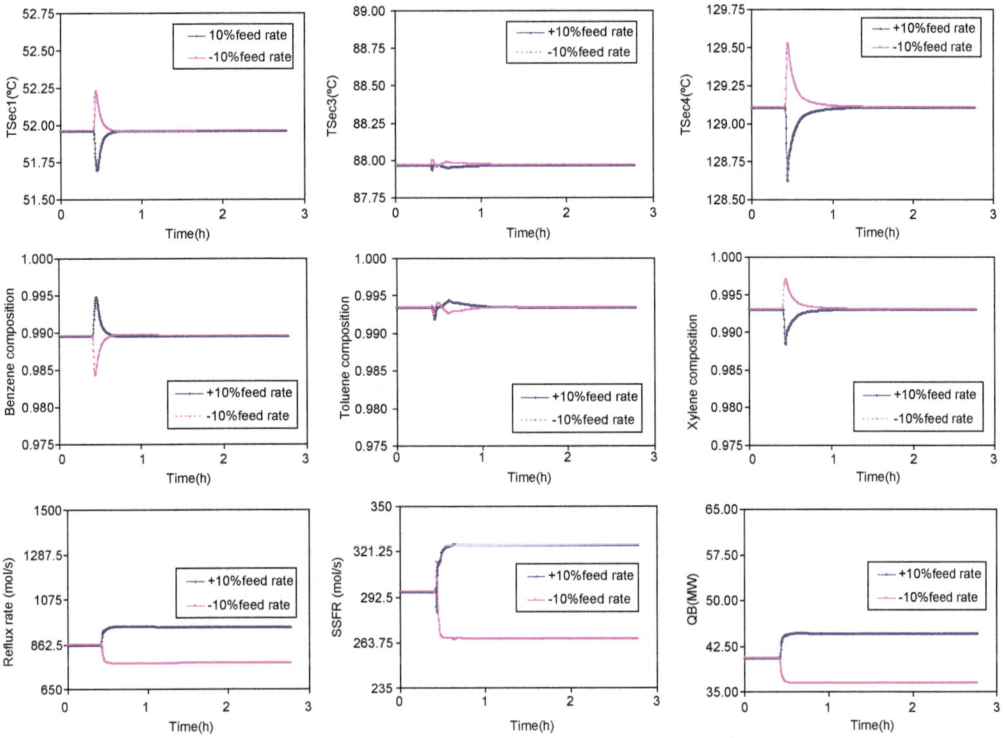


Figure 13. MPC performance for ± 10% disturbance in feed flow rate.

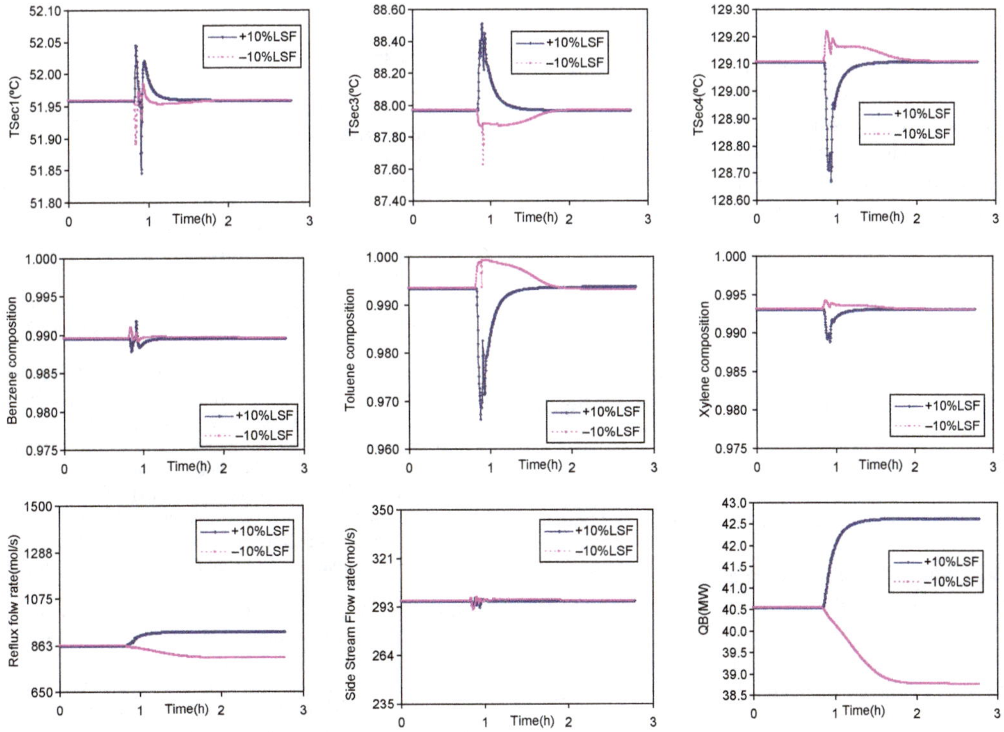


Figure 14. MPC performance for ± 10% disturbance in liquid split factor.

### 5.4. *Load change in feed flow rate*

The MPC response for ±10% load change in feed flow rate is shown in Figure 13. The temperature of all the three sections was controlled at their respective set points after maximum overshoot of 0.4 °C. The compositions were also maintained at their initial values with maximum deviation of 0.5%. PI controller stabilized in about 4–5 h (Ling and Luyben, 2010); however, MPC achieved the control performance in 1.5 h.

### 5.5. *Load change in liquid split factor*

The response for ±10% load change in liquid split factor is shown in Figure 14. The temperatures in Sections 1, 3, and 4 are controlled at their respective set points with maximum overshoot of 0.1, 0.4, and 0.4, respectively. Toluene composition shows a maximum overshoot of 0.02; other two compositions have a smaller overshoot in comparison with the toluene.

## 6. Conclusions

A mathematical model was developed for DWC, which was used as a case study for studying the control behavior of model predictive control. The controlled variables were selected as the temperatures of the 6th tray in the rectifying section, the 18th tray in the main column, and the 6th tray in the stripping section to ensure the maximum purity of benzene, toluene, and xylene, respectively. The manipulated variables selected are reflux rate, side stream flow rate, and reboiler heat duty. Model predictive controller was able to control the temperatures (and therefore product purities) of all the three sections in the presence of ±10% load changes in feed composition, flow rate, and liquid split factor. The performance of MPC was also compared with conventional controller and it is concluded that being a multivariable controller, MPC performs better than PI controller.

## Nomenclature

| | |
|---|---|
| $M_j$ | Total liquid holdup on the $j$th stage, moles |
| $x_{j,i}$ | Mole fraction of the $i$th component in the liquid phase at the $j$th stage |
| $V_j$ | Vapor flow rate from the $j$th stage, mole/s |
| $L_j$ | Liquid flow rate from the $j$th stage, mole/s |
| $y_{j,i}$ | Mole fraction of the $i$th component in the vapor phase at the $j$th stage |
| $F_j$ | Feed flow rate on the $j$th stage, mole/s |
| $Z_{j,i}$ | Mole fraction of the $i$th component in feed at the $j$th stage |
| $S_j$ | Side stream flow rate at the $j$th stage, mole/s |
| $n_k$ | Number of stages in the $k$th section ($k = 1,2,3$, and 4) |
| $n_c$ | Number of components = 3 |
| $H_{L_j}$ | Liquid enthalpy on the $j$th stage, J/mole |
| $H_{V_j}$ | Vapor enthalpy on the $j$th stage, J/mole |
| $P^{sat}_{j,i}$ | Saturation pressure of the $i$th component at the $j$th stage, Pa |
| $P_j$ | Pressure on the $j$th stage, Pa |
| $M_0$ | Liquid holdup in condenser, moles |
| $x_{D,i}$ | Mole fraction of the $i$th component in distillate |
| $L_0$ | Reflux rate, mole/s |
| $D$ | Distillate rate, mole/s |
| $H_D$ | Liquid enthalpy of distillate, joule/mol |
| $q_C$ | Condenser duty, J/s |
| $M$ | Control horizon |
| $M_R$ | Liquid holdup in reboiler, mole |
| $N_u$ | Number of manipulated variables |
| $N_y$ | Number of controlled variables |
| $p$ | Prediction horizon |
| $x_{w,i}$ | Mole fraction of the $i$th component in bottom product |
| $w$ | Bottom product flow rate, mole/s |
| $w_{u,j}$ | Weight on control move |
| $w_{y,j}$ | Weight on controlled variable deviation |
| $y_{sp}$ | Set point of controlled variables |
| $\Delta u$ | control move of manipulated variables |
| $y*$ | Reference value of controlled variables |

## Disclosure statement

No potential conflict of interest was reported by the authors.

## References


Agachi, P. S., Nagy, Z. K., Cristea, M. V., & Imre-Lucaci, A. (2006). *Model based control. Case studies in process engineering*. Weinheim: Wiley-VCH.

Asprion, N., & Kaibel, G. (2010). Dividing wall columns: Fundamentals and recent advances. *Chemical Engineering and Processing: Process Intensification*, *49*(2), 139–146.

Backx, T., Bosgra, O., & Marquardt, W. (2000). *Integration of model predictive control and optimization of processes*. Aachen: LPT, RWTH.

Buck, C., Hiller, C., & Fieg, G. (2011). Applying model predictive control to dividing wall columns. *Chemical Engineering & Technology*, *34*(5), 663–672.

De Temmerman, J., Dufour, P., Nicolaï, B., & Ramon, H. (2009). MPC as control strategy for pasta drying processes. *Computers & Chemical Engineering*, *33*(1), 50–57.

Halvorsen, I. J., & Skogestad, S. (2004). Shortcut analysis of optimal operation of Petlyuk distillation. *Industrial & Engineering Chemistry Research*, *43*(14), 3994–3999.

Hiller, C., Buck, C., Ehlers, C., & Fieg, G. (2010). Nonequilibrium stage modeling of dividing wall columns and experimental validation. *Heat and Mass Transfer*, *46*(10), 1209–1220.

Ignat, R., & Woinaroschy, A. (2011). Dynamic analysis and controllability of dividing-wall distillation columns. *Chemical Engineering*, *25*, 647–652.

Kaibel, G. (1987). Distillation columns with vertical partitions. *Chemical Engineering & Technology*, *10*(1), 92–98.

Kolbe, B., & Wenzel, S. (2004). Novel distillation concepts using one-shell columns. *Chemical Engineering and Processing: Process Intensification*, *43*(3), 339–346.

Kvernland, M., Halvorsen, I., & Skogestad, S. (2010). Model predictive control of a Kaibel distillation column. *In: Proceedings of the 9th International Symposium on Dynamics and Control of Process Systems (DYCOPS)*, 539–544.
Ling, H., & Luyben, W. L. (2009). New control structure for divided-wall columns. *Industrial & Engineering Chemistry Research*, *48*(13), 6034–6049.
Ling, H., & Luyben, W. L. (2010). Temperature control of the BTX divided-wall column. *Industrial & Engineering Chemistry Research, 49*(1), 189–203.
Rodríguez Hernández, M., & Chinea-Herranz, J. A. (2012). Decentralized control and identified-model predictive control of divided wall columns. *Journal of Process Control, 22*(9), 1582–1592.
Rewagad, R. R., & Kiss, A. A. (2012). Dynamic optimization of a dividing-wall column using model predictive control. *Chemical Engineering Science*, *68*(1), 132–142.
Schultz, M. A., Stewart, D. G., Harris, J. M., Rosenblum, S. P., Shakur, M. S., & O'Brien, D. E. (2002). Reduce costs with dividing-wall columns. *Chemical Engineering Progress*, *98*(5), 64–71.
Segovia-Hernandez, J. G., Hernandez-Vargas, E. A., & Marquez-Munoz, J. A. (2007). Control properties of thermally coupled distillation sequences for different operating conditions. *Computers & Chemical Engineering, 31*(7), 867–874.
Skogestad, S., & Postlethwaite, I. (2005). *Multivariable feedback control. Analysis and design*. 2nd ed. Chichester: John Wiley.
Sotudeh, N., & Shahraki, B. H. (2007). A method for the design of divided wall columns. *Chemical Engineering & Technology*, *30*(9), 1284–1291.
Triantafyllou, C., & Smith, R. (1992). The design and optimisation of fully thermally coupled distillation columns: Process design. *Chemical Engineering Research & Design*, *70*(A2), 118–132.
Van Diggelen, R. C., Kiss, A. A., & Heemink, A. W. (2010). Comparison of control strategies for dividing-wall columns. *Industrial & Engineering Chemistry Research*, *49*(1), 288–307.
Woinaroschy, A., & Isopescu, R. (2010). Time-optimal control of dividing-wall distillation columns. *Industrial & Engineering Chemistry Research*, *49*(19), 9195–9208.

# 4

# Design and stability analysis of CFOA-based amplifier circuits using Bode criterion

Ivailo Pandiev*

*Department of Electronics, Faculty of Electronic Engineering and Technologies, Technical University – Sofia, Sofia, 1797, Bulgaria*

In this paper the frequency stability of small-signal high-speed amplifier circuits using Bode criterion is analysed theoretically. In particular, the inverting and non-inverting amplifiers employing current-feedback operational amplifiers are under review. Based on the analysis of the operational principle, the equations for complex transfer functions of both circuits and formulas for the related electrical parameters are obtained. Moreover, using these formulas recommendations for a stable operation are given. As well, design procedure of the amplifiers with resistive and capacitive load is suggested. The efficiency of the proposed procedure and recommendations are verified by simulation modelling and experimental testing of sample electronic circuits.

**Keywords:** analogue circuits; high-speed amplifiers; operational amplifiers; CFOA; stability analysis; Bode plots; frequency response; analogue simulation

*Subject classification codes*: TSSC-IFAC-2014

## 1. Introduction

The small-signal high-speed (with bandwidth > 1 MHz) amplifiers are essential building blocks for video amplifiers, RF/IF amplifiers, high-speed A/D drivers and D/A buffers (Analog Dev., MT-060, 2008; Jung, 2002; Tietze & Schenk, 2008). In the past 10 years current-feedback operational amplifiers (CFOAs) have been basically used as active building blocks for design of high-speed amplifiers. As a kind of the monolithic operational amplifiers (op amps) family, the CFOAs have been realized to overcome the finite gain-bandwidth product of the conventional voltage-feedback operational amplifiers (VFOAs) (Jassim, 2013; Seifart, 2003). However, the CFOA-based amplifier circuits are less understood and documented in comparison with the amplifiers using VFOAs.

Stability analysis is an essential part in the design process of the analogue circuits, especially for the high-speed amplifiers, that use CFOAs. The practical interpretation of the sustainability definition is: 'An amplifier circuit is stable if all voltages and currents are reduced to zero when the input voltages and currents are zero'. Otherwise, in lack of an input signal unexpected oscillations (or self-oscillations) can occur at the output, which is unacceptable. In the theory of electronic circuits, various criteria of stability are designed. For the analysis of analogue circuits, the most commonly used are the criteria of Nyquist (Seifart, 2003) and Bode (Laker & Sansen, 1994; Nagaria, Gopi Krishna, & Singh Rakesh, 2008).

In the analysis of the Bode criteria, the behaviour of the magnitude and phase frequency characteristics or Bode plots are investigated. The Bode plots can be drawn directly from experimental data or computer simulations.

For the amplifier circuits it is assumed that the open-loop transfer function of the CFOAs is stable and their logarithmic a.c. transfer characteristics monotonously decreased with an increase in the frequency $\omega$ of the input signal. For this case the Bode criterion says: the closed-loop system containing op amp with negative feedback is stable only if the a.c. transfer characteristic of the open-loop system crosses the $x$-axis (0 dB) before the linear phase characteristic has reached $-180°$.

A relatively large number of books, publications and company application reports are devoted to the theory and the design of the amplifier circuits employing CFOAs (Jassim, 2013; Jung, 2002; Kamath, 2014; Mancini, 2001, 2002; Palumbo, 1997; Pandiev, 2012; Safari & Azhari, 2012; Schmid, 2003; Seifart, 2003; Texas Instruments, 2002; Tietze & Schenk, 2008). The authors' attention is focused on the structure and principle of operation at d.c. and large frequency range (Jung, 2002; Safari & Azhari, 2012; Seifart, 2003; Tietze & Schenk, 2008). However, for the analysis of basic amplifier circuits a simplified model of the CFOA is used. The attention of the author in Jassim (2013) is focused on the design technique, employing CMOS CFOA. Some results on the behaviour of the non-inverting amplifier are given, which also confirm the

*Email: ipandiev@tu-sofia.bg

efficiency of the designed CMOS CFOA. In the application report (Schmid, 2003), a method for measuring parasitic components in a prototype or final printed circuit board (PCB) design is discussed. A standard oscilloscope and a low-frequency waveform generator to collect valuable information for SPICE simulations are used there. The authors propose some design recommendations for the selection of the feedback resistor and gain resistor for the basic inverting and non-inverting amplifiers, employing CFOAs (Mancini, 2002; Texas Instruments, 2002). The note on the application (Mancini, 2001) has analysed the frequency stability of the inverting and non-inverting amplifier circuits, taking into account the influence of parasitic capacitance $C_N$ and $C_F$ (where $C_N$ is the capacitance connected to the inverting input and $C_F$ is the capacitance connected in parallel to the feedback resistor of the op amp). In the analysis, a simplified model of the CFOA has been used and at the same time the effect of the load in the loop-gain transfer function is not considered. An analysis of stability and compensation of CFOA is presented (Palumbo, 1997). In the theoretical analysis of the inverting amplifier, integrator and differentiator using a small-signal model of a CFOA, some of the input parasitic capacitances and the effect of the mounting capacitance are not included. Furthermore, the load is considered only as capacitance $C_L$ without taking into account the parallel connection of the load resistance $R_L$ and the load capacitance $C_L$ (i.e. $Z_L = R_L||(1/sC_L)$). This study is particularly useful for pencil-and-paper design of CFOA and takes into account both the resistive and the capacitive feedback. Recently, compensation techniques for improving amplitude and phase response of CFOA-based inverting amplifier are investigated (Kamath, 2014). The active compensation technique employing composite CFOA consisting of CFOA1 and CFOA2 is used in the place of single CFOA.

In this paper based on theoretical analysis of the high-speed amplifier circuits employing CFOAs and based on the results obtained for the amplifiers using VFOA, the frequency stability is studied by using the Bode criteria (Mancini, 2001). The results of the analysis are used to define the recommendations for improving the stability and to create a modified design procedure based on the procedure proposed (Pandiev, 2012).

The organization of the paper is as follows: in Section 2 the structure and relation between the input and the output voltages and currents for a linear high frequency model of the CFOA used in this work are presented; in Sections 3 and 4 the principles of operation of the inverting and non-inverting amplifiers at low and high frequencies are described; also in Section 4 recommendations for improving the frequency stability are defined, based on the obtained results; the proposed design procedure is given in Section 5; to illustrate the proposed theoretical analyses and the procedure, in Section 6 examples of studying the frequency stability of the inverting and non-inverting amplifiers at several voltage gains and various CFOAs are given. Finally in Section 7, the concluding remarks are given.

## 2. Current-feedback operational amplifiers

The most common CFOA (or transimpedance amplifier) is equivalent to a (positive second-generation current conveyor) CCII+ plus an output voltage buffer. These op amps have a high impedance non-inverting input $y$, a low-impedance inverting input $x$, a current output $z$ and a voltage output $o$. In some of the CFOAs the port $z$, between the first stage (CCII+) and the second stage (voltage follower), is defined as an external pin. The port is the output of the voltage buffer, where the output resistance $r_o$ is very low (several ohms magnitude). The linear model of the CFOA used in this work is presented in Figure 1.

The model in Figure 1 reflects the small-signal behaviour of the real device. It includes the following elements: input and output buffers (voltage followers); $i_x$ and $i_z$ – input and output current through the current-controlled current source (or current mirror); $r_{\text{in}}^+$ and $C_{\text{in}}^+$ – resistance and capaitance of the non-inverting input; $r_{\text{in}}^-$ and $C_{\text{in}}^-$ – resistance and capacitance of the inverting input; $r_t$ and $C_t$ – equivalent to $Z_t$ – transmission impedance and $r_o$ – output resistance.

For the linear CFOA, the ideal relations between input and output voltages and currents can be given by the following hybrid matrix:

$$\begin{bmatrix} i_y \\ u_x \\ i_z \\ u_o \end{bmatrix} = \begin{bmatrix} 1/Z_{\text{in}}^+ & 0 & 0 \\ 1 & Z_{\text{in}}^- & 0 \\ 0 & 1 & 1/Z_t \\ 0 & 0 & 1 \end{bmatrix} \begin{bmatrix} u_y \\ i_x \\ u_z \end{bmatrix}, \tag{1}$$

where $Z_{\text{in}}^+ = r_{\text{in}}^+||(1/pC_{\text{in}}^+)$, $Z_{\text{in}}^- = r_{\text{in}}^-||(1/pC_{\text{in}}^-)$ and $Z_t = r_t||(1/pC_t)$.

The matrix representation given in Equation (1) is valid only for ideal input and output voltage buffers with voltage gain equal to one.

In general, the CFOAs have several important applications. For example, the CFOAs are basically used for the design of amplifiers, active filters and oscillators. The objective in this paper is the frequency stability analysis of the basic amplifier circuits, using CFOAs.

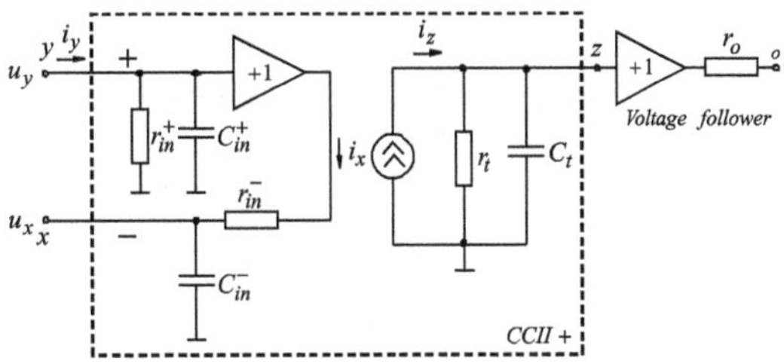


Figure 1. A linear model of the real CFOA.

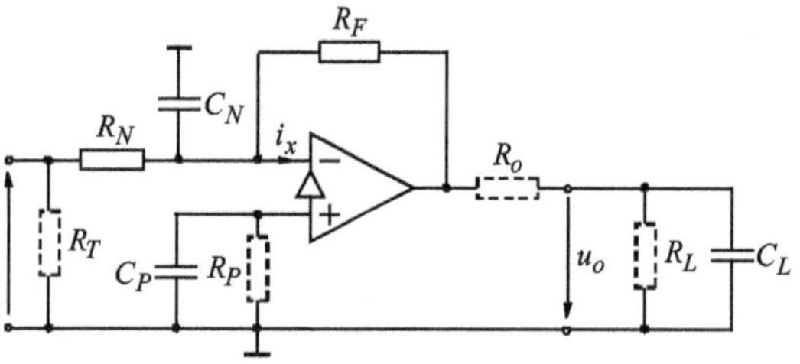


Figure 2. An inverting amplifier circuit using CFOA.

## 3. Principle of operation and basic analysis (approximately up to 50 MHz)

The objects of study are the inverting and non-inverting amplifier circuits, employing CFOAs.

### *3.1. An inverting amplifier*

The schematic structure of the high-speed inverting amplifier, employing CFOA, is shown in Figure 2. In this circuit is introduced a parallel negative feedback through the resistors $R_F$ and $R_N$. The resistor $R_P$ is used for compensation of the input bias current of the CFOA. If the circuit in Figure 2 is used as a video line driver, the best frequency response can be obtained by the addition of small resistances $R_T$ and $R_o$ at each terminal (with values of 50 or 75 Ω, for example).

For low frequencies (approximately up to 50 MHz), the influence of the parasitic capacitances $C_{in}^{+}$, $C_{in}^{-}$ and $C_L$ can be neglected, thereby the real linear model of the CFOA (Figure 1) is simplified. Including the external elements, we obtain the a.c. equivalent circuit of the analysed inverting amplifier (Figure 3). The [*Y*]-matrix of the circuit was composed using the well-known formulas (Boyanov & Shoikova, 1989), and after some transformations (using the condition $r_o \ll r_t$) we obtain the following expression for the transfer function:

$$A_U(s) = \frac{U_o(s)}{U_i(s)}$$

$$= \frac{-R_F/R_N}{R_F C_t\left(1+\frac{r_o}{R_L}\right)\left(1+\frac{r_{in}^-}{R_N}+\frac{r_{in}^-}{R_F}\right)\left[s+\frac{1+\frac{R_F}{r_t}\left(1+\frac{r_o}{R_L}\right)\left(1+\frac{r_{in}^-}{R_N}+\frac{r_{in}^-}{R_F}\right)}{R_F C_t\left(1+\frac{r_o}{R_L}\right)\left(1+\frac{r_{in}^-}{R_N}+\frac{r_{in}^-}{R_F}\right)}\right]}$$

$$= \frac{H}{s+\omega_p}. \tag{2}$$

Comparison of the left and right sides of Equation (2) results in the following formulas:

$$H = A_{U0} = \frac{-R_F/R_N}{R_F C_t\left(1+\frac{r_o}{R_L}\right)\left(1+\frac{r_{in}^-}{R_N}+\frac{r_{in}^-}{R_F}\right)} \tag{3a}$$

and

$$\omega_p = \frac{1+\frac{R_F}{r_t}\left(1+\frac{r_o}{R_L}\right)\left(1+\frac{r_{in}^-}{R_N}+\frac{r_{in}^-}{R_F}\right)}{R_F C_t\left(1+\frac{r_o}{R_L}\right)\left(1+\frac{r_{in}^-}{R_N}+\frac{r_{in}^-}{R_F}\right)}, \tag{3b}$$

where $H$ is the d.c. voltage gain and $\omega_p$ is the pole frequency, defining working frequency bandwidth $f_{-3dB}$ of the circuit.

Therefore, for $r_{in}^- \approx 0$, $r_o \ll R_L$ and $R_F \ll r_t$, the working frequency bandwidth $f_{-3dB} \approx 1/(2\pi R_F C_t)$ of the amplifier depends only on the internal capacitance $C_t$ and the external feedback resistor $R_F$. Moreover, the bandwidth $f_{-3dB}$ does not depend on the resistance $R_N$, setting the voltage gain of the circuit. The possibility for independent adjustment of gain and bandwidth is one of the main advantages of the amplifiers with CFOAs in comparison to those realized with VFOAs.

Based on the transfer function (2) for the module and the phase is obtained

$$|\dot{A}_U| = \frac{A_{U0}}{\sqrt{1+(f/f_p)^2}} \text{ and } \varphi_{A_U} = 180^\circ - \arctan\left(\frac{f}{f_p}\right).$$

### *3.2. A non-inverting amplifier*

The electronic circuit of the non-inverting amplifier is shown in Figure 4. In this circuit the CFOA is with serial negative feedback through the resistors $R_F$ and $R_N$. The resistor $R_P$ is used for compensation of the input bias current of the op amp, and the resistors $R_T$ and $R_o$ are used

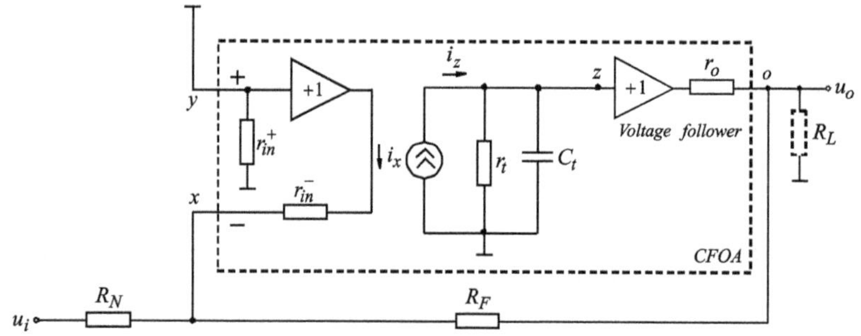


Figure 3. An equivalent circuit of the inverting amplifier using CFOA.

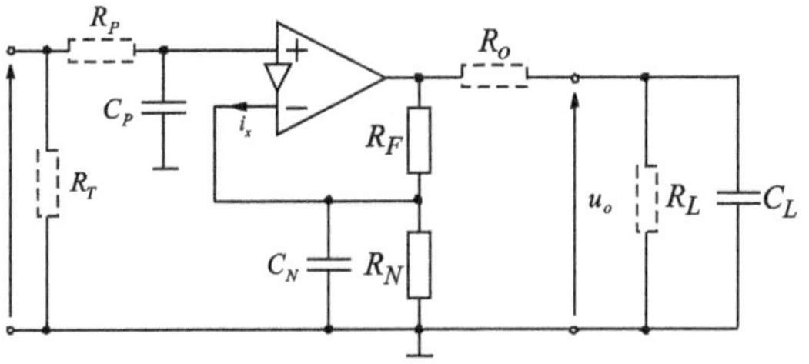


Figure 4. A non-inverting amplifier circuit using CFOA.

for the termination of each end of the circuit. The symbol analysis of the circuit in Figure 4 is implemented providing that $C_{\text{in}}^{+}$, $C_{\text{in}}^{-}$ and $C_{\text{L}}$ are neglected (see Figure 1). The a.c. equivalent circuit of the analysed non-inverting amplifier is presented in Figure 5.

In a.c. mode of operation (using the condition $r_{\text{o}} \ll r_{\text{t}}$), the transfer function can be written as

$$A_U(s) = \frac{U_{\text{o}}(s)}{U_i(s)}$$
$$= \frac{1 + R_{\text{F}}/R_{\text{N}}}{R_{\text{F}}C_{\text{t}}\left(1 + \frac{r_{\text{o}}}{R_{\text{L}}}\right)\left(1 + \frac{r_{\text{in}}^{-}}{R_{\text{N}}} + \frac{r_{\text{in}}^{-}}{R_{\text{F}}}\right)\left[s + \frac{1 + \frac{R_{\text{F}}}{r_{\text{t}}}\left(1 + \frac{r_{\text{o}}}{R_{\text{L}}}\right)\left(1 + \frac{r_{\text{in}}^{-}}{R_{\text{N}}} + \frac{r_{\text{in}}^{-}}{R_{\text{F}}}\right)}{R_{\text{F}}C_{\text{t}}\left(1 + \frac{r_{\text{o}}}{R_{\text{L}}}\right)\left(1 + \frac{r_{\text{in}}^{-}}{R_{\text{N}}} + \frac{r_{\text{in}}^{-}}{R_{\text{F}}}\right)}\right]}$$
$$= \frac{H}{s + \omega_{\text{p}}}. \tag{4}$$

Comparison of the left and right sides of Equation (4) results in the following formula for the d.c. voltage gain:

$$H = A_{U0} = \frac{1 + R_{\text{F}}/R_{\text{N}}}{R_{\text{F}}C_{\text{t}}\left(1 + \frac{r_{\text{o}}}{R_{\text{L}}}\right)\left(1 + \frac{r_{\text{in}}^{-}}{R_{\text{N}}} + \frac{r_{\text{in}}^{-}}{R_{\text{F}}}\right)}. \tag{5}$$

The formula (3b) is valid also for the non-inverting amplifier, because the denominators of the transfer functions, given in Equations (2) and (4), are in the same form.

According to formula (3b) during the implementation of amplifier circuits with a relatively large gain, the resistor $R_{\text{N}}$ has small values, while the resistor $R_{\text{F}}$ has a relatively large value, determined by the range of the working frequency bandwidth. In these cases, the resistance $R_{\text{N}}$ is close to the resistance $r_{\text{in}}^{-}$. Therefore, in the design process of amplifier circuits, the value of $r_{\text{in}}^{-}$ and the resulting ratios with $R_{\text{F}}$ and $R_{\text{N}}$ should be always given. The analysis of the datasheets of various CFOAs shows that the resistance $r_{\text{in}}^{-}$ has values from several ohms to several tenths of ohms. For example, the resistance $r_{\text{in}}^{-}$ is equal to 8 Ω for the op amp AD8009 (from Analog Devices), while for the op amp AD844 (from Analog Devices) $r_{\text{in}}^{-}$ is 50 Ω. For relatively large gains and low frequencies (up to 50 MHz) according to formula (3b), the value of the feedback resistor $R_{\text{F}}$ is obtained by $R_{\text{F}} = r_{\text{t}}(f_{\text{t}}/f_{-3\text{dB}}) - r_{\text{in}}^{-}A_{U0}$, where $f_{\text{t}} = 1/(2\pi r_{\text{t}}C_{\text{t}})$ is the transit frequency of the CFOA.

The module and the phase for the non-inverting amplifier are given as

$$|\dot{A}_U| = \frac{A_{U0}}{\sqrt{1 + \left(\frac{f}{f_{\text{p}}}\right)^2}} \text{ and } \varphi_{A_U} = \arctan\left(\frac{f}{f_{\text{p}}}\right).$$

From the analysis of the obtained formulas for the inverting and non-inverting amplifiers at low frequencies (i.e. $f \ll f_{\text{p}}$), the phase shift $\varphi_{\dot{A}_U}$ for the non-inverting amplifier is approximately equal to 0° and for the inverting amplifier is – 180°, respectively. For higher frequencies (i.e. $f > f_{\text{p}}$) the gain is decreased to $|\dot{A}_U| = A_{U0} \cdot (f_{\text{p}}/f)$ (the slope of $|\dot{A}_U|$ is approximately equal to −20 dB/dec). Furthermore, $|\varphi_{\dot{A}_U^-}| \approx 90°$, which ensures the stable operation of the circuits (the phase margin is greater than 45°).

The results of these analyses can be used to study amplifier circuits with CFOAs, whose operating frequency bandwidth is up to 50 MHz. In these cases, the frequency response is maximally flat in the pass-band. Sample CFOAs suitable for applications up to 50 MHz are THS6184 (from Texas Instruments), LT1256 (from Linear Technology) and AD844 (from Analog Devices).

## 4. Analysis at high frequencies of the amplifier circuits

At higher frequencies ( > 50 MHz), analyses of the transfer function of the input network, and thus of the overall transfer characteristic of the systems of the amplifiers affect

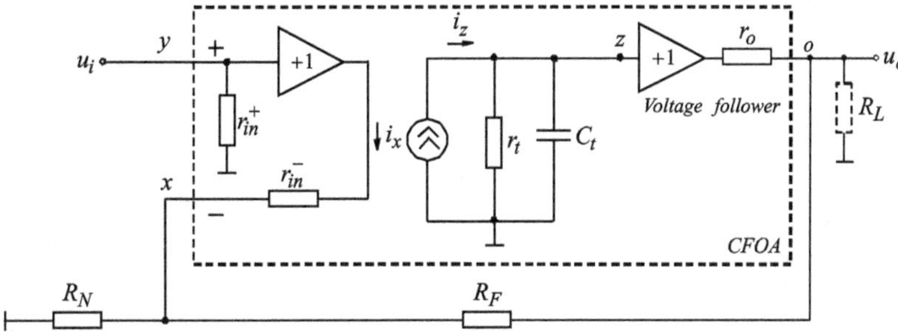


Figure 5. An equivalent circuit of the non-inverting amplifier using CFOA.

two capacitances $C_P$ and $C_N$. $C_P$ is the capacitance to non-inverting input, which is formed by the capacitance $C_{in}^{+}$ – Figure 2 plus the mounting capacitance (i.e. $C_P = C_{in}^{+} + C_M$, where $C_M$ is the parasitic board capacity with values usually up to 3 pF (Texas Instruments, 2002)). $C_N$ is the capacitance to inverting input, which is formed by $C_{in}^{-}$ and the mounting capacitance (i.e. $C_N = C_{in}^{-} + C_M$). Also the transfer characteristic of the amplifier circuits is affected by the load capacitance $C_L$, connected in parallel to the resistance $R_L$. In the following two subsections of the paper, the effects of $C_P$, $C_N$ and $C_L$ on the frequency response of the inverting and non-inverting amplifiers are examined separately.

### 4.1. *Effect of $C_P$, $C_N$ and $C_L$ on the frequency response of the non-inverting amplifier*

The analysis of the circuit in Figure 4 is performed according to the method of the nodal voltages. The CFOA is replaced by the linear model, given in Figure 1. The transfer function (using the condition $r_{in}^{-} \ll r_t$) at $Z_L = R_L||(1/sC_L)$ (the load impedance is a parallel connection of resistor $R_L$ and parasitic capacitance $C_L$) and using the condition $r_o \ll r_t$ can be found by

$$A_U(s) \approx \frac{\frac{1}{(r_{in}^{+}||R'_P)C_P}}{s + \frac{1}{(r_{in}^{+}||R'_P)C_P}}$$

$$\frac{\frac{r_{in}^{+}}{r_{in}^{+}+R'_P}\frac{C_N R_F + C_t r_o}{C_N C_t R_F r_{in}^{-}\left(1+\frac{r_o}{R_L}+\frac{C_L}{C_N}\frac{r_o}{r_{in}^{-}}\right)}\left(s+\frac{1+\frac{R_F}{R_N}}{C_N R_F + C_t r_o}\right)}{s^2 + s\frac{1}{C_N r_{in}^{-}}\frac{\left(1+\frac{r_{in}^{-}}{R_N}+\frac{r_{in}^{-}}{R_F}\right)\left(1+\frac{r_o}{R_L}\right)}{1+\frac{r_o}{R_L}+\frac{C_L}{C_N}\frac{r_o}{r_{in}^{-}}} + \frac{1}{C_N C_t R_F r_{in}^{-}\left(1+\frac{r_o}{R_L}+\frac{C_L}{C_N}\frac{r_o}{r_{in}^{-}}\right)}}$$

$$= H\frac{\omega_{p,in}}{s+\omega_{p,in}}\frac{s+\omega_z}{s^2+\frac{\omega_p}{Q_p}s+\omega_p^2}, \quad (6)$$

where $R'_P = R_G + R_P$ is the equivalent resistance to the non-inverting input and $R_G$ is the internal resistance of the input voltage source.

The equalization of the left and right sides of Equation (6) results in the following formulas for the basic parameters:

$$H = \frac{r_{in}^{+}}{r_{in}^{+}+R'_P}\frac{C_N R_F + C_t r_o}{R_F r_{in}^{-} C_t C_N}\frac{1}{\left(1+\frac{r_o}{R_L}+\frac{C_L}{C_N}\frac{r_o}{r_{in}^{-}}\right)}$$

$$\approx \frac{r_{in}^{+}}{r_{in}^{+}+R'_P}\frac{1}{r_{in}^{-}C_t\left(1+\frac{r_o}{R_L}+\frac{C_L}{C_N}\frac{r_o}{r_{in}^{-}}\right)} \quad (7)$$

is the transmission coefficient;

$$\omega_{p,in} = 1/[(r_{in}^{+}||R'_P)C_P] \quad (8)$$

is the pole angular frequency related to the effect of the resistance $R'_P$ and capacitance $C_P$ to the non-inverting input;

$$\omega_z = \frac{1+R_F/R_N}{R_F C_N + r_o C_t} \quad (9)$$

is the zero angular frequency related to the effect of the resistance $R_F$ and capacitance $C_N$ to inverting input of the CFOA;

$$\omega_p = \frac{1}{\sqrt{R_F r_{in}^{-} C_t C_N\left(1+\frac{r_o}{R_L}+\frac{C_L}{C_N}\frac{r_o}{r_{in}^{-}}\right)}} \quad (10)$$

is the pole angular frequency (self-oscillating frequency or undamped natural frequency) and

$$Q_p = \frac{\sqrt{r_{in}^{-}C_N\left(1+\frac{r_o}{R_L}+\frac{C_L}{C_N}\frac{r_o}{r_{in}^{-}}\right)}}{\sqrt{R_F C_t}\left(1+\frac{r_{in}^{-}}{R_N}+\frac{r_{in}^{-}}{R_F}\right)\left(1+\frac{r_o}{R_L}\right)}; \quad (11)$$

is the quality factor of the circuit.

After substitution of $s = j\omega$ in formula (6), based on the obtained general complex function, the module and the phase shift can be found by using

$$|A_U(j\omega)| = A_U(\omega) = \frac{\omega_{p,in}}{\sqrt{\omega^2+\omega_{p,in}^2}}\frac{H\sqrt{\omega_z^2+\omega^2}}{\sqrt{(\omega_p^2-\omega^2)^2+\left(\omega\frac{\omega_p}{Q_p}\right)^2}} \text{ and} \quad (12a)$$

$$\phi_{A_U} = \arctan\left(\frac{\omega}{\omega_z}\right) - \arctan\frac{\omega_p\omega}{Q_p(\omega_p^2-\omega^2)} - \arctan\left(\frac{\omega}{\omega_{p,in}}\right). \quad (12b)$$

To compensate the effect of the parasitic capacitance $C_P$, capacitor $C'_P$ can be placed in parallel with $R'_P$ so that $C'_P \gg C_P$ (Stoianov, 2000). The modified transfer function of the input network with capacitor $C'_P$ in parallel with $R'_P$ is

$$T_{in}(s) = \frac{r_{in}^{+}}{r_{in}^{+}+R'_P}\frac{1+sR'_P C'_P}{1+s[(r_{in}^{+}||R'_P)(C_P+C'_P)]}. \quad (13)$$

If $C'_P \gg C_P$ and $R'_P \ll r_{in}^{+}$, then

$$\frac{1+sR'_P C'_P}{1+s[(r_{in}^{+}||R'_P)(C_P+C'_P)]} \cong 1 \text{ and } u_i \cong u_y.$$

In this compensation method, the working frequency bandwidth is not narrowed and also the spikes are not reduced.

At the compensated effect of the capacitance $C_P$, the analysis of formulas (6), (12a) and (12b) shows that the transfer function is characterized by a double pole (or a complex pole) with angular frequency equal to $\omega_{\mathrm{p}}$ and one real zero with angular frequency $-\omega_z$. However, for $\omega = 0$ the voltage gain has a value $A_U(0) = H(\omega_z/\omega_{\mathrm{p}}) = 1 + (R_{\mathrm{F}}/R_{\mathrm{N}})$, while for much higher frequencies the gain decreases to zero. Therefore, the transfer characteristic of the non-inverting amplifier circuit is a low-pass type. For $Q_{\mathrm{p}} \geq 0.707$ at a frequency equal to $\omega_{\mathrm{p}}$, the denominator tends to zero and the voltage gain theoretically increases towards infinity. As a result, peaking of the frequency characteristic occurs and the phase shift between the input and output signal increases rapidly to 180°. Moreover, the circuit of the non-inverting amplifier becomes unstable. In the cases where $Q_{\mathrm{p}} < 0.707$ and $\omega_{\mathrm{p}} \ll \omega_z$, the frequency response monotonically decreases and the phase shift decreases to $-180°$ (the phase margin is less than 45°). With increase in the frequency, the first component in the formula (12b) tends to be 90°, while the second component tends to be $-180°$. From the theory of electronic circuits (Seifart, 2003), it is known that for $\omega \geq 10\omega_{\mathrm{p}}$ the phase shift increases to $-180°$. According to formula (12b), the stable operation of the amplifier can be produced at the condition that the difference between $\omega_{\mathrm{p}}$ and $\omega_z$ does not exceed 10 times. At $\omega_z < \omega_{\mathrm{p}}$ ringing in the output signal occurs, which can cause unstable operation.

Therefore, for the working bandwidth $\omega_{-3\mathrm{dB}}$, where $|A_U(j\omega)|$ decreases with 3 dB (or $1/\sqrt{2}$), the following is obtained:

$$\omega_{-3\mathrm{dB}} = \frac{\omega_{\mathrm{p}}^2}{\omega_z}\sqrt{1+\left(1-\frac{1}{2Q_{\mathrm{p}}^2}\right)\frac{\omega_z^2}{\omega_{\mathrm{p}}^2}+\sqrt{\left[1+\left(1-\frac{1}{2Q_{\mathrm{p}}^2}\right)\frac{\omega_z^2}{\omega_{\mathrm{p}}^2}\right]^2+\frac{\omega_z^4}{\omega_{\mathrm{p}}^4}}}. \quad (14)$$

At $Q_{\mathrm{p}} = 1/\sqrt{2}$ formula (14) is simplified and yields

$$\omega_{-3\mathrm{dB}} = \frac{\omega_{\mathrm{p}}^2}{\omega_z}\sqrt{1+\sqrt{1+\frac{\omega_z^4}{\omega_{\mathrm{p}}^4}}}. \quad (15)$$

For $\omega_z \gg \omega_p$ $\omega_{-3\mathrm{dB}} \approx \omega_p$. The value of the module of the general complex function (12a) at $\omega = \omega_p$ is found by the formula

$$A_U(\omega_{\mathrm{p}}) = Q_{\mathrm{p}}H\frac{\sqrt{\omega_z^2+\omega_{\mathrm{p}}^2}}{\omega_{\mathrm{p}}^2}.$$

Based on the analysis of the formulas for $\omega_{\mathrm{p}}$, $\omega_z$ and $Q_{\mathrm{p}}$ for the chosen resistance $R_{\mathrm{F}}$ and CFOA at $Q_{\mathrm{p}} < 0.707$, in the frequency response ringing can occur under the condition

$$R_{\mathrm{N}} > \frac{R_{\mathrm{F}}}{k_\omega - 1}, \quad (16)$$

where

$$k_\omega = \sqrt{\frac{R_{\mathrm{F}}C_{\mathrm{N}}}{r_{\mathrm{in}}^{-}C_{\mathrm{t}}}}\frac{1}{\sqrt{1+\frac{r_{\mathrm{o}}}{R_{\mathrm{L}}}+\frac{C_L}{C_{\mathrm{N}}}\frac{r_{\mathrm{o}}}{r_{\mathrm{in}}^{-}}}}$$

is the coefficient of peaking of the output signal.

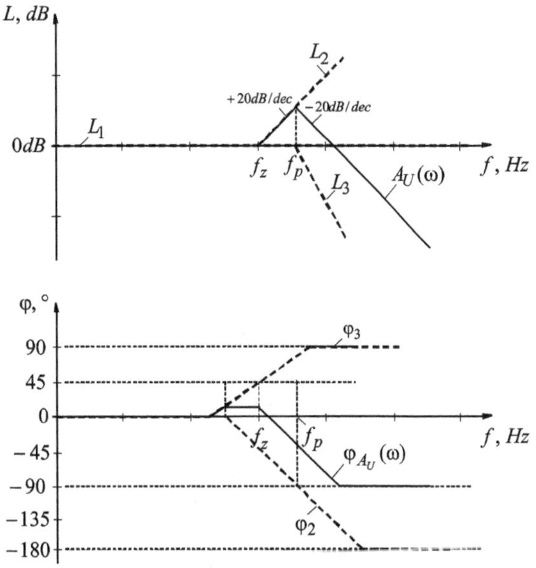


Figure 6. Magnitude and phase Bode plots for the loop gain of a voltage follower employing CFOA.

The magnitude and phase Bode plots for the loop gain of a voltage follower using CFOA are shown in Figure 6. A voltage follower or buffer with op amp is obtained from the non-inverting amplifier by removing the resistor $R_{\mathrm{N}}$. This circuit is analogous to the emitter (source) follower employing one transistor. The Bode plots are constructed by summing the corresponding logarithmic characteristics of all their sections, realizing zeros and poles:

$$L(\omega) = 20\lg A_U(\omega) = \sum_{i=1}^{n} L_i(\omega) \text{ and} \quad (17)$$

$$\varphi_{A_U}(\omega) = \sum_{i=1}^{n} \varphi_i(\omega), \quad (18)$$

where $L_i(\omega)$ are the individual logarithmic magnitude characteristics and $\varphi_i(\omega)$ are the individual phase characteristics.

Formulas (12a) and (12b), and the corresponding Bode plots for $\omega_z < \omega_{\mathrm{p}}$, given in Figure 6, show the following:

(1) In the voltage follower operation mode ($A_U(0) = 1$), there is always causes peaking in the frequency response, the coefficient $k_\omega$ is greater than unit. By increasing the d.c. voltage gain $A_{U0}$, the frequency $\omega_z$ increases. As a result, the amount of peaks in the frequency response decreases;

(2) The phase margin of the voltage follower and the non-inverting amplifier is greater than 45°. However, the stability of the circuit is worse. For a given frequency of the input signal, at which the asymptotes $L_2$ and $L_3$ intersect, the rate of change in voltage gain is about 40 dB/dec. The peaks in the magnitude plot lead to an increase in the amplitude of the output signal, which may adversely affect the next stages of the electronic devices and systems.

To compensate the effect of the capacitances $C_N$ and $C_L$, the following recommendations can be used:

(1) Reduce the value of $C_N$ by removing ground or power plane around the circuit trace to the inverting input;
(2) Reduce the value of the feedback resistor $R_F$. As a result, the operating frequency bandwidth is expanded – formula (14);
(3) Reduce the value of $C_L$ by minimizing the length of output cables;
(4) For amplifiers the condition

$$A_{U0} > \sqrt{\frac{R_F C_N}{r_{in}^- C_t}} \frac{1}{\sqrt{1 + \frac{r_o}{R_L} + \frac{C_L}{C_N}\frac{r_o}{r_{in}^-}}}$$

has to be kept, then $\omega_p < \omega_z$. As well it has to maintain the $Q_p$ to be less than $1/\sqrt{2}$, which can be achieved at $(R_F C_t/2) > r_{in}^- C_N + r_o C_L$.
(5) Connect a small resistor ($10\Omega \ldots 20\Omega$) between the output of the CFOA and the capacitive load.

For video line drivers (working up to 100 MHz) to isolate the output terminal from the load capacitance, a small resistor $R_o$, between the output of the op amp and the load, can be connected (Mancini, 2001; THS3091 – datasheet, 2014). Thus, the working bandwidth is narrowed by forming the additional pole frequency of $\omega_{pT,out} = 1/[(R_o||R_L)C_L]$ and the d.c. transmission coefficient of the output network at $R_o = R_L$ (where $R_L$ is the characteristic impedance of the cable), yielding $T_{out,dc} = R_L/(R_o + R_L)$. Furthermore, the phase margin decreases at higher frequencies.

### 4.2. *Effect of $C_P$, $C_N$ and $C_L$ on the frequency response of the inverting amplifier*

The analysis of the inverting amplifier (Figure 2) is also performed according to the method of the nodal voltages, when the op amp is replaced by the linear model, presented in Figure 1. For the corresponding transfer function (using the condition $r_{in}^- \ll r_t$) is found

$$A_U(s) = \frac{\dfrac{1}{C_N C_t R_F r_{in}^-\left(1+\frac{r_o}{R_L}+\frac{C_L}{C_N}\frac{r_o}{r_{in}^-}\right)}\left(-\frac{R_F}{R_N}\right)}{s^2 + s\dfrac{1}{C_N r_{in}^-}\dfrac{\left(1+\frac{r_{in}^-}{R_N}+\frac{r_{in}^-}{R_F}\right)\left(1+\frac{r_o}{R_L}\right)}{1+\frac{r_o}{R_L}+\frac{C_L}{C_N}\frac{r_o}{r_{in}^-}} + \dfrac{1}{C_N C_t R_F r_{in}^-\left(1+\frac{r_o}{R_L}+\frac{C_L}{C_N}\frac{r_o}{r_{in}^-}\right)}}$$

$$= \frac{H\omega_p^2}{s^2 + \frac{\omega_p}{Q_p}s + \omega_p^2}. \qquad (19)$$

The equalization of the left and right sides of Equation (19) results in the following formula for the d.c. transmission coefficient:

$$H = A_{U0} = -\frac{R_F}{R_N}. \qquad (20)$$

The formulas for the angular pole frequency and the quality factor matched with formulas (10) and (11), respectively. Furthermore, the obtained transfer function of the inverting amplifier is not affected by adding the parasitic capacitor $C_P$ to the circuit.

The analysis of formula (19) shows that the transfer function is characterized by one double pole with frequency equal to $\omega_p$. When taking into account the relation $r_o/r_t$, the transfer function is obtained with one zero with positive real part and one double pole. The value of the zero angular frequency can be found by $\omega_z = (R_F/r_o r_{in}^- C_t)$. The existence of zero with a positive real part in the transfer function makes the amplifier circuit unstable. The comparison of $\omega_p$ and $\omega_z$ for the inverting amplifier shows that $\omega_p \gg \omega_z$, as it always satisfies the condition

$$\frac{r_o}{R_F}\sqrt{\frac{r_{in}^- C_t}{R_F C_N}}\frac{1}{\sqrt{1+\frac{r_o}{R_L}}} \ll 1.$$

Furthermore, the $\omega_z$ is much larger than the unity bandwidth ($f_1 \approx 1/2\pi r_{in}^- C_t$), since $R_F \gg r_o$.

The main disadvantages of the inverting circuit compared to the non-inverting circuit are relatively small input resistance and phase inversion of the input voltage, which can be an unwanted effect for some practical applications.

## 5. Design procedure

As a result of the theoretical analysis, the aforementioned analytical formulas are the bases of the design procedure for amplifiers employing CFOAs. The schematic design for the circuits in Figures 2 and 4 is based on the following sequence:

(1) *Technical specification.* The circuit elements are calculated using predefined: amplitude of the input voltage $U_{im}$ or amplitude of the input voltage source $e_G$ with internal resistance $R_G$; input resistance $R_{iA}$; amplitude of the output voltage $U_{R_L}$ with load resistance $R_L$ and capacitance $C_L$; parasitic board capacitance $C_M$; output resistance $R_{oA}$; cut-off frequency $f_{-3dB}$; relative error $\varepsilon_{io}$ [%] defined by the input offset current and voltage, temperature drift $\varepsilon_{\Delta io}$ [%] of the $\varepsilon_{io}$ within temperature range $\Delta T$ and minimum value of a signal-to-noise (SN) ratio [dB].

(2) *An electronic circuit is selected.* An object of an analysis and design are the amplifier circuits shown in Figures 2 and 4. The inverting circuit (Figure 2) provides smaller input resistance, while the non-inverting circuit (Figure 4) is with greater input resistance. Furthermore, the inverting circuit reduces the influence of the parasitic input capacitance at high frequencies (the inverting input $x$ is a virtual ground – $u_{yx} \approx 0$).

(3) *The op amp is selected.* The main advantages of the CFOAs are the greater slew rate and the wider bandwidth compared to the VFOAs. The higher value of the slew rate is associated with a higher consumption current in a dynamic mode of operation. In order to produce low value of the power dissipation, most of the CFOAs work with supply voltages less than ±5 V. Then, at high frequencies, without additional output power stage, most CFOAs can get a maximum output current greater than 20 mA. The op amp is selected according to the following conditions:

- Maximum output voltage $U_{\rm om} \geq U_{R_{\rm L}}$ ($U_{\rm om}$ is the maximum output voltage of the op amp);
- The power supply voltage $V_{\rm CC} = -V_{\rm EE}$ is selected higher than the maximum output voltage $U_{\rm om}$, as saving the condition $V_{\rm CC\,min} < V_{\rm CC} < V_{\rm CC\,max}$;
- Maximum output current $I_{\rm o,max} > I_{\rm L}$, where $I_{\rm L} = U_{R_{\rm L}}/R_{\rm L}$;
- Small-signal bandwidth $f_1 > (5\ldots10)f_{-3\rm dB}$, where $f_1$ is the cut-off frequency at voltage gain equal to 1;
- Slew rate $\mathrm{SR_{CFOA}} > 2\pi f_{-3\rm dB} U_{R_{\rm L}}$.

(4) *The value of the equivalent quality factor* of the frequency response is obtained:

$$Q_{\rm p} = \sqrt{\omega_{-3\rm dB}}\frac{\sqrt{r_{\rm in}^{-}C_{\rm N}\left(1+\frac{r_{\rm o}}{R_{\rm L}}+\frac{C_{\rm L}}{C_{\rm N}}\frac{r_{\rm o}}{r_{\rm in}^{-}}\right)}}{1+\frac{r_{\rm o}}{R_{\rm L}}}.$$

(5.1.) *For low-frequency* (up to 50 MHz) *amplifiers*, the value of the feedback resistor $R_{\rm F}$ is calculated:

- For the inverting amplifier $R_{\rm F} = r_{\rm t}(f_{\rm t}/f_{-3\rm dB}) - (1+|A_{U0}|)r_{\rm in}^{-}$, where $f_{\rm t} = 1/(2\pi r_{\rm t}C_{\rm t})$ is the transit frequency of the CFOA and $|A_{U0}| = U_{R_{\rm L}}/U_{\rm im}$ is the voltage gain of the circuit;
- For the non-inverting amplifier $R_F = r(f_{\rm t}/f_{-3\rm dB})_t - r_{\rm in}^{-}A_{U0}$;

(5.2.) *At higher frequencies* ($>$ 50 MHz) and $\omega_{\rm p} < \omega_z$, the value of the feedback resistor $R_{\rm F}$ is calculated:

$$R_{\rm F} = r_{\rm t}\frac{f_{\rm t}f_{\rm N}}{f_{-3\rm dB}^2}\frac{1}{1+\frac{r_{\rm o}}{R_{\rm L}}}\left[\left(1-\frac{1}{2Q_{\rm p}^2}\right)\right.$$
$$\left.+\sqrt{\left(1-\frac{1}{2Q_{\rm p}^2}\right)^2+1}\,\right],$$

where $f_{\rm N} = 1/(2\pi r_{\rm in}^{-}C_{\rm N})$.

(6) *The value of the gain resistor* $R_{\rm N}$ is calculated:

- For the inverting amplifier $R_{\rm N} = R_{\rm F}/|A_{U0}|$;
- For the non-inverting amplifier $R_{\rm N} = R_{\rm F}/(A_{U0}-1)$.

The calculated values for the resistors $R_{\rm F}$ and $R_{\rm N}$ according to the aforementioned formulas have to be consistent with the values from the datasheet of the chosen CFOA.

(7) *The value of the compensation resistor* $R_{\rm P}$ is determined as $R_{\rm P} = R_{\rm N}||R_{\rm F}$.

(8) *The phase margin* is calculated: $|\varphi_m| = 180^\circ - |\varphi_{A_U}(\omega_1)|$, where $\omega_1 = 2\pi f_1$;

For the inverting amplifier, the phase shift is

$$\varphi_{A_U}(\omega) = 180 - \arctan\frac{\omega_{\rm p}\omega}{Q_{\rm p}(\omega_{\rm p}^2-\omega^2)}.$$

The obtained value for the phase margin has to be greater than 45°. Otherwise, other op amps with smaller values of the parasitic capacitances should be chosen.

(9) *The input impedance* $Z_{\rm iA}$ is calculated:

- For the inverting amplifier: $Z_{\rm iA} \approx R_{\rm N}$;
- For the non-inverting amplifier:

$$Z_{\rm iA} \approx \frac{r_{\rm in}^{+}+R'_{\rm p}}{\sqrt{1+(f/f_{\rm in}^{+})^2}},$$

where $f_{\rm in}^{+} = 1/2\pi(r_{\rm in}^{+}+R'_{\rm p})C_{\rm p}$ is the cut-off frequency of the input electrical network.

At low frequencies ($f \ll f_{\rm in}^{+}$) $Z_{\rm iA} \approx r_{\rm in}^{+}+R'_{\rm p}$.

(10) *The output resistance* $R_{\rm oA}$ is calculated: $R_{\rm oA} \approx r_{\rm o}/\beta A_{d0}$, where $\beta = R_{\rm N}/(R_{\rm N}+R_{\rm F})$ is the negative feedback coefficient and $A_{d0} \approx r_{\rm t}/r_{\rm in}^{-}$ is the d.c. open-loop voltage gain of the chosen CFOA.

(11) *The output offset voltage of the circuit is calculated.* First, the output offset voltage for room temperature – 25° (using condition $R_{\rm p} = R_{\rm F}||R_{\rm N}$) is calculated: $U_{\rm o,err} = (1+R_{\rm F}/R_{\rm N})U_{\rm io} - R_{\rm F}I_{\rm io}$, where $U_{\rm io}$ is the input offset voltage and $I_{\rm io}$ is the input offset current of the chosen CFOA. For video drivers the resistor $R_{\rm P}$ is removed and a resistor $R_{\rm T}$ in parallel to the non-inverting input of the amplifier is connected. In this case the output offset voltage is $U_{\rm o,err} = (1+R_{\rm F}/R_{\rm N})[U_{\rm io} - (R_{\rm G}||R_{\rm T})I_{\rm B}^{+} + (R_{\rm F}||R_{\rm N})I_{\rm B}^{-}]$. Then the relative error $\varepsilon_{\rm io} = (U_{\rm o,err}/U_{R_{\rm L}})100\%$ is compared to the value given in step № 1.

(12) *The output offset voltage drift is calculated.* First, the output offset voltage drift is calculated: $\Delta U_{\rm o,err} = (1+R_{\rm F}/R_{\rm N})\Delta U_{\rm io}(T) - R_{\rm F}\Delta I_{\rm io}(T)$ for the given temperature range $\Delta T$. Then the relative error $\varepsilon_{\Delta\rm io} = (\Delta U_{\rm o,err}/U_{R_{\rm L}})100\%$ is compared to the value given in step № 1. If the result does not satisfy the specification, a more precise op amp or performing new calculations for the resistances with lower values can be chosen.

13. *The SN ratio* is calculated. First, the resulting noise voltage density at the amplifier's output is calculated: $\bar{S}_{U,\rm out} = \sqrt{\sum_i S_{U_i}^2}$ for $i = 1,2,\ldots$, where $S_{U_i}$ is the individual noise components.

Then

$$\mathrm{SN} = \frac{U_{\mathrm{o,eff}}}{U_{\mathrm{oN}}} = \frac{U_{\mathrm{o,eff}}}{\bar{S}_{U,\mathrm{out}}\sqrt{B_{\mathrm{eq}}}},$$

where $B_{\mathrm{eq}} = 1.57f_{-3\mathrm{dB}}$ is the bandwidth of the circuit multiplied by the correction factor of $\pi/2 = 1.57$ and $U_{\mathrm{o,eff}}$ is the output effective value. The obtained value for the SN is compared to the value given in step $\mathcal{N}\underline{o}$ 1 of the procedure. If the result does not satisfy the specification, other op amps with lower voltage and current noise can be chosen.

## 6. Verification check, experimental testing and discussions

To verify the theoretical analysis and the proposed design procedure, in this section examples of studying the frequency stability of the inverting and non-inverting amplifiers at several voltage gains are given. A wide bandwidth ($B_1 > 300\,\mathrm{MHz}$) CFOA type AD8011 (AD8011 – datasheet, 2014) and high-current ($I_{\mathrm{o\,max}} > 200\,\mathrm{mA}$) CFOA type THS3091 (THS3091 – datasheet, 2014) are chosen as active building elements for the investigated electronic circuits. The THS3091 uses an 8-pin SOIC and the 8-pin SOIC with PowerPAD™ (Texas Instruments Incorporated, Dallas, Texas 75243, USA) packages. The package type PowerPAD™ is designed so that the thermal pad is exposed on the bottom side of the integrated circuit (IC). The thermal coefficient for the PowerPAD packages is substantially improved over the basic SOIC.

The verification check of the op amps is performed by Cadence OrCAD® (Cadence Design Systems, Inc., San Jose, CA 95134, USA), using OrCAD PSpice® program with AD8011AN PSpice macro-model (version 1.0) (AD8011 SPICE macro-model, 2014) and THS3091 PSpice macro-model (THS3091 PSpice Model, 2014). The AD8011AN model simulates the input offset voltage and current (offsets will not vary with input common-mode voltage), small-signal closed-loop gain and phase versus frequency, output current limiting and output voltage limiting, slew rate, step response performance (slew rate is based on 10–90% of step response), quiescent current at operating point, noise effects, input impedance and output impedance. The values of the modelled parameters at $V_{\mathrm{S}} = \pm 5\,\mathrm{V}$, $R_{\mathrm{L}} = 1\,\mathrm{k\Omega}$ and $T_{\mathrm{A}} = 27°\mathrm{C}$ are as follows: $U_{\mathrm{io}} = 2\,\mathrm{mV}$, $I_{\mathrm{B}}^{+} = 5.125\,\mu\mathrm{A}$, $I_{\mathrm{B}}^{-} = 5.125\,\mu\mathrm{A}$, $r_{\mathrm{in}}^{+} = 517\,\mathrm{k\Omega}$, $C_{\mathrm{in}}^{+} = 1\,\mathrm{pF}$, $r_{\mathrm{in}}^{-} = 50\,\Omega$, $C_{\mathrm{in}}^{-} = 2.3\,\mathrm{pF}$, $r_{\mathrm{t}} = 1.27\,\mathrm{M\Omega}$, $C_{\mathrm{t}} = 1.52\,\mathrm{pF}$ (or $f_{\mathrm{t}} = 82.5\,\mathrm{kHz}$), $f_1 \approx 670\,\mathrm{MHz}$ (at $A_{U0} = 1$), $\bar{S}_{Ui} = 7.53\,\mathrm{nV}/\sqrt{\mathrm{Hz}}$ (at $f = 10\,\mathrm{kHz}$), $U_{\mathrm{om}} = \pm 4\,\mathrm{V}$, $\mathrm{SR} > 1000\,\mathrm{V}/\mu\mathrm{s}$, $I_{\mathrm{o\,max}} = 60\,\mathrm{mA}$ and $r_{\mathrm{o}} = 22\,\Omega$.

The THS3091 model simulates input offset voltage, input bias currents, small-signal closed-loop gain and phase versus frequency (bandwidth is high in gains of +1 V/V and +2 V/V and low at higher gains), output voltage limiting, slew rate, step response performance (slew rate is correct at 2 V step), settling time, quiescent current, noise effects, output impedance and loading effects. The values of the modelled parameters at $V_{\mathrm{S}} = \pm 15\,\mathrm{V}$, $R_{\mathrm{L}} = 100\,\Omega$ and $T_{\mathrm{A}} = 27°\mathrm{C}$ are as follows: $U_{\mathrm{io}} = 0.9\,\mathrm{mV}$, $I_{\mathrm{B}}^{+} = -4.5\,\mu\mathrm{A}$, $I_{\mathrm{B}}^{-} = -3.5\,\mu\mathrm{A}$ (or $I_{\mathrm{iB}} = 4\,\mu\mathrm{A}$ and $I_{\mathrm{io}} = 1\,\mu\mathrm{A}$), $r_{\mathrm{in}}^{+} = 1.1\,\mathrm{M\Omega}$, $C_{\mathrm{in}}^{+} = 1.2\,\mathrm{pF}$, $r_{\mathrm{in}}^{-} = 32\,\Omega$, $C_{\mathrm{in}}^{-} = 1.4\,\mathrm{pF}$, $r_{\mathrm{t}} = 848\,\mathrm{k\Omega}$, $C_{\mathrm{t}} = 0.8\,\mathrm{pF}$ (or $f_{\mathrm{t}} = 234\,\mathrm{kHz}$), $f_1 \approx 240\,\mathrm{MHz}$ (at $A_{U0} = 1$), $\bar{S}_{Ui} < 6\,\mathrm{nV}/\sqrt{\mathrm{Hz}}$ (at $f > 10\,\mathrm{kHz}$), $U_{\mathrm{om}} \approx \pm 3.1\,\mathrm{V}$, $\mathrm{SR} > 1000\,\mathrm{V}/\mu\mathrm{s}$ (at $R_{\mathrm{F}} = 1.21\,\mathrm{k\Omega}$ and $A_{U0} = 2$) and $r_{\mathrm{o}} = 100\,\Omega$.

Table 1. Comparison between calculated parameters and simulation results for $A_{U0} = +5$, $R_{\mathrm{L}} = 50\,\Omega$ and $C_{\mathrm{L}} = 20\,\mathrm{pF}$.

| Parameter | Calculated results | Simulation results |
|---|---|---|
| $A_{U0} = +5, f_{-3\,\mathrm{dB}} = 100\,\mathrm{MHz}$, $R_{\mathrm{F}} = 1.96\,\mathrm{k\Omega}$ and $R_{\mathrm{N}} = 490\,\Omega$ ($R_{\mathrm{F}} = 1.96\,\mathrm{k\Omega} \pm 1\%$ and $R_{\mathrm{N}} = 487\,\Omega \pm 1\%$), $U_{\mathrm{im}} = 100\,\mathrm{mV}$, $R_{\mathrm{G}} = 50\,\Omega$ and $R_{\mathrm{T}} = 50\,\Omega$ | | |
| $U_{\mathrm{o,err}}$ | 19.40 mV | 19.32 mV |
| $A_{U0}$ | 5 | 5 |
| $Q_{\mathrm{P}}$ | 0.429 | 0.5 |
| $f_{-3\,\mathrm{dB}}$ | – | 100.2 MHz |
| $\varphi_{\mathrm{m}}$ | 130.5° | 126.4° |
| $A_{U0} = +5, f_{-3\,\mathrm{dB}} = 120\,\mathrm{MHz}$, $R_{\mathrm{F}} = 1.32\,\mathrm{k\Omega}$ and $R_{\mathrm{N}} = 330\,\Omega$ ($R_{\mathrm{F}} = 1.33\,\mathrm{k\Omega} \pm 1\%$ and $R_{\mathrm{N}} = 332\,\Omega \pm 1\%$), $U_{\mathrm{im}} = 100\,\mathrm{mV}$, $R_{\mathrm{G}} = 50\,\Omega$ and $R_{\mathrm{T}} = 50\,\Omega$ | | |
| $U_{\mathrm{o,err}}$ | 16.12 mV | 16.05 mV |
| $A_{U0}$ | 5 | 5 |
| $Q_{P}$ | 0.54 | 0.52 |
| $f_{-3\,\mathrm{dB}}$ | – | 121.2 MHz |
| $\varphi_{\mathrm{m}}$ | 120° | 116° |
| $A_{U0} = +5, f_{-3\,\mathrm{dB}} = 150\,\mathrm{MHz}$, $R_{\mathrm{F}} = 845\,\Omega$ and $R_{\mathrm{N}} = 211\,\Omega$ ($R_{\mathrm{F}} = 845\,\Omega \pm 1\%$ and $R_{\mathrm{N}} = 210\,\Omega \pm 1\%$), $U_{\mathrm{im}} = 100\,\mathrm{mV}$, $R_{\mathrm{G}} = 50\,\Omega$ and $R_{\mathrm{T}} = 50\,\Omega$ | | |
| $U_{\mathrm{o,err}}$ | 13.69 mV | 13.64 mV |
| $A_{U0}$ | 5 | 5 |
| $Q_{\mathrm{P}}$ | 0.602 | 0.650 |
| $f_{-3\,\mathrm{dB}}$ | – | 140.6 MHz |
| $\varphi_{\mathrm{m}}$ | 110° | 106° |

The verification check of the models for AD8011AN and THS3091 is performed by comparing the simulation results to the datasheet typical parameters of the real op amps.

In Table 1 the calculated parameters and the simulation results for three values of the working bandwidth – 100, 120 and 150 MHz – are presented. The voltage gain is chosen with value equal to $+5$ and $R_L||C_L = 50\,\Omega||20\,pF$.

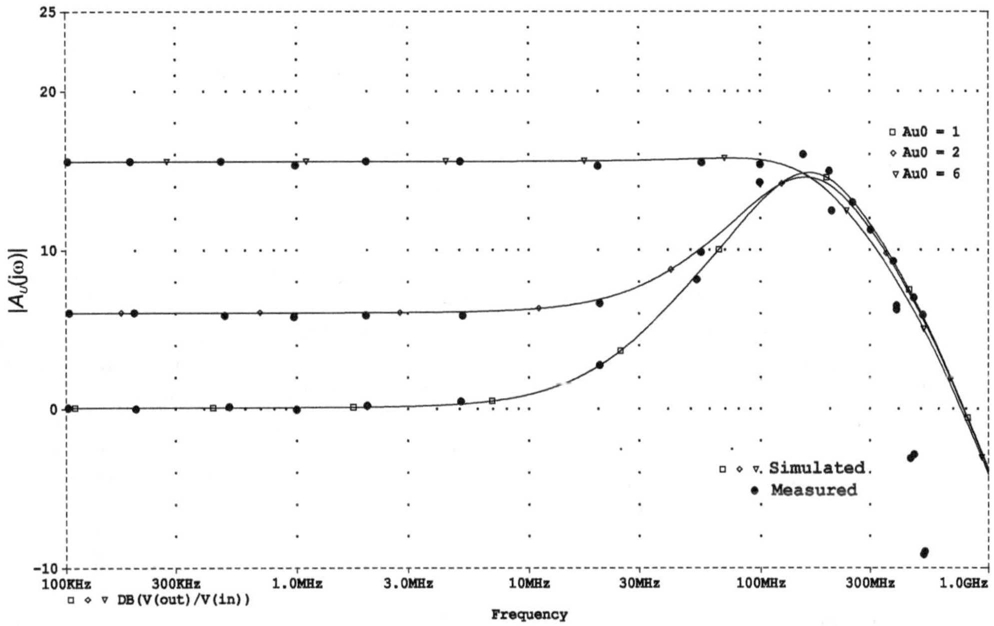


Figure 7. Module of the complex transfer function at $R_F = 1\,k\Omega$ and d.c. voltage gain $+1$, $+2$ and $+6$, respectively.

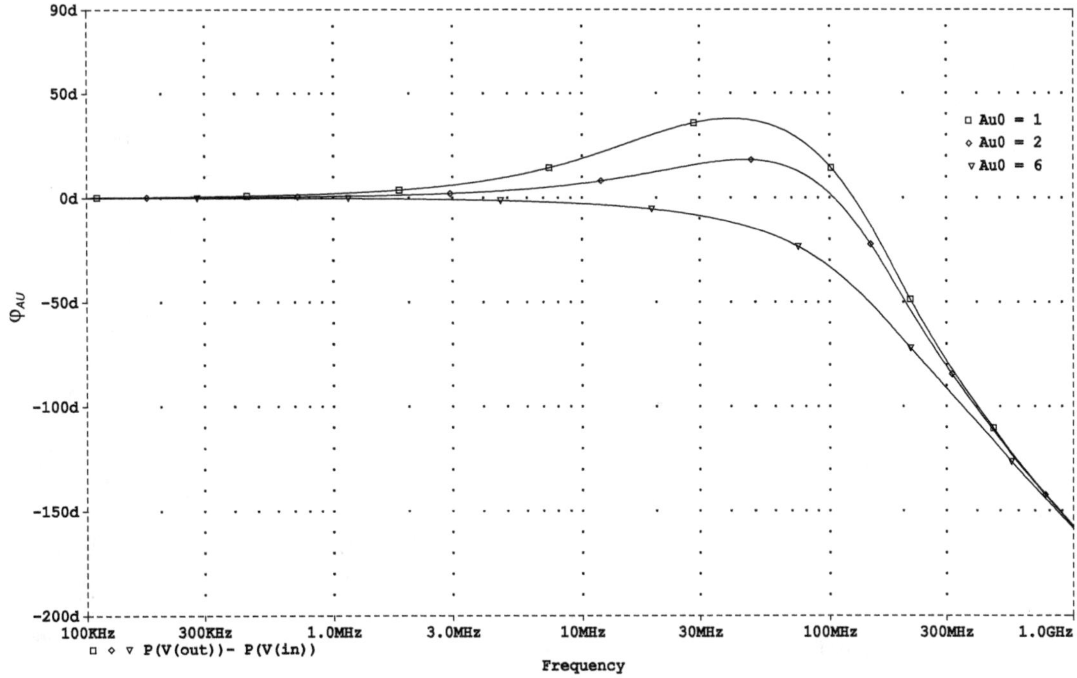


Figure 8. Phase shift of the complex transfer function at $R_F = 1\,k\Omega$ and d.c. voltage gain $+1$, $+2$ and $+6$, respectively.

Based on the proposed design procedure, values for the passive components and values for the basic dynamitic parameters were found. The maximum error between the calculated values of the electrical parameters and the simulation results is not higher than 10%. Moreover, an error of 10% is quite acceptable considering the tolerances of the technological parameters.

After implementation of a verification check by computer simulations, a combination of simulation and experimental study on various circuits of inverting and non-inverting amplifiers was performed. The tested circuits were implemented on a FR4 PCB laminate with surface-mount device passive components for the resistors and capacitors.

The study of the electronic circuits is performed in two stages. The first stage of computer simulations and experimental study is implemented for the inverting and non-inverting amplifier circuits using CFOA AD8011ARZ, biased with ±5 V supplies. The values of the chosen passive components for the investigated non-inverting amplifier circuit (Figure 4) are as follows: (1) $R_F = 1\,k\Omega$ and $R_N \to \infty$ at $A_{U0} = 1$; (2) $R_F = 1\,k\Omega$ and $R_N = 1\,k\Omega$ at $A_{U0} = 2$; (3) $R_F = 1\,k\Omega$ and $R_N = 200\,\Omega$ at $A_{U0} = 6$ with tolerances ±1%. The resistor $R_T$ is chosen equal to 51 Ω with tolerance ±1%. The coefficient $k_\omega$ of ringing of the output signal, according to the formula given in Section 4.1, is 5.53. The a.c. transfer characteristics of the circuits are obtained experimentally by using network analyser HP4195A. To measure the output signal, an active probe type HP41800A with input impedance 100kΩ/3pF is used. For the a.c. sweep analysis, the frequency is swept from 100 kHz to 1 GHz by decades, with 100 points per decade. The input voltage source is chosen with amplitude −10 dBm or 70.8 mV (with initial phase shift equal to zero), to ensure amplitude of the output voltage, not higher than 500 mV at $A_{U0} = 6$. The simulation and experimental results for the module and phase shift at three values of the d.c. voltage gain for the non-inverting amplifier are plotted in Figures 7 and 8. As can be seen for low frequencies approximately up to 10 MHz, the gains (Figure 7) are with constant value and are frequency independent. At $A_{U0}$ equal to 1 and 2 in the form of the frequency response causes peaking and for the voltage follower the amplitude reaches almost 15 dB. Moreover at gains 1 and 2, the module of the transfer function decreases with greater speed, such as for the frequency equal to 500 MHz reaches value equal to − 10 dB. The difference between the simulation and the experimental results is due to the influence of the additional parasitic poles determined by the inertial intermediate stages of the CFOA. The simulated values of the zero-pole pair at $A_{U0} = 1$ are $f_z = 30$ MHz and $f_p = 250$ MHz, respectively. For $A_{U0} = 2$, $f_z = 60$ MHz and $f_p = 250$ MHz. At the voltage gain equal to 6, $f_z = 180$ MHz and $f_p = 250$ MHz, the amount of peaking is small, because the frequencies are close and as a result, the transfer characteristic monotonically decreases to unity. The phase margin (Figure 8) for the three gains is greater than 45°, which means that the

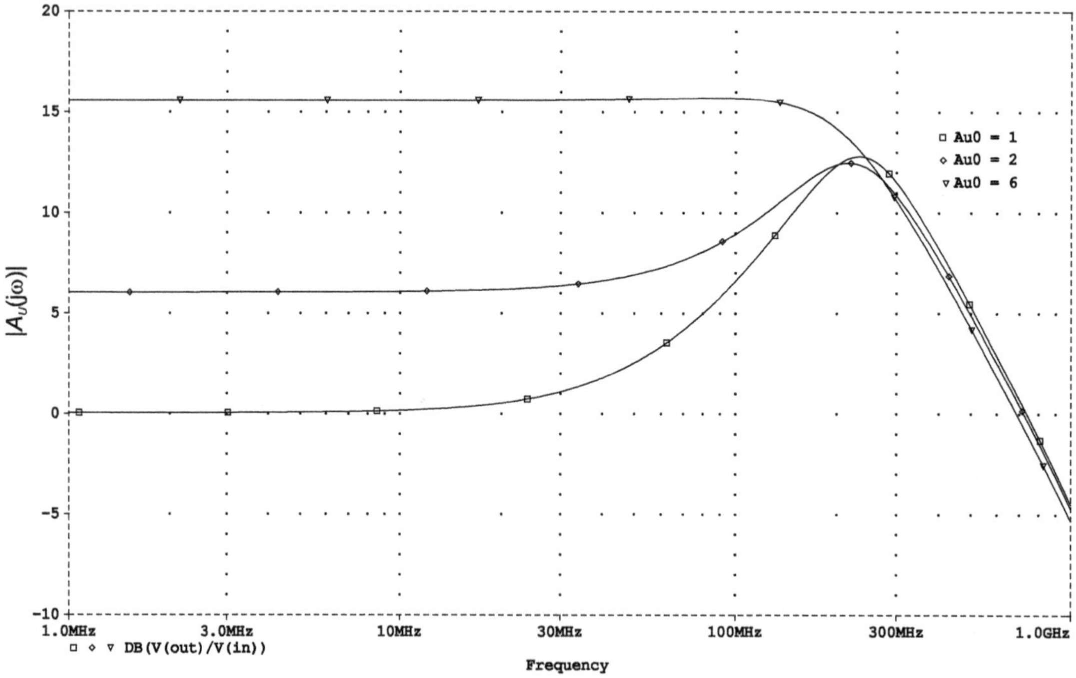


Figure 9. Alternating current transfer characteristics at $R_F = 500\,\Omega$ and d.c. voltage gain + 1, + 2 and + 6, respectively.

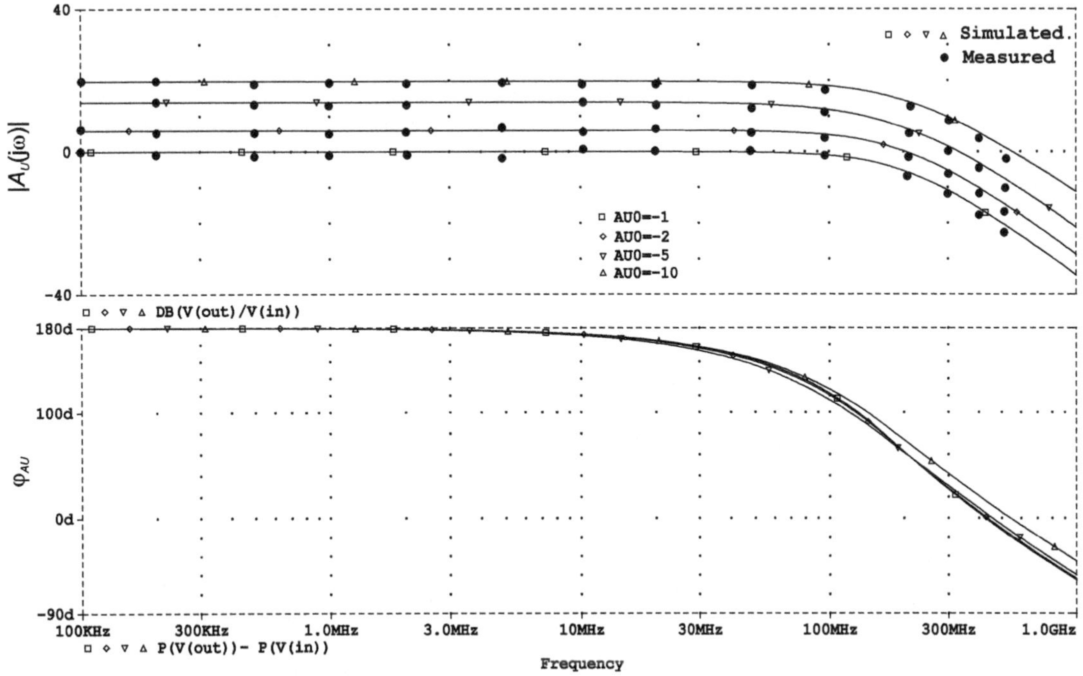


Figure 10. Alternating current transfer characteristics at d.c. voltage gain $-1$, $-2$, $-5$ and $-10$, respectively.

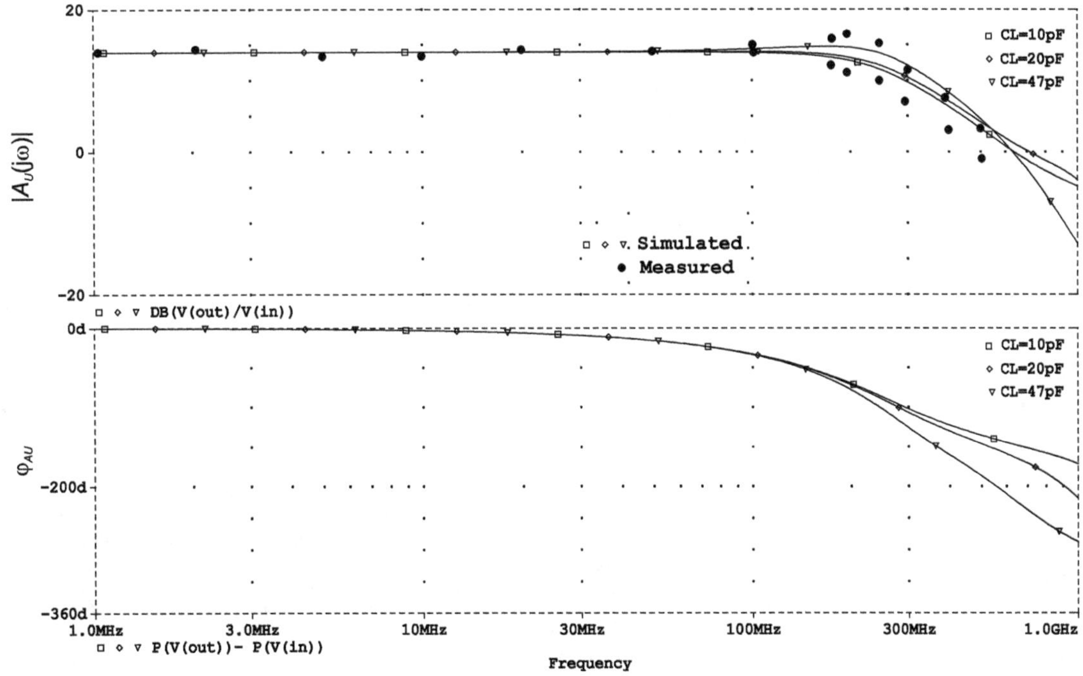


Figure 11. Alternating current transfer characteristics at $R_F = 1000\,\Omega$ and $C_L$ with values 10, 20 and 47 pF.

amplifier circuits are stable according to the Bode criterion. At gains equal to 1 and 2, the phase shift between the input and the output signal is positive, because $\omega_z < \omega_p$. This additional phase shift has a value less than 50°, which does not affect the stability of the circuits.

To verify the results of the theoretical analysis at $R_F = 500\,\Omega$, new non-inverting amplifiers were implemented with the following passive components: (1) $R_F = 500\,\Omega$ and $R_N \to \infty$ at $A_{U0} = 1$; (2) $R_F = 500\,\Omega$ and $R_N = 500\,\Omega$ at $A_{U0} = 2$; (3) $R_F = 500\,\Omega$ and $R_N = 100\,\Omega$ at $A_{U0} = 6$ with tolerances $\pm 1\%$. In this case the coefficient $k_\omega$ is equal to 3.91. The simulation results for the transfer characteristics at the three values of the d.c. voltage gain are shown in Figure 9. The comparative analysis of the transfer characteristics in Figure 7 shows that the circuit with the lower value of $R_F$ is with expanded value of working bandwidth and the amplitude of the peaking at voltage gain equal to 1 and 2 is smaller. These simulations and experimental results confirm the correctness of the theoretical analyses and effectiveness of the obtained formulas.

The inverting amplifier circuit (Figure 2) using CFOA AD8011ARZ is studied for four d.c. voltage gains. The values of the calculated passive components are as follows: (1) $R_F = 1\,\text{k}\Omega$ and $R_N = 1\,\text{k}\Omega$ at $A_{U0} = -1$; (2) $R_F = 1\,\text{k}\Omega$ and $R_N = 499\,\Omega$ at $A_{U0} = -2$; (3) $R_F = 1\,\text{k}\Omega$ and $R_N = 200\,\Omega$ at $A_{U0} = -5$ and (4) $R_F = 499\,\Omega$ and $R_N = 51\,\Omega$ at $A_{U0} = -10$. All resistors were chosen with tolerance $\pm 1\%$. The resistor $R_T = 51\Omega \pm 1\%$ is used for gains $-1$, $-2$ and $-5$. The a.c. transfer characteristic at the four values of the voltage gains is presented in Figure 10. As can be seen, all frequency characteristics are not ringing ($Q_p < 0.707$) and decreased monotonically with increase in the frequency of the input signal. Furthermore, the zero frequency $f_z$ is much greater than the frequency of the double pole. The phase shift between the input and the output signal varies from $+180°$ to $-180°$, as for the frequency approximately equal to 400 MHz the phase shift is 0°. For gains $-1$ and $-2$ the phase margin is greater than 45°. At gains equal to $-5$ and $-10$ for frequencies greater than 240 MHz, the phase margin is less than 45° and the circuits are unstable. The phase shift of the output signal varies approximately up to $-90°$. This phase shift is determined by the influence of additional parasitic poles in the frequency response of the op amp.

The second series of computer simulations and experimental study was performed for the non-inverting amplifier circuits (with $R_F = 1\,\text{k}\Omega$, $R_N = 250\,\Omega$ and $A_{U0} = 5$) at complex (active-capacitive) load employing CFOA THS3091D (using an 8-pin SOIC with PowerPAD™), biased with $\pm 15$ V supplies. The a.c. transfer characteristic at amplitude of the output voltage equal to 4 V and for three values (10, 20 and 47 pF) of the capacitive load of the amplifier is plotted in Figure 11. The $R_L$ was chosen equal to $100\,\Omega$, as the maximum output current is 40 mA. For frequencies up to 30 MHz, the module is with constant value, approximately equal to 5.0 (or $\approx 14$ dB). At capacitive load equal to 10 and 20 pF, the quality factor, according to formula (11), is equal to 0.5 and 0.71, respectively. In these cases, the peaks are not observed and the transfer characteristic decreases monotonically with an increase in the frequency of the input signal. At capacitive load equal to 50 pF, the quality factor becomes larger than 0.707 and in the form of transfer characteristic ringing occurs. Furthermore, the phase shift between the input and the output signal increases and at frequency equal to 330 MHz it becomes 135°. At further increase of the frequency, the phase margin becomes smaller than 45°, which decreases the stability of the amplifier.

## 7. Conclusion

In this paper, a study of the frequency stability analysis for high-speed inverting and non-inverting amplifiers using CFOAs has been presented. Based on the analysis of the principle of operation, equations for the complex transfer functions are obtained, as well as recommendations for improving the stability at complex load and design procedure are defined. The proposed procedure can be useful for the analysis and design of high-speed amplifier circuits employed in various analogue and mixed-signal circuits, such as video amplifiers, line drivers and analogue switches. The results obtained by the theoretical analyses are validated through simulation and experimental testing of sample electronic circuits using monolithic op amps AD8011 and THS3091. The maximum error between calculated values and the simulation results for the basic electrical parameters is not higher than 10%.

## Acknowledgements

The AD8011 and THS3091 CFOAs, used in this work, were provided by Analog Devices and Texas Instruments, respectively.

## Disclosure statement

No potential conflict of interest was reported by the authors.

## References

AD8011 300 MHz current feedback amplifier – datasheet. (2014). Analog devices. Retrieved from http://www.analog.com/static/imported-files/data_sheets/AD8011.pdf

AD8011 SPICE macro-model, rev. A, 9/97. (2014). Analog devices. Retrieved from http://www.analog.com/en/all-operational-amplifiers-op-amps/operational-amplifiers-op-amps/ad8011/products/product.html#product-designtools

Boyanov, J., & Shoikova, E. (1989). *Electronic circuits theory*. Sofia: Tehnika.

Choosing Between Voltage Feedback (VFB) and Current Feedback (CFB) Op Amps. (2008). Analog devices. Retrieved from http://www.analog.com/media/en/training-seminars/tutorials/MT-060.pdf.

Jassim, H. (2013). A new design technique of CMOS current feedback operational amplifier (CFOA). *Circuits and Systems*, *4*(1), 11–15. doi:10.4236/cs.2013.41003

Jung, W. G. (2002). *Op Amp applications handbook.* Analog devices. Retrieved from http://www.analog.com/library/analogDialogue/archives/39-05/op_amp_applications_handbook.html.

Kamath, D. (2014). Bandwidth enhancement of inverting amplifier using composite CFOA block. *International Journal of Innovative Research in Electrical, Electronics, Instrumentation and Control Engineering, 2*(4), 1387–1390.

Laker, K., & Sansen, W. (1994). *Design of analog integrated circuits and systems.* New York, NY: McGraw-Hill.

Mancini, R. (2001). *Current feedback amplifier analysis and compensation* (Application Report - SLOA021A). Dallas, TX: Texas Instruments Inc.

Mancini, R. (2002). *Op amps for everyone. Design guide* (Rev. B). Dallas, TX: Texas Instruments Inc

Nagaria, R. K., Gopi Krishna, M., & Singh Rakesh, Kr. (2008). Comparative performance of low voltage CMOS – CFOA suitable for analog VLSI. *WSEAS Transactions on Electronics, 5*(6), 226–231.

Palumbo, G. (1997). *Current feedback amplifier: Stability and compensation.* Proceedings of the 40th Midwest Symposium on Circuits and Systems, Sacramento, CA, 249–252.

Pandiev, I. (2012). *Analysis and design of CFOA-based high-speed amplifier circuits.* TELFOR, Proceedings of papers, Serbia, Belgrade, 979–982. doi:10.1109/TELFOR.2012.6419373.

Safari, L., & Azhari, S. (2012). A high performance fully differential pure current mode operational amplifier and its applications. *Journal of Engineering, Science and Technology, 7*(4), 471–486.

Schmid, R. (2003). *Measuring board parasitics in high-speed analog design* (Application Report – SBOA094). Dallas, TX: Texas Instruments Inc..

Seifart, M. (2003). *Analoge Schaltungen. 6 Auflage.* Berlin: Verlag Technik.

Stoianov, I. (2000). *Electronic measurement systems.* Sofia: TU-Sofia.

Texas Instruments. (2002). *High speed analog design and application seminar.* Dallas, TX: Texas Instruments Inc.

THS3091 current-feedback operational amplifier – datasheet. (2014). Texas Instruments. Retrieved from http://www.ti.com/lit/ds/symlink/ths3091.pdf

THS3091 PSpice Model. (2014). Texas Instruments. Retrieved from http://www.ti.com/product/THS3091/toolssoftware

Tietze, V., & Schenk, Ch. (2008). *Electronic circuits* (2nd ed.). Berlin: Springer-Verlag.

# Dynamic modelling and experimental validation of an automotive windshield wiper system for hardware in the loop simulation

Mark Dooner[a], Jihong Wang[a]* and Alexandros Mouzakitis[b]

*[a]School of Engineering, The University of Warwick, Coventry, UK; [b]Jaguar Land Rover Product Development Centre, Gaydon, Warwickshire, UK*

In order to remain competitive, automotive companies use advanced simulation methods to assist in product development. Hardware in the loop (HIL) simulation is one such technique. To use HIL in the development of automotive electronic control units (ECU), accurate simulation models of the ECU's sensors and actuators are needed. In this work, a full dynamic mathematical model of an automotive windshield wiper system is developed and validated. In the modelling phase, the wiper motor is analysed and a unique mathematical model is developed to capture the devices two speed operation. A multi-body dynamic model of the linkages is implemented using the MathWorks' SimMechanics software. The model is validated experimentally and its parameters are identified using genetic algorithms. The model is then simplified to allow it to be simulated in real time, making it suitable for HIL simulation. The HIL compatible model is used in the development of Automotive ECUs.

**Keywords:** windshield-wiper; hardware-in-the-loop; real-time simulation; genetic algorithm

## 1. Introduction

The Global Automotive Report 2013 compiled by Clearwater Corporate Finance LLP estimated the value of the automotive industry at $800bn (Clearwater Corporate Finance, 2013). In addition, worldwide passenger car sales in 2012 exceeded 60 million units and sales have tended to increase over the last decade (http://www.oica.net). The total number of passenger cars produced in 2012 was also greater than 60 million – representing an increase of 5.3% in production from 2011 (http://www.oica.net). New vehicles must meet stringent safety and environmental requirements, with industry standards such as ISO26262 being widely adhered to (Jeon, Cho, Jung, Park, & Han, 2011) whilst maintaining acceptable comfort and performance standards to make the product viable. It follows that automotive companies capable of producing quality and safe products in relatively short production times will benefit from the highly lucrative global automotive industry.

The development process for new products in the automotive industry follows the classic V model (Robert Bosch GmbH, 2007). Figure 1 shows an adapted version of this, highlighting the use of models in the production process. Modern product development processes use model-based design, development and testing tools to improve the tractability of requirements, the speed of development and the validity of early testing. A particularly important simulation-based testing procedure is hardware in the loop (HIL) simulation.

HIL is an advanced real-time simulation technique in which a purely simulated system has certain aspects replaced by hardware components (Hu & Azarnasab, 2013). In the automotive industry, HIL simulation is used extensively in the development and testing of electronic control units (ECU) (Ganesh, 2005; Schuette & Ploeger, 2007) because they allow extensive tests to be carried out before the final design and manufacture of components interfacing with the ECUs. Most of the innovations in modern luxury vehicles are in the electronic/software domains, with electronic systems replacing traditionally mechanical and pneumatic systems (Von Tils, 2006) and there are in excess of 100 ECUs in a modern luxury car (Waltermann, 2009). The purpose of the model developed here is to provide a simulation model of a windscreen wiper system to be used in the HIL testing of an ECU by Jaguar Land Rover.

The benefits of using a simulation model such as the one developed in this paper are as follows: (1) reduced need for a hardware prototype which takes up space and resources, (2) models can be quickly updated to incorporate design changes, whereas up to date prototypes are often unavailable, (3) tests done with simulation models have higher repeatability because all variables can be controlled and (4) testing for the development of ECUs

*Corresponding author. Email: jihong.wang@warwick.ac.uk

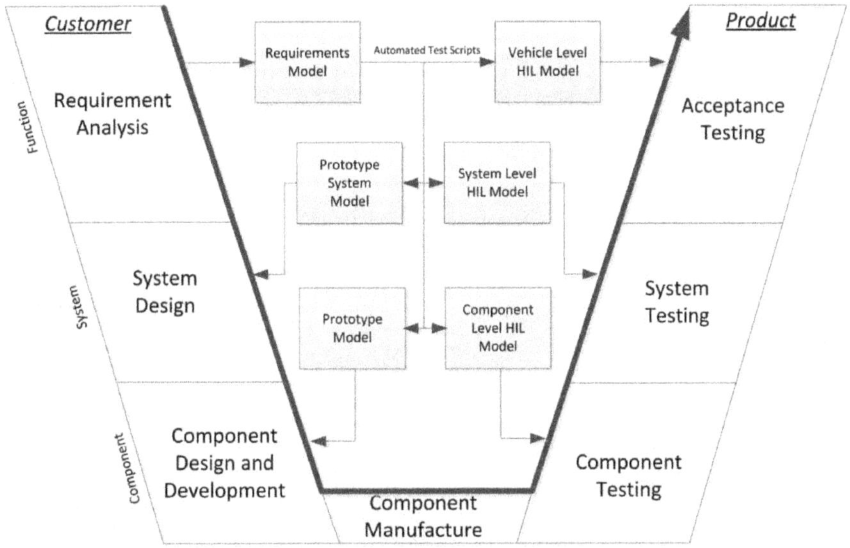


Figure 1. V model for product development.

can be carried out sooner than tests relying on hardware prototypes.

This paper first describes the wiper system to be modelled by showing its physical structure and operation principles. Then the modelling process of each element of the system is shown. Genetic algorithms (GA) are then used to identify the unknown parameters in the model. The model is then validated by comparing its performance against real data. Once this is complete, the model is simplified to allow it to be simulated in real time, making it suitable for HIL simulation.

## 2. Existing windscreen wiper models

Models of windscreen wiper systems in the existing literature tend to concentrated on one particular aspect of the system and are not suitable for real-time simulation. Some examples of previous work are given to illustrate this point: In Chang and Lin (2004) a mathematical model of a wiper system is developed in order to investigate, and attempt to control, chaotic behaviour present the system. A mathematical model of the mechanics of a wiper blade is developed in Okura, Sekiguchi, and Oya (2000) to investigate the vibrational behaviour of the system during reversal (i.e. when the blades change direction). Significant noise can be heard at this point and the model was used to identify design parameters which could be modified to reduce this noise. Chatter vibration was investigated in Grenouillat and Leblanc (2002) using a simulation model of the blade on the windscreen. It was found that the mechanical configuration of the wipers had a large effect on the chatter noise. Models such as the three identified above which deal with vibrations are generally too complicated for HIL simulation and are unsuitable for developing ECUs.

More similar research to that presented here is given in Xiaoyu, Yanfeng, and Yengjie (2011) where physical modelling techniques are used to create a mechanical model of a wiper system to assist in the design of future components; however the model was not validated. Finally, previous publications of this work Wei, Mouzakitis, Wang, and Sun (2011) and Dooner, Wang, and Mouzakitis (2013) develop a model for HIL simulation which includes the motor and linkages.

## 3. Wiper system modelling

The model developed in the section simulates different behaviours from the models discussed in Section 2. The important elements of the wiper system that this model captures are (1) the dynamic torque load of the linkages driven by the motor and how this affects the motor current, (2) the switching transients on the motor current when switching between ON and OFF modes as well as FAST and SLOW, (3) such behaviour must be modelled across the systems entire operating speed. The model does not take into account effects such as vibration of the mechanical element.

In addition to capturing the system dynamics and coupling between the motor and linkages, the model must be capable of being simulated in real time (for use in HIL simulation) and be easy to update to incorporate design changes since the model will be used in the early stages of product development. To meet these goals a physical modelling approach has been employed using the MathWorks physical modelling tool, Simscape.

The system can be split into two elements for modelling: the wiper motor and the linkage system, as shown in

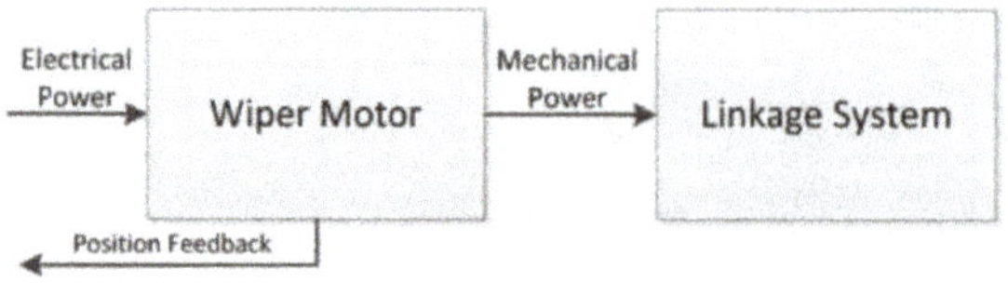


Figure 2. System block diagram.

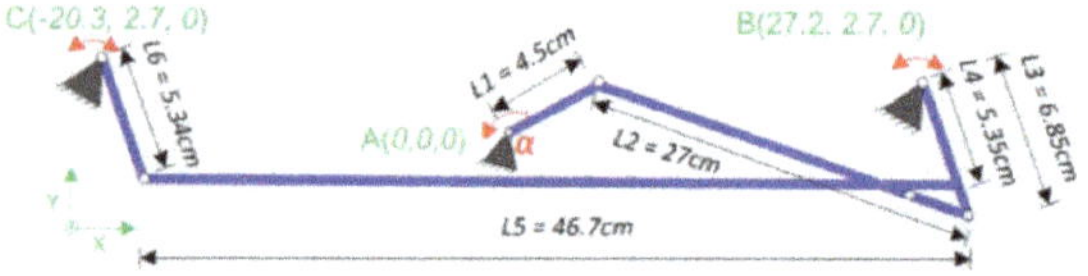


Figure 3. Linkages operation.

Figure 2. These are both initially modelled separately and then coupled together into one model.

### 3.1. *Linkage system modelling*

The linkages used in a windshield wiper system are designed such that unidirectional rotational motion from a Permanent Magnet Direct Current (PMDC) motor can be translated into the oscillatory motion of the two wipers. The design and operation of the linkage system is demonstrated in Figure 3. The linkages have initially been modelled using the MathWorks tool SimMechanics. This was chosen because it simplifies the modelling and simulation process of Multi-body dynamic systems by using embedded equations and modular design (Wood & Kennedy, 2003). SimMechanics has also been successfully employed in similar projects (Xiaoyu et al., 2011).

The linkages are defined in the SimMechanics modelling environment as shown in Figure 4. Each individual linkage is represented by a 'Solid' block which defines the shape and material of the linkage. Two transformation blocks are used to define the length of the linkage and assign a co-ordinate system to either end. Each end is then attached to a revolute joint. The full five bar system is modelled using this method and is approximated as being in a plane. The complete model of the linkages is shown in Figure 5.

The measured parameters of the model are shown in Figure 3, adapted from Wei et al. (2011). The model is designed to allow any of the parameters to be easily changed so that the model can represent any physically viable design of the linkage system shown in Figure 3, allowing for design updates to be easily implemented. Also included in the parameters is the density of the material, 2.7 g/cm$^3$, and the initial angle, ($\alpha$), of the crank, determining the park position of the wipers. The park position refers to the wipers rest position when not in operation, that is, at the bottom of the windscreen.

### 3.2. *Wiper motor modelling*

The brushed PMDC wiper motor is connected directly to the battery meaning that the speed of the motor, and thus

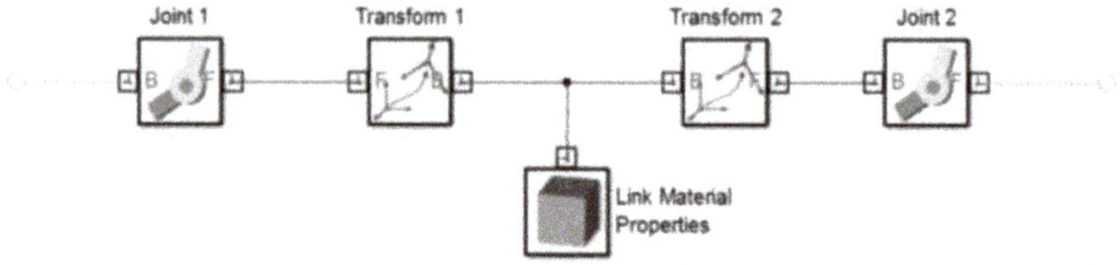


Figure 4. SimMechanics tool.

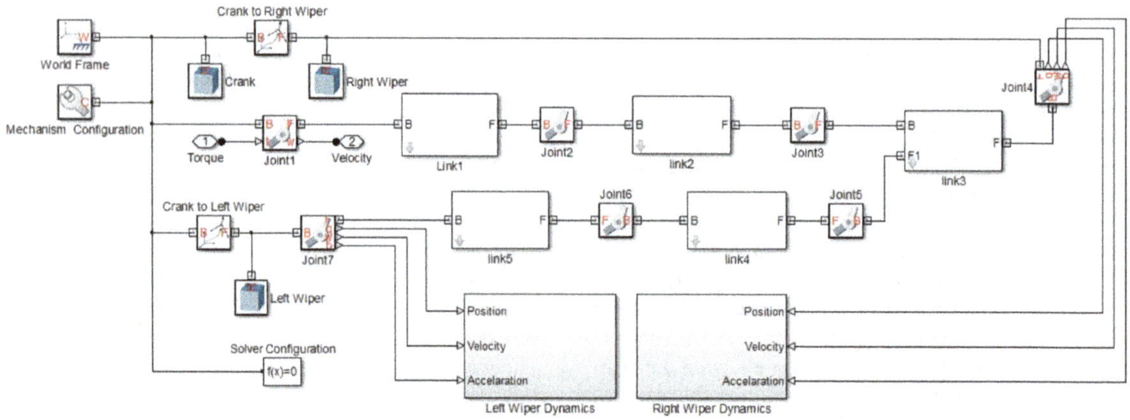


Figure 5. Complete linkages model.

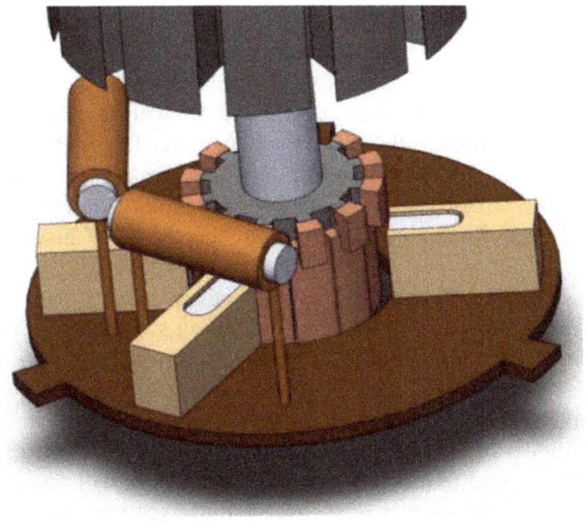

Figure 6. Wiper motor structure.

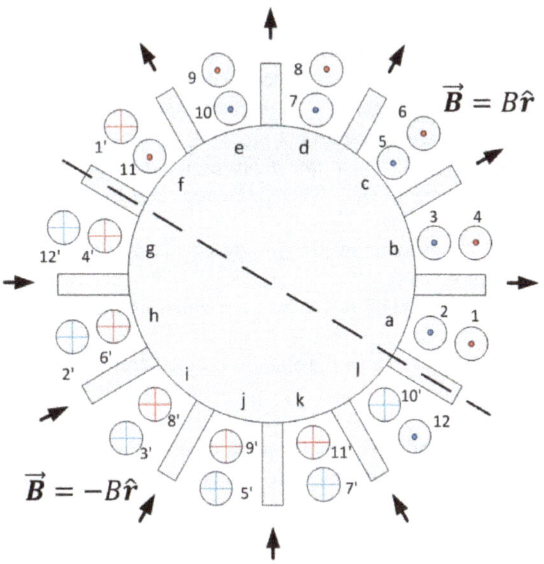


Figure 7. Slow speed armature winding.

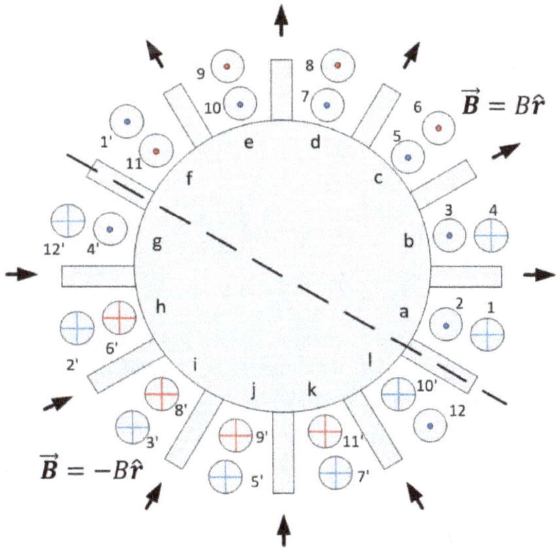


Figure 8. Fast speed armature winding.

the wipers, cannot be controlled by changing the input voltage. In order to control the speed of the motor a third brush is added, offset from the magnetic neutral line, giving the motor two electrical inputs and one common ground output (Hameyer & Belmans, 1996). The structure can be seen in Figure 6.

The brush directly opposite to the common brush in the magnetic neutral line is connected to the slow speed input. Applying a voltage to this brush will cause the motor to operate as a standard two brush PMDC motor. Applying a voltage to the brush offset from the magnetic neutral line will cause the motor to operate in its fast mode. This increases the rotational velocity of the motor shaft and also the armature current, with a reduction in efficiency (Hillier, 1987).

No model of a wiper motor capturing the fast and slow speed operation, and the switching between them, could be found. Therefore a unique model of the wiper motor is developed here using classical electromagnetic theory.

To understand the two brush operation of the wiper motor the difference in the current paths around the armature windings in the two different states are considered. The current paths in the slow and fast operations are shown in Figures 7 and 8, respectively where the 12 armature slots are labelled '*a*' to '*l*' and contain two windings each. The 12 windings are labelled '1' to '12' with a tick representing the return path of the winding. The instantaneous direction of the current is represented using a cross for current into the page and a dot for current out of the page and the current paths are colour coded, with blue representing current path 1 and red representing current path 2. The arrows show the direction of the magnetic field caused by the permanent magnet. The magnetic field can be described in terms of the flux density vector, $\vec{\boldsymbol{B}}$, or the scalar denoting the magnitude of the magnetic flux density, $B$, and the vector, $\hat{\boldsymbol{r}}$, which points radially away from the centre of the armature.

It can be seen that for the slow operation the current paths are symmetrical around the magnetic neutral line, as in a normal PMDC motor. However, in fast operation the current paths are not symmetrical, that is, they are unbalanced. The value of the current in each current path the same in slow mode, however it is different in fast mode due to the different physical lengths of the current paths.

An analysis is carried out based on Chapter 1 in Chiasson (2005) on the torque developed in each armature slot.

The magnitude of the force in each slot is a function of the dimensions of the armature, the magnitude and direction of the $\vec{B}$ field, and the magnitude and direction of the current. The unbalanced nature of the current paths in the wiper motor complicates the determination of the total torque (which is usually the torque in one armature slot multiplied by the number of slots. There are three cases to consider:

- The currents in the slot are equal and in the same direction.
- The currents in the slot are equal and in opposite directions.
- The currents in the slot are unequal and in the same direction.

In the first case, that is, the normal case as in slot '$a$' Figure 6, the torque generated per slot can be shown to be,

$$\vec{T}_a = l_1 l_2 B i \hat{z} = K_t i \hat{z},$$

where $\vec{T}_a$ is the torque in slot '$a$', $l_1$ is the length of the armature, $l_2$ is the diameter of the armature, $i$ is the current, $B$ magnetic flux density and $K_t$ is the torque constant of the motor. Unit vector $\hat{z}$ points along the axis of rotation.

In the second case (as in slot '$f$' in Figure 7), the torque produced by the two windings in the slot cancel each other out. In the case of slot '$f$':

$$\vec{T}_{11} = l_1 \left(\frac{l_2}{2}\right) B i \hat{z},$$

$$\vec{T}_{1'} = -l_1 \left(\frac{l_2}{2}\right) B i \hat{z},$$

$$\vec{T}_{11} + \vec{T}_{1'} = \vec{T}_f = 0.$$

In the third case (as in slot '$c$' of Figure 8), the torques produced by the two windings sum, but are of different magnitudes. In the case of slot '$c$':

$$\vec{T}_5 = l_1 \left(\frac{l_2}{2}\right) B i_1 \hat{z},$$

$$\vec{T}_6 = l_1 \left(\frac{l_2}{2}\right) B i_2 \hat{z},$$

$$\vec{T}_5 + \vec{T}_6 = \vec{T}_c = l_1 l_2 B (i_1 + i_2) \hat{z} = K_t (i_1 + i_2) \hat{z},$$

where $i_1$ and $i_2$ are the currents through the separate current paths.

Expressions for the torque produced by the motor in slow and fast modes can now be derived by summing the torque produced in each individual slot, depending on current pattern,

$$T_{\text{slow}} = T_a + T_b + \cdots + T_l = 10 K_t i_{\text{slow}}, \quad (1)$$

$$T_{\text{fast}} = T_a + T_b + \cdots + T_l = 8 K_t i_{\text{fast}} \quad (2)$$

in scalar form where $T_{\text{slow}}$ and $T_{\text{fast}}$ are the torques produced in the motor's slow and fast modes, respectively. Likewise, $i_{\text{slow}}$ and $i_{\text{fast}}$ are the input currents in its slow and fast operations, respectively, with $i_{\text{fast}}$ equalling $(i_1 + i_2)$.

When analysing the back electromotive force (EMF) produced by the motor, it is simpler to analyse armature winding loops, rather than armature slots. The following shows an analysis of the EMF produced by a single loop, which will then be applied to the wiper motor under investigation; the analysis is based on Chapter 1 in Chiasson (2005).

Figure 9 represents a single coil of wire in the armature. The transparent section represents the air gap and thus the flux surface of the magnetic field, $S$, in the motor. Lengths $l_1$ and $l_2$ are the height and width of the armature, respectively, meaning that the flux surface is approximately a half cylinder of radius $l_2/2$ and length $l_1$. Figure 9 shows that the positive direction of travel around the coil is taken to be anti-clockwise, in accordance with the vector $\vec{r}$ pointing radially away from the centre of the armature. The magnetic field in the air gap generated by the permanent magnet is known to be approximately radially directed with a constant magnitude of $B$. Therefore an expression of the vector $\vec{B}$ is given as:

$$\vec{B} = \begin{cases} +B\hat{r} & \text{for} \quad 0 < \theta < \pi, \\ -B\hat{r} & \text{for} \quad \pi < \theta < 2\pi. \end{cases}$$

A surface element, $\mathrm{d}\vec{S}$, of the flux surface defined in Figure 9 is shown in Figure 10; its direction is always outward of the cylinder. From Figure 10, an expression for $\mathrm{d}\vec{S}$ can be derived as:

$$\mathrm{d}\vec{S} = \left(\frac{l_2}{2}\right) \mathrm{d}\theta \, \mathrm{d}z \vec{r}, \quad (3)$$

where $\theta$ is the position of the coil, with $\theta = 0$ on the magnetic neutral line between slots $g$ and $f$ in Figures 7 and 8. The flux can be derived by integrating the flux density over the air gap shown in Figure 9, making use of the expression

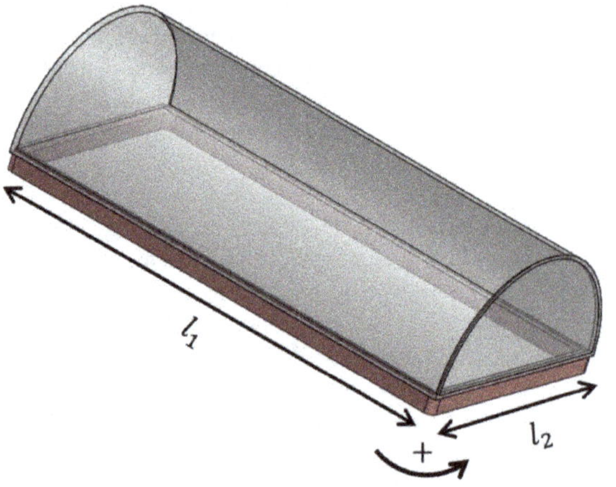


Figure 9. Flux surface of a single coil.

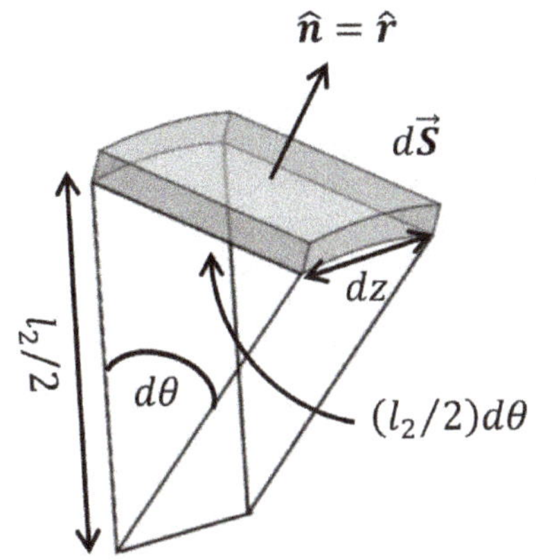


Figure 10. Surface element.

for the surface element area in Equation (3). Thus, using Equation (3), an expression for the flux, $\phi$, can be derived:

$$\phi(\theta_R) = \int_S \vec{B} \cdot d\vec{S},$$

$$\phi(\theta_R) = \int_0^{l_1} \int_{\theta=\theta_R}^{\theta=\pi} (B\hat{r}) \cdot \left(\frac{l_2}{2} d\theta \, dz\hat{r}\right)$$

$$+ \int_0^{l_1} \int_{\theta=\pi}^{\theta=\pi+\theta_R} (-B\hat{r}) \cdot \left(\frac{l_2}{2} d\theta \, dz\hat{r}\right),$$

$$\phi(\theta_R) = -l_1 l_2 B\left(\theta_R - \frac{\pi}{2}\right) \quad \text{for } 0 < \theta_R < \pi.$$

Similarly, it can be shown that the flux for $\pi < \theta_R < 2\pi$ is:

$$\phi(\theta_R) = -l_1 l_2 B\left(\theta_R - \frac{\pi}{2} - \pi\right).$$

The induced EMF in the rotor loop can therefore be calculated as:

$$\xi = -\frac{d\phi}{dt} = (l_1 l_2 B)\frac{d\theta_R}{dt} = K_e \omega_R,$$

where $\xi$ is the EMF, $K_e$ is the back EMF constant and equals $l_1 l_2 B$ and $\omega_R$ is the rotor's angular velocity.

Using the above expression, an expression for the EMF produced by each coil can be derived for both the slow and fast modes of operation. These are then summed to get an expression for the total EMF produced by the motor in each mode. Referring to Figure 7, it can be seen that the forward and return paths of coil 1 both lie in the positive direction of $\vec{B}$ when the motor is operating in its slow mode. Likewise, coil 12's forward and return paths both lie in the negative direction of $\vec{B}$. This means that these two coils will produce no back EMF. This is also true for the fast mode. Figure 8 shows that the current in coil 4 in the fast mode is in the reverse direction to the so called positive direction defined in Figure 9. This means that the EMF produced by the will be negative. Therefore, the total EMF in the motors slow and fast modes are:

$$\xi_{slow} = 10K_e\omega_r, \tag{4}$$

$$\xi_{fast} = 8K_e\omega_r. \tag{5}$$

When a voltage is applied to either of the inputs, the current will see two paths to ground. In the slow operation the resistance of the paths is virtually equal. However in the fast operation, one path is physically shorter than the other and thus its resistance is lower. Considering the two current paths as two resistors in parallel, the overall armature resistance of the motor in its fast mode is smaller than in its slow mode. A similar analysis can be done for the inductance, with the same conclusion.

The eight parameter dynamic simulation model of the wiper motor is implemented as shown in Figure 11. Each separate DC motor represents a separate speed and implements the model with the following equations (Chiasson, 2005):

$$V - R_a I_a - L\frac{dI_a}{dt} - K_e\frac{d\theta}{dt} = 0,$$

$$K_t I_a - J\frac{d^2\theta}{dt^2} - b\frac{d\theta}{dt} - T_L = 0,$$

where $V$ is input voltage, $R_a$ is armature resistance, $I_a$ is armature current, $L$ is armature inductance, $K_e$ is the EMF constant, $\theta$ is the rotor position, $K_t$ is the torque constant, $J$ is the motor inertia, $b$ is the damping and $T_L$ is the torque load.

The using Equations (1)–(5), along with the analysis of the resistance and inductance of the motor, it is concluded that to represent the fast mode, the resistance, inductance and $K$ parameters will be lower than the slow mode. The mechanical parameters remain the same for both modes.

The model also includes a gear box to model the worm and wheel gears built into the motor. The park switch is

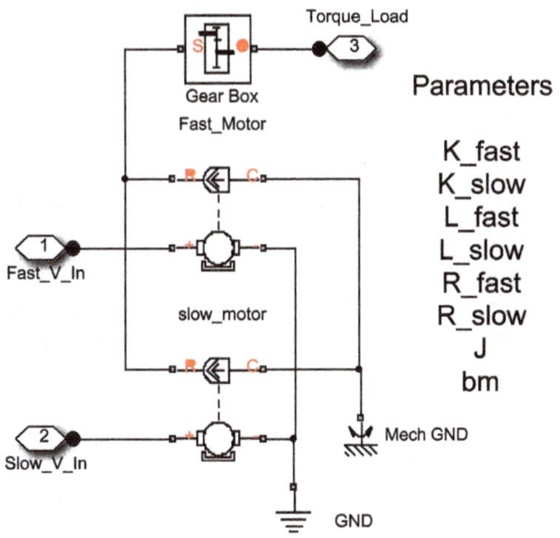


Figure 11. Wiper motor model implementation.

generated by measuring the position of the crank shaft and outputting a pulse when the wipers are in their park position.

## 4. Modelling for real-time simulation

The model presented in Section 3 is not optimized for real-time simulation. A powerful enough machine could simulate the model in real time; however there are steps that can be taken to make its simulation less computationally intensive.

The off-line model was simulated on a Desk-top PC with 4 GB of RAM and a 3.4 GHz processor. The solver used was a fixed step (ode14x) solver with a step size of 1 ms. Simulating 60 s of data took 86 s, giving a time per simulation step of 1.43 ms, that is, one simulation step of 1 ms takes 1.43 ms to simulate. This means that to simulate the model in real time for HIL, significant increases in code efficiency would be required.

To move from off-line to online real-time simulation there are four areas in which changes can be made (Miller & Wendlandt 2013):

- Increase the step size of the simulation; however the step size is fixed at 0.001 s in this case.
- Utilize the local solvers supplied by SimScape. The local solvers greatly increases the simulation speed but cannot be used when a SimMechanics model is connected to the physical network.
- Reduce the number of iterations the solver makes, reducing accuracy but increasing speed.
- Decrease the overall model fidelity by removing insignificant or irrelevant elements.

After considering the available options, the decision was made to remove the SimMechanics element from the model and replace it with a mathematical model so that local solvers could be used. Therefore the linkage model needed to be replaced, without a significant loss of accuracy.

### 4.1. *Position of the wipers*

To simulate the position of the wipers, a simple look-up table was used. This method was chosen because there is a direct relationship between the position of the crank and the position of the wipers. Two look-up tables were used per wiper to represent the forward and backward sweeps. This approximation introduced virtually no error into the simulation.

### 4.2. *Torque load*

Using a look-up table to represent the torque load applied to the motor by the linkages would have been unwieldy because, not only is the instantaneous torque value dependent of the position of the rotor, it is also dependent on the motor's angular velocity.

The fundamental shape of the torque produced by the linkages is the same for each revolution, or 'wipe'. The shape of the torque produced by a forward and backward sweep of the wipers has been modelled separately with two polynomials, $f(x)$ and $g(x)$, respectively, where $x$ is the angle of the motor's rotor and is between 0 and $2\pi$. The polynomials are:

$$\begin{aligned} f(x) &= -0.19224x^4 + 0.021796x^3 - 0.15435x^2 \\ &\quad - 0.077812x + 0.68626, \\ g(x) &= 0.4184x^7 - 0.30315x^6 - 0.42441x^5 + 0.12634x^4 \\ &\quad + 0.21484x^3 - 0.10093x^2 - 0.29859x + 0.63276. \end{aligned}$$

Both the average and peak-to-peak values of the torque increase with velocity. This was modelled using Simulink to represent the following equation:

$$\tau = \begin{cases} (C_2 * v) * (f(x) + (C_1 * v)) & \text{if } \cos x \text{ is } + \\ (C_2 * v) * (g(x) + (C_1 * v)) & \text{if } \cos x \text{ is } - \end{cases}, \tag{6}$$

where $\tau$ is the torque, $v$ is the velocity of the motor and $C_1$ and $C_2$ are unknown unit-less constants to be identified. $C_1$ determines the average value of the torque and $C_2$ determines the peak-to-peak value of the fundamental shape.

To estimate $C_1$ and $C_2$ a GA was used. A GA is an advanced optimization tool based on the principles of biological evolution and natural selection. Typically, a GA will attempt to minimize a cost function by changing the values of its population using the following operators: selection, mating and mutation. In this case the population consists of the two constants to be identified, $C_1$ and $C_2$. For more information on GAs refer to Goldberg (1989) and Haupt and Haupt (2004).

The cost function used in the GA is defined below.

$$\sigma_i = \sum_{1}^{N} |\tau - \tau'|, \tag{7}$$

$$\text{cost} = \frac{1}{N}\sum_{i=1}^{5} \sigma_i, \tag{8}$$

where $\sigma_i$ is the result of the absolute error between the torque, $\tau$, calculated from the off-line SimMechanics model and the torque, $\tau'$, calculated by the simplified approximated model described in Equation (6) at a specific speed. $N$ is the number of sampled points, in this case $N = 1001$. The value of *cost* is the sum of $\sigma_i$ for $i = (1, 2, \ldots, 5)$, that is, the torque load at five different speeds, representing a range of speeds of the wiper system. The result of *cost* is the value to be minimized by the GA.

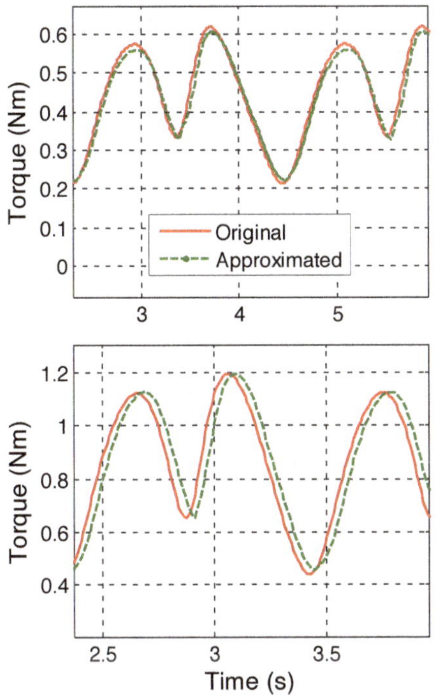


Figure 12. Torque load approximation.

The GA identified the constants as $C_1 = 0.0035$ and $C_2 = 0.2738$. The performance of the torque approximation at the slowest and fastest speeds is shown in Figure 12. It can be seen that the approximated model accurately matches the original torque.

The look-up tables and torque approximation model replace the SimMechanics linkages in the model.

### *4.3. Performance*

With the SimMechanics linkages removed and replaced with a simplified model, local solvers can be used. The modified model was simulated under the same conditions as the off-line model and was measured as being able to simulate 60 s of simulation time in 4 s of real time, meaning each simulation step of 1 ms can be simulated in 0.067 ms. The model is now suitable for simulation in real time and ready for implementation in an HIL test facility.

## 5. Identifying the model parameters

Motor parameterization could be achieved in a number of ways, for example: Measuring the parameters directly, datasheets or by using optimization algorithms. A GA was chosen for the following reasons: (1) Although some parameters can be measured directly, many cannot because the motor and linkages are a single unit and thus no load or constant load tests (including step inputs) are difficult to do. Also, data used in a GA may be available where a physical prototype is not, (2) Because the motor model is unique, datasheets do not contain the relevant information and (3) For off-line parameterization GA is ideal because it is superior at finding a global maximum solution over local optimizers but takes more time, which is not a problem for off-line parameterization.

The eight parameters to identify are: K_fast, K_slow, L_fast, L_slow, R_fast. R_slow, J and bm. The cost function for this GA is more complicated than Equation (7) because it is attempting to minimize the error in both the angle of the wipers and the motor current, hence the cost function is a multi-objective function. The cost function used is:

$$\mathrm{Cost}_{\mathrm{cur}} = W_{\mathrm{cur}} \sum_{j=1}^{N} \sum_{i=1}^{4} \{\mu(j,i) - \mu'(j,i)\}^2, \tag{9}$$

$$\mathrm{Cost}_{\mathrm{pos}} = W_{\mathrm{pos}} \sum_{j=1}^{N} \sum_{i=1}^{4} \{\theta(j,i) - \theta'(j,i)\}^2, \tag{10}$$

$$\mathrm{Cost}_{\mathrm{total}} = \frac{1}{N}(\mathrm{Cost}_{\mathrm{cur}} + \mathrm{Cost}_{\mathrm{pos}}). \tag{11}$$

Equations (9) and (10) compare the simulated current, $\mu'$, and simulated wiper angle, $\theta'$, with measured current, $\mu$, and position, $\theta$, data, respectively. $W_{\mathrm{cur}}$ and $W_{\mathrm{pos}}$ are the weights applied to Equations (9) and (10), respectively, to bring the errors of each equation into a similar range. $W_{\mathrm{cur}} = 8$ and $W_{\mathrm{pos}}$ is 1, which were chosen to bring the errors in the current and position to with a similar absolute value so one cost function does not dominate in Equation (11). In this case $N$, which is the number of data points sampled, is 501. Equation (11) sums the value of Equations (9) and (10) and divides it by $N$ to calculate the accumulated cost of the current and angle. In around 50 generations the algorithm found an optimal set of parameters. Table 1 shows the values of the parameters identified by the GA. The parameters were identified using four data sets, representing the full operating range of the system. The results were then validated against a fifth data set which was not used for parameter identification.

The performance of the model in representing the motor current and wiper angle of the real system is shown in Figures 13 and 14 at the slowest and fastest speeds,

Table 1. Wiper motor equivalent parameters.

| Parameter | Value | Unit |
|---|---|---|
| K_fast | 0.0423 | V/rpm |
| K_slow | 0.065 | V/rpm |
| L_fast | $9.41 \times 10^{-10}$ | H |
| L_slow | $9.6 \times 10^{-10}$ | H |
| R_fast | 0.58 | Ω |
| R_slow | 0.6 | Ω |
| *J* | 0.004 | kgm$^2$ |
| *bm* | 0.0185 | Nm/(rad/s) |

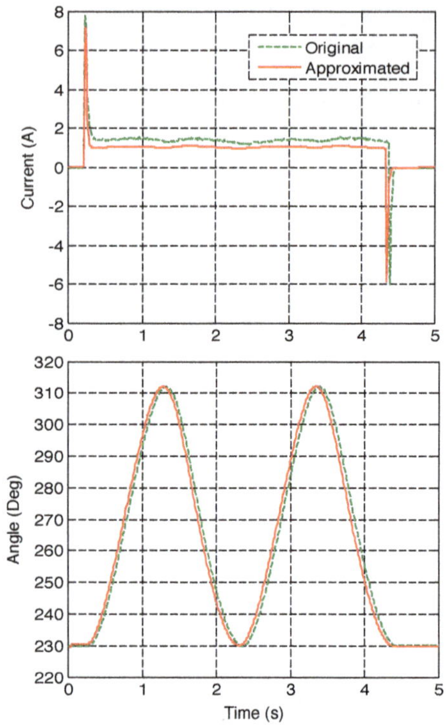


Figure 13. Model performance (slow).

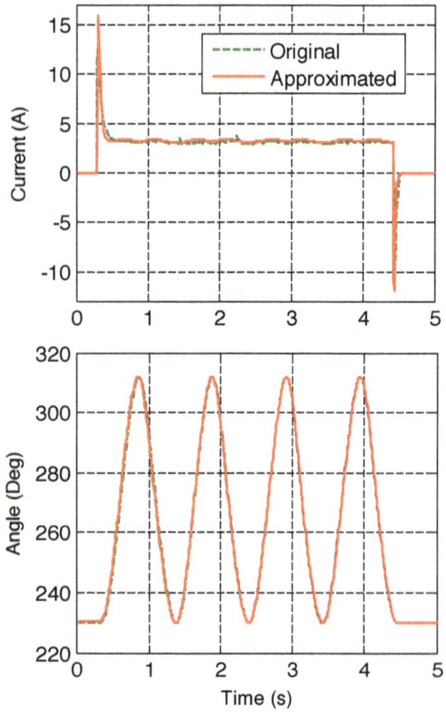


Figure 14. Model performance (fast).

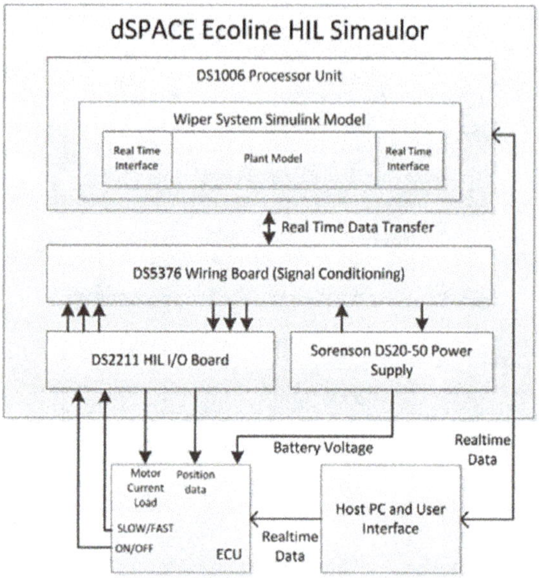


Figure 15. HIL test rig configuration.

respectively. It can be seen that the model accurately recreates the output of the real system, particularly in the fast mode of operation. The largest error, and thus the worst-case performance of this model, is the steady state current error in the slow operation. This error is likely to be due to the simplified friction model implemented in the motor and linkages. The accuracy model could be improved by increasing the complexity and fidelity of the friction model, however this will add extra parameters to be identified by the GA.

## 6. Implementing the model in an HIL facility

A diagram of the HIL facility in which the wiper model is implemented is shown in Figure 15.

The model is integrated into an existing control system for the wipers and real-time executable code is generated using the MathWorks real-time workshop tool. This code is transferred to the DS1006 processor board and can be simulated in real time whilst interacting with hardware. The results of the simulation can now be used to test the ECU's control system.

## 7. Conclusion

In the automotive industry ECUs are developed with the aid of HIL simulation. Accurate, real-time capable simulation models of the loads and actuators connected to the ECUs are required in order to carry out this testing. An off-line physical model of a windscreen wiper motor and linkage system has been developed whose parameters can be updated to capture design changes in the development process. The physical model has then been simplified using system identification methods to decrease its simulation time, making it suitable for HIL simulation. GAs were used to identify the unknown parameters of the model. The model has been implements in an HIL rig used for the testing of ECUs.

Further work on this project will be to add the wiper blades and windscreen elements to the model.


## Acknowledgements

The authors would like to thank the support in test facilities from the Advantage West Midlands and the European Regional Development Fund for Birmingham Science City Energy Efficiency and Demand Reduction project.


## Disclosure statement

No potential conflict of interest was reported by the authors.

## References


Chang, S. C., & Lin, H. P. (2004). Chaos attitude motion and chaos control in an automotive wiper system. *International Journal of Solids and Structures*, *41*, 3491–3504. doi:10.1016/j.ijsolstr.2004.02.005

Chiasson, J. (2005). *Modeling and high-performance control of electric machines*. Hoboken, New Jersey: John Wiley & Sons.

Clearwater Corporate Finance, Global Automotive Report. (2013). Retrieved from http://www.clearwatercf.com/

Dooner, M., Wang, J., & Mouzakitis, A. (2013). *Development of a simulation model of a windshield wiper system for Hardware in the Loop simulation*. 19th International Conference on Automation and Computing (ICAC), London.

Ganesh, B. (2005). *Hardware in the loop simulation (HIL) for vehicle electronics systems testing and validation*. SAE Technical Paper 2005-26-304. doi:10.4271/2005-26-304

Goldberg, G. (1989). *Genetic algorithms in search, optimization, and machine learning*. Addison-Wesley Publishing Company.

Grenouillat, R., & Leblanc, C. (2002). *Simulation of chatter vibrations for wiper systems*. SAE Technical Paper 2002-01-1239. doi:10.4271/2002-01-1239

Hameyer, K., & Belmans, R. J. M. (1996). Permanent magnet excited brushed DC motors. *IEEE Transactions on Industrial Electronics*, *43*, 247–255. doi:10.1109/41.491348

Haupt, R., & Haupt, S. (2004). *Practical genetic algorithms*. Hoboken, New Jersey: John Wiley & Sons.

Hillier, V. (1987). *Fundamentals of automotive electronics*. London: Hutchinson Education.

Hu, X., & Azarnasab, E. (2013). Progressive simulation-based design for networked real-time embedded systems. In K. Popovici, & P. J. Mosterman (Eds.), *Real-time simulation technologies: Principles, methodologies and applications* (pp. 181–198). Boca Raton, FL: Taylor Francis Group.

Jeon, S. H., Cho, J. H., Jung, Y., Park, S., & Han, T. M. (2011). *Automotive hardware development according to ISO 26262*. 13th International Conference on Advanced Communication Technology (ICACT), Seoul.

Miller, S., & Wendlandt, J. (2013). Real-time simulation of physical systems using simscape™. In K. Popovici & P. J. Mosterman (Eds.), *Real-time simulation technologies: Principles, methodologies and applications* (pp. 581–597). Boca Raton, FL: Taylor Francis Group.

Okura, S., Sekiguchi, T., & Oya, T. (2000). *Dynamic analysis of blade reversal behavior in a windshield wiper system*. SAE Technical Paper 2000-01-0127. doi:10.4271/2000-01-0127

Robert Bosch GmbH. (2007). *Automotive electrics automotive electronics*. Sussex: John Wily and Sons Ltd.

Schuette, H., & Ploeger, M. (2007). *Hardware-in-the-loop testing of engine control units – a technical survey*. SAE Technical Paper 2007-01-0500. doi:10.4271/2007-01-0500

Von Tils, V. (2006). *Trends and challenges in automotive electronics*. IEEE International Symposium on Power Semiconductor Devices and IC's, Naples.

Waltermann, P. (2009). *Hardware-in-the-loop: The technology for testing electronic controls in automotive engineering*. Translation of 6th Paderborn Workshop 'Designing Mechatronic Systems', Paderborn.

Wei, J., Mouzakitis, A., Wang, J., & Sun, H. (2011). *Vehicle windscreen wiper mathematical model development and optimisation for model based hardware-in-the-loop simulation and control*. 17th International Conference on Automation and Computing (ICAC), Huddersfield.

Wood, G., & Kennedy, D. (2003). *Simulating mechanical systems in Simulink with SimMechanics*. Retrieved from http://www.mathworks.com

Xiaoyu, Z., Yanfeng, X., & Yengjie, L. (2011). *Based on Matllab electrically operated windshield wiper systems design method research*. Third International Conference on Measuring Technology and Mechatronics Automation, Shanghai.

# 6

# A novel artificial intelligent control system to suppress the vibration of a FGM Plate

Jalal Javadi Moghaddam* and Ahmad Bagheri

*Department of Mechanical Engineering, Faculty of Engineering, University of Guilan, PO Box 3756, Rasht, Iran*

In this paper, an adaptive neuro-fuzzy sliding-mode-based genetic algorithm (ANFSGA) control system is proposed to control functionally graded material (FGM) plates. The model of the FGM plate is considered by the finite element method based on the classical laminated plate theory. Moreover, to show the performance of the proposed ANFSGA intelligent control system, a traditional sliding-mode control (SMC) system and an adaptive neuro-fuzzy (ANF) SMC system are designed to suppress the vibrations of the FGM plate as a comparison. The proposed genetic algorithm control system uses the ANF SMC system in the crossover and mutation operation. In this way, the online learning ability can be used by adjusting the control parameters to deal with external disturbance. The control objective is to drive the system state to the original equilibrium point and thus, the asymptotically stability of the proposed control system can be achieved.

**Keywords:** neuro fuzzy; FGM plates; sliding mode; on line; genetic algorithm

## 1. Introduction

FGMs offer good possibilities for optimizing engineering structures to achieve high performance and material efficiency. It is well known that functionally gradient materials are used in high-temperature conditions. That is why they have successful applications as electronic devices, optical films, anti-wear and anticorrosion coatings and biomaterials.

In the solid mechanical field, many researchers have investigated the application of piezoelectric materials as the sensors and actuators for the purpose of monitoring and controlling which is used in active structural systems.

Advanced reinforced composite structures incorporating piezoelectric sensors and actuators are increasingly becoming important due to the development of adaptive structures.

These structures offer potential benefits in a wide range of engineering applications such as vibration and noise suppression, shape control and precision positioning. Parashkevova, Ivanova, and Bontcheva (2004) developed an optimal design and defined two cost functions of functionally graded plates. Turteltaub (2002) introduced a numerical procedure to determine an optimal material layout of a functionally graded material (FGM) within the context of a transient phenomenon. Numerous approaches have been introduced into the analysis of plates with bonded piezoelectric sensors and actuators. Piezoelectric materials are more promising to use in structural mechanics in a way to develop adaptive structures. This is due to their coupled mechanical and electrical properties.

Reddy (2004) and Robbins and Reddy (1991) investigated the transformed piezoelectric and dielectric coefficients, and presented the finite element model of a piezoelectrically actuated beam by using four different displacement-based one-dimensional beam theories. He, Ng, Sivashanker, and Liew (2001) presented a finite element formulation based on the classical laminated plate theory for the shape and vibration control of the FGM plates with integrated piezoelectric sensors and actuators. Moita, Correia, Martins, Mota Soares, and Mota Soares (2006) presented a finite element formulation based on the classical laminated plate theory for laminated structures with integrated piezoelectric layers or patches, acting as sensors and actuators. Based on the first-order shear deformation theory, Liew, He, and Kitipornchai (2004) developed a generic finite element formulation to account for the coupled mechanical and electrical responses of FGM shells with piezoelectric sensors and actuator layers. The large-scale shell structures with distributed piezoelectric components of complicated geometrical configurations are approximated by the hybrid strain or mixed formulation based on lower order triangular shell finite elements investigated by To and Chen (2007).

In the control field, variable structure control with sliding mode, which is commonly known as sliding-mode control (SMC), is a nonlinear control strategy that is well

*Corresponding author. Email: jalaljavadimoghaddam@gmail.com

known for its robustness characteristics. Many methods based on sliding mode have been developed to control the dynamic systems; in particular, Bagheri and Moghaddam (2009) developed decoupled adaptive neuro-fuzzy SMC system methods for the chaos control problem in a system without precise system model information. Wai and Lee (2004) investigated a double-inductance double-capacitance resonant driving circuit and a sliding-mode fuzzy-neural-network control system for the motion control of an linear piezoelectric ceramic motor. Bagheri and Javadi Moghaddam (2010) developed artificial intelligence control system for underwater vehicle. Lin and Wai (2003) presented adaptive and fuzzy-neural-network sliding-mode controllers for the motor-quick-return servomechanism.

A total sliding-mode-based genetic algorithm control system for a linear piezoelectric ceramic motor driven by a newly designed hybrid resonant inverter is discussed by Wai and Tu (2007).

In this paper, the traditional SMC and the adaptive neuro-fuzzy (ANF) SMC and also the adaptive neuro-fuzzy sliding-mode-based genetic algorithm (ANFSGA) control system are presented to control the FGM plate in a vibration problem. It can be understood that the proposed control system can be easily used to other mechanical and electrical systems.

## 2. Model description

A cantilevered (CFFF) FGM plate with the integrated sensors and actuators is shown in Figure 1. The piezoelectric actuator layer and the piezoelectric sensor layer are distributed uniformly on the top and bottom layers of the laminated plate, respectively. The region between the two surfaces is made of the combined aluminum oxide and Ti-6A1-4 V materials. The material properties of the FGM plate are graded through the thickness direction according to a volume fraction power law distribution. The material properties can be easily found in the literature (Liew et al., 2004; Touloukian, 1967).

### 2.1. *Mathematical model using the classical laminated plate theory (CLPT)*

In the CLPT theory, the displacement field is presented by the following form:

$$\{u\} = \begin{Bmatrix} u_1 \\ u_2 \\ u_3 \end{Bmatrix} = \begin{Bmatrix} u_0 \\ v_0 \\ w_0 \end{Bmatrix} - \begin{Bmatrix} z\dfrac{\partial w_0}{\partial x} \\ z\dfrac{\partial x_0}{\partial y} \\ 0 \end{Bmatrix} = [H]\{\bar{u}\}, \quad (1)$$

$$\{\bar{u}\} = \left\{u_0,\ v_0, w_0, \frac{\partial w_0}{\partial x}, \frac{\partial w_0}{\partial y}\right\}^{\mathrm{T}}, \quad (2)$$

$$[H] = \begin{bmatrix} 1 & 0 & 0 & -z & 0 \\ 0 & 1 & 0 & 0 & -z \\ 0 & 0 & 1 & 0 & 0 \end{bmatrix}, \quad (3)$$

where $\{\bar{u}\}$ is the midplane displacement.

$u_0, v_0, w_0$ are displacements in the $x$, $y$ and $z$ directions, and $\partial w_0/\partial x, \partial w_0/\partial y$ are rotations of the $yz$ and $xz$ planes due to bending.

The strains according to the displacement field in Equation (1) are given by

$$\begin{Bmatrix} \varepsilon_1 \\ \varepsilon_2 \\ \varepsilon_6 \end{Bmatrix} = \begin{Bmatrix} \dfrac{\partial u_0}{\partial x} \\ \dfrac{\partial v_0}{\partial y} \\ \dfrac{\partial u_0}{\partial y} + \dfrac{\partial v_0}{\partial x} \end{Bmatrix} - z \begin{Bmatrix} \dfrac{\partial^2 w_0}{\partial x^2} \\ \dfrac{\partial^2 w_0}{\partial y^2} \\ 2\dfrac{\partial^2 w_0}{\partial x \partial y} \end{Bmatrix}. \quad (4)$$

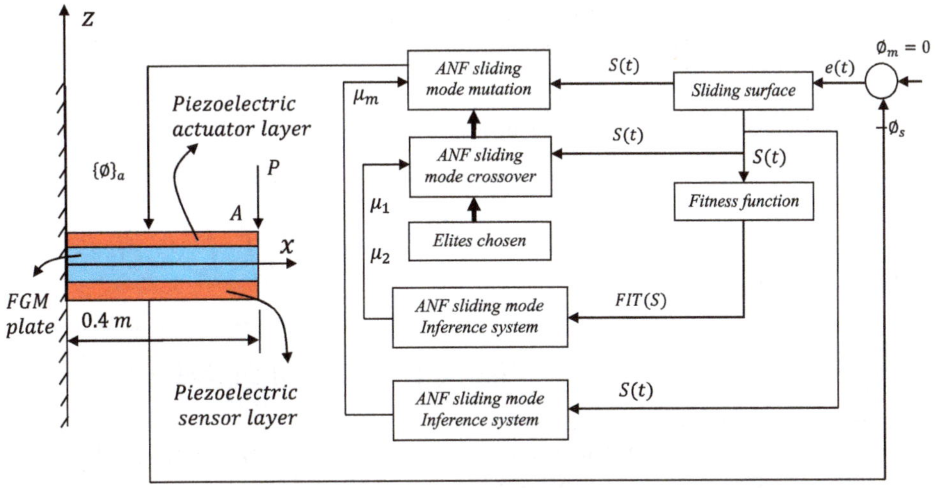


Figure 1. The block diagram of an ANFSGA control system and 2D plot of the FGM plate with distributed piezoelectric layer as an actuator on top and a sensor on bottom.

The equations of equilibrium and electrostatics are given as follows:

$$\sigma_{ij,j} + f_{bi} = \rho \ddot{u}_i, \tag{5}$$

$$D_{i,i} = 0. \tag{6}$$

In the quasi-static and plane stress formulation analysis, the constitutive relationship for the FGM lamina in the principal material coordinates of the lamina can be considered by the following form:

$$\sigma_{ij} = c_{ijkl}\varepsilon_{kl} - e_{ijk}E_k, \tag{7}$$

$$D_k = e_{ijk}\varepsilon_{ij} + k_{kl}E_l, \tag{8}$$

where $E_i = -\Phi_{,i}$ and $\Phi$ is the electric potential, $\sigma_{ij}$ denotes stress, $\varepsilon_{ij}$, $E_i$ and $D_i$ are the strain, electric field and the electric displacements respectively. $c_{ijkl}$ is the elastic coefficients, $[e]$ and $[k]$ are accordingly, the piezoelectric stress constants and the dielectric permittivity coefficients for a constant elastic strain. The symbol $\rho$ is the density of the plate which varies according to the following form:

$$\rho(z) = (\rho_T - \rho_A)\left(\frac{(2z+h)}{2h}\right)^n + \rho_A. \tag{9}$$

The relationship between piezoelectric stress constants and the piezoelectric strain can be obtained by the following form:

$$e_{31} = d_{31}c_{11} + d_{32}c_{12}, \tag{10}$$

$$e_{32} = d_{31}c_{12} + d_{32}c_{22}. \tag{11}$$

In the present paper, the effective mechanical properties' definitions of the plate are assumed to vary through the thickness of the uniform plate and can be written as

$$c_{ij}(z) = \left(c_{ij}^{T} - c_{ij}^{A}\right)\left(\frac{(2z+h)}{2h}\right)^n + c_{ij}^{A}. \tag{12}$$

Here, the simple power law distribution method is used, where $c_{ij}^T$ and $c_{ij}^A$ are the corresponding elastic properties of the Ti-6A1-4 V and aluminum oxide, $n$ and $h$ are the power law index and thickness of the plate, respectively.

According to Hamilton's principle and by using the above equations, the variational form of the equations of motion for the FGM plate can be written as

$$\int_{t_0}^{t_1}\int_{v}(-\rho\ddot{u}_i\delta u_i - \sigma_{ij}\delta\varepsilon_{ij} + D_i\delta E_i)\,dv\,dt + \int_{t_0}^{t_1}\int_{v} f_{bi}\delta u_i dv\,dt + \int_{t_0}^{t_1} f_{ci}\delta u_i dt + \int_{t_0}^{t_1}\int_{s}(f_{si}\delta u_i + q\delta\varnothing)\,ds\,dt = 0 \tag{13}$$

here $q$ is the surface charge, $t_0$ and $t_1$ are arbitrary time intervals, the symbols 'v' and 's' represent the volume and surface of the solid, respectively. $f_{bi}$, $f_{ci}$ and $f_{si}$ denote the body force, concentrated load and specified traction, respectively.

### 2.2. *Finite element model*

In this section, a finite element model of the FGM plate as a plant is introduced. The displacements and electric potential at the element level can be defined in terms of nodal variables by the following form:

$$\{u\} = [H][N_u]\{u^e\}, \tag{14}$$

$$\{\varnothing\} = [N_\varnothing]\{\varnothing^e\}, \tag{15}$$

where $[N_u]$ and $[N_\varnothing]$ are the shape functions, which include linear interpolation functions and non-conforming Hermite cubic interpolation functions. These shape functions can be found in the literature (He et al., 2001; Reddy, 2004). $\{u^e\}$ is the generalized nodal displacements and $\{\varnothing^e\}$ is the nodal electric potentials.

The infinitesimal engineering strains that are associated with the displacements are given by

$$\{\varepsilon\} = [B_u]\{u^e\}, \tag{16}$$

where the strain matrices $[B_u] = [[B_{u1}][B_{u2}][B_{u3}][B_{u4}]] = [A_u] - z[C_u]$ and $[B_{ui}] = [A_{ui}] - z[C_{ui}]$ are for $i = 1, 2, 3$ and 4.

$[A_{ui}]$ and $[C_{ui}]$ are derivative matrixes of linear and non-conforming Hermite cubic interpolation functions, respectively (Reddy, 2004).

The electric field vector $\{E\}$ can be expressed in terms of nodal variables as

$$\{E\} = -\nabla\varnothing = -[B_\varnothing][\varnothing^e], \tag{17}$$

where $[B_\varnothing] = \nabla[N_\varnothing]$. Substituting Equations (7), (8), (14), (16) and (17) into Equation (13), and assembling the element equations yield

$$[M_{uu}]\{\ddot{u}\} + [K_{uu}]\{u\} + [K_{u\varnothing}]\{\varnothing\} = \{F_m\}, \tag{18}$$

$$[K_{\varnothing u}]\{u\} - [K_{\varnothing\varnothing}]\{\varnothing\} = \{F_q\}. \tag{19}$$

The matrices and vectors are given by

$$[M_{uu}] = \sum_{\text{elem}}\sum_{K=1}^{NL}\int_{-1}^{1}\int_{-1}^{1}[N_u]^T[I][N_u]|J|\,d\xi\,d\eta, \tag{20}$$

$$[K_{uu}] = \int_{v}[B_u]^T[C][B_u]\,dv = \sum_{\text{elem}}\sum_{K=1}^{NL}\int_{-1}^{1}\int_{-1}^{1}([A_u]^T[A][A_u] - [A_u]^T[B][C_u] - [C_u]^T[B][A_u] + [C_u]^T[Q][C_u])|J|\,d\xi\,d\eta, \tag{21}$$

$$[K_{u\varnothing}] = \int_{v}[B_u]^T[e]^T[B_\varnothing]\,dv = \sum_{\text{elem}}\sum_{K=1}^{NL}(z_{K+1} - z_K)\int_{-1}^{1}\int_{-1}^{1}([A_u][e]^T[B_\varnothing] - \frac{1}{2}(z_{K+1} + z_K)[C_u]^T[e]^T[B_\varnothing])|J|\,d\xi\,d\eta, \tag{22}$$

$$[K_{\varnothing u}] = \int_V [B_\varnothing]^T[e][B_u]\,dv$$
$$= \sum_{\text{elem}}\sum_{K=1}^{NL}(z_{K+1}-z_K)\int_{-1}^{1}\int_{-1}^{1}([B_\varnothing]^T[e][A_u]$$
$$-\frac{1}{2}(z_{K+1}+z_K)[B_\varnothing]^T[e][C_u])|J|\,d\xi\,d\eta, \quad (23)$$

$$[K_{\varnothing\varnothing}] = \int_V [B_\varnothing]^T[k][B_\varnothing]\,dv$$
$$= \sum_{\text{elem}}\sum_{K=1}^{NL}(z_{K+1}-z_K)$$
$$\times \int_{-1}^{1}\int_{-1}^{1}[B_\varnothing]^T[k][B_\varnothing]|J|\,d\xi\,d\eta, \quad (24)$$

$$\{F_m\} = \int_V [N]^T[H]^T[f_b]\,dv + \int_{sf}[N]^T[H]^T[f_s]\,ds$$
$$+[N]^T[H]^T[f_c] \quad (25)$$

$$\{F_q^e\} = \int_{S_q}[N_\varnothing]^T\{q\}\,ds, \quad (26)$$

where

$$[I] = \left(\int_{z_K}^{z_{K+1}}(\rho_T-\rho_A)\left(\frac{(2z+h)}{2h}\right)^n[H]^T[H]\right.$$
$$\left.+\rho_A[H]^T[H]\right)dz, \quad (27)$$

$$[A,\,B,\,Q] = \int_{z_K}^{Z_{K+1}}\left(([C]^T-[C]^A)\left(\frac{(2z+h)}{2h}\right)^n(1,\,z,\,z^2)\right.$$
$$\left.+[C]^A(1,\,z,\,z^2)dz\right). \quad (28)$$

Substituting Equation (19) into Equation (18), one can obtain

$$[M_{uu}]\{\ddot{u}\} + ([K_{uu}]+[K_{u\varnothing}][K_{\varnothing\varnothing}]^{-1}[K_{\varnothing u}])\{u\}$$
$$= \{F_m\} + [K_{u\varnothing}][K_{\varnothing\varnothing}]^{-1}\{F_q\}. \quad (29)$$

where $\{F_q\}$ for the sensor and actuator layer can be written as

$$\{F_q\} = \begin{Bmatrix}\{F_q\}_s\\ \{F_q\}_a\end{Bmatrix} = \begin{bmatrix}[K_{\varnothing u}]_s & 0\\ 0 & [K_{\varnothing u}]_a\end{bmatrix}\begin{Bmatrix}\{u\}_s\\ \{u\}_a\end{Bmatrix}$$
$$-\begin{bmatrix}[K_{\varnothing\varnothing}]_s & 0\\ 0 & [K_{\varnothing\varnothing}]_a\end{bmatrix}\begin{Bmatrix}\{\varnothing\}_s\\ \{\varnothing\}_a\end{Bmatrix}. \quad (30)$$

Here, the subscript 's' denotes the sensors and subscript 'a' represents the actuators.

For the sensor layer, the applied charge $\{F_q\}$ is zero and the converse piezoelectric effect is assumed negligible. Using Equation (19), the sensor output is

$$\{\varnothing\}_s = [K_{\varnothing\varnothing}]_s^{-1}[K_{\varnothing u}]_s\{u\}_s \quad (31)$$

and the sensor charge due to deformation from Equation (19) is

$$\{F_q\}_s = [K_{\varnothing u}]_s\{u\}_s. \quad (32)$$

For the actuator layer, from Equation (19), $\{F_q\}_a$ can be written in the following form:

$$\{F_q\}_a = [K_{\varnothing u}]_a\{u\}_a - [K_{\varnothing\varnothing}]_a\{\varnothing\}_a. \quad (33)$$

As mentioned above and substituting Equations (32) and (33) into Equation (30), thus Equation (30) can be expressed as

$$\{F_q\} = [K_{\varnothing u}]\{u\} - [K_{\varnothing\varnothing}]_a\{\varnothing\}_a \quad (34)$$

substituting Equation (34) into Equation (29) and by using some mathematics operations one can obtain

$$[M_{uu}]\{\ddot{u}\} + [C_s]\{\dot{u}\} + [K_{uu}]\{u\} = \{F_m\} - [K_{u\varnothing}]_a\{\varnothing\}_a, \quad (35)$$

where $[C_s] = a[M_{uu}] + b[K_{uu}]$ is the damping matrix, a and b are Rayleigh's coefficients.

## 3. Control system

In this section the control objective is to find a control law $\{\varnothing\}_a$ so that the desired sensor output $\{\varnothing\}_m(t)$ as a state can be tracked by the sensor output $\{\varnothing\}_s(t)$. Here, to suppress the vibration, the fact that the state (mode shape) of the plate goes to equilibrium point should be considered, therefore $\{\varnothing\}_m(t) = 0$ and the corresponding sensor output $\{\varnothing\}_s(t) \to 0$. It is noted that the proposed control systems can be used in forced vibrations problem by selecting a proper $\{\varnothing\}_m(t)$. Based on this strategy, the mode shapes of the plate can be held on arbitrary trajectory or desired mode shape.

### 3.1. *Traditional sliding mode*

In this section a traditional sliding-mode control (TSMC) system is designed and fabricated to suppress vibrations of the FGM plate. To achieve the control objective, the following tracking error vector can be defined $e(t) = \{\varnothing\}_m(t) - \{\varnothing\}_s(t)$, Moreover, the sliding surface can be expressed as

$$S(t) = \left(\frac{d}{dt}+\lambda\right)^2\int_0^t e(\tau)\,d\tau, \quad (36)$$

where $\lambda$ is a positive constant. Note that since the function $S(t) = 0$ when $t = 0$, there is no reaching phase as in the traditional sliding-mode control (Lin & Hsu, 2004; Slotine

& Li, 1991). Differentiating $S(t)$ with respect to time and using Equation (31), one can obtain:

$$\dot{S} = \ddot{e}(t) + 2\lambda\dot{e}(t) + \lambda^2 e(t), \tag{37}$$

$$\dot{S} = \ddot{\varnothing}_m - [K_{\varnothing\varnothing}]_s^{-1}[K_{\varnothing u}]_s\{\ddot{u}\}_s + 2\lambda(\dot{\varnothing}_m - \dot{\varnothing}_s) + \lambda^2(\varnothing_m - \varnothing_s), \tag{38}$$

Now, $\{\ddot{u}\}$ can be expressed as

$$\{\ddot{u}\} = [M_{uu}]^{-1}(\{F_m\} - [K_{u\varnothing}]_a\{\varnothing\}_a - [C_s]\{\dot{u}\} - [K_{uu}]\{u\}) \tag{39}$$

substituting Equation (39) in Equation (38) one can obtain

$$\dot{S} = \ddot{\varnothing}_m - [K_{\varnothing\varnothing}]_s^{-1}[K_{\varnothing u}]_s[M_{uu}]^{-1}(\{F_m\} - [K_{u\varnothing}]_a\{\varnothing\}_a - [C_s]\{\dot{u}\} - [K_{uu}]\{u\}) + 2\lambda(\dot{\varnothing}_m - \dot{\varnothing}_s) + \lambda^2(\varnothing_m - \varnothing_s). \tag{40}$$

The aforementioned tracking problem is to design a control law $\{\varnothing\}_a$ so that the state remains on the surface $S(t) = 0$ for all times. In designing the sliding-mode control system, first of all the equivalent control law $\{\varnothing\}_{aeq}$, which will determine the dynamic of the system on the sliding surface, can be found. The equivalent control law is derived from recognizing

$$\dot{S}|_{\{\varnothing\}_a=\{\varnothing\}_{aeq}} = 0. \tag{41}$$

Substituting Equation (44) into Equation (43) and rearranging yield

$$\{\varnothing\}_{aeq} = -\left([K_{\varnothing\varnothing}]_s^{-1}[K_{\varnothing u}]_s[M_{uu}]^{-1}[K_{u\varnothing}]_a\right)^{-1}\left(\ddot{\varnothing}_m - [K_{\varnothing\varnothing}]_s^{-1}[K_{\varnothing u}]_s[M_{uu}]^{-1}(\{F_m\} - [C_s]\{\dot{u}\} - [K_{uu}]\{u\}) + 2\lambda(\dot{\varnothing}_m - \dot{\varnothing}_s) + \lambda^2(\varnothing_m - \varnothing_s)\right). \tag{42}$$

Thus, given $\dot{S}(t) = 0$, the dynamics of the system on the sliding surface for $t > 0$ is given by

$$\ddot{e}(t) + 2\lambda\dot{e}(t) + \lambda^2 e(t) = 0. \tag{43}$$

In this controller if the system parameters are perturbed or unknown, the equivalent control design cannot guarantee the performance specified by Equation (43). Moreover, the stability of the controlled system may be destroyed. To ensure the system performance designed by Equation (43) under the existence of the uncertainties, a robust controller $\{\varnothing\}_{aR}$ is designed by the following form:

$$\{\varnothing\}_{aR} = -k_k \operatorname{sign}(S), \tag{44}$$

where $k_k$ is the gain control. Finally, the TSMC control system can be obtained as

$$\{\varnothing\}_a = \{\varnothing\}_{aeq} + \{\varnothing\}_{aR}. \tag{45}$$

### 3.2. *ANF sliding mode*

The architecture diagram of the neuro-fuzzy inference mechanism is depicted in Figure 2. The ANF sliding-mode controller is composed of a neuro-fuzzy network with the online learning algorithm.

Let ***input*** $= [S(t), \dot{S}(t)]$ and ***output*** $= \{\varnothing\}_a$ be the input and output variables to the ANF sliding-mode system, respectively.

#### 3.2.1. *Description of ANF*

In the proposed controller, the four-layer NN is used (Figure 2). Layers I–IV represent the inputs to the network, the membership functions, the fuzzy rule base and the outputs of the network, respectively.

#### 3.2.2. *Layer I: input layer*

Inputs and outputs of nodes in this layer are represented as

$$\text{net}_1^1 = S(t),\ y_1^1 = f_1^{\ 1}(\text{net}_1^1) = \text{net}_1^1 = S(t), \tag{46}$$

$$\text{net}_1^1 = \dot{S}(t),\ y_2^1 = f_2^{\ 1}(\text{net}_2^1) = \text{net}_2^1 = \dot{S}(t), \tag{47}$$

where $y_1^1$ and $y_2^1$ are outputs of the input layer. In this layer, the weights are unity and fixed.

#### 3.2.3. *Layer II: membership layer*

In this layer, each node performs a fuzzy set and the Gaussian function is adopted as a membership function

$$\text{net}_{1,j}^{\text{II}} = -\frac{\left(x_{1,j}^{\text{II}} - m_{1,j}^{\text{II}}\right)^2}{\left(\sigma_{1,j}^{\text{II}}\right)^2},\quad y_{1,j}^{\text{II}} = f_{1,j}^{\text{II}}\left(\text{net}_{1,j}^{\text{II}}\right) = \exp\left(\text{net}_{1,j}^{\text{II}}\right), \tag{48}$$

$$\text{net}_{1,k}^{\text{II}} = -\frac{\left(x_{2,k}^{\text{II}} - m_{2,k}^{\text{II}}\right)^2}{\left(\sigma_{2,k}^{\text{II}}\right)^2},\quad y_{2,k}^{\text{II}} = f_{2,k}^{\text{II}}\left(\text{net}_{2,k}^{\text{II}}\right) = \exp\left(\text{net}_{2,k}^{\text{II}}\right), \tag{49}$$

where $m_{1,j}^{\text{II}}$, $m_{2,k}^{\text{II}}$ and $\sigma_{1,j}^{\text{II}}$, $\sigma_{2,k}^{\text{II}}$ are the mean and the standard deviation of the Gaussian function, respectively. The variables $x_{1,j}^{\text{II}}$ and $x_{2,k}^{\text{II}}$ are the outputs of layer I.

#### 3.2.4. *Layer III: rule layer*

This layer includes the rule base used in the fuzzy logic control. Each node in this layer which multiplies the input signals and outputs can be expressed as follows:

$$\text{net}_{jk}^{\text{III}} = \left(x_{1,j}^{\text{III}} \times x_{2,k}^{\text{III}}\right),\quad y_{jk}^{\text{III}} = f_{jk}^{\text{III}}\left(\text{net}_{jk}^{\text{III}}\right) = \text{net}_{jk}^{\text{III}} \tag{50}$$

here $x_{1,j}^{\text{III}}$ and $x_{2,k}^{\text{III}}$ are the outputs of layer II. The values of link weights between the membership layer and rule base layer are unity.

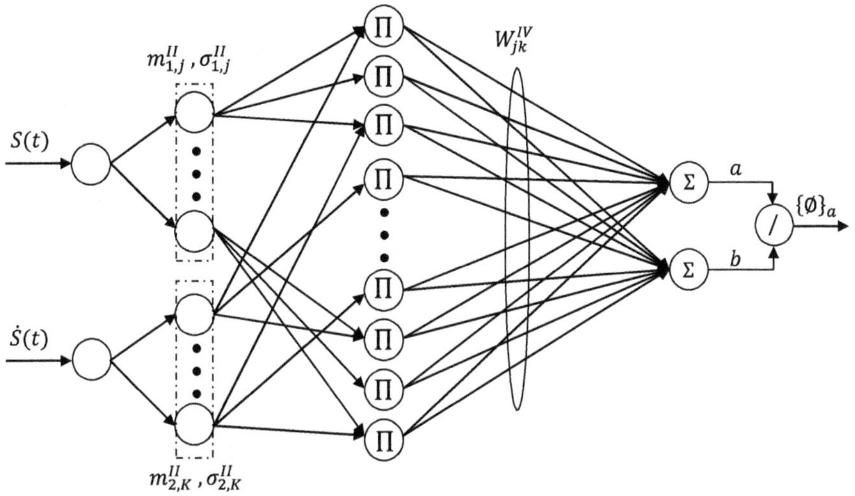


Figure 2. Schematic diagram of the neuro-fuzzy network.

### 3.2.5. *Layer IV: output layer*

This layer represents the inference and defuzzification which are used in the fuzzy logic system. For defuzzification, the center of area method is used. Therefore, the following form can be obtained:

$$a_i = \sum_j \sum_k W^{IV}_{jk} y^{III}_{jk}, \quad b_i = \sum_j \sum_k y^{III}_{jk},$$

$$\text{net}^{IV}_{0\,i} = \frac{a_i}{b_i}, \quad y^{IV}_{0\,i} = f_0^{\,IV}(\text{net}^{IV}_{0\,i}) = \frac{a_i}{b_i}, \tag{51}$$

where $y^{III}_{jk}$ is the output of the rule layer, $a_i$ and $b_i$ are the numerator and the denominator of the function used in the center of area method according to each degree and $W^{IV}_{jk}$ is the center of the output membership functions used in the fuzzy logic system, respectively. The aim of the learning algorithm is to adjust the weights of $W^{IV}_{jk}$, $m^{II}_{1,j}$, $m^{II}_{2,k}$ and $\sigma^{II}_{1,j}$, $\sigma^{II}_{2,k}$. Finally, $y^{IV}_0$ is the output of the proposed inference system.

The online learning algorithm is a gradient descent search algorithm in the space of network parameters. The Lyapunov function is chosen as $(1/2)S^2(t)$. The aim is to minimize the derivative of the Lyapunov function with respect to time or $S(t)\dot{S}(t)$.

### 3.2.6. *Online learning algorithm*

The error expression for the input of Layer IV can be expressed as follows:

$$\delta^{IV}_{0\,i} = -\frac{\partial S(t)\dot{S}(t)}{\partial y^{IV}_{0\,i}} \frac{\partial y^{IV}_{0\,i}}{\partial\, \text{net}^{IV}_{i}} = \varsigma_1 S(t), \tag{52}$$

where $\varsigma_1$ is the learning rate for $W^{IV}_{jk}$. Therefore, the changing of $W^{IV}_{jk}$ is written as

$$\dot{W}^{IV}_{jk} = -\frac{\partial S(t)\dot{S}(t)}{\partial\, \text{net}^{IV}_{0\,i}} \frac{\partial\, \text{net}^{IV}_{0\,i}}{\partial a_i} \frac{\partial a_i}{\partial W^{IV}_{jk}} = \frac{1}{b_i} \delta^{IV}_{0\,i} y^{III}_{jk}. \tag{53}$$

Since the weights in the rule layer are unified, only the approximated error term needs to be calculated and propagated by the following equation:

$$\delta^{III}_{jk\,i} = -\frac{\partial S(t)\dot{S}(t)}{\partial\, \text{net}^{IV}_{0\,i}} \frac{\partial\, \text{net}^{IV}_{0\,i}}{\partial y^{III}_{1,j}} \frac{\partial y^{III}_{1,j}}{\partial\, \text{net}^{III}_{jk\,i}} = \frac{1}{b_i} \delta^{IV}_{0\,i} \left( W^{IV}_{jk} - \partial y^{IV}_{0\,i} \right). \tag{54}$$

The error received from Layer III is computed as

$$\delta^{II}_{1,j\,i} = \sum_k \left[ \left( -\frac{\partial S(t)\dot{S}(t)}{\partial\, \text{net}^{III}_{jk\,i}} \right) \frac{\partial\, \text{net}^{III}_{jk\,i}}{\partial y^{II}_{1,j}} \frac{\partial y^{II}_{1,j\,i}}{\partial\, \text{net}^{II}_{1,j\,i}} \right] = \sum_k \delta^{III}_{jk\,i} y^{III}_{jk\,i}, \tag{55}$$

$$\delta^{II}_{2,k\,i} = \sum_j \left[ \left( -\frac{\partial S(t)\dot{S}(t)}{\partial\, \text{net}^{III}_{jk\,i}} \right) \frac{\partial\, \text{net}^{III}_{jk\,i}}{\partial y^{II}_{2,k}} \frac{\partial y^{II}_{2,k_i}}{\partial\, \text{net}^{II}_{2,k_i}} \right] = \sum_j \delta^{III}_{jk\,i} y^{III}_{jk\,i}. \tag{56}$$

The update laws of $m^{II}_{1,j}$, $m^{II}_{2,k}$ and $\sigma^{II}_{1,j}$, $\sigma^{II}_{2,k}$ also can be obtained by the gradient decent search algorithm, it means:

$$\dot{m}^{II}_{1,j\,i} = -\frac{\partial S(t)\,\dot{S}(t)}{\partial\, \text{net}^{II}_{1,j\,i}} \frac{\partial\, \text{net}^{II}_{1,j\,i}}{\partial m^{II}_{1,j\,i}} = \varsigma_2 \delta^{II}_{1,j\,i} \frac{2\left(x^{II}_{1,j} - m^{II}_{1,j\,i}\right)}{\left(\sigma^{II}_{1,j\,i}\right)^2}, \tag{57}$$

$$\dot{m}^{\mathrm{II}}_{2,k\,i} = -\frac{\partial S(t)\,\dot{S}(t)}{\partial\,\mathrm{net}^{\mathrm{II}}_{2,k_i}}\frac{\partial\,\mathrm{net}^{\mathrm{II}}_{2,k_i}}{\partial m^{II}_{2,k_i}} = \varsigma_3\delta^{\mathrm{II}}_{2,k\,i}\frac{2\left(x^{\mathrm{II}}_{2,k} - m^{\mathrm{II}}_{2,k_i}\right)}{\left(\sigma^{\mathrm{II}}_{2,k_i}\right)^2}, \tag{58}$$

$$\dot{\sigma}^{\mathrm{II}}_{1,j\,i} = -\frac{\partial S(t)\,\dot{S}(t)}{\partial\,\mathrm{net}^{\mathrm{II}}_{1,j\,i}}\frac{\partial\,\mathrm{net}^{\mathrm{II}}_{1,j\,i}}{\partial \sigma^{\mathrm{II}}_{1,j\,i}} = \varsigma_4\delta^{\mathrm{II}}_{1,j\,i}\frac{2\left(x^{\mathrm{II}}_{1,j} - m^{\mathrm{II}}_{1,j\,i}\right)^2}{\left(\sigma^{\mathrm{II}}_{1,j\,i}\right)^3}, \tag{59}$$

$$\dot{\sigma}^{\mathrm{II}}_{2,k\,i} = -\frac{\partial S(t)\,\dot{S}(t)}{\partial\,\mathrm{net}^{\mathrm{II}}_{2,k_i}}\frac{\partial\,\mathrm{net}^{\mathrm{II}}_{2,k_i}}{\partial m^{\mathrm{II}}_{2,k_i}} = \varsigma_5\delta^{\mathrm{II}}_{2,k\,i}\frac{2\left(x^{\mathrm{II}}_{2,k} - m^{\mathrm{II}}_{2,k_i}\right)^2}{\left(\sigma^{\mathrm{II}}_{2,k_i}\right)^3}, \tag{60}$$

where $\varsigma_2$, $\varsigma_3$, $\varsigma_4$, and $\varsigma_5$ are the learning-rate parameters of the mean and the standard deviation of the Gaussian functions.

### *3.3. ANFSGA control system*

In this section a control law-based genetic algorithm is designed to the FGM plate for tracking mode shapes, suppurating vibration and external disturbance rejection. The proposed control system is included in the SMC concept and the neuro-fuzzy sliding-mode-based evolutionary procedure. In order to achieve the control object, the evolutionary spirit of GA is embedded. The neuro-fuzzy approach is used to further ensure the correct evolutionary direction and decide the appropriate evolutionary step. In this section, the control law is made as the chromosome in GA with floating point coding can be considered. This process, which is a real one, is to be replaced by the ANF sliding-mode crossover method. An ANF sliding-mode mutation is used just after selecting the chromosomes. The first step after creating a generation is to calculate the fitness function of each member in the population. If the evolutionary direction is correct, the fittest control action can be obtained. In order to achieve the correct evolutionary direction and to ensure the stable system dynamic, the concept of the SMC system is embedded in the genetic operators to form the direction-based operators with the ANF sliding-mode evolutionary procedure.

Now, a fitness function is defined as an exponential term by the following form (Wai & Tu, 2007):

$$\mathrm{FIT}(S) = \exp[-\zeta \times (S(t)^2 + \dot{S}(t)^2)] \in [0, 1], \tag{61}$$

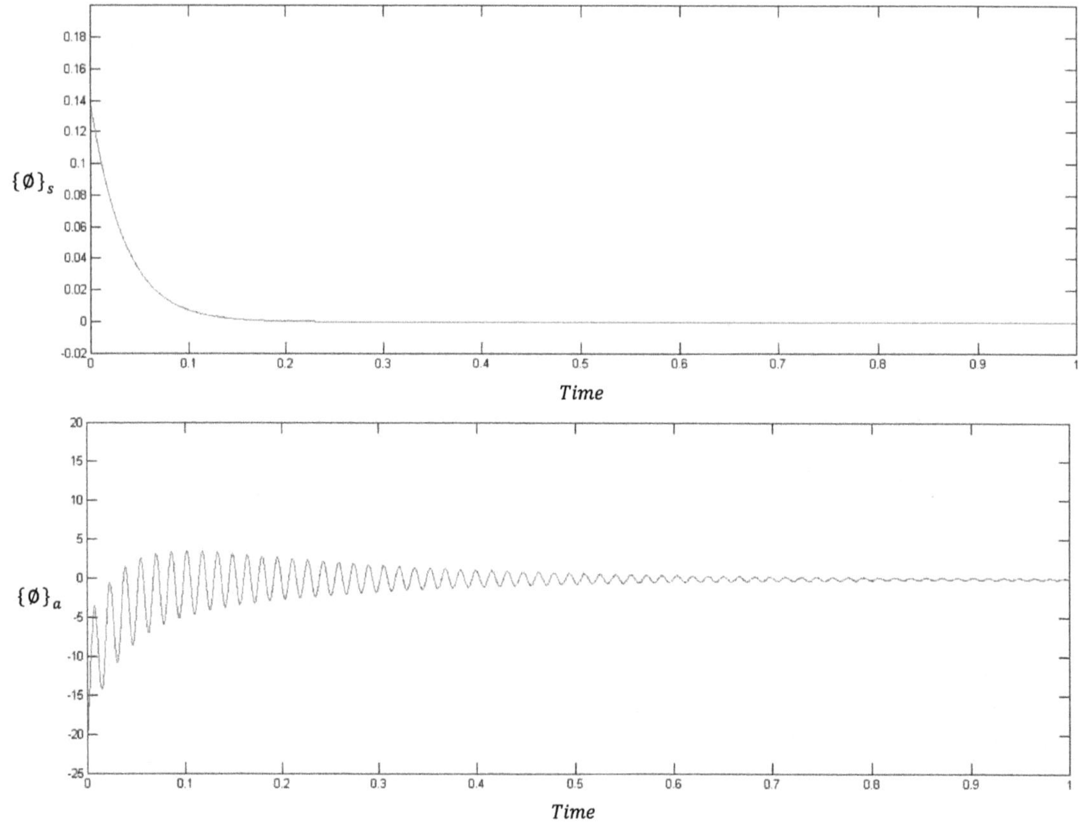


Figure 3. Simulation result of piezoelectric sensor and actuator due to the traditional sliding mode with robust controller.

where $\zeta$ is a positive constant, $S$ is the sliding surface and $\dot{S}$ is the first derivative of $S$ which is defined as Equation (36). The next step after evaluation is to create a new population from the current generation. The selection operation determines which chromosome participates in producing offspring for the next generation. Initially, the population is selected randomly, which means that several control actions are randomly selected from the operational region $[\{\varnothing\}_{a_{\min}}, \{\varnothing\}_{a_{\max}}]$. After comparing the fitness values of all the individuals, the best one is regarded as the elite. If the fitness value of the new control action is higher than all the previous ones, it will become the new elite.

Crossover operation is used to reshape the GA system, which can produce offspring by charging the features of the parent. In this study, the sliding surface is combined with the crossover operation by the following form:

$$\{\varnothing\}_{a_{GA,new}} = \{\varnothing\}_{a_{GA,\,old}} + \mu_1 \times S + \mu_2 \times \dot{S}, \tag{62}$$

where $\{\varnothing\}_{a_{GA,new}}$ is the generated offspring, $\{\varnothing\}_{a_{GA,old}}$ is the selected elitist chromosome of the last generation, $\mu_1$ and $\mu_2$ are the positive tuning parameters of $S$ and $\dot{S}$, respectively. Here, the important problem is selecting the tuning parameters. The small tuning step may not satisfy the stability conditions. Therefore, an ANF sliding-mode system is used to produce the tuning coefficients. In this section the ANF sliding-mode mechanism of the previous section is considered to produce $\mu_1$ and $\mu_2$. Let ***input*** $= \mathrm{FIT}(S)$ for both ANF sliding-mode mechanisms and ***output***$_1 = \mu_1$ and ***output***$_2 = \mu_2$. For the two systems, different means and the standard deviations of the Gaussian function are used.

To avoid the problem of local optimization an ANF sliding-mode mechanism is used in mutation operation. Traditional mutation methods are not useful to produce better offspring in an online learning ability. Therefore, the stability of the system may be destroyed. If the control action cannot let the system dynamic stay on the sliding surface after fuzzy sliding-mode crossover, the mutation operation will further compel the system dynamic to close the sliding surface by using the fuzzy sliding-mode inference mechanism.

The offspring after mutation operation can be expressed as

$$\{\varnothing\}^{\Delta}_{a_{GA,new}} = \{\varnothing\}_{a_{GA,new}} + \mu_m, \tag{63}$$

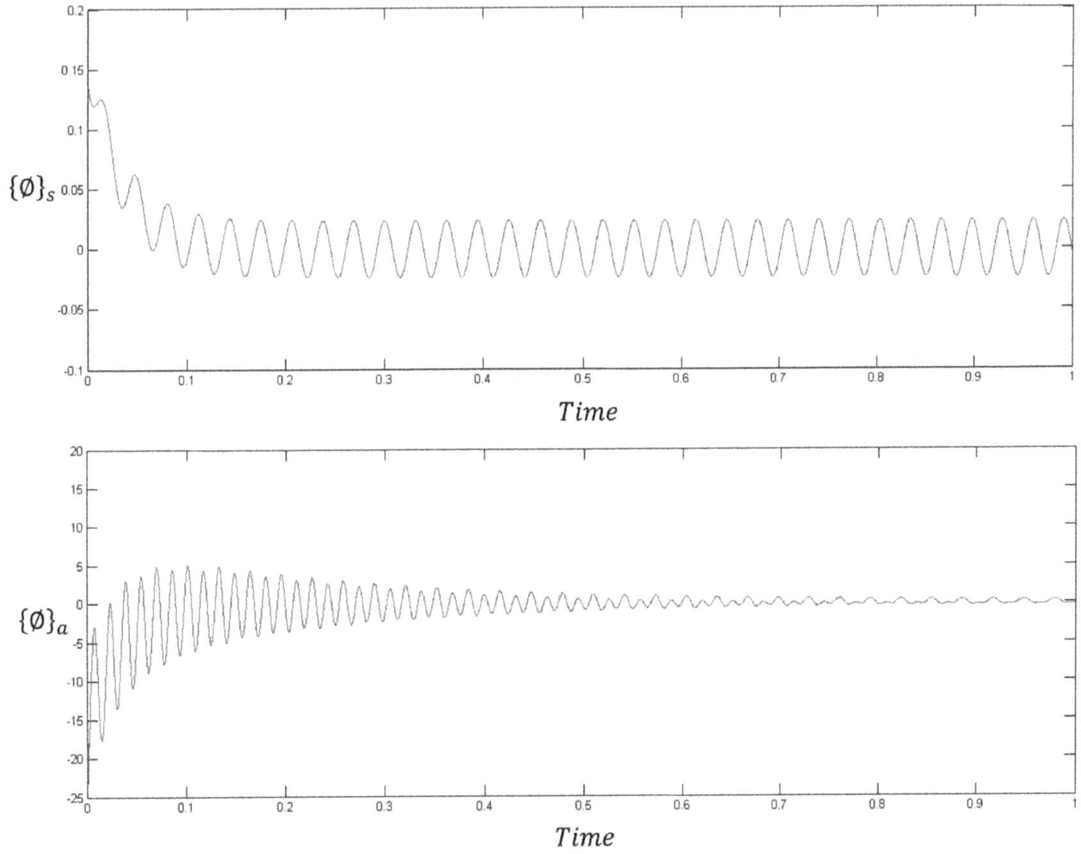


Figure 4. Simulation result of piezoelectric sensor and actuator due to the traditional sliding mode with robust controller in the disturbance condition.

where $\mu_m$ is the adjustment of mutation operation. $\{\varnothing\}^{\Delta}_{\mathrm{aGA,new}}$ is the offspring after mutation operation which is produced by the ANF sliding-mode inference mechanism. In this situation, the input to the ANF sliding-mode system is the sliding surface or ***input*** $= S(t)$. If the fitness value is lower than a specified value ($\mathrm{FIT}_B$), mutation occurs. On the other hand, if the fitness value is higher than the specified value, the mutation idles.

The main process of the proposed GA-based controller is represented by the following pseudo-code:

*Step 1* Select the size of population [$N$] and the fitness function [FIT($S$)].
*Step 2* Generate the initial population.
*Step 3* Evaluate the fitness value via (61) and sort the sequence to choose the elite $\{\varnothing\}_{\mathrm{aGA,old}}$.
*Step 4* Do ANF sliding-mode crossover operation to generate $\{\varnothing\}_{aGA,\mathrm{new}}$ via (62).
*Step 5* Compare the fitness value with the specified value ($\mathrm{FIT}_B$), if it is not lower, then go to Step 7, otherwise follow the chart.
*Step 6* Do ANF sliding-mode mutation operation to generate the $\{\varnothing\}^{\Delta}_{\mathrm{aGA,new}}$ via (63).
*Step 7* Output control action.
*Step 8* Program complete? If yes then it is the end, if not go to Step 3.

It is noted that in the proposed controller, for the crossover and mutation operations Equations (46)–(51) are used. The adaptive laws and the online learning algorithm are used in Equations (52)–(60).

The chattering phenomenon is a particular problem in the control algorithms. The chattering problem can reduce the control accuracy and destroy the stability of system. To find the smooth control action and reduce chattering phenomena, the following soft limit switching function $f_{\mathrm{SL}}$ is presented as

$$f_{\mathrm{SL}}(S) = \frac{S(t)^2}{1+S(t)^2}\tanh(S(t)). \qquad (64)$$

## 4. Simulation results

The finite element model for the FGM plate is based on the general concept of solid mechanics and to ensure the accuracy, it is validated with different values of volume fraction

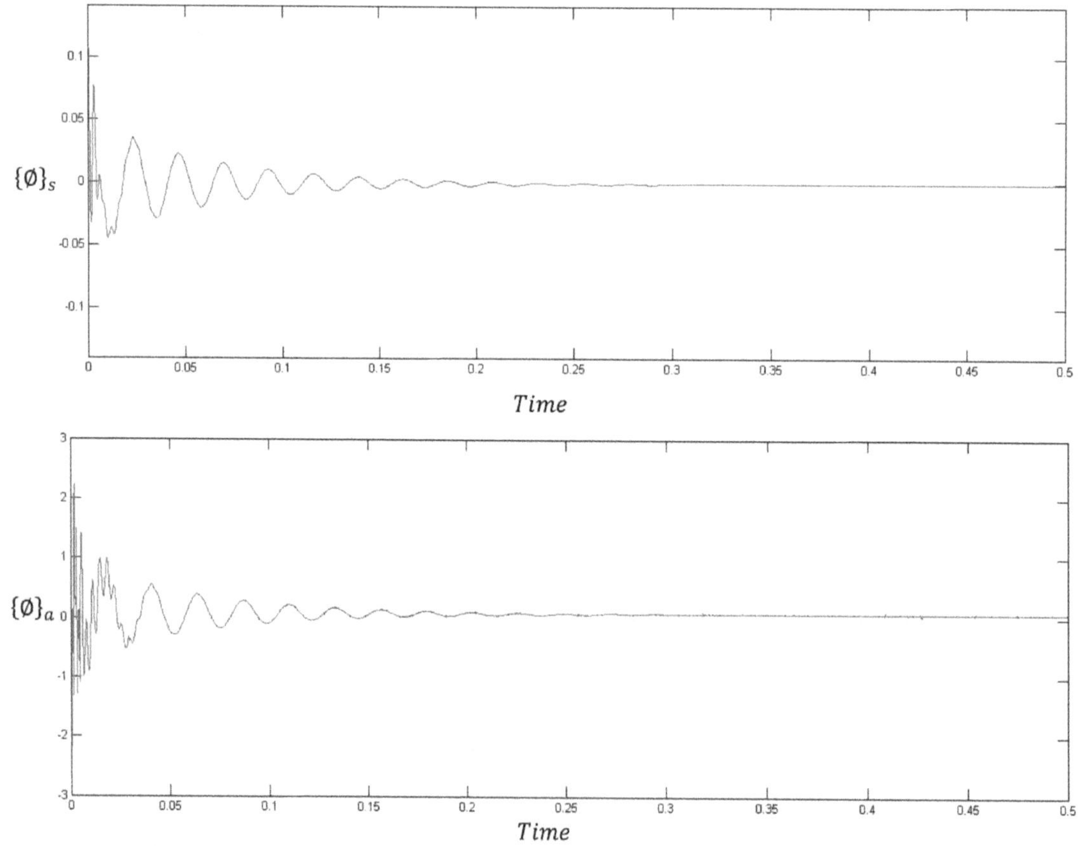


Figure 5. Simulation result of piezoelectric sensor and actuator due to the ANF SMC system.

power law exponent $n$ and compared with the results of Bishop (1979) and Praveen and Reddy (1998).

The G-1195N piezoelectric films bond both the top and bottom surfaces of the FGM plate as shown in Figure 1. The plate is square with both length and width set as 0.4 m. The thickness of the plate is set as 5 mm, and each G-1195N piezoelectric layer has a thickness equal to 0.1 mm. The material properties of piezoelectric materials are elastic modulus $E = 63 \times 10^9\ \mathrm{N/m^2}$, Poison's ratio $\upsilon = 0.3$, density $\rho = 7600\ \mathrm{kg/m^3}$, piezoelectric constant $d_{31} = 254 \times 10^{-12}(\mathrm{m/V})$, piezoelectric constant $d_{32} = 254 \times 10^{-12}(\mathrm{m/V})$ and dielectric coefficients $k_{33} = 15 \times 10^{-9}(\mathrm{F/m})$. The material constants of the constituent of the FGM plate are listed as follows: for aluminum oxide, $E = 3.2024 \times 10^{11}\ \mathrm{N/m^2}$, $\upsilon = 0.2600$, density $\rho = 3750\ \mathrm{kg/m^3}$ and for Ti-6Al-4 V $E = 1.0570 \times 10^{11}\ \mathrm{N/m^2}$, $\upsilon = 0.2981$, density $\rho = 4429\ \mathrm{kg/m^3}$.

The cantilevered (CFFF) plate is considered as the boundary condition. For the vibration control analysis, $64(8 \times 8)$ elements are used to model the FGM plate and to simplify the vibration analysis, the modal superposition algorithm is used and the first six modes are considered in this modal space analysis. An initial modal damping for each mode has been assumed to be 0.8%. A unit of force is imposed at point A of the FGM plate (Figure 1) in the vertical direction and is subsequently removed to generate motion from the initial displacement. Power law exponent for FGM plate is selected as $n = 5$.

In the design of proposed control systems, the effect of external disturbance are modeled as

$$\Delta\mathrm{Dis} = Am\,[\bar{K}_{uu}]\begin{bmatrix}\sin(\omega_d) & \cos(\omega_d) & \sin(\omega_d) & \cos(\omega_d) & \sin(\omega_d) & \cos(\omega_d)\end{bmatrix}^{\mathrm{T}}, \quad (65)$$

where $Am = 0.000001$ is the amplitude of disturbance, $[\bar{K}_{uu}]$ is the normalized matrix of $[K_{uu}]$ and $\omega_d = 200$ is the frequency of disturbance. Therefore, Equation (35) in the disturbance condition can be rewritten as

$$[\bar{M}_{uu}]\{\ddot{u}\} + [\bar{C}_{\mathrm{s}}]\{\dot{u}\} + [\bar{K}_{uu}]\{u\} = \{F_m\} - [\bar{K}_{u\varnothing}]_{\mathrm{a}}\{\varnothing\}_{\mathrm{a}} + \Delta\mathrm{Dis} \quad (66)$$

here $[\bar{M}_{uu}]$, $[\bar{C}_{\mathrm{s}}]$ and $[\bar{K}_{u\varnothing}]_{\mathrm{a}}$ are the normalized matrices of $[M_{uu}]$, $[C_{\mathrm{s}}]$ and $[K_{u\varnothing}]_{\mathrm{a}}$.

The simulation results are shown in Figures 3–8.

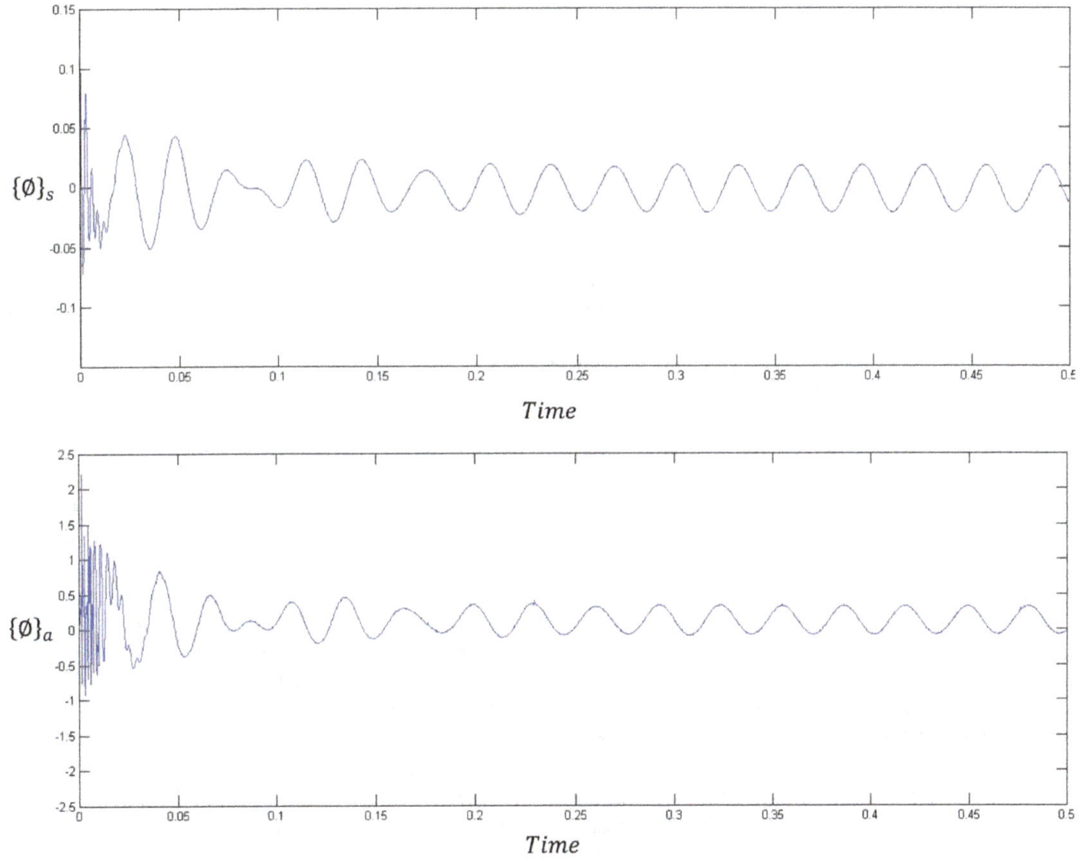


Figure 6. Simulation result of piezoelectric sensor and actuator due to the ANF SMC system in the disturbance condition.

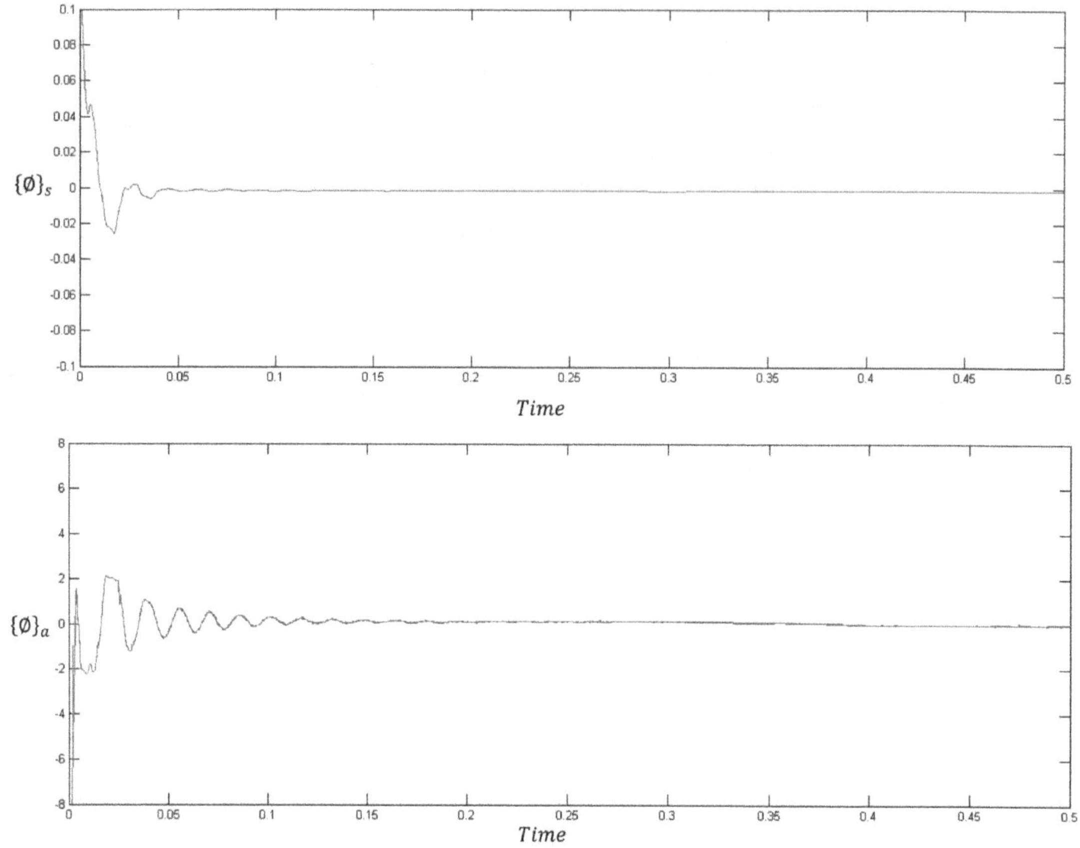


Figure 7. Simulation result of piezoelectric sensor and actuator due to the ANFSGA control system.

The effectiveness of the TSMC system is depicted in Figure 3 with $\lambda = 14.8$, Figure 4 shows that the ability of the TSMC system is improved by selecting $k_k = 1.5$ for the robust term in the disturbance condition.

The plots of the ANF SMC system (Figures 5 and 6 with control parameters $\varsigma_1 = 1.5$, $\varsigma_2 = \varsigma_3 = \varsigma_4 = \varsigma_5 = 0.05$ and $\lambda = 1.2$) show that the rate of the voltage which is applied on the actuator layer is smaller than the TSMC system.

In this study, 10 initial populations are randomly chosen from the reasonable region $[\{\varnothing\}_{a_{\min}} = -1, \{\varnothing\}_{a_{\max}} = 1]$ for the ANFSGA control system. The ability to suppress the vibration and reduce the external disturbance of the ANFSGA control system rather than the ANF sliding mode and also the TSMC system are demonstrated in Figures 7 and 8. Figure 5 shows that the settling time in the response of the ANF SMC system is nearly 0.17, but Figure 7 shows that the response of the ANFSGA control system is nearly 0.05.

In the proposed control system, a small voltage can be used to drive the system states to the equilibrium point. Therefore, it is superior to the ANF sliding mode and TSMC system to suppress the vibrations. The control parameters of the ANFSGA control system in the crossover operation to produce $\mu_1$ are $\varsigma_1 = 0.0001$, $\varsigma_2 = \varsigma_3 = \varsigma_4 = \varsigma_5 = 0.0001$ and also to produce $\mu_2$ are $\varsigma_1 = 0.00005$, $\varsigma_2 = \varsigma_3 = \varsigma_4 = \varsigma_5 = 0.0001$. In the mutation operation, the control parameters of the ANFSGA control system are $\varsigma_1 = 0.0003$, $\varsigma_2 = \varsigma_3 = \varsigma_4 = \varsigma_5 = 0.0002$. The sliding surface parameter is selected as $\lambda = 1.2$. The threshold value to activate the mutation operation is applied as $\mathrm{FIT}_B = 0.1$ and the parameter fitness value $\zeta = -16.3$ is used. It can be regarded that the associated fuzzy sets with the Gaussian function for each input signal are divided into NE (negative), ZE (zero) and PO (positive). Moreover, the means of the Gaussian functions are set as $-0.5$, 0, 0.5 and the standard deviations of the Gaussian functions are set as 0.3 for the NE, ZE and PO neurons.

The reasonable region $[\{\varnothing\}_{\mathrm{a}_{\min}} = -1, \{\varnothing\}_{\mathrm{a}_{\max}} = 1]$ bounds the power fluctuation. The reasonable region only makes the system use $\{\varnothing\}_{\mathrm{a}}$ of interval $[\{\varnothing\}_{\mathrm{a}_{\min}} = -1, \{\varnothing\}_{\mathrm{a}_{\max}} = 1]$ and then the values close to 1 and $-1$ are produced by crossover and the mutation algorithm. Moreover, stability of the system is kept in this operation. Therefore, based on this method, the operator can set the reasonable region without tuning the control parameters to have a desired response.

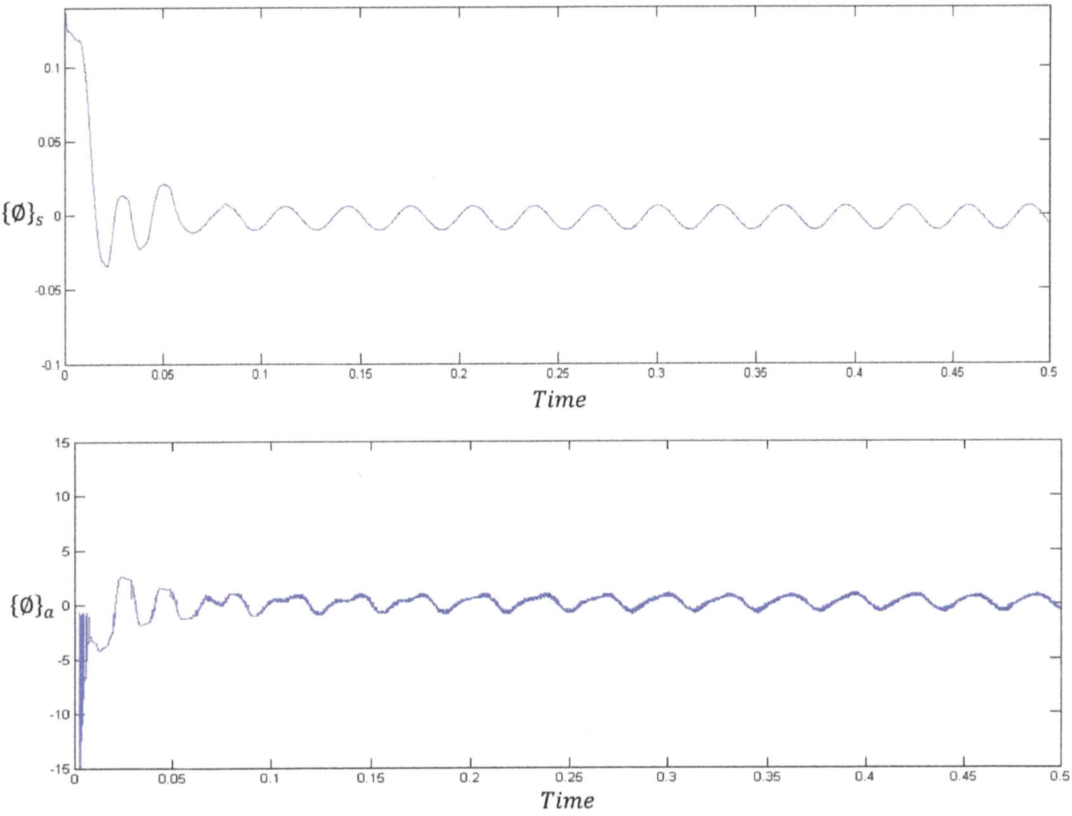


Figure 8. Simulation result of piezoelectric sensor and actuator due to the ANFSGA control system in the disturbance condition.

## 5. Conclusion

A general finite element model of the FGM plate has been introduced in this paper. A TSMC system has been designed to suppress the vibration of the FGM plate in the normal and disturbance conditions. The ANF sliding mode and the ANFSGA control system as the intelligent control methods have been successfully designed and effectively used to reduce the disturbance and eliminate the vibrations for the FGM plate. It is noted that, in the proposed controller, no constrained conditions and prior knowledge of the controlled plant have been used in the design process. Therefore, any information of the FGM plate is not utilized in the ANFSGA control system. The proposed controller is a flexible kind of control systems and it can be applied in another engineering applications.

## Disclosure statement

No potential conflict of interest was reported by the authors.

## References

Bagheri, A., & Javadi moghaddam, J. (2010). An adaptive neuro-fuzzy sliding mode based genetic algorithm control system for under water remotely operated vehicle. *Expert Systems with Applications, Expert Systems with Applications*, *37*, 647–660.

Bagheri, A., & Moghaddam, J. J. (2009). Decoupled adaptive neuro-fuzzy (DANF) sliding mode control system for a Lorenz chaotic problem. *Expert Systems with Applications*, *36*, 6062–6068.

Bishop, R. E. D. (1979). *The mechanics of vibration*. New York, NY: Cambridge University Press.

He, X. Q., Ng, T. Y., Sivashanker, S., & Liew, K. M. (2001). Active control of FGM plates with integrated piezoelectric sensors and actuators. *International Journal of Solids and Structures*, *38*, 1641–1655.

Liew, K. M., He, X. Q., & Kitipornchai, S. (2004). Finite element method for the feedback control of FGM shells in the frequency domain via piezoelectric sensors and actuators. *Computer Methods in Applied Mechanics and Engineering*, *193*, 257–273.

Lin, C. M., & Hsu, C. F. (2004). Adaptive fuzzy sliding-mode control for induction servomotor systems. *IEEE Transactions on Energy Conversion*, *19*(2), 362–368.

Lin, F.-J., & Wai, R.-J. (2003). Adaptive and fuzzy neural network sliding-mode controllers for motor-quick-return servomechanism. *Mechatronics*, *13*, 477–506.

Moita, J. M. S., Correia, V. M. F., Martins, P. G., Mota Soares, C. M. M., & Mota Soares, C. A. (2006). Optimal design in vibration control of adaptive structures using a simulated annealing algorithm. *Composite Structures*, *75*, 79–87.

Parashkevova, L., Ivanova, J., & Bontcheva, N. (2004). Optimal design of functionally graded plates with thermo-elastic plastic behaviour. *Comptes Rendus Mécanique*, *332*, 493–498.

Praveen, G. N., & Reddy, J. N. (1998). Nonlinear transitent thermoelastic analysis of functionally graded ceramic-metal plates. *International Journal of Solids and Structures*, *35*, 4457–4476.

Reddy, J. N. (2004). *Mechanics of laminated composite plates and shells: Theory and analysis* (2nd ed.). Boca Raton: CRC Press.

Robbins, D. H., & Reddy, J. N. (1991). Analysis of piezoelectrically actuated beams using a layer wise displacement theory. *Computers and Structures*, *41*, 265–279.

Slotine, J. J. E., & Li, W. (1991). *Applied nonlinear control*. EnglewoodCli4s, NJ: Prentice-Hall.

To, C. W. S., & Chen, T. (2007). Optimal control of random vibration in plate and shell structures with distributedpiezoelectric components. *International Journal of Mechanical Sciences*, *49*, 1389–1398.

Touloukian, Y. S. (1967). *Thermophysical properties of high temperature solid materials*. New York, NY: Macmillian.

Turteltaub, S. (2002). Functionally graded materials for prescribed field evolution. *Computer Methods in Applied Mechanics and Engineering*, *191*, 2283–2296.

Wai, R.-J., & Lee, J.-D. (2004). Intelligent motion control for linear piezoelectric ceramic motor drive. *IEEE Transactions on Systems, Man, and Cybernetics – Part B: Cybernetics*, *34*(5), 2100–2111.

Wai, R.-J., & Tu, C.-H. (2007). Design of total sliding-mode-based genetic algorithm control for hybrid resonant-driven linear piezoelectric ceramic motor. *IEEE Transactions on Power Electrics*, *22*(2), 563–575.

# 7

# Sliding mode control of a permanent magnet synchronous generator for variable speed wind energy conversion systems

Jacob Hostettler and Xin Wang*

*Department of Electrical and Computer Engineering, Sourthern Illinois University Edwardsville, Edwardsville, IL, USA*

Difficulties in achieving the maximum level of efficiency in power extraction from available wind resources warrant the collective attention of modern control and power systems engineers. A strong movement towards sustainable energy resources, and advances in control system methodologies make previously unattainable levels of efficiency possible. One such promising method is sliding mode control. This control method, touted for its robustness given un-modelled dynamics present in the system, provides ideal characteristics for application in the control of permanent magnet synchronous generators employed in variable speed wind energy conversion systems. Application of this method for control using dynamic models of the $d$-axis and $q$-axis currents, as well as those of the high-speed shaft rotational speed results in a high-level efficiency in power extraction from a varying wind resource.

**Keywords:** permanent magnet synchronous generator; wind energy conversion system; sliding mode control

**Nomenclature**

| | |
|---|---|
| $\tau_m$ | Mechanical torque |
| $\rho$ | Air density |
| $R_t$ | Rotor radius |
| $C_p(\cdot)$ | Power coefficient |
| $\lambda$ | Tip speed ratio |
| $\beta$ | Blade pitch angle |
| $v$ | Wind speed |
| $\omega_l$ | Low-speed shaft rotational speed |
| $u_d$ | $d$-axis voltage |
| $u_q$ | $q$-axis voltage |
| $R_s$ | Stator resistance |
| $i_d$ | $d$-axis current |
| $i_q$ | $q$-axis current |
| $L_d$ | $d$-axis inductance |
| $L_q$ | $q$-axis inductance |
| $\omega_e$ | Generator rotational frequency |
| $\Psi_m$ | Generator flux linkage |
| $\omega_r$ | High-speed shaft rotational speed |
| $\tau_e$ | Generator torque |
| $B$ | Stiffness Coefficient |
| $J$ | High-speed shaft inertia |

## 1. Introduction

The primary generator control challenge inherent in the use of variable speed wind energy conversion systems (WECS) involves coping with a consistently unpredictable and uncontrollable energy resource (Munteanu, Bratcu, Cutululis, & Ceangă, 2008). More accurately, the goal of control lies in achieving the maximum efficiency in extraction of available power from the wind, known as Maximum Power Point Tracking (MPPT) (Errami, Ouassaid, & Maaroufi, 2013; Orlando, Liserre, Mastromauro, & Dell'Aquila, 2013). A restriction on the amount power available for extraction, as well as the nonlinear nature of mechanical components compounds the severity of this problem (Garcia-Sanz & Houpis, 2012). Greater global implementation of WECS technology emphasizes the need to develop more advanced methods of control in order to achieve desired levels of efficiency, and to allow for the creation of WECS with the ability to compete with more traditional forms of power generation (Alizadeh & Yazdani, 2013; Hirth, 2013).

A growing trend away from the use of asynchronous machines such as doubly fed induction generators (DFIGs), brought on by the implementation of permanent magnet synchronous generators (PMSGs) further compounds the need for new control strategies (Beltran, El Hachemi Benbouzid, & Ahmed-Ali, 2012, Orlando et al., 2013). The increased use of PMSGs in WECS stems in part from improved performance attributes such as high reliability, high efficiency, a high energy-to-weight ratio, and greater power factor (Sharma & Singh, 2012). The smaller size and reduced costs of PMSGs make them ideal for use in WECS (Qiao, Qu, & Harley, 2009). PMSGs eliminate the necessity of a gearbox, which further reduces costs associated with maintenance by allowing

*Corresponding author. Email: xwang@siue.edu

for direct coupling of the shafts of the generator and the rotor (Alizadeh & Yazdani, 2013; Chinchilla, Arnaltes, & Burgos, 2006). A variety of proposed control schemes exist for achieving MPPT in PMSG-based WECS (Tan & Islam, 2004). These methods are often concerned with control of different aspects, such as speed and torque (Alizadeh & Yazdani, 2013; Rajaei, Mohamadian, & Yazdian Varjani, 2013; Yan et al., 2013).

One proposed control method that shows promise in helping us to achieve high efficiency, robustness, and stability is sliding mode control (SMC). For WECS employing DFIGs, second-order SMC applied to torque control demonstrates high MPPT with low variations in torque (Beltran et al., 2012). The usefulness of this control method extends into synchronous machines as well. Research into the use of SMC in PMSGs effectively demonstrates the potential power of this control method (Yan et al., 2013). The strength of SMC comes from the ability to control high-order systems, while exhibiting resiliency against disturbances and variations in model parameters (Utkin, Guldner & Shi, 2009). Research into SMC for PMSGs illustrates the potential of this control method, and accentuate SMCs promise as a candidate for achieving the goal increasing the efficiency level in regards to MPPT (Valenciaga & Puleston, 2008). Results demonstrated by research involving the use of SMC with permanent magnet synchronous motors, and second-order SMC applied to PMSGs continue to demonstrate the strength of SMC as a viable method for effective control of synchronous machines (Benelghali, El Hachemi Benbouzid, Charpentier, Ahmed-Ali, & Munteanu, 2011; Vu, Yu, Choi, & Jung, 2013; Zhang, Sun, Zhao, & Sun, 2013).

Although the positive attributes of SMC make it seem an ideal control method, it does not exist without fault. The phenomenon known as chattering invites an understandable level of criticism. Due to the nonideailties in switching devices, the response of the system under SMC oscillates about the desired reference known as the sliding surface (Young, Utkin, & Ozguner, 1999).

This work aims to expand on existing research into second-order SMC architecture, in an attempt to improve MPPT. A novel third-order SMC architecture employing the dynamics associated with the rotational speed of the generator along with the standard $d, q$ model of a PMSG is developed. Active control of the $q$-axis reference current of the PMSG model provides the main distinction for this control design in comparison to existing architectures. Simulations of this design involve the application of a highly varying wind input to the developed system.

The structure of this paper is as follows: Section 2 develops and outlines models for the aerodynamics of the wind turbine and the PMSG. Section 3 presents the design of the third-order sliding mode controller. Section 4 gives the results of simulations conducted in MATLAB and SIMULINK. Lastly, Section 5 concludes this work with comments on the effectiveness of the proposed control method.

## 2. Wind turbine and generator modelling

The architecture considered is that of a variable speed, fixed pitch, rigid drive train WECS employing a PMSG (Garcia-Sanz & Houpis, 2012; Munteanu et al., 2008).

### *2.1. Aerodynamics*

Mechanical torque developed by the rotor, and the tip-speed ratio are described by Garcia-Sanz & Houpis (2012):

$$\tau_m = \frac{\rho \pi R_t^2 C_p(\lambda, \beta) v^3}{2\omega_l} \tag{1}$$

$$\lambda = \frac{\omega_l R_t}{v} \tag{2}$$

$\tau_m$ is dependent on the air density $\rho$, the blade length of the turbine $R_t$, the wind speed $v$, and the low-speed shaft speed $\omega_l$. Also important is the power efficiency coefficient, which denotes the percentage of power extracted from the available power in the wind. The power coefficient can be approximated as Garcia-Sanz & Houpis (2012):

$$C_p(\lambda, \beta) = c_1 \left( \frac{c_2}{\lambda_i} - c_3\beta - c_4 \right) \exp \frac{-c_5}{\lambda_i} \tag{3}$$

where

$$\lambda_i = \left( \frac{1}{\lambda + c_6\beta} - \frac{c_7}{\beta^3 + 1} \right)^{-1} \tag{4}$$

An ideal limit on the amount of power that can be extracted from the wind is known as the Betz Limit. Under ideal assumptions, approximately 59.3% of the available power can be extracted (Munteanu et al., 2008). Upon consideration of the non-ideal nature of the turbine, this value will decrease. Figure 1 shows an estimation of a $C_p$ versus $\lambda$

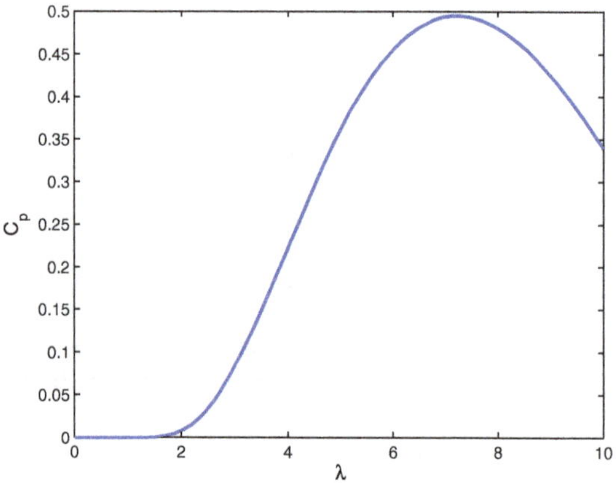


Figure 1. Power Coefficient v.s. Tip Speed Ratio.

curve, where $\beta = 0^o$, $c_1 = 0.39, c_2 = 116, c_3 = 0.4, c_4 = 5, c_5 = 16.5, c_6 = 0.089$, and $c_7 = 0.035$. The peak gives the max power coefficient ($C_{p,\max}$) and the optimal tip speed ratio ($\lambda_o$) (Garcia-Sanz & Houpis, 2012). Figure 1 gives $C_{p,\max} = 0.4953$, and $\lambda_o = 7.2$.

### 2.2. *PMSG Modelling*

Application of the Park's Transformation to the $(a, b, c)$ coordinate model of synchronous generator yields the $(d, q)$ coordinate model of a PMSG. The equations for the direct ($d$) axis and quadrature ($q$) axis voltages are given as Munteanu et al., 2008:

$$u_d = R_s i_d + L_d \frac{\mathrm{d}i_d}{\mathrm{d}t} - L_q i_q \omega_e \tag{5}$$

$$u_q = R_s i_q + L_q \frac{\mathrm{d}i_q}{\mathrm{d}t} + (L_d i_d + \Psi_m)\omega_e \tag{6}$$

Considering further the use of a surface-mounted PMSG $L_d = L_q$. Therefore, both inductances will become $L$. Thus, the re-arrangement of (5) and (6) yields the following second-order system (Munteanu et al., 2008):

$$\frac{\mathrm{d}i_d}{\mathrm{d}t} = -\frac{R_s}{L} i_d + \omega_e i_q - \frac{1}{L} u_d \tag{7}$$

$$\frac{\mathrm{d}i_q}{\mathrm{d}t} = -\frac{R_s}{L} i_q - \omega_e i_d - \frac{1}{L} u_q + \frac{1}{L} \Psi_m \omega_e \tag{8}$$

This model is expanded to a third-order system through consideration of the equation of motion for the high-speed shaft (HSS) rotational speed (Munteanu et al., 2008):

$$\frac{\mathrm{d}\omega_r}{\mathrm{d}t} = \frac{\tau_m}{J} - \frac{\tau_e}{J} - \frac{B\omega_r}{J} \tag{9}$$

Considering $\omega_e = (P/2)\omega_r$, where $P$ is the number of stator poles, $\tau_e = K_t i_q$, and $K_t = \frac{3}{4} P \Psi_m$, the complete third-order model used in control design is as follows Munteanu et al., 2008:

$$\frac{\mathrm{d}i_d}{\mathrm{d}t} = -\frac{R_s}{L} i_d + \frac{P}{2} i_q \omega_r - \frac{1}{L} u_d \tag{10}$$

$$\frac{\mathrm{d}i_q}{\mathrm{d}t} = -\frac{R_s}{L} i_q - \frac{P}{2}\left(i_d - \frac{\Psi_m}{L}\right)\omega_r - \frac{1}{L} u_q \tag{11}$$

$$\frac{\mathrm{d}\omega_r}{\mathrm{d}t} = \frac{\tau_m}{J} - \frac{K_t i_q}{J} - \frac{B\omega_r}{J} \tag{12}$$

## 3. SMC design

SMC design is applied to Equations (10) and (11), and expanded to include (12) in order to create a new SMC architecture for the WECS using a PMSG (Perruquetti & Barbot, 2002; Utkin et al., 2009).

### 3.1. *Sliding surfaces*

The sliding surfaces are to be defined as

$$s_d(t) = [i_d(t) - i_d^*(t)] = 0 \tag{13}$$

$$s_q(t) = [i_q(t) - i_q^*(t)] = 0 \tag{14}$$

$$s_{\omega_r}(t) = [\omega_r(t) - \omega_r^*(t)] = 0 \tag{15}$$

$i_d^*(t)$, $i_q^*(t)$, and $\omega_r^*(t)$ are the reference values for their respective surfaces. Due to the nature of field-oriented control, the $d$-axis stator current reference $i_d^*(t) = 0$. The speed reference $\omega_r^*(t) = i(\lambda v/R_t)$, where $i$ is the WECS fixed drive train multiplying ratio (Garcia-Sanz & Houpis, 2012). The $q$-axis stator current reference $i_q^*(t)$ is a dynamic value, and will be revealed as the resulting output of the control law developed for Equation (12).

### 3.2. *Reachability*

The reachability conditions for Equations (13)–(15) are given respectively as

$$s_d(t)\dot{s}_d(t) < 0 \tag{16}$$

$$s_q(t)\dot{s}_q(t) < 0 \tag{17}$$

$$s_{\omega_r}(t)\dot{s}_{\omega_r}(t) < 0 \tag{18}$$

These inequalties ensure that the trajectories will remain driven towards their respective sliding surfaces (Perruquetti & Barbot, 2002).

### 3.3. *Parameter variations*

Possible un-modelled dynamics present in Equations (10)–(12) are taken into consideration by the $R_s = \hat{R}_s + \Delta R_s$, where $\hat{R}_s$ is the nominal value, and $\Delta R_s$ is a bounded disturbance. The same reasoning is applied to $L = \hat{L} + \Delta L$, $\Psi_m = \hat{\Psi}_m + \Delta\Psi_m$, $\tau_m = \hat{\tau}_m + \Delta\tau_m$, $J = \hat{J} + \Delta J$, and $B = \hat{B} + \Delta B$.

### 3.4. *Direct axis current control design*

In order to develop the $d$-axis control, Equation (16) must be satisfied. From Equation (10), this inequality can be re-written as

$$\frac{s_d(t)}{L}\left[-R_s i_d(t) + \frac{P}{2} L i_q(t)\omega_r(t) - u_d(t) - L\frac{\mathrm{d}i_d^*(t)}{\mathrm{d}t}\right] < 0 \tag{19}$$

Denoting the $d$-axis control law as

$$u_d(t) = u_{d,eq}(t) + u_{d,N}(t) \tag{20}$$

where $u_{d,eq}(t)$ is the equivalent control, and $u_{d,N}(t)$ is the switching control. The equivalent control is given as follows:

$$u_{d,eq}(t) = -\hat{R}_s i_d(t) + \frac{P}{2}\hat{L} i_q(t)\omega_r(t) - \hat{L}\frac{\mathrm{d}i_d^*(t)}{\mathrm{d}t} \qquad (21)$$

Hence, $\dot{s}_d$ from inequality (19) becomes

$$\dot{s}_d(t) = \frac{1}{\Delta L}\left[-\Delta R_s i_d(t) - \Delta L\left(\frac{\mathrm{d}i_d^*(t)}{\mathrm{d}t} - \frac{P}{2} i_q(t)\omega_r(t)\right)\right] - \frac{1}{\Delta L} u_{d,N}(t) \qquad (22)$$

Due to the bounded nature of the uncertainties discussed in III C, along with the variables $\mathrm{d}i_d^*(t)/\mathrm{d}t$, $i_d(t)$, $i_q(t)$, and $\omega_r(t)$ there exists a positive constant $u_{do}$ such that

$$u_{do} > \left|-\Delta R_s i_d(t) + \Delta L\frac{P}{2} i_q(t)\omega_r(t) - \Delta L\frac{\mathrm{d}i_d^*(t)}{\mathrm{d}t}\right|. \qquad (23)$$

From this, the switching portion of $u_d(t)$ is determined to be:

$$u_{d,N}(t) = -u_{do}\mathrm{sgn}(s_d(t)) \qquad (24)$$

where

$$\mathrm{sgn}(s) = \begin{cases} 1, & \text{if } s > 0 \\ 0, & \text{if } s = 0 \\ -1, & \text{if } s < 0 \end{cases}$$

The design of this controller should ensure that $i_d(t)$ is driven to $i_d^*(t) = 0$, and will remain there despite disturbances.

### 3.5. *Quadrature axis control design*

Following the same method, and considering the $q$-axis control, the inequality (17) must be satisfied. Re-writting $\dot{s}_q(t)$ in terms of Equations (11) and (14), Equation (17) can be re-written as

$$\frac{s_q(t)}{L}\left[-R_s i_q(t) - \frac{P}{2}(L i_d(t) - \Psi_m)\omega_r(t)\right] + \frac{s_q(t)}{L}\left[-u_q(t) - L\frac{\mathrm{d}i_q^*(t)}{\mathrm{d}t}\right] < 0. \qquad (25)$$

Denoting the $q$-axis control law as the sum of the equivalent and switching controls, respectively:

$$u_q(t) = u_{q,eq}(t) + u_{q,N}(t). \qquad (26)$$

The equivalent control can be given as

$$u_{q,eq}(t) = -\hat{R}_s i_q(t) + \frac{P}{2}(\hat{L} i_d(t) - \hat{\Psi}_m)\omega_r(t) - \hat{L}\frac{\mathrm{d}i_q^*(t)}{\mathrm{d}t}. \qquad (27)$$

Now $\dot{s}_q$ from inequality (25) is written as

$$\dot{s}_q(t) = \frac{1}{\Delta L}\left[-\Delta R_s i_q(t) - \frac{P}{2}(\Delta L i_d(t) - \Delta\Psi_m)\omega_r(t)\right] - \frac{1}{\Delta L}\left[\Delta L\frac{\mathrm{d}i_q^*(t)}{\mathrm{d}t} - u_{q,N}(t)\right]. \qquad (28)$$

Again, due to the bounded nature of the uncertainties and variables discussed in Section 3.3, there exists a positive constant $u_{qo}$ such that

$$u_{qo} > \left|-\Delta R_s i_q(t) - \frac{P}{2}(\Delta L i_q(t) + \Delta\Psi_m)\omega_r(t) - \Delta L\frac{\mathrm{d}i_d^*(t)}{\mathrm{d}t}\right|. \qquad (29)$$

From this, the switching portion of $u_q(t)$ is determined to be

$$u_{q,N}(t) = -u_{qo}\mathrm{sgn}(s_q(t)). \qquad (30)$$

The result of this control design should be that $i_q(t)$ is maintained at the reference $i_q^*(t)$.

### 3.6. *Control design based on rotational speed dynamics*

Again, making use of the similar method used for the $d$- and $q$-axis controls, and ensuring that inequality (18) is satisfied, $\dot{s}_{\omega_r}(t)$ can be rewritten from Equations (12) and (15). Equation (18) now becomes

$$\frac{s_{\omega_r}(t)}{K_t}\left[\tau_m - K_t i_q(t) - B\omega_r(t) - J\frac{\mathrm{d}\omega_r^*(t)}{\mathrm{d}t}\right] < 0. \qquad (31)$$

For the rotational speed controller, the control variable becomes the quadrature axis reference current. It follows that the control law is designated

$$i_q^*(t) = i_{q,eq}^*(t) + i_{q,N}^*(t). \qquad (32)$$

With equivalent control

$$i_{q,eq}^*(t) = \frac{1}{K_t}\left[\hat{\tau}_m - \hat{B}\omega_r(t) - \hat{J}\frac{\mathrm{d}\omega_r^*(t)}{\mathrm{d}t}\right]. \qquad (33)$$

$\dot{s}_q$ from inequality (31) is now re-written as

$$\dot{s}_{\omega_r}(t) = \frac{1}{K_t}\left[\Delta\tau_m - \Delta B\omega_r(t) - \Delta J\frac{\mathrm{d}\omega_r^*(t)}{\mathrm{d}t}\right]. \qquad (34)$$

Based on previous reasoning for the $d$- and $q$-axis control design, there exists a postive constant $i_{qo}$ such that

$$i_{qo} = \left|\frac{1}{K_t}\left[\Delta\tau_m - \Delta B - \Delta J\frac{\mathrm{d}\omega_r^*(t)}{\mathrm{d}t}\right]\right|. \qquad (35)$$

It then follows that

$$i_{q,N}^*(t) = -i_{qo}\mathrm{sgn}(s_{\omega_r}(t)). \qquad (36)$$

This controller will actively control $i_q^*(t)$. Therefore the output is the input to the $q$-axis control developed previously.

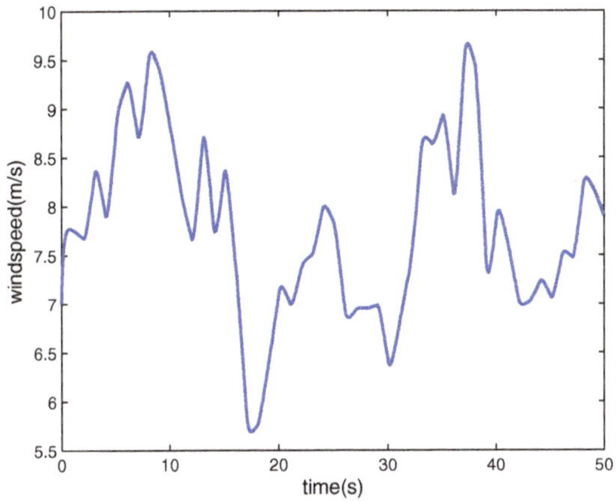


Figure 2. SIMULINK Model.

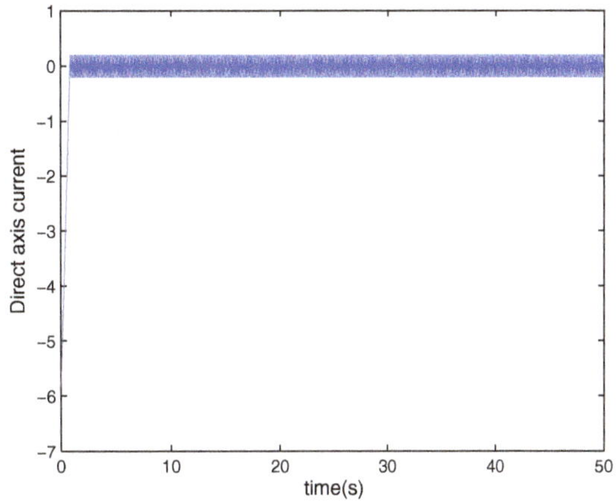


Figure 3. Wind Input.

Table 1. Turbine and generator parameters.

| | |
|---|---|
| Rotor radius | $R_t = 3\,\text{m}$ |
| Statorresistance | $R_s = 3.5\Omega$ |
| d-axis inductance | $L_q = 35\,\text{mH}$ |
| q-axis inductance | $L_d = 35\,\text{mH}$ |
| Flux linkage | $\Psi_m = 0.3\,\text{Wb}$ |
| Poles | $P = 6$ |
| Inertia | $J = 1$ |
| Stiffness | $B = 0.001$ |

## 4. Simulation results

The implementation of the wind turbine aerodynamics, PMSG dynamics, and SMC control architecture in SIMULINK is shown in Figure 2. The nominal PMSG parameters used are listed in Table 1. Parameter variations were implemented within the Generator Dynamics block. The SIMULINK wind model in Figure 2 is capable of producing varying degrees of wind intensity and speed variations. Figure 3 shows the wind input used for the remainder of the simulation results listed. This particular wind input was chosen to demonstrate the effectivness of the control design because of its highly varying nature, and range of wind speeds.

The first result shown is that of $i_d(t)$. As can be seen by Figure 4, the $d$-axis current accurately tracks the reference value $i_d^*(t) = 0$. Chattering about the reference is due to switching delays. However, despite constant fluctuations in wind speed, the $d$-axis control scheme effectively resists the effects of disturbances and parameter variations.

One of the key components of the proposed control scheme is the active control of the quadrature axis current reference $i_q^*(t)$ through the implementation of a sliding mode controller based on the rotational speed dynamics of the WECS. Figures 5 and 6 show the speed reference $\omega_r^*(t)$ and the rotational speed $\omega_r(t)$, respectively. Inspection of these figures show that accurate speed control is achieved. $i_q^*(t)$ is a direct result of the new control design. Figure 7 shows the value for $i_q^*(t)$, and Figure 8 shows the actual value of $i_q(t)$, accurately tracking the defined reference.

The final and probably the most important result regards MPPT. Figure 9 shows the power coefficient throughout the course of the simulation. Immediately following the initial approach towards $C_{p,\max}$, it can be seen that the value of $C_p$ is held at or near $C_{p,\max}$ despite all fluctuations in the wind and parameters.

The proposed control algorithm has been realized in hardware for controlling the PMSG-based wind energy

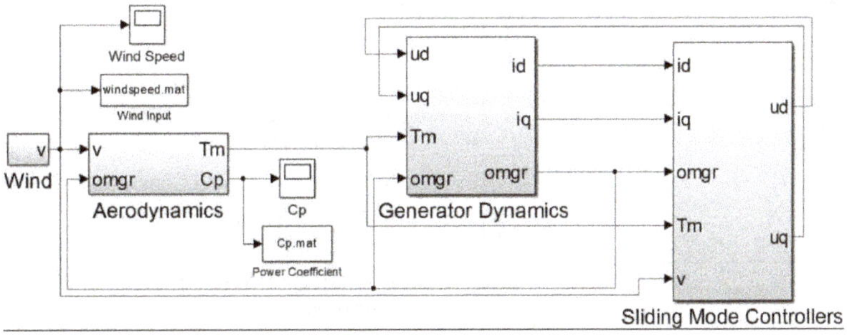


Figure 4. Actual Direct Axis Current $i_d(t)$.

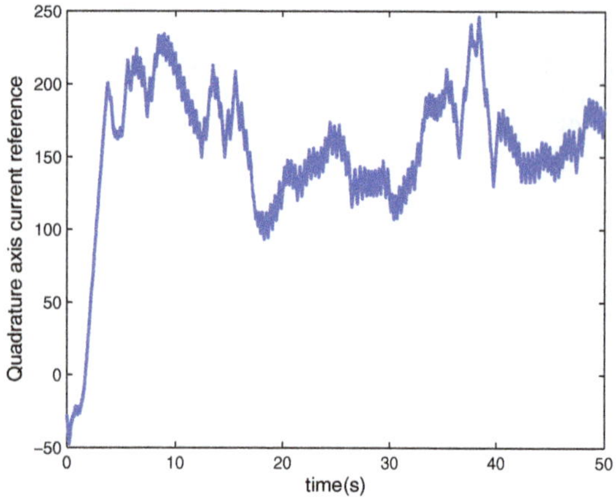


Figure 5. Quadrature Axis Current Reference $i_q^*(t)$.

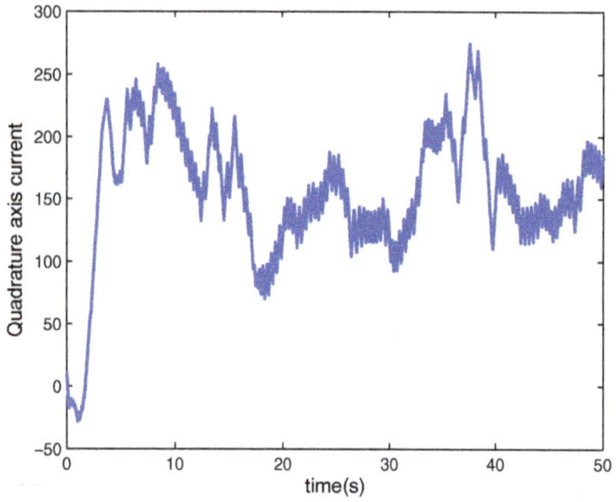


Figure 6. Quadrature Axis Current $i_q(t)$.

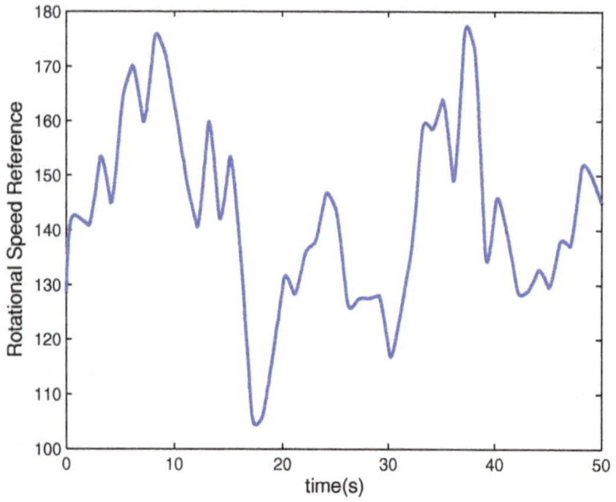


Figure 7. Rotational Speed Reference $\omega_r^*(t)$.

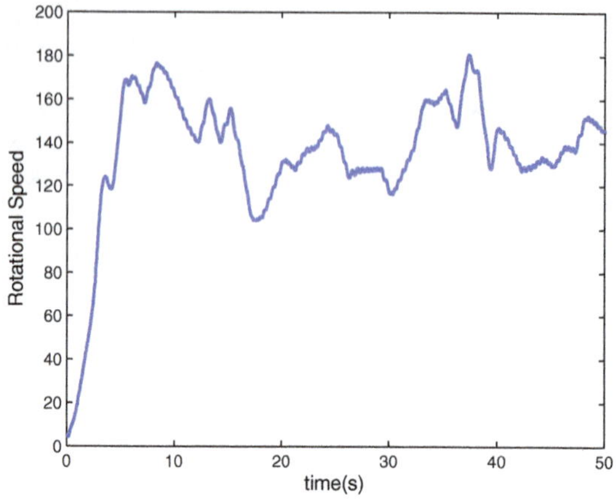


Figure 8. Actual Rotational Speed $\omega_r(t)$.

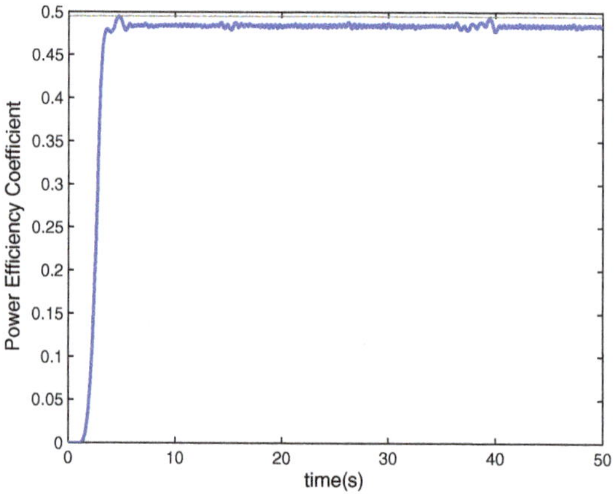


Figure 9. Power Coefficient.

system in real-time using dSPACE DS1103 PPC control system. Both simulation and hardware implementation have shown the efficacy of the proposed SMC approach.

In future research, the proposed SMC approach for PMSG-based variable speed WECSs will be tested on the wind turbines with the dynamic prescribed vortex wake model (Currin, Coton, & Wood, 2008; Qi & Barltrop, 2015).

## 5. Conclusion

The proposed sliding mode controller designed for a variable-speed WECS employing a surface-mounted PMSG has been designed and simulated using MATLAB/SIMULINK. Preliminary testing of this novel control design results in a system with low sensitivity to disturbances, and the ability to maintain a relatively constant level of power extraction efficiency in the presence of a highly varying wind input, and bounded model parameter variations. Initial research into this method shows promise for active control of the quadrature axis current reference based on the use of an SMC control design for the dynamics of the high-speed shaft rotational speed of the system. Research involving system parameter variation, chattering reduction, and comparison to other methods requires further investigation. However, the initial effectiveness of this control topology shows that it merits future research.

## Disclosure statement

No potential conflict of interest was reported by the authors.

## References

Alizadeh, O., & Yazdani, A. (2013). A strategy for real power control in a direct-drive PMSG-based wind energy conversion system. *IEEE Transactions on Power Delivery*, *28*(3), 1297–1305.

Beltran, B., El Hachemi Benbouzid, M., & Ahmed-Ali, T. (2012). Second-order sliding mode control of a doubly fed induction generator driven wind turbine. *IEEE Transactions on Energy Conversion*, *27*(2), 261–269.

Benelghali, S., El Hachemi Benbouzid, M., Charpentier, J. F., Ahmed-Ali, T., & Munteanu, I. (2011). Experimental validation of a marine current turbine simulator: application to a permanent magnet synchronous generator-based system second-order sliding mode control. *IEEE Transactions on Industrial Electronics*, *58*(1), 118–126.

Chinchilla, M., Arnaltes, S., & Burgos, J. C. (2006). Control of permanent-magnet generators applied to variable-speed wind-energy systems connected to the grid. *IEEE Transactions on Energy Conversion*, *21*(1), 130–135.

Currin, H. D., Coton, F. N., & Wood, B. (2008). Dynamic prescribed vortex wake model for AeroDyn. *ASME J. Solar Energy Engineering*, *130*(3), 031007-1–031007-7.

Errami, Y., Ouassaid, M., & Maaroufi, M. (2013). Control of a PMSG based Wind Energy Generation System for Power Maximization and Grid Fault Conditions. *Energy Procedia*, *42*, 220–229.

Garcia-Sanz, M., & Houpis, C. (2012). Aerodynamics and mechanical modeling of wind turbines. *Wind energy systems control engineering design*. Boca Raton, FL: Taylor & Francis Group LLC.

Hirth, L. (2013). The market value of variable renewables: The effect of solar wind power variability on their relative price. *Energy Economics*, *38*, 218–236.

Munteanu, I., Bratcu, A., Cutululis, N., & Ceangă, E. (2008). *Optimal control of wind energy systems towards a global approach*. London: Springer.

Orlando, N. A., Liserre, M., Mastromauro, R. A., & Dell'Aquila, A. (2013). A survey of control issues in PMSG-based small wind-turbine systems. *IEEE Transactions on Industrial Informatics*, *9*(3), 1211–1221.

Perruquetti, W., & Barbot, J. P. (2002). *Sliding mode control in engineering*. Boca Raton, FL: CRC Press.

Qi, Q., & Barltrop, N. (2015, Jun). Unsteady aerodynamics of offshore floating wind turbines using free vortex wake model. *Proc. of the 25th International Ocean and Polar Engineering Conference*, Kona, HI.

Qiao, W., Qu, L., & Harley, R. G. (2009). Control of IPM synchronous generator for maximum wind power generation considering magnetic saturation. *IEEE Transactions on Industry Applications*, *45*(3), 1095–1105.

Rajaei, A., Mohamadian, M., & Yazdian Varjani, A. (2013). Vienna-rectifier-based direct torque control of PMSG for wind energy application. *IEEE Transactions on Industrial Electronics*, *60*(7), 2919–2929.

Sharma, S., & Singh, B. (2012). Control of permanent magnet synchronous generator-based stand-alone wind energy conversion system. *IET Power Electronics*, *5*(8), 1519–1526.

Tan, K., & Islam, S. (2004). Optimum control strategies in energy conversion of PMSG wind turbine system without mechanical sensors. *IEEE Transactions on Energy Conversion*, *19*(2), 392–399.

Utkin, V. I., Guldner, J., & Shi, J. (2009). *Sliding mode control in electromechanical systems* (2nd ed.). Boca Raton, FL: CRC Press.

Valenciaga, F., & Puleston, P. F. (2008). High-order sliding control for a wind energy conversion system based on a permanent magnet synchronous generator. *IEEE Transactions on Energy Conversion*, *23*(3), 860–867.

Vu, N., Yu, D. Y., Choi, H. H., & Jung, J. W. (2013). T-S fuzzy-model-based sliding-mode control for surface-mounted permanent-magnet synchronous motors considering uncertainties. *IEEE Transactions on Industrial Electronics*, *60*(10), 4281–4291.

Yan, J., Lin, H., Feng, Y., Guo, X., Huang, Y., & Zhu, Z. Q. (2013). Improved sliding mode model reference adaptive system speed observer for fuzzy control of direct-drive permanent magnet synchronous generator wind power generation system. *IET Renewable Power Generation*, *7*(1), 28–35.

Young, K. D., Utkin, V. I., & Ozguner, U. (1999). A control engineer's guide to sliding mode control. *IEEE Transactions on Control Systems Technology*, *7*(3), 328–342.

Zhang, X., Sun, L., Zhao, K., & Sun, L. (2013). Nonlinear speed control for PMSM system using sliding-mode control and disturbance compensation techniques. *IEEE Transactions on Power Electronics*, *28*(3), 1358–1365.

# 8

## Design of optimal disturbance cancellation controllers via modified loop transfer recovery

Tadashi Ishihara[a]* and Hai-Jiao Guo[b]

[a]*Faculty of Science and Technology, Fukushima University, Fukushima, Japan;* [b]*Department of Electrical and Information Engineering, Tohoku Gakuin University, Tagajo, Japan*

The authors proposed a modified loop transfer recovery (LTR) method for designing the disturbance cancellation controllers where the disturbances were assumed to be step functions and the optimality of the controllers was not considered. This paper discusses the extension of their work to a more general class of disturbances with optimality consideration. It is assumed that the plant is minimum phase and the disturbances entering the plant input side are generated by function generators with unknown initial conditions. A quadratic performance index explicitly representing the disturbance cancellation requirement is introduced. The optimal disturbance cancellation controller minimizing the performance index is constructed by the separation principle. As a target for the LTR design, the optimal disturbance cancellation controller based on the measurement of the plant state is chosen. It is shown that the target feedback property can be recovered in the output feedback controller by a simple modification of the standard LTR procedure. A numerical example is presented to illustrate the effectiveness of the proposed optimal design.

**Keywords:** linear system; optimal disturbance cancellation; polynomial disturbances; sinusoidal disturbances; loop transfer recovery

### 1. Introduction

The authors discussed a loop transfer recovery (LTR) design of the disturbance cancellation controllers (Guo, Ishihara, & Takeda, 1996) where the disturbances were assumed to be step functions and the optimality of the controllers was not considered. It was pointed out that the standard LTR theory (e.g. Saberi, Chen, & Sannuti, 1993; Stein & Athans, 1987) could not directly be applied but that a procedure obtained by modifying the standard LTR procedure could be used for the disturbance cancellation controllers.

The extension of our earlier work to a more general class of disturbances has initially been discussed in our conference paper (Ishihara & Guo, 2012) with the optimality consideration. In this paper, the extension is discussed in more detail with a design example illustrating the effectiveness of the proposed optimal design. It is assumed that the plant is minimum phase and the disturbances, which enter the plant input side, are generated by function generators with unknown initial conditions. The class of disturbances includes sinusoids with known frequencies as well as polynomial functions of time such as steps and ramps. To guarantee the optimality, a quadratic performance index explicitly representing the disturbance cancellation requirement is introduced as in Ishihara and Guo (2008).

Assuming that the state of the plant and that of the disturbance model are perfectly measurable, we obtain the optimal control law by reducing the optimal disturbance cancellation problem to a standard linear-quadratic (LQ) problem (Anderson & Moore, 1990). The optimal output feedback controller is constructed by the separation principle with the use of the Kalman filter jointly estimating the state of the plant and that of the disturbance model. As a target for the LTR design, we choose the optimal disturbance cancellation controller including the optimal disturbance estimator based on the measurement of the plant state. The use of the optimal disturbance estimator ensures that the target has guaranteed large stability margins. It is shown that the target can be recovered by an extended version of the modified LTR procedure proposed in Guo et al. (1996).

It should be noted that, in the LTR design of LQG (Linear-Quadratic-Gaussian) controllers (e.g. Stein & Athans, 1987), the covariance parameters in the stochastic model and the weighting coefficients of the performance index are used as tuning parameters to achieve desired feedback property. Note also that, as in our earlier work (Guo et al., 1996), we use sensitivity matrices instead of loop transfer matrices to discuss the LTR.

This paper is organized as follows. The plant description is given in Section 2. The optimal output disturbance cancellation controller is constructed in Section 3. Section 4 discusses the modified LTR design. In Section 5, an illustrative numerical example is presented. Concluding remarks are given in Section 6.

*Corresponding author. Email: ishihara@sss.fukushima-u.ac.jp

## 2. Plant description

Consider a plant given by

$$\dot{x}_p(t) = A_p x_p(t) + B_p[u(t) + d(t)], \quad y(t) = C_p x_p(t), \quad (1)$$

where $x_p(t)$ is an $n_p$ dimensional state vector, $u(t)$ is an $m$ dimensional control input, $y(t)$ is an $m$ dimensional output and $d(t)$ is an $m$ dimensional disturbance vector described by

$$d(t) = C_d x_d(t), \quad \dot{x}_d(t) = A_d x_d(t), \quad (2)$$

where $x_d(t)$ is an $n_d$ dimensional state vector. The disturbance model (2) can be used to describe a fairly general class of persistent disturbances including steps, ramps and sinusoids with known frequency.

Define

$$G_p(s) \triangleq C_p(sI - A_p)^{-1}B_p. \quad (3)$$

For the system given by Equations (1) and (2), we assume the following conditions:

A1: $(A_p, B_p, C_p)$ is a minimal realization and $G_p(s)$ is non-singular for almost all $s$.
A2: $(A_p, B_p, C_p)$ is a minimum phase, that is, $G_p(s)$ has no zero in the closed right half plane.
A3: $(C_d, A_d)$ is an observable pair.
A4: All the eigenvalues of $A_d$ are in the closed left plane.

## 3. Optimal disturbance cancellation

### 3.1. *Optimal controller*

The optimal disturbance cancellation controller for the plant (1) is constructed based on the separation principle. First, we give the optimal disturbance cancellation control law under the assumption that the state and the disturbance are perfectly measurable.

PROPOSITION 1 *Assume that the state vectors $x_d(t)$ and $x_p(t)$ in Equations (1) and (2) are measurable. Consider the quadratic performance index:*

$$J_d \triangleq \int_0^{\infty} \{y'(t)Qy(t) + [u(t) + d(t)]'R[u(t) + d(t)]\}dt, \quad (4)$$

*where $Q$ and $R$ are positive definite matrices. Then the optimal controller minimizing (4) is given by*

$$u(t) = -Fx_p(t) - d(t), \quad (5)$$

*where $F$ is the optimal feedback gain matrix of the optimal regulator problem for the plant $(A_p, B_p, C_p)$ with the standard quadratic performance index:*

$$J \triangleq \int_0^{\infty} [y'(t)Qy(t) + u'(t)Ru(t)]dt, \quad (6)$$

*where $Q$ and $R$ are the same as the weighting matrices in Equations (4).*

*Proof* Define a new control input as

$$u_c(t) \triangleq u(t) + d(t). \quad (7)$$

Using the new control input, we can rewrite the plant dynamics (1) as

$$\dot{x}_p(t) = A_p x_p(t) + B_p u_c(t), \quad y(t) = C_p x_p(t). \quad (8)$$

In addition, we can write the performance index (4) as

$$J_d \triangleq \int_0^{\infty} [y'(t)Qy(t) + u'_c(t)Ru_c(t)]dt. \quad (9)$$

Note that $u_c(t)$ can take an arbitrary value by an appropriate choice of $u(t)$ since the disturbance $d(t)$ is perfectly measurable. Consequently, the optimal control problem for the plant (1) with the disturbance (2) can be reduced to the standard quadratic optimal problem under the assumption that $x_d(t)$ and $x_p(t)$ are perfectly measurable. The optimal control input $u_c(t)$ is given as

$$u_c(t) = -Fx_p(t), \quad (10)$$

where $F$ is the optimal feedback gain matrix of the plant $(A_p, B_p, C_p)$ minimizing the performance index (6). It follows from Equations (7) and (10) that the optimal control input $u(t)$ for the disturbance cancellation problem is given by (5). ■

By the separation principle (e.g. Anderson & Moore, 1990), the output feedback disturbance cancellation controller for Equations (1) and (2) is obtained from Proposition 1 as

$$u(t) = -F\hat{x}_p(t) - C_d\hat{x}_d(t), \quad (11)$$

where $\hat{x}_p(t)$ and $\hat{x}_d(t)$ are optimal estimates of $x_p(t)$ and $x_d(t)$, respectively. The estimates are obtained by constructing a Kalman filter for the extended stochastic model:

$$\dot{x}(t) = Ax(t) + Bu(t) + \bar{B}w(t), \quad y(t) = Cx(t) + v(t), \quad (12)$$

where

$$x(t) \triangleq \begin{bmatrix} x_d(t) \\ x_p(t) \end{bmatrix}, \quad A \triangleq \begin{bmatrix} A_d & 0 \\ B_pC_d & A_p \end{bmatrix}, \quad B \triangleq \begin{bmatrix} 0 \\ B_p \end{bmatrix}, \quad (13)$$

$v(t)$ and $w(t)$ are mutually independent white noise processes with covariance matrices $V$ and $W$, respectively, and $\bar{B}$ is chosen such that $(A, \bar{B})$ is controllable. The choice of the matrix $\bar{B}$ will be discussed in Section 4.

The structure of the output feedback controller is shown in Figure 1

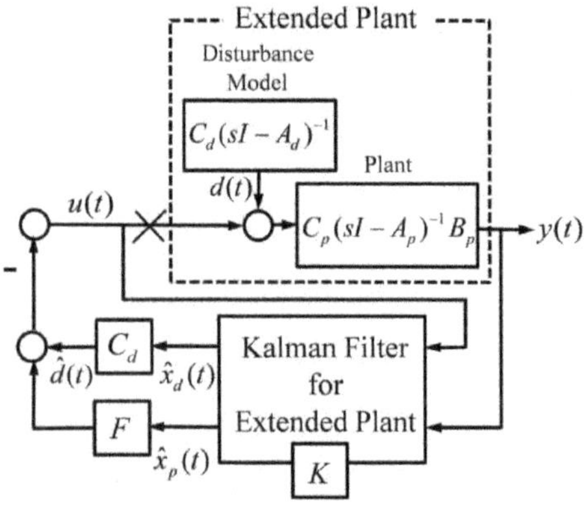


Figure 1. The optimal disturbance cancellation controller.

### *3.2. Sensitivity property*

First, we give the transfer function matrix of the controller (11) in left factorization form as follows.

LEMMA 1 *Consider the controller (11) with the Kalman filter for the stochastic model (12). Let $C_{uy}(s)$ denote the controller transfer function matrix of the output feedback controller from the output $y(t)$ to the control input $u(t)$. Define the partition of the Kalman filter gain matrix $K$ as*

$$K = \begin{bmatrix} K_d \\ K_p \end{bmatrix}. \tag{14}$$

*Then the controller transfer function matrix $C_{uy}(s)$ can be written as*

$$C_{uy}(s) = -M^{-1}(s)N(s), \tag{15}$$

*where*

$$M(s) \triangleq [I + F(sI - A_p + K_pC_p)^{-1}B_p][I + C_d(sI - A_d)^{-1}K_d \times C_p(sI - A_p + K_pC_p)^{-1}B_p]^{-1}, \tag{16}$$

$$\begin{aligned} N(s) \triangleq\ & F[I + (sI - A_p + K_pC_p)^{-1}B_pC_d(sI - A_d)^{-1}K_dC_p]^{-1} (sI - A_p + K_pC_p)^{-1} \\ & \times [K_p + B_pC_d(sI - A_d)^{-1}K_d] - C_d(sI - A_d)^{-1}K_d \\ & \times \{I - C_p[I + (sI - A_p + K_pC_p)^{-1}B_pC_d(sI - A_d)^{-1} \\ & \times K_dC_p]^{-1} (sI - A_p + K_pC_p)^{-1} \\ & \times [K_p + B_pC_d(sI - A_d)^{-1}K_d]\}. \end{aligned} \tag{17}$$

*Proof* The expression can easily be obtained by straightforward matrix calculation. ■

Using the above lemma, we can obtain the following result on the sensitivity property.

PROPOSITION 2 *Consider the output feedback controller (11) for the plant (1) and (2). Let $\Sigma(s)$ denote the sensitivity matrix at the plant input side. The sensitivity matrix can be factored as*

$$\Sigma(s) = S(s)M(s), \tag{18}$$

*where*

$$S(s) \triangleq [I + F(sI - A_p)^{-1}B_p]^{-1} \tag{19}$$

*is the sensitivity matrix of the standard LQ regulator and $M(s)$ is defined in Equation (16).*

*Proof* The sensitivity matrix at the plant input side is defined as

$$\Sigma(s) \triangleq [I - C_{uy}(s)G_p(s)]^{-1}, \tag{20}$$

where $G_p(s)$ is defined in Equation (3). It follows from Equations (15) and (20) that

$$\Sigma(s) = [M(s) + N(s)G_p(s)]^{-1}M(s). \tag{21}$$

Simple but somewhat tedious matrix calculation using Equations (16) and (17) yields

$$M(s) + N(s)G_p(s) = I + F(sI - A_p)^{-1}B_p. \tag{22}$$

The factorization (18) follows from Equations (21) and (22). ■

*Remark 1* Note that the eigenvalue of the matrix $A_d$ for the disturbance model (2) appears as a zero of $[I + C_d(sI - A_d)^{-1}K_dC_p(sI - A_p + K_pC_p)^{-1}B_p]^{-1}$ in the denominator matrix of Equation (16) provided $K_d \neq 0$. In the case that the matrix $A_d$ has an eigenvalue on the imaginary axis, the eigenvalue appears as a zero of the sensitivity matrix (18) under the assumptions A1–A4, which shows that the controller has the internal model for the disturbances. It is worth pointing out that the internal model is not assumed a priori but emerges as a result of minimization of the performance index (4).

## 4. Modified LTR design

The design of the optimal output feedback disturbance cancellation controller requires the determination of $F$, $K_d$ and $K_p$ such that design specifications are satisfied. Since the stabilizability of the extended model (12) consisting of the plant and the disturbance model is not guaranteed, the classical LTR method cannot directly be applied. To overcome the difficulty, the modified LTR method has been proposed in Guo et al. (1996) for the step disturbances. However, they have not considered the use of the optimal target. In this section, we discuss the modified LTR design using the optimal target for a more general class of disturbances. The target and the recovery procedure of the modified LTR design are discussed in the following subsections.

### 4.1. *Target controller design*

As a target of the design, we choose the optimal disturbance cancellation controller for the case where the plant state $x_p(t)$ is measurable but the disturbance state $x_d(t)$ is not. By the separation principle, the target controller is obtained from Equation (5) as

$$u(t) = -Fx_p(t) - C_d\bar{x}_d(t), \tag{23}$$

where $F$ is the optimal feedback gain matrix minimizing the performance index (6) and $\bar{x}_d(t)$ is the optimal estimate of $x_d(t)$ based on the observation of $x_p(t)$.

The optimal estimate $\bar{x}_d(t)$ is obtained from the relation:

$$\dot{x}_d(t) = A_d x_d(t), \tag{24}$$

$$\dot{x}_p(t) - A_p x_p(t) - B_p u(t) = B_p C_d x_d(t), \tag{25}$$

where the left side of Equation (25) is available from the measurement of $x_p(t)$. Note that the observation relation (25) includes redundant rows. Multiplying the both sides of Equation (25) by $(B_p'B_p)^{-1}B_p'$ from the left, we obtain

$$(B_p'B_p)^{-1}B_p'\dot{x}_p(t) - (B_p'B_p)^{-1}B_p'A_p x_p(t) - u(t) = C_d x_d(t). \tag{26}$$

From Equations (24) and (26), we can construct the stochastic model as

$$\begin{aligned}\dot{x}_d(t) &= A_d x_d(t) + B_d w_d(t),\\ (B'_pB_p)^{-1}B'_p\dot{x}_p(t) &- (B'_pB_p)^{-1}B'_pA_p x_p(t) - u(t) \\ &= C_d x_d(t) + v_d(t),\end{aligned} \tag{27}$$

where $w_d(t)$ and $v_d(t)$ are mutually independent zero-mean white noise processes with the covariance matrices $W_d$ and $V_d$, respectively, and $B_d$ is chosen such that $(A_d, B_d)$ is controllable. Using the Kalman filter theory, we can obtain the optimal estimate $\bar{x}_d(t)$ by the optimal disturbance estimator based on the observation of $x_p(t)$ as

$$\begin{aligned}\dot{\bar{x}}_d(t) = A_d\bar{x}_d(t) &+ \bar{K}_d[(B_p'B_p)^{-1}B_p'\dot{x}_p(t)\\ &- (B_p'B_p)^{-1}B_p'A_p x_p(t) - u(t) - C_d\bar{x}_d(t)],\end{aligned} \tag{28}$$

where $\bar{K}_d$ is the optimal estimator gain matrix given by

$$\bar{K}_d = \bar{P}_d C_d' V_d^{-1}, \tag{29}$$

with $\bar{P}_d$ satisfying the Riccati equation:

$$A_d\bar{P}_d + \bar{P}_d A_d' - \bar{P}_d C_d' V_d^{-1} C_d \bar{P}_d + B_d W_d B_d' = 0. \tag{30}$$

The structure of the target control system is shown in Figure 2.

*Remark 2* Note that the target controller is used only in the design process but is not used as a real-time controller. Although the optimal disturbance estimator (28) includes time derivative of $x_p(t)$, the differentiation is acceptable in the design process. In software such as Simulink, the time derivative of $x_p(t)$ is easily obtained from the input of the integrator generating $x_p(t)$ in the block diagram representation of the target control system.

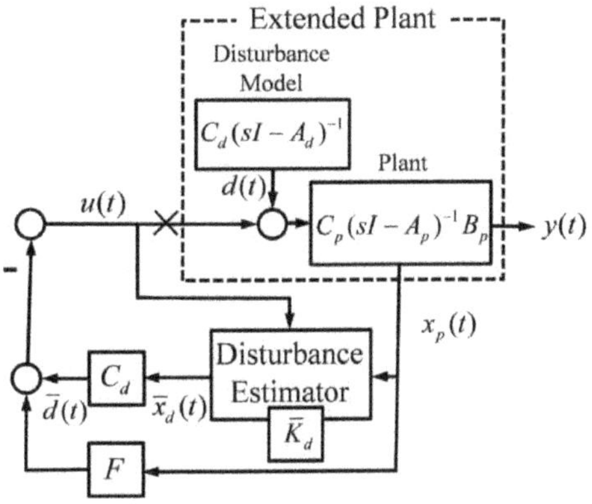


Figure 2. The target control system.

For the target controller (23), we have the following result.

LEMMA 2 *Consider the controller (23) with the optimal disturbance estimator (28). The controller transfer function matrix $\bar{C}_{uy}(s)$ from $x_p(t)$ to $u(t)$ can be written as*

$$\bar{C}_{uy}(s) = -\bar{M}^{-1}(s)\bar{N}(s), \tag{31}$$

*where*

$$\bar{M}(s) \triangleq I - C_d(sI - A_d + \bar{K}_d C_d)^{-1}\bar{K}_d, \tag{32}$$

$$\begin{aligned}\bar{N}(s) \triangleq F &+ C_d(sI - A_d + \bar{K}_d C_d)^{-1}\bar{K}_d(B_p'B_p)^{-1}\\ &\times B_p'(sI - A_p).\end{aligned} \tag{33}$$

*Proof* The expression can easily be obtained by straightforward matrix calculation. ■

Using the above lemma, we can obtain the following result for the target sensitivity property.

PROPOSITION 3 *Consider the target control system consisting of the plant (1) and the optimal controller (23) with the optimal disturbance state estimator (28). Let $\bar{\Sigma}(s)$ denote the sensitivity matrix at the plant input side. Then the sensitivity matrix is factored as*

$$\bar{\Sigma}(s) = S(s)\bar{M}(s), \tag{34}$$

*where $S(s)$ is the sensitivity matrix defined in Equation (19) and $\bar{M}(s)$, which is given in Equation (32), can be regarded as the sensitivity matrix for the optimal disturbance estimator (28).*

*Proof* The target sensitivity matrix at the plant input side is defined as

$$\bar{\Sigma}(s) \triangleq [I - \bar{C}_{uy}(s)(sI - A_{\rm p})^{-1}B_{\rm p}]^{-1}, \quad (35)$$

where $\bar{C}_{uy}(s)$ is defined in Equation (31). It follows from Equations (31) and (35) that

$$\bar{\Sigma}(s) = [\bar{M}(s) + \bar{N}(s)(sI - A_{\rm p})^{-1}B_{\rm p}]^{-1}\bar{M}(s). \quad (36)$$

It readily follows from Equations (32) and (33) that

$$\bar{M}(s) + \bar{N}(s)(sI - A_{\rm p})^{-1}B_{\rm p} = I + F(sI - A_{\rm p})^{-1}B_{\rm p}. \quad (37)$$

The factorization (34) follows from Equations (36) and (37). By the matrix inversion lemma, we can write $\bar{M}(s)$ as

$$\bar{M}(s) = [I + C_{\rm d}(sI - A_{\rm d})^{-1}\bar{K}_{\rm d}]^{-1}, \quad (38)$$

which shows that Equation (38) is the sensitivity matrix for the estimation error dynamics of the optimal disturbance estimator (28). ■

Using the above result, we can show that the target controller provides guaranteed robustness property as in the target for the standard LTR design.

PROPOSITION 4 *For the single-input-single-output (SISO) case ($m = 1$), the target sensitivity function (34) satisfies:*

$$|\bar{\Sigma}(j\omega)| \leq 1 \quad \text{for all } \omega, \quad (39)$$

*which guarantees that the target control system has the large gain margins (infinite gain margin and phase margin more than 60 degrees) for any choice of the optimal feedback gain matrix $F$ and the optimal estimator gain matrix $\bar{K}_{\rm d}$.*

*Proof* Note that the optimal disturbance estimator (28) is constructed using the Kalman filter theory. By the well-known Kalman inequality for optimal LQ regulators and Kalman filters, it is guaranteed that

$$|S(j\omega)| \leq 1 \text{ and } |\bar{M}(j\omega)| \leq 1 \text{ for all } \omega. \quad (40)$$

The inequality (39) follows from Equations (34) and (40). ■

### 4.2. *Recovery procedure*

The target feedback property is recovered in the output feedback controller by the following recovery procedure.

PROPOSITION 5 *Assume that the target controller is determined, that is, the optimal feedback gain matrix $F$ and the optimal disturbance estimator matrix $\bar{K}_{\rm d}$ are fixed. Consider the optimal disturbance cancellation controller using the Kalman filter gain matrix $K$ determined for the stochastic model (12) and (13) with the covariance matrices $V = I$ and $W = \sigma^2 I$, where $\sigma$ is a positive scalar, and*

$$\bar{B} \triangleq \begin{bmatrix} \bar{K}_{\rm d} \\ B_{\rm p} \end{bmatrix}. \quad (41)$$

*Then, as $\sigma$ tends to infinity, the sensitivity matrix $\Sigma(s)$ defined in Equation (18) for the output feedback controller approaches the target sensitivity matrix $\bar{\Sigma}(s)$ defined in Equation (34).*

*Proof* Let $K(\sigma)$denote the Kalman filter gain matrix for the stated stochastic model. Using the Popov-Belevitch-Hautus test (e.g. Anderson & Moore, 1990), we can show that the pair $(A, \bar{B})$ is stabilizable, $(C, A)$ is observable and that the invariant zero of the realization $(A, \bar{B}, C)$ are in the open left plane under the condition A1–A4. These results guarantee that the Kalman filter gain matrix $K(\sigma)$ satisfies the asymptotic property

$$\lim_{\sigma\to\infty} \sigma^{-1}K(\sigma) = \bar{B}, \quad (42)$$

which justifies that the Kalman filter gain matrix $K(\sigma)$ for sufficiently large $\sigma$ can be written as

$$K(\sigma) = \begin{bmatrix} K_{\rm d}(\sigma) \\ K_{\rm p}(\sigma) \end{bmatrix} = \sigma \begin{bmatrix} \bar{K}_{\rm d} \\ B_{\rm p} \end{bmatrix}. \quad (43)$$

Using the above expression, we can obtain the asymptotic expressions of the two transfer function matrices in $M(s)$ defined in Equation (16) as

$$\begin{aligned} &[sI - A_{\rm p} + K_{\rm p}(\sigma)C_{\rm p}]^{-1}B_{\rm p} \\ &\quad = (sI - A_{\rm p})^{-1}B_{\rm p}[I + \sigma C_{\rm p}(sI - A_{\rm p})^{-1}B_{\rm p}]^{-1} \\ &\quad \to 0\,(\sigma \to \infty), \end{aligned} \quad (44)$$

$$\begin{aligned} &K_{\rm d}(s)[sI - A_{\rm p} + K_{\rm p}(\sigma)C_{\rm p}]^{-1}B_{\rm p} \\ &\quad = \sigma\bar{K}_{\rm d}(sI - A_{\rm p})^{-1}B[I + \sigma C(sI - A)^{-1}B]^{-1} \\ &\quad \to \bar{K}_{\rm d}\,(\sigma \to \infty). \end{aligned} \quad (45)$$

From Equations (16), (18), (32) and (34), we have $M(s) \to \bar{M}(s)$ as $\sigma \to \infty$, which implies that $\Sigma(s) \to \bar{\Sigma}(s)$ as $\sigma \to \infty$. ■

*Remark 3* Note that the modified recovery procedure differs from the original (e.g. Saberi et al., 1993) in the point that it requires the optimal estimator gain matrix $\bar{K}_{\rm d}$ used in the target. Despite the difference, the proposed procedure retains the philosophy of the original LTR design.

## 5. Illustrative example

In this section, a numerical example is presented to illustrate the design procedure proposed in the preceding

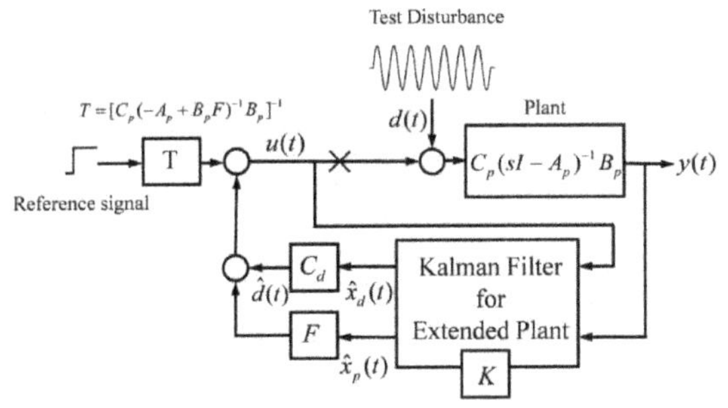


Figure 3. Control system set-up for the example.

section. Consider a SISO plant described by Equation (1) with

$$A_P = \begin{bmatrix} -2.0 & -1.0 & -0.5 \\ 2.0 & 0 & 0 \\ 0 & 1.0 & 0 \end{bmatrix},$$

$$B_P = \begin{bmatrix} 0.5 \\ 0 \\ 0 \end{bmatrix}, \quad C_P = \begin{bmatrix} 0 & 0 & 1 \end{bmatrix}, \tag{46}$$

which is a minimal realization of the transfer function given by

$$G(s) = \frac{1}{(s+1)(s^2+s+1)}. \tag{47}$$

It is assumed that the sinusoidal disturbance with the angular frequency $\omega = \pi$ (rad/sec) enters the plant input side. The disturbance model is given by Equation (2) with

$$A_d = \begin{bmatrix} 0 & -\pi^2 \\ 1 & 0 \end{bmatrix}, \quad C_d = \begin{bmatrix} 0 & 1 \end{bmatrix}. \tag{48}$$

The set-up of the control system for this example is shown in Figure 3 where the reference input is inserted with the pre-compensator:

$$T = [C_p(-A_p + B_pF)^{-1}B_p]^{-1}. \tag{49}$$

In the subsequent discussion on the time response, the unit step signal is applied as a reference at $t = 0$ and the test disturbance signal

$$d(t) = \begin{cases} 0 & (0 \le t < 10), \\ 10\sin \pi t & (10 \le t \le 24) \\ 0 & (t > 24) \end{cases}, \tag{50}$$

is injected to check the disturbance rejection capability.

Note that the control system shown in Figure 3 has the two-degree-of-freedom structure: the response to the reference input is independent of the choice of the Kalman filter gain matrix $K$ while the disturbance cancellation capability depends on $K$ as well as $F$.

### 5.1. *Target controller design*

First, we determine the target controller by appropriate choice of the optimal feedback gain matrix $F$ and the optimal gain matrix $\bar{K}_d$ for the disturbance estimator. The reference input with pre-compensator (49) is included as in the output feedback case shown in Figure 3.

Note that the target controller has the two-degree-of-freedom structure as in the output feedback case. It is reasonable to determine $F$ first setting $\bar{K}_d = 0$ and then determine $\bar{K}_d$. Let us assume that the optimal feedback gain matrix $F$ corresponding to the performance index (6) with $R = 1$ and $Q = 100$ provides satisfactory response to the step reference signal. Then $\bar{K}_d$ is determined considering the disturbance cancellation capability. To determine $\bar{K}_d$ by Equations (29) and (30), we set:

$$B_d = \begin{bmatrix} 1 \\ 0 \end{bmatrix}, \quad W_d = q, \quad V_d = 1, \tag{51}$$

where $q$ is a scalar tuning parameter.

For $q = 10$, $10^3$ and $10^4$, the magnitude characteristics of the target sensitivity function (34) are shown in Figure 4. Note that a sharp dip exists at the angular frequency of the disturbance, which reflects that the target controller has the internal model of the disturbance. The sensitivity is reduced at the low-frequency region as $q$ increases.

The time response of the target to the unit step reference and the test disturbance (50) is shown in Figure 5. It is seen that the disturbance cancellation capability is improved as $q$ is increased.

The stability margins of the target are summarized in Table 1, which confirms that the target has the infinite gain margin and the phase margin more than 60 degrees irrespective of $q$. For this illustrative example, we choose $\bar{K}_d$ corresponding to $q = 10^3$ as the disturbance estimator gain matrix for the target.

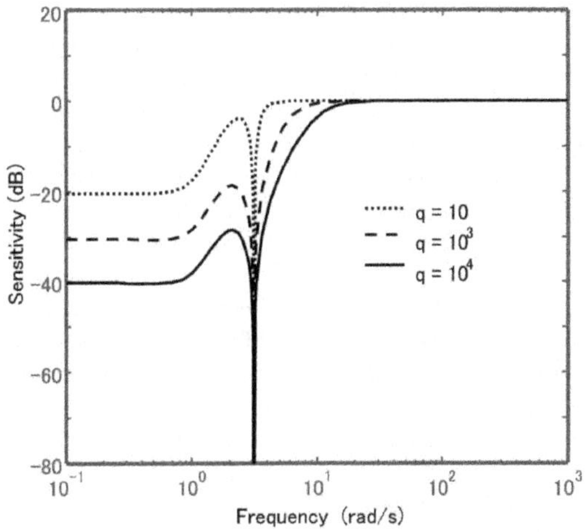


Figure 4. The target sensitivity function.

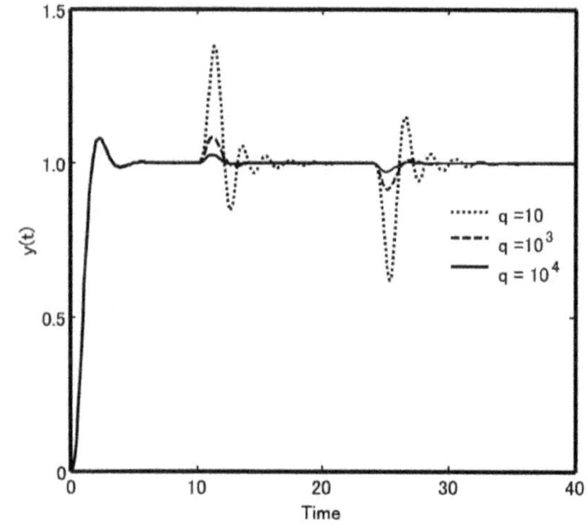


Figure 5. The time–response of the target.

Table 1. Target stability margins.

| $Q$ | Gain margin (dB) | Phase margin (degree) |
|---|---|---|
| 10 | $\infty$ | 63.9 |
| $10^3$ | $\infty$ | 63.8 |
| $10^4$ | $\infty$ | 84.0 |

Table 2. Stability margins of the output feedback case.

| $\rho$ | Gain margin (dB) | Phase margin (degree) |
|---|---|---|
| $10^3$ | 6.83 | 37.7 |
| $10^6$ | 10.1 | 51.8 |
| $10^{12}$ | 15.9 | 79.3 |

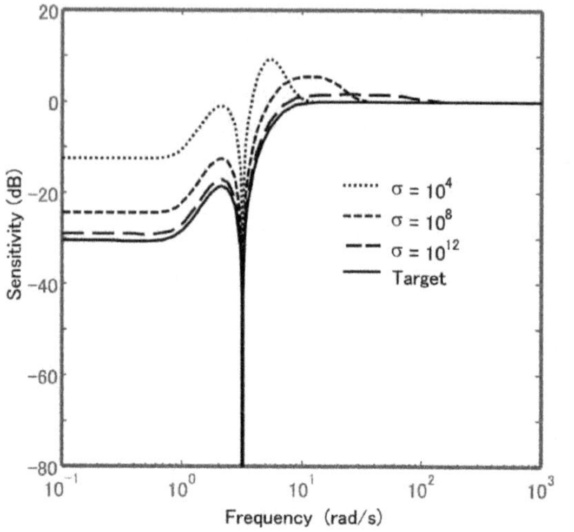


Figure 6. The sensitivity function for the output feedback controller.

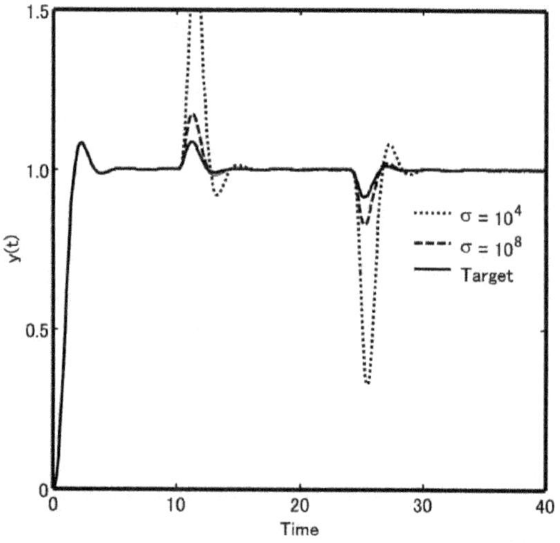


Figure 7. The time–response of the output feedback controller.

### 5.2. *Target recovery*

The second step determines the Kalman filter gain matrix $K$ by the formal procedure using the $\sigma$ introduced in Proposition 5. Consider the output feedback disturbance cancellation controller (11). The feedback gain matrix $F$ is the same as in the target and the estimate of the extended state $x(t)$ is obtained as the Kalman filter gain matrix for the stochastic model (12) where the matrix $\bar{K}_{\mathrm{d}}$ determined in the first step is included in $\bar{B}$ defined in Equation (41). For $\sigma = 10^4,\ 10^8$ and $10^{12}$, the magnitude characteristics of the sensitivity matrix (18) and the time response for the unit step reference and the test disturbance (50) are shown in Figures 6 and 7, respectively. It is confirmed numerically that the sensitivity characteristics and the time response approach those of the target as $\sigma$ is increased. The stability margins of the output feedback controller are summarized in Table 2. It is seen that the stability margins are improved as $\sigma$ is increased.

As shown above, the proposed design procedure provides flexible design with a small number of tuning parameters taking account of stability margins.

## 6. Conclusions

Our earlier work (Guo et al., 1996) has been extended to a more general class of disturbances with optimality consideration. The optimal disturbance cancellation controller based on the measurement of the plant state is chosen as the target with the guaranteed stability margins. It has been shown that the target feedback property can be recovered in the output feedback controller by the extended version of the modified LTR procedure proposed in our earlier work. A numerical example has been presented to show that the proposed optimal design provides efficient tuning of the disturbance cancellation capability with attention to the stability margins. The result of this paper can be extended to non-minimum phase plants using the partial LTR technique used in Ishihara, Guo, and Takeda (2005).

For the disturbance cancellation at the plant output side, a completely different approach is required. Some fundamental issues have been discussed in Ishihara and Guo (2011, 2013a) for step disturbances. Recently, it has been extended to sinusoidal and polynomial disturbances in the conference papers (Ishihara & Guo, 2013b, 2013c).

It is an interesting future problem to develop a new type of predictive controllers (Maciejowski, 2002; Mosca, 1995) based on the disturbance cancellation technique with additional practical constraints.

## Disclosure statement

No potential conflict of interest was reported by the authors.

## References

Anderson, B. D. O., & Moore, J. B. (1990). *Optimal control, linear quadratic methods*. Englewood Cliffs, NJ: Prentice Hall.

Guo, H.-J., Ishihara, T., & Takeda, H. (1996). Design of discrete-time servosystems using disturbance estimators via LTR technique. *Transactions of the Society of Instrument and Control Engineers*, *32*, 646–652.

Ishihara, T., & Guo, H.-J. (2008). LTR design of integral controllers for time-delay plants using disturbance cancellation. *International Journal of Control*, *81*, 1027–1034.

Ishihara, T., & Guo, H.-J. (2011, August 28–September 2). *Partial LTR design of optimal output disturbance cancellation controllers for non-minimum phase plants*. Proc. of 18th IFAC World Congress, Milano, Italy, pp. 7903–7908.

Ishihara, T., & Guo, H.-J. (2012, August 20–23). *Design of optimal disturbance cancellation controllers for a class of disturbances via loop transfer recovery*. Proc. of SICE Annual Conference 2012, Akita, Japan, pp. 1609–1612.

Ishihara, T., & Guo, H.-J. (2013a). Design of optimal output disturbance cancellation controllers via loop transfer recovery. *Systems Science and Control Engineering: An Open Access Journal*, *1*, 57–66.

Ishihara, T., & Guo, H.-J. (2013b, June 23–26). *Design of optimal output disturbance cancellation controllers for sinusoidal output disturbances via loop transfer recovery*. Proc. of Asian Control Conference 2013, Istanbul, Turkey, TuA5.2.

Ishihara, T., & Guo, H.-J. (2013c, September 14–17). *Design of optimal output disturbance cancellation controllers for polynomial output disturbances via loop transfer recovery*. Proc. of SICE Annual Conference 2013, Nagoya, Japan, pp. 1223–1230.

Ishihara, T., Guo, H.-J., & Takeda, H. (2005). Integral controller design based on disturbance cancellation: Partial LTR approach for non-minimum phase plants. *Automatica*, *41*, 2083–2089.

Maciejowski, J. M. (2002). *Predictive control with constraints*. London: Pearson Education.

Mosca, E. (1995). *Optimal, predictive, and adaptive control*. Englewood Cliffs, NJ: Prentice Hall.

Saberi, A., Chen, B. M., & Sannuti, P. (1993). *Loop transfer recovery: Analysis and design*. New York, NY: Springer-Verlag.

Stein, G., & Athans, M. (1987). The LQG/LTR procedure for multivariable feedback control design. *IEEE Transactions on Automatic Control*, *32*, 105–114.

# 9

# Static alignment of inertial navigation systems using an adaptive multiple fading factors Kalman filter

Behrouz Safarinejadian* and Mojtaba Yousefi

*Control Engineering Department, Shiraz University of Technology, Modarres Blvd., P.O. Box 71555-313, Shiraz, Iran*

Kalman filter is not an optimal estimation method for the systems without any exact model. Therefore, a multiple fading factors matrix has been used as a multiplier for the covariance matrices in these systems. In this paper, a novel method, named adaptive multiple fading factor Kalman filter, is proposed for the systems without initial alignment of strap down inertial navigation systems. By applying this algorithm to different channels of Kalman filter, different coefficients of fading factors are computed. The simulation results show a noticeable increment in alignment's precision and alignment's speed, and a noticeable decrement in sensitivity to unknown noises.

**Keywords:** strap down; initial alignment; inertial navigation systems; multiple fading factor filter

## 1. Introduction

Inertial navigation system (INS) is an autonomous system in small intervals of time which has high precision and good coverage. Although, a strap down inertial navigation system (SINS) is the main hardware of the INS, it does not have the problems of a traditional INS. Most importantly, the SINS has some other advantages such as reducing the total system's complexity, costs and power consumption.

Recently, these systems have grown rapidly in all fields of application. Nowadays, SINS is widely used for positioning and navigating in aircrafts, ships, vehicles, etc. (Dzhashitov et al., 2014; Silson, 2011). Furthermore, SINS is used as a powerful means in constant tracking of a position, direction and velocity of any moving object without referring to any external reference (Wu & Pan, 2013; Yin, Sun, & Wang, 2013).

Notwithstanding, SINS has some drawbacks. One of its main fallibilities is its initial alignment error that has a negative effect on the system velocity and position error. Initial alignment in SINS is defined as measuring coordinates of transformation matrix from body frame to navigation frame and aligning misaligned angles to zero. More precisely, SINS's initial alignment contains two stages, coarse alignment and the precise alignment, each of which has its own purposes (Jiong, Lei, Rong, & Jianyu, 2011).

The purpose of the coarse alignment is estimation of transferring matrix coordinates from body frame to navigation frame by taking the best advantages of the vector of the Earth's gravity $g$ and measuring the value of the Earth's rotation rate $\omega_{ie}$. In contrast, the purpose of precise alignment is computation of the small misaligned angles between the navigation frame and body frame accurately through processing those data obtained from various sensors. In this section, the constant drift of inertial sensors is omitted and the accuracy of alignment is enhanced.

However, precise alignment complicates system's mechanism design and decreases system's accuracy and reliability. Initial alignment in SINS should be done with a lot of care and attention at the shortest possible time. Moreover, there are some consequences which make both alignments really complicated and time consuming.

For initial alignment in SINSs, Kalman filter has a high applicability when the system's model is accurate and the system's noise is white Gaussian and uncorrelated. For decades, random estimation techniques, especially Kalman filtering (KF) and its more developed forms, extended Kalman filter (EKF) and unscented Kalman filter (UKF), have been widely used in INSs (Grewal, Henderson, & Miyasako, 1991; Ladetto, 2000; Wang & Chen, 2010). Besides, Kalman filter makes it possible to combine measurements of nearly most navigation system's sensors and reach more precise estimation of position and velocity.

With an unauthentic model of the system and unavailable exact statistical data of the system's noise, relying on the Kalman filter's estimations may lead to inaccurate results. Sometimes, it may lead to the diversion of Kalman filter's estimations from the true state. Therefore, to prevent these problems, a vast amount of research has been done and more efficient models of KF have been proposed such as adaptive KF algorithm (Wang & Chen, 2010),

*Corresponding author. Email: safarinejad@sutech.ac.ir

$H_\infty$ filtering algorithm (Yue & Yuan, 2001), fading factor algorithms (Gao, Miao, & Ni, 2010a), etc.

Descriptively, the fading algorithm of Kalman filter is such an adaptive algorithm which wipes previous data out by exponent $\alpha_k$, a method that limits the memory of the Kalman filter (Fagin, 1964; Sorenson & Sacks, 1971). In Ydstie and Co (1985), an adjustable algorithm for fading factor has been suggested, wherein fading factors are determined based on 'memory length'. A fast fading occurs when the system model is not accurate and slow fading is used for accurate models. Few years later, an optimal fading factors KF algorithm had been offered, in which an exponential weight changing approach was used to balance the model errors and unknown drifts (Gao, Yang, Cui, & Zhang, 2006; Özbek & Alive, 1998; Xia, Rao, Ying, & Shen, 1994; Xu, Qin, & Peng, 2004). Zhang, Jin, and Tian (2003) show the use of adaptive filtering techniques to develop the speed of the dynamic alignment of a micro-electro-mechanical system inertial measurement unit (MEMS IMU) with a real-time kinematic global positioning system (RTK GPS) for a nautical function. In Mohamed and Schiwarz (1999), real-time adaptive algorithms are used for GPS information processing. To prevent the covariance matrix of estimate error from being asymmetric, an enhanced algorithm has been suggested, which reaches use of the filter residual. Along with statistical progression of the filter residuals using a chi-square test, the fading factors have been calculated distinctly to increase the predicted variance modules of the state vector (Geng & Wang, 2008; Shi, Miao, & Ni, 2010). In Kim, Jee, Park, and Lee (2009), the stability of the adaptive fading EKF has been studied. Furthermore, an adaptive UKF with several fading factors based on gain adjustment has been presented and examined on the estimation system approach of a Pico-satellite by simulation in Söken and Hajiyev (2009). Also, a new strong tracking square root Cubature Kalman Filter with suboptimal multiple fading factors is used, which can adjust the structure parameters of the filter and improve the performance of target state tracking in Li, Zhu, and Zhang (2014). In Gao et al. (2010a), two types of fading factor Kalman filter algorithms are suggested by computing unbiased estimations of the innovative sequence.

In this paper, an adaptive multiple fading factors Kalman filter is suggested that exploits the two types of fading factors algorithms mentioned in Gao et al. (2010a). The proposed algorithm calculates a fading factor for each channel separately. Additionally, by applying an averaging method, another fading factor is calculated for all channels. Then, the computed fading factor for each channel will be compared with the mentioned fading factor obtained from applying the averaging method. Consequently, the bigger fading factor will be chosen and multiplied by an estimation covariance matrix. By using this strategy, the bigger fading factor values are obtained. This will result in a fast fading since the system model is not accurate.

This paper is organized as follows. Section 2 contains a brief introduction of the aligning error models in SINS. In Section 3, our adaptive multiple fading factors Kalman filter is proposed. Simulation results are presented in Section 4. And consequently, in Section 5, the total overview of the present paper and its conclusions will be manifested.

## 2. SINS error model for alignment

### 2.1. *Coordinate frames*

Studying coordinate frames related to the initial alignment of SINS can be described as follows (Qin, 2006):

- Inertial frame (I frame): its origin is in the centre of earth; Page equator $z$-axis, the axis of $x$ in Page Equator hidden. Direction is arbitrarily to be chosen, $y$-axis complementary to the right system.
- Navigation frame (n frame): it is a local geographic coordinate frame; $x$-axis towards the East, $y$-axis towards the north, $z$-axis perpendicular to the local level, the local positions.
- Body frame (b frame): origin is of body centre; $x$-axis is along the transverse axis, $y$-axis is the longitudinal axis, $z$-axis perpendicular to the longitudinal symmetry page.
- Earth fixed frame (e frame): its origin is in the centre. $z$-axis coincides with the axis of the earth's polar axis, while the other two are fixed between the two tropics. Detailed information is available in Gao, Miao, and Shen (2010c) and Qin (2006).

### 2.2. *SINS error model*

Assuming that the navigation system is fixed to the ground, to study the behaviour of an INS, a suitable error model must be chosen. Using a linear error model for this situation is quite a good approximation. In the SINS error model, the state vector consists of 12 state variables, for example, 3 velocity errors, 3 attitude angle errors, 3 accelerometer biases and 3 gyroscope drifts. Navigation mode equations are presented as follows:

$$\dot{X} = AX + Gw, \tag{1}$$

where

$$X = [\delta V_E \quad \delta V_N \quad \delta V_u \quad \phi_E \quad \phi_N \quad \phi_u \quad \nabla_x \quad \nabla_y \quad \nabla_z \quad \varepsilon_x \quad \varepsilon_y \quad \varepsilon_z]^T,$$

$$w = [w_{ax} \quad w_{ay} \quad w_{az} \quad w_{gx} \quad w_{gy} \quad w_{gz}]^T,$$

$$A = \begin{bmatrix} A_1 & A_2 & C_b^n & 0_{3\times3} \\ 0_{3\times3} & A_3 & 0_{3\times3} & -C_b^n \\ 0_{3\times3} & 0_{3\times3} & 0_{3\times3} & 0_{3\times3} \\ 0_{3\times3} & 0_{3\times3} & 0_{3\times3} & 0_{3\times3} \end{bmatrix},$$

$$A_1 = \begin{bmatrix} 0 & 2\omega_{ie}\sin L & -2\omega_{ie}\cos L \\ -2\omega_{ie}\sin L & 0 & 0 \\ 2\omega_{ie}\cos L & 0 & 0 \end{bmatrix},$$

$$A_2 = \begin{bmatrix} 0 & -g & 0 \\ g & 0 & 0 \\ 0 & 0 & 0 \end{bmatrix}, \quad G = \begin{bmatrix} C_b^n & 0_{3\times3} \\ 0_{3\times3} & C_b^n \\ 0_{3\times3} & 0_{3\times3} \end{bmatrix},$$

$$A_3 = \begin{bmatrix} 0 & \omega_{ie}\sin L & -\omega_{ie}\cos L \\ -\omega_{ie}\sin L & 0 & 0 \\ \omega_{ie}\cos L & 0 & 0 \end{bmatrix},$$

where $\omega_{ie}$ is the Earth rotation rate, $g$ is the local acceleration gravity and $L$ is the local latitude. $\nabla_x$, $\nabla_y$ and $\nabla_z$ are the biases of the three accelerometers, $\varepsilon_x, \varepsilon_y$ and $\varepsilon_z$are the drifts of the three gyros and $\delta v_{\mathrm{E}}, \delta v_{\mathrm{N}}$ and $\delta v_{\mathrm{u}}$ are the east, north and vertical velocity errors, respectively. $w$ is the noise vector of the system, $C_{\mathrm{b}}^n$ is the transformation matrix from the body frame to the navigation frame and $\phi_{\mathrm{E}}$, $\phi_{\mathrm{N}}$ and $\phi_{\mathrm{u}}$ are the east, north and azimuth misalignment angles, respectively. Note that the velocity error is considered as the output and the output equation will be represented as follows:

$$z = HX + v, \tag{2}$$

where $H = \begin{bmatrix} I_{3\times3} & 0_{3\times9} \end{bmatrix}$ and $v = \begin{bmatrix} v_{\mathrm{E}} & v_{\mathrm{N}} & v_{\mathrm{u}} \end{bmatrix}$ are the output noise vector.

## 3. Adaptive several fading factors Kalman filter

### *3.1. Development of single fading factor*

Discrete linear system state and output equations will be represented as follows:

$$x = \phi_{k-1}x_{k-1} + \Gamma_{k-1}w_{k-1}, \tag{3}$$

$$z_k = H_k x_k + v_k. \tag{4}$$

In this system, $x_k$ is the state vector at time instant $k$, $\phi_{k-1}$ is a state transition matrix, $\Gamma_{k-1}$ is a system perturbation matrix, $z_k$ is the output at time $k$, $H_k$ is the output matrix, $w_{k-1}$ is the process noise vector and finally, $v_k$ is the output noise vector.

When the system noise and the measurement noise are uncorrelated and white Gaussian, the Kalman filter equations will be presented as follows:

$$\hat{x}_k^- = \phi_{k-1}\hat{x}_{k-1}^+, \tag{5}$$

$$\hat{x}_k^+ = \hat{x}_k^- + k_k(z_k - H_k\hat{x}_k^-), \tag{6}$$

$$k_k = p_k^- H_k^T (H_k p_k^- H_k^T + R_k)^{-1}, \tag{7}$$

$$p_k^- = \phi_{k-1}p_{k-1}^+\phi_{k-1}^T + \Gamma_{k-1}Q_{k-1}\Gamma_{k-1}^T, \tag{8}$$

$$p_k^+ = (I - k_kH_k)p_K^-(I - k_kH_k)^T + k_kR_kk_k^T. \tag{9}$$

In these equations, $\hat{x}_k^-$ is the propagation state vector, $p_k^-$ represents the covariance matrix for $\hat{x}_k^-$, $k_k$ is the gain matrix, $\hat{x}_k^+$ is the estimated state vector, $p_k^+$ is the covariance matrix for $\hat{x}_k^+$, $Q_k$ is the system noise covariance matrix, $R_k$ represents the measurement noise covariance matrix and $I$ manifests the identity matrix.

When the system model is accurate, and noise characteristics are under Gaussian conditions, white and known accurately, Kalman filter is an optimal estimator. However, satisfying the aforementioned conditions is usually very difficult in the actual system. As mentioned earlier, Kalman filter performance highly depends on the previous measurements and also system model accuracy. If the measurements are inaccurate, Kalman filter estimations might be inaccurate or even incorrect. To eliminate these problems, the $\lambda_k$ factor is used to omit exponential fading of previous data, which is a technique to limit the memory of the Kalman filter (Fagin, 1964; Sorenson & Sacks, 1971). The equations of a single fading factor Kalman filter are similar to the standard Kalman filter. The only difference is the time propagation equation of the error covariance that is expressed as follows:

$$p_k^- = \lambda_k\phi_{k-1}p_{k-1}\phi_{k-1}^T + \Gamma_{k-1}Q_{k-1}\Gamma_{k-1}^T, \tag{10}$$

where $\lambda_k > 1$ is the fading factor.

When there is an unpredictable disturbance in the system, reaching the optimal performance using a fixed exponential fading factor is usually difficult. Therefore, a variable fading factor algorithm has been suggested. Fading factors are verified based on 'memory length' (Ydstie & Co, 1985). If the obtained data have an inappropriate fit with the system model, a fast fading will happen. Otherwise, a slow fading will happen if we have an appropriate fit. An optimal fading factor KF algorithm has been proposed in Xia et al. (1994). That can be expressed as follows:

ALGORITHM 1

$$A_k = \lambda_k H_K \phi_{K-1} p_{K-1}^+ \phi_{K-1}^T H_k^T,$$

$$M_k = H_k\phi_{k-1}p_{k-1}^+\phi_{k-1}^T H_k^T,$$

$$B_k = H_k\Gamma_{k-1}Q\Gamma_{k-1}^T H_k^T + R_k,$$

$$N_k = \hat{\delta}_{V_k} - B_k,$$

$$J_k = \phi_{k-1}p_{k-1}^+\phi_{k-1}^T,$$

where $\hat{\delta}_{v_k}$ is the estimation of the innovation sequence covariance and optimal fading factor is obtained as

$$\lambda_k = \max\{1, \mathrm{tr}(N_k M_k^{-1})/m\}.$$

In this case, $m$ is a dimension of the output vector.

In this algorithm, only a single fading factor is multiplied in the estimation error covariance matrix, which can change at any time step. This method may not give optimal performance of filters and in some situations, it leads to the divergence of the filter.

As it was discussed, in most cases, the single fading factor is not sufficient and has different rates for different channels. When matrix $p_k^-$ is not asymmetric, the time propagation equation of error covariance is offered as follows:

$$p_k^- = \Lambda_k \phi_{k-1} p_{k-1} \phi_{k-1}^T \Lambda_k^T + \Gamma_k Q_{k-1} \Gamma_{k-1}^T. \quad (11)$$

In this equation, $\Lambda_k$ is a diagonal fading factor matrix. Here, the second algorithm which in this paper is called Algorithm 2 will be described. Assuming that the output matrix satisfies the following equation:

$$H_k = [s_{m\times m} \quad 0_{m\times(n-m)}]_{m\times n}, \quad (12)$$

where $s_{m\times m} = \mathrm{diag}(s_1, s_2, \ldots, s_m)$, $m < n$.

In this case, Algorithm 2 can be described as

$$\begin{aligned} A_k &= H_k \Lambda_k \phi_{k-1} p_{k-1}^+ \phi_{k-1}^T \Lambda_k^T H_k^T, \\ B_k &= H_k \Gamma_{k-1} Q_{k-1} \Gamma_{k-1}^T H_k^T + R_k, \\ N_k &= \hat{\delta}_{v_k} - B_k, \\ J_k &= \phi_{k-1} p_{k-1} \phi_{k-1}^T. \end{aligned}$$

Now, the optimal fading factors can be extracted as $\Lambda_k = (\lambda_1, \lambda_2, \ldots, \lambda_m, 1, 1, \ldots, 1)$, and $\lambda_i$ can be calculated as follows:

$$\lambda_i = \begin{cases} \max\left\{1, \sqrt{\dfrac{\hat{\delta}_{v_k}^{ii} - b_{ii}}{s_i^2 j_{ii}}}\right\}, & \hat{\delta}_{v_k}^{ii} > b_{ii} \\ 1, & \hat{\delta}_{v_k}^{ii} < b_{ii} \end{cases}.$$

In this algorithm, only $\lambda_1, \lambda_2, \ldots, \lambda_m$ can be adaptively estimated, and other matrix components cannot be estimated which are determined by the dimension of the output.

### 3.2. *The proposed adaptive multiple fading factors Kalman filter*

In this section, an adaptive multiple fading factors algorithm is presented. As mentioned, Algorithm 1 cannot save the filter optimums completely, and sometimes ends in the divergence of the filter. Furthermore, Algorithm 2 cannot provide the optimal solution. It is because of its method of calculating fading factors $(\lambda_1, \lambda_2, \ldots, \lambda_m)$, which in some cases results in $\lambda_k = 1$. Here, an algorithm will be provided, which based on simulation results has more efficient results than the mentioned algorithms.

The innovative sequence is a part of measurements that includes new information about the state of the system. Indeed, it is a Gaussian white process with zero mean and covariance $H_k p_k^- H_k^T + R_k$. Considering Equation (11), the following result is obtained:

$$\begin{aligned} \delta_{v_k} &= H_k(\Lambda_k \phi_{k-1} p_{k-1}^+ \phi_{k-1}^T \Lambda_k^T + \Gamma_{k-1} Q_{k-1} \Gamma_{k-1}^T) H_k^T + R_k \\ &= H_k \Lambda_k \phi_{k-1} p_{k-1}^+ \phi_{k-1}^T \Lambda_k^T H_k^T + H_k \Gamma_{k-1} Q_{k-1} \Gamma_{k-1}^T H_k^T \\ &+ R_k = A_k + B_k, \end{aligned} \quad (13)$$

where

$$\begin{aligned} A_k &= H_k \Lambda_k \phi_{k-1} p_{k-1}^+ \phi_{k-1}^T \Lambda_k^T H_k^T, \\ B_k &= H_k \Gamma_{k-1} Q_{k-1} \Gamma_{k-1}^T H_k^T + R_k. \end{aligned}$$

The unbiased estimate of the innovation sequence covariance can be extracted as

$$\hat{\delta}_{v_k} = \frac{1}{k-1} \sum_{i=1}^{k} v_i v_i^T, \quad (14)$$

where $v_i$ is the innovation sequence, therefore

$$\hat{\delta}_{v_k} = A_k + B_k. \quad (15)$$

Generally, the output matrix equation in navigation and positioning applications can be considered as Equation (12). If the output matrix $H_K$ satisfies Equation (12), then we have

$$\begin{aligned} A_k &= H_k \Lambda_k \phi_{k-1} p_{k-1}^+ \phi_{k-1}^T \Lambda_k^T H_k^T = H_k \Lambda_k J_k \Lambda_k^T H_k^T \\ &= [s_{m\times m} \quad 0_{m\times(n-m)}] \begin{bmatrix} \Lambda_{m\times m}^1 & 0_{m\times(n-m)} \\ 0_{(n-m)\times m} & \Lambda_{(n-m)\times(n-m)}^2 \end{bmatrix} \\ &\times \begin{bmatrix} J_{m\times m}^{11} & J_{m\times(n-m)}^{12} \\ J_{(n-m)\times m}^{21} & J_{(n-m)\times(n-m)}^{22} \end{bmatrix} \\ &\times \begin{bmatrix} \Lambda_{m\times m}^1 & 0_{m\times(n-m)} \\ 0_{(n-m)\times m} & \Lambda_{(n-m)\times(n-m)}^2 \end{bmatrix} \begin{bmatrix} s_{m\times m} \\ 0_{(n-m)\times m} \end{bmatrix} \\ &= [\Lambda_{m\times m}^1 \quad 0_{m\times(n-m)}] \begin{bmatrix} s_{m\times m} & 0_{m\times(n-m)} \\ 0_{(n-m)\times m} & 0_{(n-m)\times(n-m)} \end{bmatrix} \\ &\times \begin{bmatrix} J_{m\times m}^{11} & J_{m\times(n-m)}^{12} \\ J_{(n-m)\times m}^{21} & J_{(n-m)\times(n-m)}^{22} \end{bmatrix} \\ &\times \begin{bmatrix} s_{m\times m} & 0_{m\times(n-m)} \\ 0_{(n-m)\times m} & 0_{(n-m)\times(n-m)} \end{bmatrix} \begin{bmatrix} \Lambda_{m\times m}^1 \\ 0_{(n-m)\times m} \end{bmatrix} \\ &= [\Lambda_{m\times m}^1 \quad 0_{m\times(n-m)}] \begin{bmatrix} s_{m\times m} J_{m\times m}^{11} s_{m\times m} & 0_{m\times(n-m)} \\ 0_{(n-m)\times m} & 0_{(n-m)\times(n-m)} \end{bmatrix} \\ &\times \begin{bmatrix} \Lambda_{m\times m}^1 \\ 0_{(n-m)\times m} \end{bmatrix} = \Lambda_{m\times m}^1 s_{m\times m} J_{m\times m}^{11} s_{m\times m} \Lambda_{m\times m}^1. \end{aligned}$$

Therefore, the aforementioned equation can be rewritten as

$$a_{ii} = \lambda_i^2 s_i^2 j_{ii} \quad (i = 1, 2, \ldots, m). \quad (16)$$

where $a_{ii}$ and $j_{ii}$ are the $i$th diagonal elements of matrix $A_K$ and $J_K$, respectively.

By considering the obtained results of the aforementioned equation and according to Equation (15), the following conclusion is achieved:

$$\hat{\delta}_{v_k}^{ii} = a_{ii} + b_{jj}. \tag{17}$$

In this conclusion, $\hat{\delta}_{v_k}^{ii}$ is the $i$th diagonal element of matrix $\hat{\delta}_{v_k}$ and $b_{ii}$ is the $i$th diagonal element of matrix $B_K$.

Based on the earlier discussions, the fading factors are calculated using the following equation:

$$\lambda_i^2 = \frac{\hat{\delta}_{v_k}^{ii} - b_{ii}}{s_i^2 j_{ii}} \quad (i = 1, 2, \ldots, m). \tag{18}$$

Here, we will propose a new formula to calculate $\lambda_k$. The fading factors may satisfy the following condition:

$$\lambda_k = d_k, \tag{19}$$

$$M_k = H_k \phi_{k-1} p_{k-1}^{+} \phi_{k-1}^{T} H_k^{T},$$

$$N_k = \hat{\delta}_{v_k} - B_k,$$

$$d_k = \max\{1, \operatorname{abs}(\operatorname{tr}(N_k M_k^{-1}))\}. \tag{20}$$

Here, $d_k > 1$ is also considered as a fading factor.

Considering the sign of square root $\lambda_i$, the optimal fading factors matrix can be expressed as

$$\lambda_i = \begin{cases} \max\left\{d_k, \sqrt{\dfrac{\hat{\delta}_{v_k}^{ii} - b_{ii}}{s_i^2 j_{ii}}}\right\}, & \hat{\delta}_{v_k}^{ii} > b_{ii}, \\ d_k, & \hat{\delta}_{v_k}^{ii} < b_{ii}. \end{cases} \tag{21}$$

Now, the proposed algorithm can be described based on the aforementioned equations as follows. If the output matrix satisfies Equation (12), we have

$$A_k = H_k \Lambda_k \phi_{k-1} p_{k-1}^{+} \phi_{k-1}^{T} \Lambda_k^{T} H_k^{T},$$

$$B_k = H_k \Gamma_{k-1} Q_{k-1} \Gamma_{k-1}^{T} H_k^{T} + R_k,$$

$$M_k = H_k \phi_{k-1} p_{k-1}^{+} \phi_{k-1}^{T} H_k^{T},$$

$$N_k = \hat{\delta}_{v_k} - B_k,$$

$$J_k = \phi_{k-1} p_{k-1} \phi_{k-1}^{T}.$$

In the next phase, the fading factors matrix can be expressed as follows:

$$\Lambda_k = \operatorname{diag}\{\lambda_1, \lambda_2, \lambda_3, d_k, d_k, \ldots, d_k\},$$

where $\lambda_i$, $d_i$ can be calculated from Equations (20) and (21).

The calculation of the fading factor for each channel is independent of other channels. In other words, the fading factors matrix is made up of some independent fading factors. As known, the fading factors matrix is used to set the covariance matrix $p_k^-$ in order to adjust the gain matrix $k_k$. In practice, $\hat{\delta}_{v_k}$ is estimated as follows:

$$\hat{\delta}_{v_k} = \frac{1}{N} \sum_{J=0}^{N} v_{k-j} v_{k-j}^{T}, \tag{22}$$

where $N$ is the window width.

Determination of the window width properly is difficult. With a very low $N$, some of the information may be lost and the unbiased estimate of the innovation sequence covariance cannot be calculated. On the other hand, if $N$ is chosen to be very large, the volume of information will be too large, and the short-range characteristics of the innovation sequence covariance are hard to reflect.

When the velocity error is considered as the output, the output matrix of the SINS alignment can be modelled as follows:

$$H_K = [I_{3\times 3} \quad 0_{3\times 9}].$$

And the optimal fading factors matrix can be as follows:

$$\Lambda_k = \operatorname{diag}\{\lambda_1, \lambda_2, \lambda_3, d_k, d_k, \ldots, d_k\},$$

$$\lambda_i = \begin{cases} \max\left\{d_k, \dfrac{\sqrt{\hat{\delta}_{v_k}^{ii} - b_{ii}}}{s_i^2 j_{ii}}\right\}, & \hat{\delta}_{v_k}^{ii} > b_{ii} \\ d_k, & \hat{\delta}_{v_k}^{ii} < b_{ii} \end{cases} \quad (i = 1, 2, 3).$$

## 4. Simulation results

In this section, the proposed method has been evaluated based on simulation. Both the constant and drifts of each gyro are selected as 0.02(°)/h; and both the constant and random biases of each accelerometer are selected as $1 \times 10^{-4}$ g; the true attitude angles of the system are 0°, 0°, 0° and the local latitude of SINS place is 39.96°. After the coarse alignment, the horizontal accuracy is 0.1°, and the azimuth accuracy is 0.3°, both of which meet the applicability demand of linear model. In the presence of external interference, to demonstrate the capabilities of the multipurpose fading factors Kalman filter algorithm, we assume the range to be 100–150 and 200–250 s and system noise to be $Q' = 10Q$ and $Q' = 12Q$, respectively. Here, standard Kalman filter, Algorithm 1, Algorithm 2 and proposed algorithm are used in SINS initial alignment individually.

The initial state vector is assumed to be

$$X(0) = [0 \quad 0 \quad 0 \quad 0 \quad 0 \quad 0 \quad 0 \quad 0 \quad 0 \quad 0 \quad 0 \quad 0]^{\mathrm{T}}.$$

The initial state covariance matrix is set as

$$p_0 = \operatorname{diag}\{(0.1\,\mathrm{m/s})^2 \quad (0.1\,\mathrm{m/s})^2 \quad (0.1\,\mathrm{m/s})^2 \quad (1^\circ)^2 (1^\circ)^2 \times (1^\circ)^2 \quad (1 \times 10^{-4}\,\mathrm{g})^2 (1 \times 10^{-4}\,\mathrm{g}), (1 \times 10^{-4}\,\mathrm{g})^2 \times (0.02(^\circ)/\mathrm{h})^2 \quad (0.02(^\circ)/\mathrm{h})^2 \quad (0.02(^\circ)/\mathrm{h})^2\}.$$

System noise is set as

$$Q = \text{diag}\{(1 \times 10^{-4}\,\text{g})^2, (1 \times 10^{-4}\,\text{g})^2, (1 \times 10^{-4}\,\text{g})^2, \times (0.02(^\circ)/\text{h})^2, (0.02(^\circ)/\text{h})^2, (0.02(^\circ)/\text{h})^2\}.$$

The measurement noise is also set as $R = \text{diag}\{(0.1\,\text{m/s})^2, (0.1\,\text{m/s})^2, (0.1\,\text{m/s})^2\}$. When using Algorithms 1 and 2 and also the proposed algorithm, window widths are set as $N = 100$.

If the estimation of a state by each of the presented algorithms is convergent, the state will be observable, but if the estimation of a state is not convergent, the state will not be observable. Moreover, if the rank of the observability matrix is equal to the order of the system, the system will be completely observable; on the contrary, if the rank of the observability matrix is less than the order of the system, the difference between the order of the system and the rank of the observability matrix will be the number of the unobservable states. The observability matrix can be written as

$$O = \begin{bmatrix} \varphi & \varphi H & \varphi H^2 & \ldots & \varphi H^{11} \end{bmatrix}^{\mathrm{T}}.$$

Since $\text{Rank}(O) = 9$, there are only nine observable states, while the other three states are unobservable. Based on the observability analysis of the SINS error model on a stationary base (Gao, Miao, & Ni, 2010b; Gao et al., 2010c; Qin, 2006), observable misalignment angles are essential. The unobservable states combination could be $\nabla_N$, $\nabla_E$ and $\varepsilon_E$ or $\nabla_E, \varepsilon_E$ and $\varepsilon_U$. However, the most acceptable choice of three unobservable states is $\nabla_E$, $\nabla_N$ and $\varepsilon_E$ for the initial

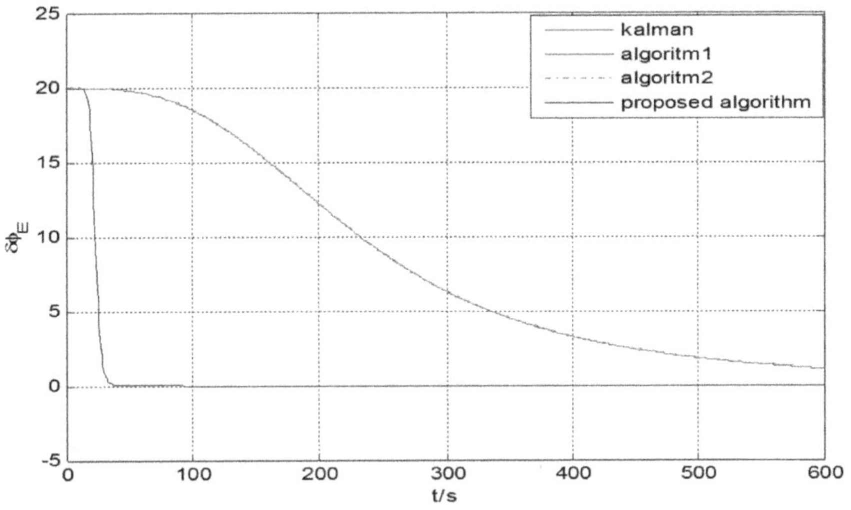


Figure 1. Estimation error of $\phi_{\text{E}}$.

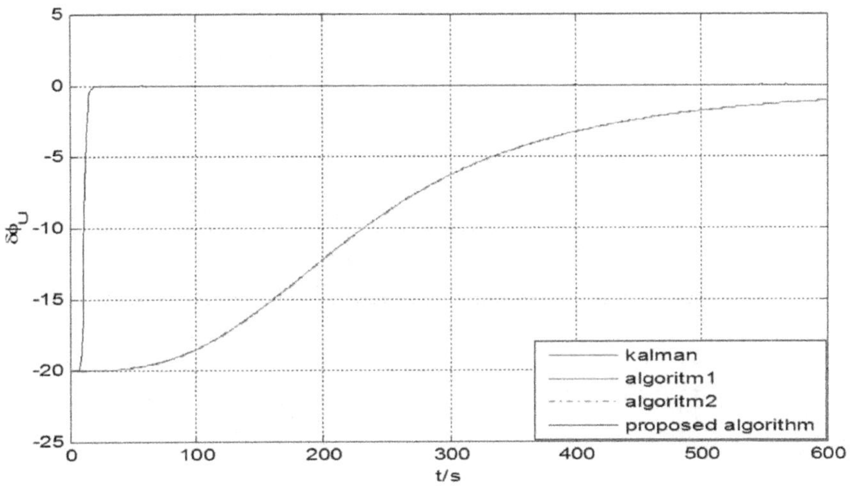


Figure 2. Estimation error of $\phi_{\text{u}}$.

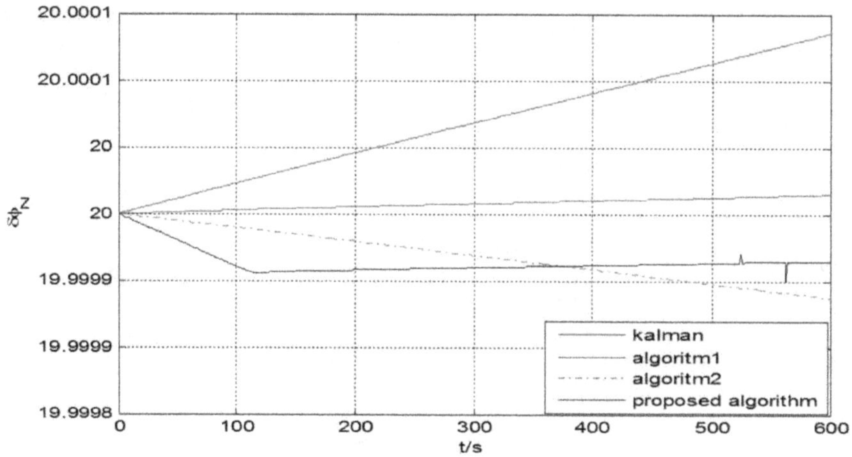


Figure 3. Estimation error of $\phi_{\mathrm{N}}$.

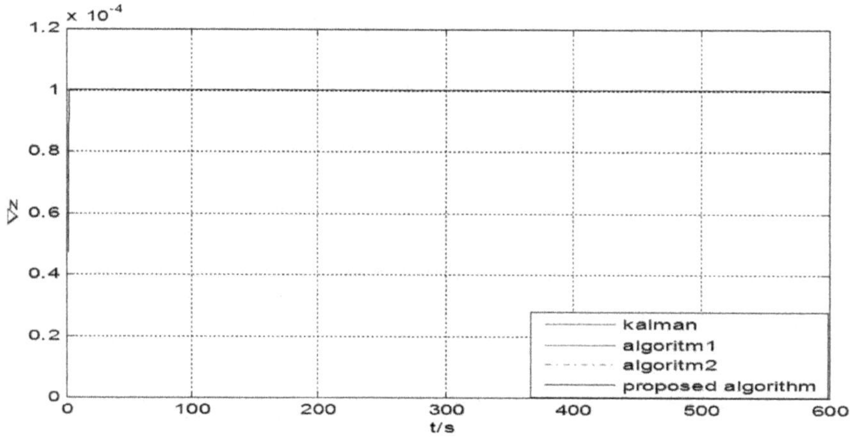


Figure 4. Estimation of $\nabla_z$.

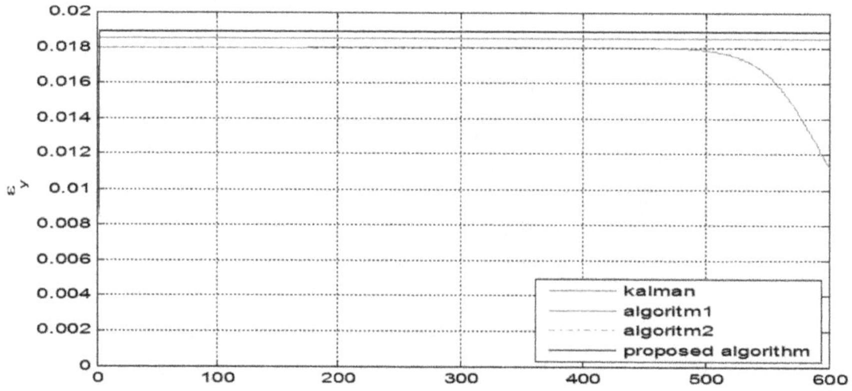


Figure 5. Estimation of $\varepsilon_y$.

alignment process (Ali & Ushaq, 2009). Observability of the system is solely characterized by the system model and not by the different filtering algorithms. In other words, the filtering algorithm does not have any influence on observability of the system. Curves of the states $\nabla_Z, \varepsilon_y, \varphi_E, \varphi_N$ and $\varphi_U$ are shown which are observable according to the results of the observability analysis. The estimation of $\nabla_U, \varepsilon_N$ and the estimation error of $\delta\varphi_E, \delta\varphi_N$ and $\delta\varphi_U$ are shown in Figures 1, 2 and 3, respectively. As it is shown, the speed convergence of estimation error of the proposed algorithm is faster than any other algorithm and also, $\nabla_U$ and $\varepsilon_N$ are approximations to $\nabla_Z$ and $\varepsilon_y$ are shown in Figures 4 and 5, respectively. From Figures 1 to 5, the following results can be obtained:

- Algorithm 1 is not superior to standard Kalman filter and in the presence of external noise, the standard Kalman filter and Algorithm 1 are not good but Algorithm 2 and the proposed algorithm have good performances.
- As can be seen in the curves, the proposed algorithm has a quicker response than any other algorithm. Besides, it is more stable.
- When the proposed algorithm is applied for initial alignment of SINS, the final estimated accuracy of misalignment angles is much more accurate than other algorithms which prove its authenticity. However, using the standard KF algorithm and Algorithm 1, the final estimation accuracy is determined by the constant drifts of the inertial sensors. The proposed algorithm can still have a perfect estimation effect. The azimuth alignment accuracy can be improved up to 50% intensity.

## 5. Conclusions

In this paper, a novel algorithm has been proposed for initial alignment in SINS. This algorithm can compute several fading factors based on the innovation sequence. According to the theoretical analysis, the proposed method not only has good convergence properties but also provides accurate estimations. In other words, this algorithm is more efficient than the single fading factor KF algorithm. Furthermore, computer simulation results have manifested that the new algorithm has strong robustness and adaptability.

### Disclosure statement

No potential conflict of interest was reported by the authors.

### References


Ali, J., & Ushaq, M. (2009). A consistent and robust Kalman filter design for in-motion alignment of inertial navigation system. *Measurement*, *42*(4), 577–582.

Dzhashitov, V. E., Pankratov, V. M., Golikov, A. V., Nikolaev, S. G., Kolevatov, A. P., Plonikov, A. D., Koffer K. V. (2014). Hierarchical thermal models of FOG-based strapdown inertial navigation system. *Gyroscopy and Navigation*, *5*, 162–173.

Fagin, S. L. (1964). Recursive linear regression theory, optimal filter theory and error analysis of optimal systems. *IEEE International Convention Record*, *12*, 216–240.

Gao, W. X., Miao, L. J., & Ni, M. L. (2010a). Multiple fading factors Kalman filter for SINS static alignment application. *Journal of Aeronautics*, *24*(1), 476–483.

Gao, W. X., Miao, L. J., & Ni, M. L. (2010b). A fast initial alignment with gyros angular rate information. [In Chinese]. *Journal of Astronautics*, *31*(6), 1596–1601.

Gao, W. X., Miao, L. J., & Shen, J. (2010c). A method of resistance to disturbance for SINS initial alignment. [In Chinese]. *Transactions of Beijing Institute of Technology*, *30*(1), 190–194.

Gao, W. G., Yang, Y. X., Cui, X. Q., Zhang, S. (2006). Application of adaptive Kalman filtering algorithm in IMU/GPS integrated navigation system. [In Chinese]. *Geometrics and Information Science of Wuhan University*, *31*(5), 466–469.

Geng, Y. R., & Wang, J. L. (2008). Adaptive estimation of multiple fading factors in Kalman filter for navigation applications. *GPS Solutions*, *12*(4), 273–279.

Grewal, M. S., Henderson, V. D., & Miyasako, R. S. (1991). Application of Kalman filtering to the calibration and alignment of inertial navigation systems. *IEEE Transactions on Automatic Control*, *36*, 3–13.

Jiong, Y., Lei, Z., Rong, S., & Jianyu, W. (2011). Initial alignment for SINS based on low-cost IMU. *Journal of Computers*, *6*(6), 1080–1085.

Kim, K. H., Jee, G. I., Park, C. G., Lee J.-G. (2009). The stability analysis of the adaptive fading extended Kalman filter using the innovation covariance. *International Journal of Control, Automation, and Systems*, *7*(1), 49–56.

Ladetto, Q. (2000, September). *On foot navigation: Continuous step calibration using both complementary recursive prediction and adaptive Kalman filtering*. Proceedings of the 13th international technical meeting of the satellite division of the institute of navigation (ION GPS 2000), Salt Lake City, UT (pp. 1735–1740).

Li, N., Zhu, R., & Zhang, Y. (2014, July 4–6). *A strong tracking square root CKF algorithm based on multiple fading factors for target tracking*. 2014 Seventh international joint conference on Computational Sciences and Optimization (CSO), Beijing (pp. 16–20).

Mohamed, A. H., & Schiwarz, K. P. (1999). Adaptive Kalman filtering for INS/GPS. *Journal of Geodesy*, *73*(4), 193–203.

Özbek, L., & Alive, F. A. (1998). Comment on "adaptive fading Kalman filter with an application". *Automatica*, *34*(12), 1663–1664.

Qin, Y. Y. (2006). *Inertial navigation. [In Chinese]*. Beijing: Science Press.

Shi, J., Miao, L. J., & Ni, M. L. (2010). An outlier rejecting and adaptive filter algorithm applied in MEMS-SINS/GPS. [In Chinese]. *Journal of Astronautics*, *31*(12), 2711–2716.

Silson, P. M. G. (2011). Coarse alignment of a ship's strapdown inertial attitude reference system using velocity loci. *IEEE Transactions on Instrumentation and Measurement*, *60*, 1930–1941.

Söken, H. E., & Hajiyev, C. (2009, June 11–13). *Adaptive unscented Kalman filter with multiple fading factors for pico satellite attitude estimation*. Proceedings of the 4th international conference on recent advances in space technologies, Istanbul (pp. 541–546).

Sorenson, H. W., & Sacks, J. E. (1971). Recursive fading memory filtering. *Information Sciences*, *3*(2), 101–119.

Wang, J.-H., & Chen, J.-B. (2010, July 11–14). *Adaptive unscented Kalman filter for initial alignment of strapdown inertial navigation systems*. International conference on machine learning and cybernetics (ICMLC), Qingdao (1384–1389).

Wu, Y., & Pan, X. (2013). Velocity/position integration formula part II: application to strapdown inertial navigation computation. *IEEE Transactions on Aerospace and Electronic Systems*, *49*, 1024–1034.

Xia, Q., Rao, M., Ying, Y., Shen X. (1994). Adaptive fading Kalman filter with an application. *Automatica*, *30*(8), 1333–1338.

Xu, J. S., Qin, Y. Y., & Peng, R. (2004). New method for selecting adaptive Kalman filter fading factor. [In Chinese]. *Systems Engineering and Electronics*, *26*(11), 1552–1554.

Ydstie, B. E., & Co, T. (1985). Recursive estimation with adaptive divergence control. *IEE Proceedings D Control Theory and Applications*, *132*(3), 124–130.

Yin, X., Sun, Y., & Wang, C. (2013). Positioning errors predicting method of strapdown inertial navigation systems based on PSO-SVM. *Abstract and Applied Analysis*, *2013*, 1–7.

Yue, X.-K., & Yuan, J.-P. (2001). $H_\infty$ filtering algorithm and its application in GPS/SINS integrated navigation system[J]. *Acta Aeronautica Et Astronautica Sinica*, *4*, 019.

Zhang, J., Jin, Z. H., & Tian, W. F. (2003). A suboptimal Kalman filter with fading factors for DGPS/MEMS-IMU/magnetic compass integrated navigation. *IEEE Intelligent Transportation Systems Proceedings*, *2*, 1229–1234.

# 10

## Scale-corrected minimal skew simplex sampling UKF for BLDCM sensorless control

Zhugang Ding, Guoliang Wei*, Xueming Ding and Haidong Lv

*Shanghai Key Lab of Modern Optical System, Department of Control Science and Engineering, University of Shanghai for Science and Technology, Shanghai 200093, People's Republic of China*

In this paper, a scale-corrected minimal skew simplex sampling unscented Kalman filter (UKF) algorithm for the permanent magnet (PM) brushless DC motors (BLDCM) sensorless control has been studied to cancel the position sensor by the use of a systematical and analytical approach. Compared with the general UKF, the sampling method with the least Sigma points called minimal skew simplex sampling is adopted to reduce amount of computation and increase the estimation precision. Moreover, the scale-corrected strategy is introduced into the minimal skew simplex sampling UKF to overcome the nonlocal effects. On the other hand, for more easily calculating the value of back-EMF, the shape function of counter electromotive force is approximated by a series of sine and cosine functions based on the law of Fourier series. The purpose of the problem addressed is, by the method of scale-corrected minimal skew simplex sampling UKF, to properly estimate the rotor speed and position without installing encoders. Finally, the effectiveness of the proposed sensorless control technique is verified by simulation in MATLAB/Simulink.

**Keywords:** unscented Kalman filter; sensorless control; scale-corrected minimal skew simplexsampling; BLDCM

### 1. Introduction

With the dramatic improvement of power electronics, inverter control technologies and permanent magnet (PM) materials, brushless DC motor (BLDCM) has undergone great developments in communication equipments, industrial automation systems, space industries and medical devices, which is also ascribed to its advantages, such as high efficiency, low power consumption, low maintenance and compact structure. In the actual BLDCM control system, it requires that the currents in the windings of BLDCM must be synchronized to the instantaneous position of the rotor, therefore, resolvers or encoders may be used to measure rotor position. However, the applications of position sensors inevitably increase the cost of system and decrease the reliability. In addition, position sensors are quite sensitive to the environments, and they even cannot work in the condition of high temperature and high humidity.

Hence, it would be a great deal of sense to obtain the rotor position signal indirectly with the easily available information of stator terminal voltage, stator current, counter electromotive force (EMF) and motor parameters instead of installing resolvers or encoders. Over the past decades, the elimination of rotor position or velocity sensors has been the focus of intensive research, which mainly includes the back-EMF method, freewheel diode method, inductance testing method, rotor flux method, state observer method and some other special methods (Acarnley & Watson, 2006). Among them, the back-EMF method (Shao, Nolan, Teissier, & Swanson, 2003) is the most technically matured, the simplest and most popular for detecting the rotor position in the trapezoidal brushless DC motor. However, in practice, there still exist some obstacles and challenges to detect the zero point information; therefore, some other indirect methods have been put forward, such as the terminal voltage method (Kim, Lee, & Kwon, 2006; Lai & Lin, 2011), third harmonic method (Moreira, 1996) and back-EMF integration method (Jahns, Becerra, & Ehsani, 1991). The freewheeling diode method was first proposed in 1991 by Ogasawara and Akagi (1991), which detected current flowing through a freewheeling diode in the silent phase to determine the rotor position. The freewheeling diode method is essentially based on the principle of the back-EMF zero-crossing, so it also has a position error of commutation points in the transient state. In Kulkarni and Ehsani (1992), Rodriguez and Emadi (2007) and Jang, Sul, Ha, Ide, and Sawamura (2003), the inductance-based method has been proposed for the motor with severe saliency such as switched reluctance motors and interior PM motors, which utilized the relationship between the motor winding inductance and the rotor position to calculate rotor position. The rotor flux method has been investigated in Tatematsu, Hamada, Uchida, Wakao, and Onuki (2000) and Kim and

*Corresponding author. Email: guoliang.wei1973@gmail.com

Ehsani (2004), and by virtue of measured voltages and currents, the flux linkage is estimated. This method, however, requires a large amount of computation and is sensitive to the variation of motor parameters. State observer methods (Zhu, Kaddouri, Dessaint, & Akhrif, 2001) contain the sliding-mode observer, extended Kalman filter (EKF) method (Bolognani, Oboe, & Zigliotto, 1999; Terzic & Jadric, 2001) and model reference adaptive system observer. Most of the state observer methods take advantage of the known information, such as the stator current and terminal voltage, to estimate the rotor position in real time.

In the past few decades, the filtering problems have been extensively investigated (Dong, Wang, & Gao, 2013; Hu, Wang, Shen, & Gao, 2013; Liang, Sun, & Liu, 2014; Shen, Wang, Ding, & Shu, 2013). Accordingly, the filter theory has been successfully applied in many branches of practical domains, such as computer vision, communications, navigation and tracking systems. It is well known that the traditional Kalman filter (KF) serves as an optimal filter in the least mean square sense for linear systems with the assumption that the system model is exactly known. In the case that the system model is nonlinear and/or uncertain, there has been an increasing research effort to improve KF with hope to enhance their capabilities of handling nonlinearities. The EKF has been shown to be an effective way for tackling the nonlinear system estimation problems. In fact, EKF has recently gained particular research attention with promising application potentials, see, e.g. Hu, Wang, Gao, and Stergioulas (2012), Kallapur, Petersen, and Anavatti (2009) and Kluge, Reif, and Brokate (2010). Unfortunately, the EKF also have three well-known drawbacks: (1) linearisation can produce highly unstable filter performance, if the timestep intervals are not sufficiently small. (2) The derivations of the Jacobian matrices are nontrivial in most applications and often lead to significant implementation difficulties. (3) Sufficiently small timestep intervals usually imply high computational overhead as the number of calculations demanded for the generation of the Jacobian and the predictions of state estimate and covariance are large. Therefore, in Julier and Uhlmann (2004), put forward a new filter, unscented Kalman filter (UKF), which approximated the Gaussian distribution associated with each state variable rather than approximated nonlinear function transformation. The core idea of UKF lies in unscented transformation (UT) which propagates the mean and covariance through a nonlinear function based on a set of chosen sample points, known as sigma points, and preserves the nonlinear nature of the system. Considering development up to now, the UT sampling strategies include: basic UT, general UT, simplex UT, spherical UT and high-order UT. In the above several sampling methods, the numbers of the sigma points of the basic and general UT are $2n$ and $2n + 1$ (Julier, Uhlmann, & Durrant-Whyte, 2000), however, in order to reduce computing cost, the minimal skew simplex UT (Julier, 2003; Julier & Uhlmann, 2002) with the least sigma points is applied into UKF in this paper. Meanwhile, the scale-corrected method (Julier, 2002) is introduced into minimal skew simplex UT to overcome the nonlocal effects caused by the increase in states or to decrease the error of high-order item, which allows any set of sigma points to be scaled by an arbitrary scaling factor. While a few practical applications of the UKF have been studied by scholars, mainly in radar tracking, signal processing and robotics. However, up to now, there is still little research involved in the scale-corrected minimal skew simplex sampling UKF for the BLDCM sensorless control.

Motivated by the above discussions, in this paper, the scale-corrected minimal skew simplex sampling UKF sensorless control technology is designed for the BLDCM drive. The main contributions are as follows: (1) UKF approach is designed to estimate the position sensors for the BLDCM sensorless control; (2) the minimal skew simplex UT is used to replace the basic UT for reducing the computing cost; (3) to overcome the nonlocal effects, the scale-corrected method of UT is introduced into the minimal skew simplex sampling UKF; and (4) the back-EMF is represented by a shape function combining with Fourier series without judging the angle range.

## 2. BLDCM model description

A typical block diagram of BLDCM drive system is shown in Figure 1 consisting primarily of PM motor ontology, detectors and power switch device (inverter). In this paper, we assume that the magnetic flux of the motor is not saturated, the motor three-phase stator windings are a star connection, the influence of the magnetic hysteresis is negligible and the back-EMF of BLDCM is trapezoidal. Then, we have the following PM BLDCM model in the three-phase stationary frame.

$$\begin{bmatrix} U_{ag} \\ U_{bg} \\ U_{cg} \end{bmatrix} = \begin{bmatrix} R & 0 & 0 \\ 0 & R & 0 \\ 0 & 0 & R \end{bmatrix} \begin{bmatrix} i_a \\ i_b \\ i_c \end{bmatrix} + \begin{bmatrix} L_s - M & 0 & 0 \\ 0 & L_s - M & 0 \\ 0 & 0 & L_s - M \end{bmatrix} \frac{\mathrm{d}}{\mathrm{d}t} \begin{bmatrix} i_a \\ i_b \\ i_c \end{bmatrix} + \begin{bmatrix} e_a \\ e_b \\ e_c \end{bmatrix} + \begin{bmatrix} U_n \\ U_n \\ U_n \end{bmatrix}, \tag{1}$$

where $U_{ag}$, $U_{bg}$ and $U_{cg}$ are the terminal voltages of three-phase windings, $i_a, i_b$ and $i_c$ are the stator three-phase currents, $e_a, e_b$ and $e_c$ are the three-phase back-EMFs, $U_n$ is the neutral-point voltage, $R$ is the phase resistance, $L_s$ is the self-inductance of three-phase windings and $M$ is the mutual inductance of three-phase windings.

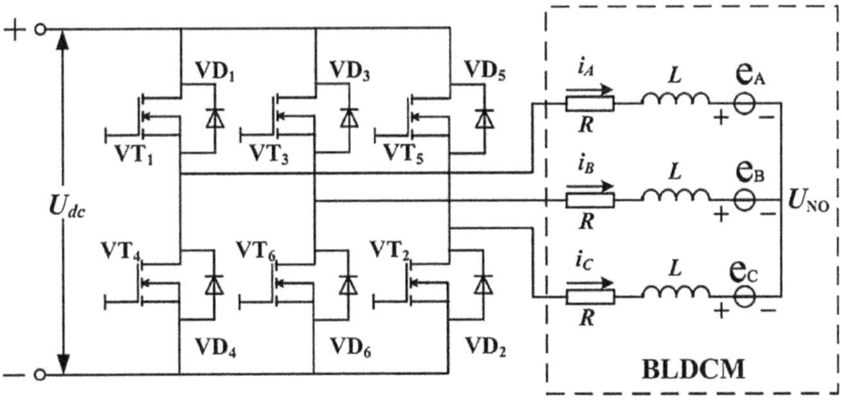


Figure 1. Block diagram of the BLDCM drive system.

For the symmetrically distributed star connected three-phase windings, the three-phase currents obey

$$i_a + i_b + i_c = 0. \tag{2}$$

Then, from Equations (1) and (2), $U_n$ can be represented as

$$U_n = \frac{[(U_{ag} + U_{bg} + U_{cg}) - (e_a + e_b + e_c)]}{3}. \tag{3}$$

Substituting Equation (3) into the BLDCM model (1), we can have

$$\frac{\mathrm{d}}{\mathrm{d}t}\begin{bmatrix} i_a \\ i_b \\ i_c \end{bmatrix} = \frac{-R}{L_s - M}\begin{bmatrix} i_a \\ i_b \\ i_c \end{bmatrix} + \frac{1}{3(L_s - M)}\begin{bmatrix} U_{ab} - U_{ca} \\ U_{bc} - U_{ab} \\ U_{ca} - U_{bc} \end{bmatrix} + \frac{1}{3(L_s - M)}\begin{bmatrix} e_b + e_c - 2e_a \\ e_a + e_c - 2e_b \\ e_a + e_b - 2e_c \end{bmatrix}, \tag{4}$$

where $U_{ab}$, $U_{bc}$ and $U_{ca}$ are line-to-line voltages, and $U_{ab} = U_{ag} - U_{bg}$, $U_{bc} = U_{bg} - U_{cg}$, $U_{ca} = U_{cg} - U_{ag}$.

The electromagnetic torque is calculated as follows:

$$T_e = \frac{e_a i_a + e_b i_b + e_c i_c}{w_r}, \tag{5}$$

where $w_r$ is the mechanical angular velocity of motor.

From Equation (5), we conclude that, to produce a steady electromagnetic torque, the sum of $e_a i_a$, $e_b i_b$ and $e_c i_c$ has to be constant as far as a certain speed is concerned. We assume that the air gap magnetic field distribution of BLDCM is an ideal trapezoidal wave, then the distribution of the magnetic induction intensity and the back-EMF are consistent. Therefore, to ensure a constant torque, the armature current with ideal square waveform should be in phase with back-EMF waveform, as shown in Figure 2. Nevertheless, the back-EMF cannot be directly measured in the running process of the motor. To solve this problem, considering the relationship of the rotor position, speed and back-EMF, the back-EMF can be expressed in the following form:

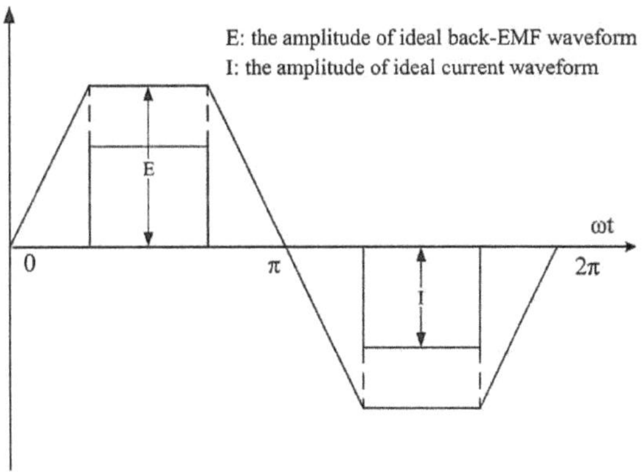


Figure 2. Ideal current and back-EMF waveforms in BLDCM.

$$e = w_e K_e B_g(\theta), \tag{6}$$

where $K_e$ is the back-EMF coefficient which is a constant, $B_g(\theta)$ is a shape function that depends on the rotor position. As shown in Figure 3, $B_g(\theta)$ can be written as the following piecewise function:

$$B_g(\theta) = \begin{cases} \frac{6\theta}{\pi}, & 0 \le \theta \le \frac{\pi}{6}, \\ 1, & \frac{\pi}{6} \le \theta \le \frac{5\pi}{6}, \\ -\frac{6(\theta - \pi)}{\pi}, & \frac{5\pi}{6} \le \theta \le \frac{7\pi}{6}, \\ -1, & \frac{7\pi}{6} \le \theta \le \frac{11\pi}{6}, \\ \frac{6(\theta - 2\pi)}{\pi}, & \frac{11\pi}{6} \le \theta \le 2\pi. \end{cases} \tag{7}$$

*Remark 1* For convenience, combining with the principle of Fourier series, the piecewise function $B_g(\theta)$ which

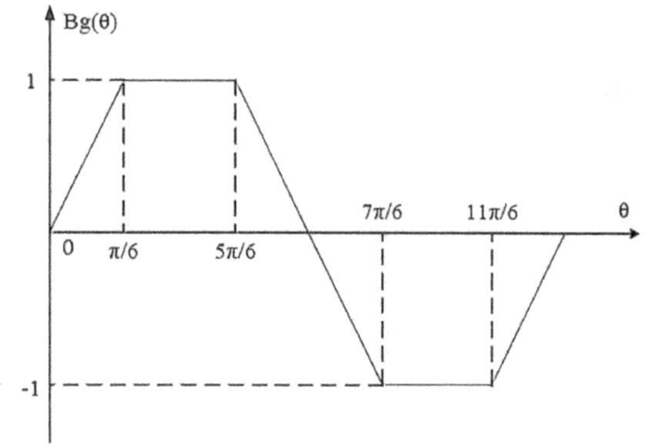


Figure 3. Shape function of $B_g(\theta)$.

takes $2\pi$ radians as one cycle can be approximated by an infinite series composed of sine and cosine functions. Then, Equation (7) can be replaced by the approximate mathematical expression (8),

$$B_g(\theta) = 1.215\sin\theta + 0.270\sin 3\theta + 0.049\sin 5\theta - 0.025\sin 7\theta - 0.030\sin 9\theta \cdots. \quad (8)$$

Three phases $A, B$ and $C$ are similar only with the difference of 120 electrical degrees. Then, three-phase back-EMFs can be expressed as

$$\begin{bmatrix} e_a \\ e_b \\ e_c \end{bmatrix} = w_e K_e \begin{bmatrix} B_g(\theta) \\ B_g\left(\theta - \frac{2\pi}{3}\right) \\ B_g\left(\theta - \frac{4\pi}{3}\right) \end{bmatrix}. \quad (9)$$

From Equations (9), (4) and (5) can be rewritten as

$$\frac{\mathrm{d}}{\mathrm{d}t}\begin{bmatrix} i_a \\ i_b \\ i_c \end{bmatrix} = \frac{-R}{L_s - M}\begin{bmatrix} i_a \\ i_b \\ i_c \end{bmatrix} + \frac{1}{3(L_s - M)}\begin{bmatrix} U_{ab} - U_{ca} \\ U_{bc} - U_{ab} \\ U_{ca} - U_{bc} \end{bmatrix} + \frac{w_e K_e}{3(L_s - M)} \times \begin{bmatrix} B_g\left(\theta - \frac{2\pi}{3}\right) + B_g\left(\theta - \frac{4\pi}{3}\right) - 2B_g(\theta) \\ B_g(\theta) + B_g\left(\theta - \frac{4\pi}{3}\right) - 2B_g\left(\theta - \frac{2\pi}{3}\right) \\ B_g(\theta) + B_g\left(\theta - \frac{2\pi}{3}\right) - 2B_g\left(\theta - \frac{4\pi}{3}\right) \end{bmatrix} \quad (10)$$

and

$$T_e = pK_e\left[B_g(\theta)i_a + B_g\left(\theta - \frac{2\pi}{3}\right)i_b + B_g\left(\theta - \frac{4\pi}{3}\right)i_c\right], \quad (11)$$

where $p = w_e/w_r$, which is the number of pole pairs.

Based on the mechanical equation of BLDCM

$$J\frac{\mathrm{d}w_r}{\mathrm{d}t} = T_e - T_L - Bw_r$$

and Equation (11), we can obtain

$$\frac{\mathrm{d}w_e}{\mathrm{d}t} = \frac{pT_e - pT_L - Bw_e}{J} = \frac{p^2K_e[B_g(\theta)i_a + B_g(\theta - 2\pi/3)i_b + B_g(\theta - 4\pi/3)i_c] - pT_L - Bw_e}{J}, \quad (12)$$

where $J$ is the rotor equivalent inertia, $B$ is the viscous friction coefficient and $T_\mathrm{L}$ is the load torque.

## 3. Scale-corrected minimal skew simplex sampling UKF

UKF provides an alternative estimation methodology for nonlinear applications which does not rely upon any linearization procedure. The principle of UKF is based on performing the state estimation by approximating the probability distribution instead of approximating the nonlinearity itself. To approximate the probability distribution, a set of sigma points is chosen and propagated through the nonlinear function, then, the state mean and covariance could be obtained by these sigma points. The detailed process of this algorithm is introduced in the following.

Consider the following discrete-time nonlinear system and measurement model:

$$\begin{cases} \boldsymbol{x}_{\kappa+1} &= \boldsymbol{f}_\kappa(\boldsymbol{x}_\kappa, \boldsymbol{u}_\kappa) + \boldsymbol{\omega}_\kappa \\ z_{\kappa+1} &= \boldsymbol{h}_{\kappa+1}(\boldsymbol{x}_{\kappa+1}) + \boldsymbol{v}_{\kappa+1}, \end{cases} \quad (13)$$

where $\boldsymbol{x}_\kappa \in \mathbb{R}^{n\times 1}$ is the state vector, $z_\kappa \in \mathbb{R}^{m\times 1}$ is the measurement vector, $\boldsymbol{f}_\kappa(\boldsymbol{x}_\kappa, \boldsymbol{u}_\kappa)$ is a known nonlinear state transition vector, $\boldsymbol{h}_{\kappa+1}(\boldsymbol{x}_{\kappa+1})$ is a known nonlinear measurement transition vector. The $\boldsymbol{\omega}_\kappa$ is system noise and $\boldsymbol{v}_{\kappa+1}$ is measurement noise. They are both additive white Gaussian noise with zero mean, and both obey the following statistical properties:

$$\begin{cases} \mathbb{E}(\boldsymbol{\omega}_\kappa) = 0, & \mathrm{Cov}(\boldsymbol{\omega}_\kappa, \boldsymbol{\omega}_j) = \boldsymbol{Q}_\kappa\delta_{\kappa j}, \\ \mathbb{E}(\boldsymbol{v}_\kappa) = 0, & \mathrm{Cov}(\boldsymbol{v}_\kappa, \boldsymbol{v}_j) = \boldsymbol{R}_\kappa\delta_{\kappa j}, \\ \mathrm{Cov}(\boldsymbol{\omega}_\kappa, \boldsymbol{v}_j) = 0, & \end{cases}$$

where $\delta_{\kappa j}$ is the *kronecker* $-\,\delta$ function.

### 3.1. *Minimal skew simplex UT*

The minimal skew simplex sigma points are chosen by the following algorithm:

(1) Choose the weight $0 \le W_0 < 1$.

(2) Choose the remaining weights as follows:

$$W_i = \begin{cases} \dfrac{(1 - W_0)}{2^j}, & i = 1, 2, \\ 2^{i-1} W_1, & i = 3, \dots, j + 1. \end{cases} \tag{14}$$

(3) Initialize the $\xi_i^j$ vectors (state dimension $j = 1$):

$$\xi_0^1 = [0], \xi_1^1 = \left[\frac{-1}{\sqrt{2W_1}}\right], \quad \xi_2^1 = \left[\frac{1}{\sqrt{2W_1}}\right].$$

(4) When the state dimension $j = 2, 3, 4, \dots, n$, the iterative formula $\xi_i^j$ is

$$\xi_i^j = \begin{cases} \begin{bmatrix} \xi_0^{j-1} \\ 0 \end{bmatrix}, & i = 0, \\ \begin{bmatrix} \xi_i^{j-1} \\ \dfrac{-1}{\sqrt{2W_{j+1}}} \end{bmatrix}, & i = 1, 2, \dots, j, \\ \begin{bmatrix} \mathbf{0}_{j-1} \\ \dfrac{1}{\sqrt{2W_{j+1}}} \end{bmatrix}, & i = j + 1, \end{cases}$$

where $\mathbf{0}_{j-1} \in \mathbb{R}^{(j-1)\times 1}$ is the zero vector.

(5) Determine the sigma points as

$$\xi_{i,\kappa-1} = \hat{\boldsymbol{x}}_{\kappa-1} + \sqrt{\boldsymbol{P}_{\kappa-1}}\xi_i^j, \quad i = 0, 1, 2, \dots, j + 1.$$

*Remark 2* The numbers of the sigma points for the basic and general UTs are $2n$ and $2n + 1$ ($n$ is the dimension of $\boldsymbol{x}$), respectively. It is generally known that the computational efficiency of the UT depends on the number of sigma points required to estimate the next states, and the low computational burden is particularly important in electric drive applications or real-time control. Fortunately, the minimal skew simplex UT can only use $n + 1$ sigma points which is the least of all UTs to capture the mean and covariance of an $n \times 1$ vector so as to reduce the computational load. Moreover, the minimal skew simplex sigma points can estimate more accurately than other UT methods, because these sigma points are chosen by matching the first two moments of $\boldsymbol{x}$ and to minimize the third-order moments.

The minimal skew simplex UT has the least $n + 1$ sigma points, so it is a sampling method with minimum calculation. Nevertheless, it is not recommended for high-dimensional nonlinear systems, because the weights of minimal skew simplex UT sigma points in (14) increase geometrically. This will cause numerical problems in high-dimensional systems. Fortunately, the dynamic system mentioned in this paper is not a high-dimensional system.

### 3.2. *Scaled UT*

However, as the dimension of the state space increases, the distance between the chosen sigma points and the centre $\hat{\boldsymbol{x}}_\kappa$ increases as well. For many kinds of nonlinearities, such as exponents or trigonometric functions, this can lead to significant difficulties. To avoid this problem, the modified approach called scaled UT is introduced into the minimal skew simplex sampling method.

As a consequence, the state weights of this transformed sequence are

$$W_i^m = \begin{cases} W_0/\alpha^2 + (1 - 1/\alpha^2), & i = 0, \\ (1 - W_0)/2^j \alpha^2, & i = 1, 2, \\ 2^{i-2} W_1/\alpha^2, & i = 3, \dots, j + 1, \end{cases}$$

and the covariance weights are

$$W_i^c = \begin{cases} W_0^m + 1 + \beta - \alpha^2, & i = 0, \\ W_i^m, & i = 1, 2, \dots, j + 1, \end{cases}$$

where $\alpha$ is a positive scaling parameter which can be made arbitrarily small to minimize higher order effects, and in general, $10^{-4} \le \alpha \le 1$. $\beta$ is a noise distribution parameter which is used to introduce the higher order term information, and if the distributions are Gaussian, $\beta$ is usually 2.

Therefore, the sigma points are modified as

$$\xi_{i,\kappa-1} = \hat{\boldsymbol{x}}_{\kappa-1} + \alpha\sqrt{\boldsymbol{P}_{\kappa-1}}\xi_i^j, \quad i = 0, 1, 2, \dots, j + 1.$$

### 3.3. *Whole process of algorithm*

We assume that the initial values of state variable $\boldsymbol{x}_0$ and the noise $\boldsymbol{\omega}_\kappa$, $\boldsymbol{v}_\kappa$ are mutually independent. The initial state statistical property is assumed as follows:

$$\begin{cases} \hat{\boldsymbol{x}}_0 & = \mathbb{E}(\boldsymbol{x}_0), \\ \boldsymbol{P}_0 & = \mathrm{Var}(\boldsymbol{x}_0) = \mathbb{E}[(\boldsymbol{x}_0 - \hat{\boldsymbol{x}}_0)(\boldsymbol{x}_0 - \hat{\boldsymbol{x}}_0)^{\mathrm{T}}], \end{cases}$$

(1) *Time update*:

Project the sigma points in time using the following nonlinear transformation:

$$\hat{\chi}_{i,\kappa|\kappa-1} = \boldsymbol{f}_{\kappa-1}(\xi_{i,\kappa-1}, \boldsymbol{u}_{\kappa-1}), \quad i = 0, 1, 2, \dots, j + 1.$$

Calculate the predicted states mean

$$\hat{\boldsymbol{x}}_{\kappa|\kappa-1} = \sum_{i=0}^{j+1} W_i^m \hat{\chi}_{i,\kappa|\kappa-1}$$

and the predicted error covariance

$$\boldsymbol{P}_{\kappa|\kappa-1} = \sum_{i=0}^{j+1} W_i^c (\hat{\chi}_{i,\kappa|\kappa-1} - \hat{\boldsymbol{x}}_{\kappa|\kappa-1})(\hat{\chi}_{i,\kappa|\kappa-1} - \hat{\boldsymbol{x}}_{\kappa|\kappa-1})^{\mathrm{T}} + \boldsymbol{Q}_{\kappa-1}.$$

(2) *Measurement update*:

Solution of the classical form of KF is adapted in UKF, so measurement update algorithm is performed in the similar way to that in classical KF and also requires output covariance $\boldsymbol{P}_{\hat{z}_\kappa}$ and cross-covariance matrix $\boldsymbol{P}_{\hat{x}_\kappa \hat{z}_\kappa}$.

Recalculate the sigma points

$$\xi_{i,\kappa|\kappa-1} = \hat{\boldsymbol{x}}_{\kappa|\kappa-1} + \alpha\sqrt{\boldsymbol{P}_{\kappa|\kappa-1}}\xi_i^j.$$

Project the sigma points through the following observation function:

$$\hat{\gamma}_{i,\kappa|\kappa-1} = \boldsymbol{h}_\kappa(\xi_{i,\kappa|\kappa-1}), \quad i = 0, 1, 2, \ldots, j+1.$$

Calculate the predicted measurements mean

$$\hat{z}_{\kappa|\kappa-1} = \sum_{i=0}^{j+1} W_i^m \hat{\gamma}_{i,\kappa|\kappa-1}.$$

The predicted covariance matrix of observation

$$\boldsymbol{P}_{\hat{z}_\kappa} = \sum_{i=0}^{j+1} W_i^c (\hat{\gamma}_{i,\kappa|\kappa-1} - \hat{z}_{\kappa|\kappa-1})(\hat{\gamma}_{i,\kappa|\kappa-1} - \hat{z}_{\kappa|\kappa-1})^{\mathrm{T}} + \boldsymbol{R}_\kappa.$$

The cross-covariance matrix

$$\boldsymbol{P}_{\hat{x}_\kappa \hat{z}_\kappa} = \sum_{i=0}^{j+1} W_i^c (\hat{\chi}_{i,\kappa|\kappa-1} - \hat{\boldsymbol{x}}_{\kappa|\kappa-1})(\hat{\gamma}_{i,\kappa|\kappa-1} - \hat{z}_{\kappa|\kappa-1})^{\mathrm{T}}.$$

Correct the predicted states and covariance matrix

$$\begin{cases} \boldsymbol{K}_\kappa = \boldsymbol{P}_{\hat{x}_\kappa \hat{z}_\kappa} \boldsymbol{P}_{\hat{z}_\kappa}^{-1}, \\ \hat{\boldsymbol{x}}_\kappa = \hat{\boldsymbol{x}}_{\kappa|\kappa-1} + \boldsymbol{K}_\kappa (z_\kappa - \hat{z}_{\kappa|\kappa-1}), \\ \boldsymbol{P}_\kappa = \boldsymbol{P}_{\kappa|\kappa-1} - \boldsymbol{K}_\kappa \boldsymbol{P}_{\hat{z}_\kappa} \boldsymbol{K}_\kappa^{\mathrm{T}}. \end{cases}$$

## 4. BLDCM discrete-time model

To implement the above algorithm on a simulation computer, the system dynamic equation should be transformed into a discrete-time state equation. Assume that the controller has a short sampling time, hence, the rotor electrical angular velocity can be regarded as a constant within a sampling period,

$$\theta(\kappa+1) = T_s\omega_e(\kappa) + \theta(\kappa),$$

where $T_s$ is the sampling time.

The model (10) and (12) must be discretized for the convenience to implement the algorithm on the computer.

$$\begin{cases} i_a(\kappa+1) &= \left(1 - \dfrac{T_sR}{L-M}\right) i_a(\kappa) \\ &\quad + \dfrac{T_s}{3(L-M)}[u_{ab}(\kappa) - u_{ca}(\kappa)] \\ &\quad + \dfrac{T_sK_e[B_g(\theta_\kappa - 2\pi/3) + B_g(\theta_\kappa - 4\pi/3) - 2B_g(\theta_\kappa)]}{3(L-M)}\omega_e(\kappa), \\ i_b(\kappa+1) &= \left(1 - \dfrac{T_sR}{L-M}\right) i_b(\kappa) \\ &\quad + \dfrac{T_s}{3(L-M)}[u_{bc}(\kappa) - u_{ab}(\kappa)] \\ &\quad + \dfrac{T_sK_e[B_g(\theta_\kappa) + B_g(\theta_\kappa - 4\pi/3) - 2B_g(\theta_\kappa - 2\pi/3)]}{3(L-M)}\omega_e(\kappa), \\ i_c(\kappa+1) &= \left(1 - \dfrac{T_sR}{L-M}\right) i_c(\kappa) \\ &\quad + \dfrac{T_s}{3(L-M)}[u_{ca}(\kappa) - u_{bc}(\kappa)] \\ &\quad + \dfrac{T_sK_e[B_g(\theta_\kappa) + B_g(\theta_\kappa - 2\pi/3) - 2B_g(\theta_\kappa - 4\pi/3)]}{3(L-M)}\omega_e(\kappa), \\ \omega_e(\kappa+1) &= \dfrac{p^2K_eT_s}{J}\left[B_g(\theta_\kappa)i_a(\kappa) + B_g\left(\theta_\kappa - \dfrac{2\pi}{3}\right)i_b(\kappa) \right. \\ &\quad \left. + B_g(\theta_\kappa - \dfrac{4\pi}{3})i_c(\kappa)\right] \\ &\quad + \left(1 - \dfrac{BT_s}{J}\right)\omega_e(\kappa) - \dfrac{pT_s}{J}T_L, \\ \theta(\kappa+1) &= T_s\omega_e(\kappa) + \theta(\kappa). \end{cases}$$

Furthermore, Equation (15) can be rewritten in the form of the UKF algorithm

$$\begin{cases} \boldsymbol{x}_{\kappa+1} &= \boldsymbol{F}_\kappa(\boldsymbol{x}_\kappa)\boldsymbol{x}_\kappa + \boldsymbol{G}_\kappa\boldsymbol{u}_\kappa + \boldsymbol{\omega}_\kappa, \\ z_{\kappa+1} &= \boldsymbol{H}\boldsymbol{x}_{\kappa+1} + \boldsymbol{v}_{\kappa+1}, \end{cases} \tag{15}$$

where $\boldsymbol{x}_\kappa = [i_a(\kappa) \;\; i_b(\kappa) \;\; i_c(\kappa) \;\; w_e(\kappa) \;\; \theta(\kappa)]^{\mathrm{T}}$, $z_{\kappa+1} = [i_a(\kappa+1) \;\; i_b(\kappa+1) \;\; i_c(\kappa+1)]^{\mathrm{T}}$, $\boldsymbol{u}_\kappa = [u_{ab}(\kappa) - u_{ca}(\kappa)$

$u_{bc}(\kappa) - u_{ab}(\kappa)u_{ca}(\kappa) - u_{bc}(\kappa)\ T_L]^{\mathrm{T}}$,

$$F_\kappa(\boldsymbol{x}_\kappa) = \begin{bmatrix} 1-\frac{RT_s}{L-M} & 0 & 0 & \frac{T_sK_e[B_g(\theta_\kappa-2\pi/3)+B_g(\theta_\kappa-4\pi/3)-2B_g(\theta_\kappa)]}{3(L-M)} & 0 \\ 0 & 1-\frac{RT_s}{L-M} & 0 & \frac{T_sK_e[B_g(\theta_\kappa)+B_g(\theta_\kappa-4\pi/3)-2B_g(\theta_\kappa-2\pi/3)]}{3(L-M)} & 0 \\ 0 & 0 & 1-\frac{RT_s}{L-M} & \frac{T_sK_e[B_g(\theta_\kappa)+B_g(\theta_\kappa-2\pi/3)-2B_g(\theta_\kappa-4\pi/3)]}{3(L-M)} & 0 \\ \frac{p^2K_eT_sB_g(\theta_\kappa)}{J} & \frac{p^2K_eT_sB_g(\theta_\kappa-2\pi/3)}{J} & \frac{p^2K_eT_sB_g(\theta_\kappa-4\pi/3)}{J} & 1-\frac{BT_s}{J} & 0 \\ 0 & 0 & 0 & T_s & 1 \end{bmatrix}$$

$$G_\kappa = \begin{bmatrix} \frac{T_s}{3(L-M)} & 0 & 0 & 0 \\ 0 & \frac{T_s}{3(L-M)} & 0 & 0 \\ 0 & 0 & \frac{T_s}{3(L-M)} & 0 \\ 0 & 0 & 0 & -\frac{pT_s}{J} \\ 0 & 0 & 0 & 0 \end{bmatrix},$$

$$H = \begin{bmatrix} 1 & 0 & 0 & 0 & 0 \\ 0 & 1 & 0 & 0 & 0 \\ 0 & 0 & 1 & 0 & 0 \end{bmatrix}.$$

## 5. Simulation results

To verify the state estimation performance of the scale-corrected minimal skew simplex sampling UKF method, simulations have been carried out for different operating conditions by the Matlab/Simulink software. Moreover, for the computation time comparison between the new algorithm and the traditional algorithm, the general UT UKF and the scale-corrected minimal skew simplex UT UKF have been operated in the same simulation model. The simulation model is given in Figure 4. The block in yellow is the embedded simulation program (M-file) with scale-corrected minimal skew simplex UT, and the red is the embedded simulation program (M-file) with general UT.

In this simulation model, the parameters of BLDCM are as follows:

| Parameters | | Value |
|---|---|---|
| Phase resistance | $R$ | 0.62 Ω |
| Stator inductance of three-phase windings | $L_s - M$ | $10^{-3}$ mH |
| Back-EMF coefficient | $K_e$ | 0.066 |
| Number of pole pairs | $p$ | 4 |
| Rotor equivalent inertia | $J$ | $3.62 \times 10^{-4}\ \mathrm{kg \cdot m^2}$ |
| Viscous friction coefficient | $B$ | $9.444 \times 10^{-5}\ \mathrm{N \cdot m \cdot s}$ |

By iterative experimentations and previous studies, good transient response and steady-state performance can be obtained by selecting the appropriate system noise covariance matrix $\boldsymbol{Q}_\kappa$ and the measurement noise covariance matrix $\boldsymbol{R}_\kappa$,

$$\boldsymbol{Q}_\kappa = \mathrm{diag}[0.1 \quad 0.1 \quad 1\times10^{-6} \quad 1\times10^{-6} \quad 0],$$

$$\boldsymbol{R}_\kappa = \mathrm{diag}[0.1 \quad 0.1 \quad 1\times10^{-3}],$$

and $\alpha = 0.009, \beta = 2$.

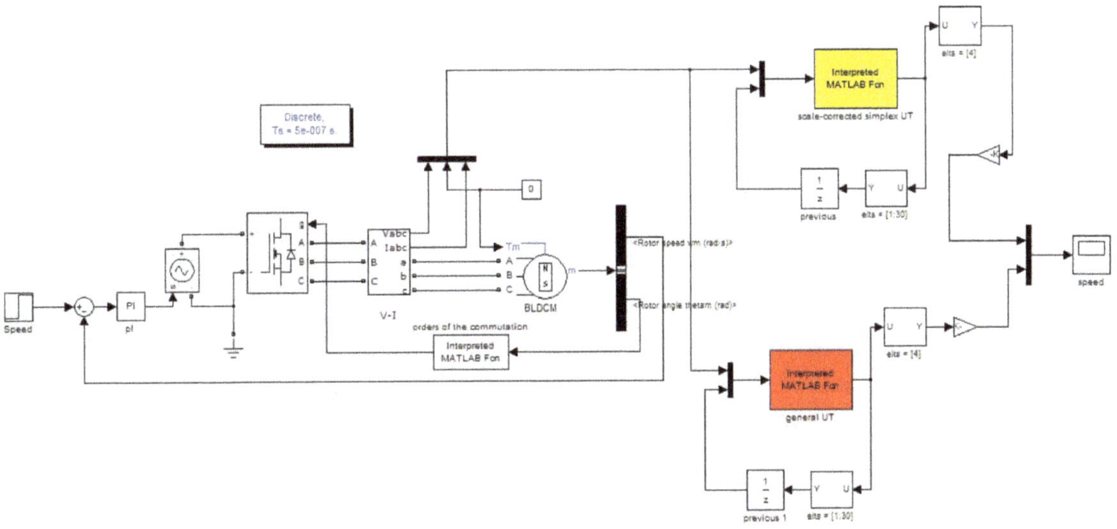


Figure 4. Simulation diagram of motor control system.

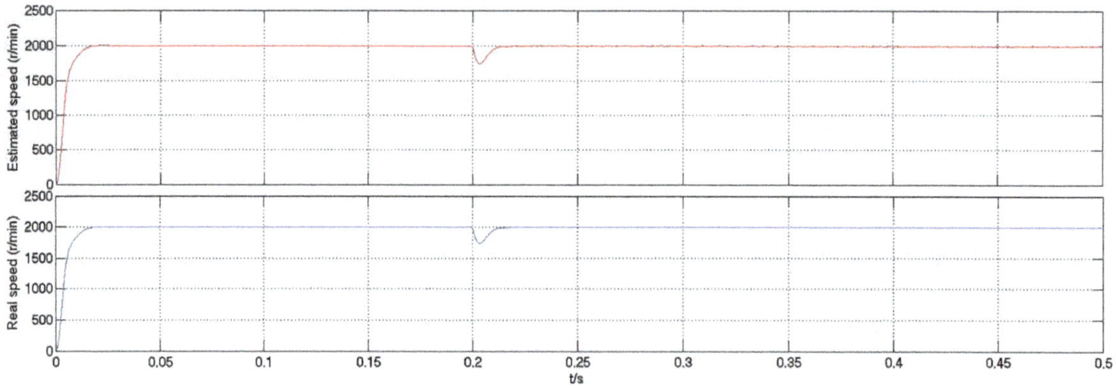


Figure 5. Speed estimated performance under abrupt load variation (scale-corrected minimal skew simplex UT).

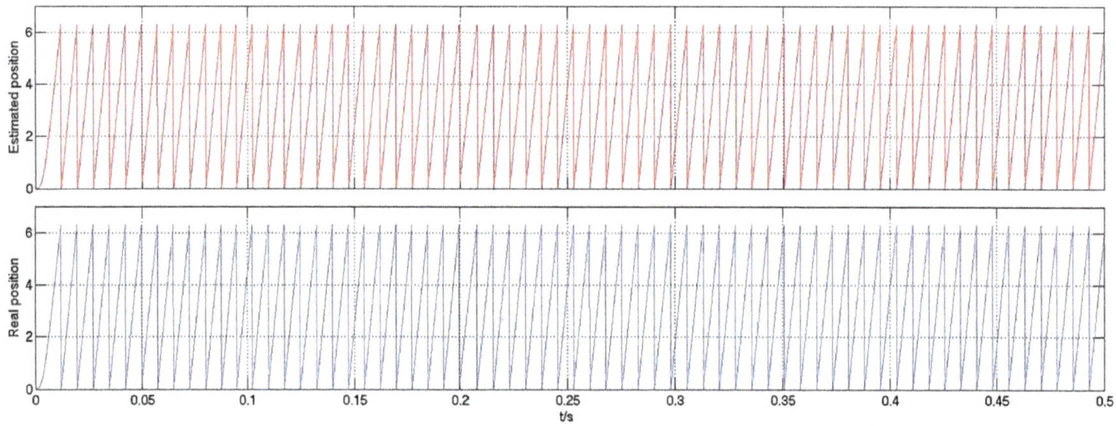


Figure 6. Position estimated performance under abrupt load variation (scale-corrected minimal skew simplex UT).

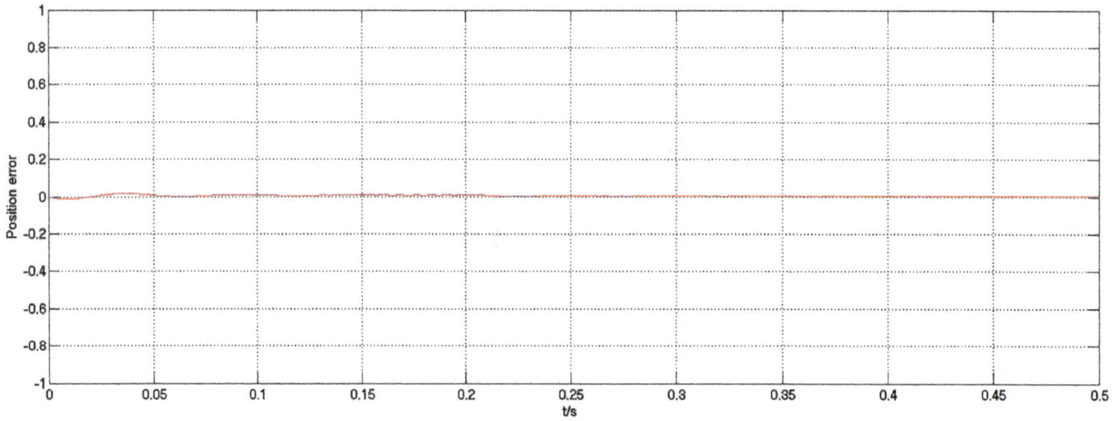


Figure 7. Position error under abrupt load variation (scale-corrected minimal skew simplex UT).

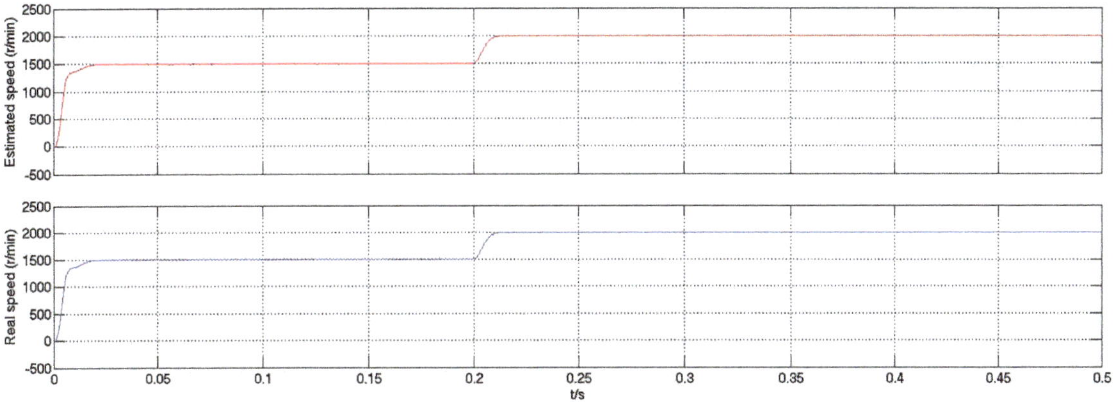


Figure 8. Speed estimated performance when reference speed varies (scale-corrected minimal skew simplex UT).

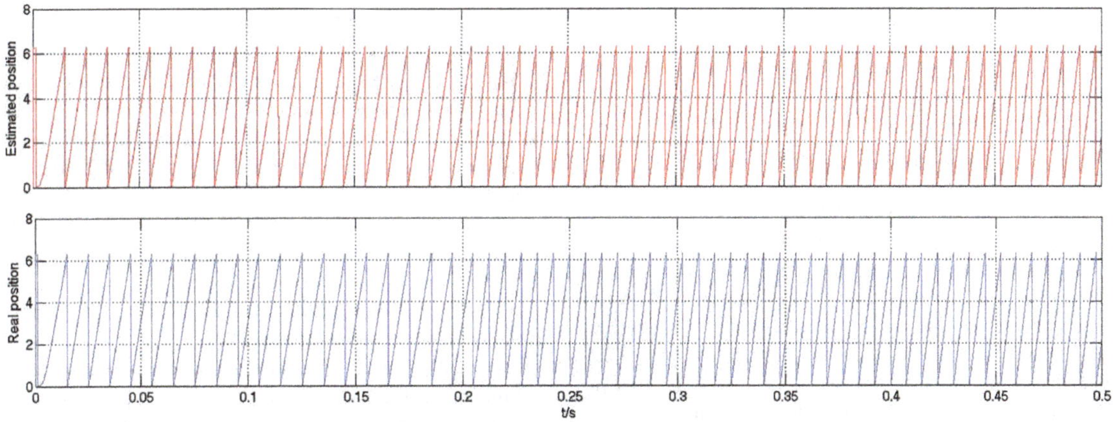


Figure 9. Position estimated performance when reference speed varies (scale-corrected minimal skew simplex UT).

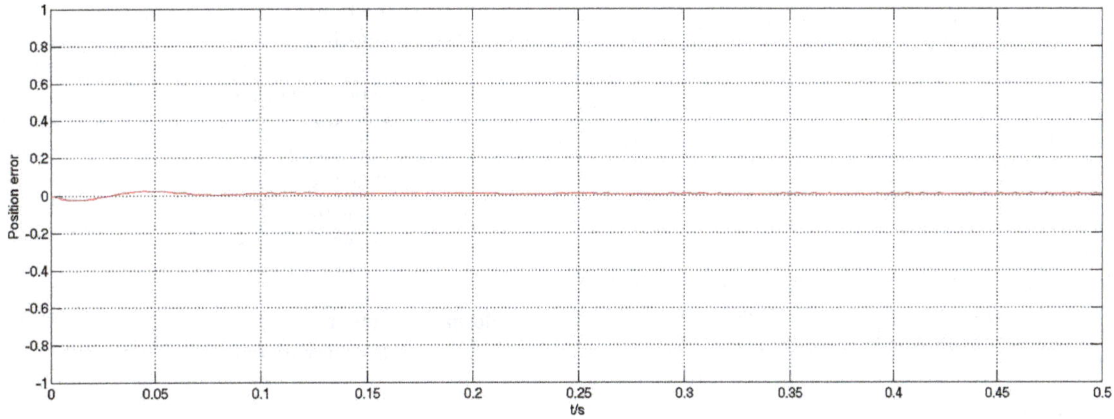


Figure 10. Position error when reference speed varies (scale-corrected minimal skew simplex UT).

Table 1. general UT.

| Times | Elapsed time($s$) |
|---|---|
| 1 | 0.000278 |
| 2 | 0.000271 |
| 3 | 0.000292 |
| 4 | 0.000284 |
| 5 | 0.000269 |
| 6 | 0.000340 |
| 7 | 0.000296 |
| 8 | 0.000308 |
| 9 | 0.000277 |
| 10 | 0.000295 |
| ⋮ | ⋮ |
| $n$ | |
| Total($n = 10,000$) | 3.431525 |

Table 2. scale-corrected minimal skew simplex UT.

| Times | Elapsed time($s$) |
|---|---|
| 1 | 0.000227 |
| 2 | 0.000234 |
| 3 | 0.000233 |
| 4 | 0.000227 |
| 5 | 0.000218 |
| 6 | 0.000240 |
| 7 | 0.000232 |
| 8 | 0.000225 |
| 9 | 0.000226 |
| 10 | 0.000229 |
| ⋮ | ⋮ |
| $n$ | |
| Total($n = 10,000$) | 2.510272 |

State estimation starts with turning the motor system on, and initial values of state estimation are chosen as

$$\hat{\boldsymbol{x}}_0 = \mathbb{E}(\boldsymbol{x}_0) = [0\ \ 0\ \ 0\ \ 0\ \ 0]^{\mathrm{T}},$$

$$\begin{aligned}\boldsymbol{P}_0 &= \mathrm{Var}(\boldsymbol{x}_0) = \mathbb{E}[(\boldsymbol{x}_0 - \hat{\boldsymbol{x}}_0)(\boldsymbol{x}_0 - \hat{\boldsymbol{x}}_0)^{\mathrm{T}}] \\ &= \mathrm{diag}[0.01\ \ 0.01\ \ 0.01\ \ 0.01\ \ 0.01].\end{aligned}$$

Then, the effectiveness and practicability of the speed estimated method for BLDCM are proved by two different groups of simulation experiments with the changes of reference speed and load torque. Figures 5–7 show the estimations of the motor speed, position when the load torque changes from 0 to 5 $\mathrm{N \cdot m}$ at $t = 0.2\,\mathrm{s}$ under the condition that the reference speed is always $2000\,\mathrm{r/min}$. Figures 8–10 show the estimations of the motor speed, position when the reference speed changes from 1500 to 2000 $\mathrm{r/min}$ at $t = 0.2\,\mathrm{s}$ under the condition that the load torque $T_{\mathrm{L}}$ is always $2\,N \cdot m$.

As shown in Figures 5 and 8, the speed observer designed in this paper can accurately estimate the motor real speed and track the speed quickly when the reference speed changes. When the outside load torque changes, the motor speed can be fed into a stable state in a very short time. Figures 6, 7, 9 and 10 verify that this method has a high accuracy of estimation of rotor position.

About the computation time of the two different UTs (the two blocks in colour ), we give Tables 1 and 2 and the single running time and the total running time are also listed in them. From the tables, it is obviously that the computation time of the scale-corrected minimal skew simplex UT is less than the general UT.

## 6. Conclusions

The sensorless control technique can effectively enhance the reliability and cut down the hardware cost. Therefore, this paper is concerned with the problem of BLDCM sensorless control by the use of scale-corrected minimal skew simplex sampling UKF. In this algorithm, we use the minimal skew simplex UT which has the least number of sigma points to capture the mean and covariance so as to save computational time. In addition, to overcome the drawback of nonlocal effects caused by the increasein state dimensions, the scaled UT is introduced into this algorithm. On the flip side, we use a back-EMF shape function based on Fourier series instead of the back-EMF function described by a piecewise linear function in order to streamline program codes. In the end, simulation results of the UKF for BLDCM driven with noise illustrate that the proposed method is effective and practical.

Recently, under pressure from high-energy prices, immediate actions should be taken to strengthen the study of motor control methods for BLDC motor to save energy. In addition, in the most recent sensorless drive methods, rotor position estimation precision depends on motor parameters and measured quantities. However, the ambient environment and motor ageing will greatly affect the accuracies of parameters and measurements that lead to the poor control performance. Therefore, adapting the artificial intelligence control which can compensate parameter variations is necessary. Also, benefiting from the rapid development of the microprocessor technology, it is possible to accomplish the complicated control algorithms such as neural network, model predictive control (Fu, Aghezzaf, and Keyser, 2014) and fuzzy logic with the fast DSP chips.

### Disclosure statement

No potential conflict of interest was reported by the authors.

### Funding

This work was supported in part by the National Natural Science Foundation of China under Grant (61374039, 61203143), Shanghai Pujiang Program under Grant 13PJ1406300, Shanghai Natural Science Foundation of China under Grant 13ZR1428500,

Innovation Program of Shanghai Municipal Education Commission 14YZ083 and the Innovation Fund Project for Graduate Student of Shanghai under Grant JWCXSL1302.

## References

Acarnley, P. P., & Watson, J. F. (2006). Review of position-sensorless operation of brushless permanent-magnet machines. *IEEE Transactions on Industrial Electronics*, *53*(2), 352–361.

Bolognani, S., Oboe, R., & Zigliotto, M. (1999). Sensorless full-digital PMSM drive with EKF estimation of speed and rotor position. *IEEE Transactions on Industrial Electronics*, *46*(1), 184–191.

Dong, H. L., Wang, Z. D., & Gao, H. J. (2013). Distributed H-infinity filtering for a class of Markovian jump non-linear time-delay systems over lossy sensor networks. *IEEE Transactions on Industrial Electronics*, *60*(10), 4665–4672.

Fu, D., Aghezzaf, E.-H., & Keyser, R. D. (2014). A model predictive control framework for centralised management of a supply chain dynamical system. *Systems Science and Control Engineering: An Open Access Journal*, *2*(1), 250–260.

Hu, J., Wang, Z. D., Gao, H. J., & Stergioulas, L. K. (2012). Extended Kalman filtering with stochastic nonlinearities and multiple missing measurements. *Automatica*, *48*(9), 2007–2015.

Hu, J., Wang, Z. D., Shen, B., & Gao, H. J. (2013). Quantised recursive filtering for a class of nonlinear systems with multiplicative noises and missing measurements. *International Journal of Control*, *86*(4), 650–663.

Jahns, T. M., Becerra, R. C., & Ehsani, M. (1991). Integrated current regulation for a brushless ECM drive. *IEEE Transactions on Power Electronics*, *6*(1), 118–126.

Jang, J. H., Sul, S. K., Ha, J. I., Ide, K., & Sawamura, M. (2003). Sensorless drive of surface-mounted permanent-magnet motor by high-frequency signal injection based on magnetic saliency. *IEEE Transactions on Industry Applications*, *39*(4), 1031–1039.

Julier, S. J. (2002). *The scaled unscented transformation*. Proceedings of the 2002 American control conference, Jefferson City, Vol. 6, pp. 4555–4559.

Julier, S. J. (2003). *The spherical simplex unscented transformation*. Annual American Control Conference (ACC 2003), Denver, CO, Vol. 1–6, pp. 2430–2434.

Julier, S. J., & Uhlmann, J. K. (2002). *Reduced sigma point filters for the propagation of means and covariances through nonlinear transformations*. Paper presented at the 20th Annual American Control Conference (ACC), Anchorage, AK, Vol. 1-6, pp. 887–892.

Julier, S. J., & Uhlmann, J. K. (2004). Unscented filtering and nonlinear estimation. *Proceedings of the IEEE*, *92*(3), 401–422.

Julier, S. J., Uhlmann, J. K., & Durrant-Whyte, H. F. (2000). A new method for the nonlinear transformation of means and covariance in filters and estimators. *IEEE Transactions on Automatic Control*, *45*(3), 477–482.

Kallapur, A. G., Petersen, I. R., & Anavatti, S. G. (2009). A discrete-time robust extended Kalman filter for uncertain systems with sum quadratic constraints. *IEEE Transactions on Automatic Control*, *54*(4), 850–854.

Kim, T. H., & Ehsani, M. (2004). Sensorless control of the BLDC motors from near-zero to high speeds. *IEEE Transactions on Power Electronics*, *19*(6), 1635–1645.

Kim D.-K., Lee K.-W., & Kwon, B. (2006). Commutation torque ripple reduction in a position sensorless brushless DC motor drive. *IEEE Transactions on Power Electronics*, *21*(6), 1762–1768.

Kluge, S., Reif, K., & Brokate, M. (2010). Stochastic stability of the extended Kalman filter with intermittent observations. *IEEE Transactions on Automatic Control*, *55*(2), 514–518.

Kulkarni, A. B., & Ehsani, M. (1992). A novel position sensor elimination technique for the interior permanent-magnet synchronous motor drive. *IEEE Transactions on Industry Applications*, *28*(1), 144–150.

Lai Y.-S., & Lin Y.-K. (2011). A unified approach to zero-crossing point detection of back EMF for brushless DC motor dives without current and hall sensors. *IEEE Transactions on Power Electronics*, *26*(6), 1704–1713.

Liang, J. L., Sun, F. B., & Liu, X. H. (2014). Finite-horizon $H_\infty$ filtering for time-varying delay systems with randomly varying nonlinearities and sensor saturations. *Systems Science and Control Engineering: An Open Access Journal*, *2*(1), 108–118.

Moreira, J. C. (1996). Indirect sensing for rotor flux position of permanent magnet AC motors operating over a wide speed range. *IEEE Transactions on Industry Applications*, *32*(6), 1394–1401.

Ogasawara, S., & Akagi, H. (1991). An approach to position sensorless drive for brushless DC motors. *IEEE Transactions on Industry Applications*, *27*(5), 928–933.

Rodriguez, F., & Emadi, A. (2007). A novel digital control technique for brushless DC motor drives. *IEEE Transactions on Industrial Electronics*, *54*(5), 2365–2373.

Shao, W. J., Nolan, D., Teissier, M., & Swanson, D. (2003). A novel microcontroller-based sensorless brushless DC (BLDC) motor drive for automotive fuel pumps. *IEEE Transactions on Industry Applications*, *39*(6), 1734–1740.

Shen, B., Wang, Z., Ding, D., & Shu, H. (2013). H-infinity state estimation for complex networks with uncertain inner coupling and incomplete measurements. *IEEE Transactions on Neural Networks and Learning Systems*, *24*(12), 2027–2037.

Tatematsu, K., Hamada, D., Uchida, K., Wakao, S., & Onuki, T. (2000). New approaches with sensorless drives. *IEEE Industry Applications Magazine*, *6*(4), 44–50.

Terzic, B., & Jadric, M. (2001). Design and implementation of the extended Kalman filter for the speed and rotor position estimation of brushless DC motor. *IEEE Transactions on Industrial Electronics*, *48*(6), 1065–1073.

Zhu, G. C., Kaddouri, A., Dessaint, L. A., & Akhrif, O. (2001). A nonlinear state observer for the sensorless control of a permanent-magnet AC machine. *IEEE Transactions on Industrial Electronics*, *48*(6), 1098–1108.

# A non-continuum lumped-parameter dynamic model applied to Indian traffic

Ajitha Thankappan[a], Amritha Sunny[b], Lelitha Vanajakshi[b]* and Shankar C. Subramanian[b]

[a]*Department of Civil Engineering, Government College of Engineering, Kannur, Kerala 670 563, India;* [b]*Department of Civil Engineering, Indian Institute of Technology Madras, Chennai 600 036, India*

Dynamic traffic flow models are essential for obtaining information about the time evolution of variables describing the traffic flow phenomena and have a critical role in the development and implementation of real-time applications such as Intelligent Transportation Systems. Macroscopic traffic flow models that treat the traffic as a continuum are preferred for such applications. But, existing macroscopic models characterize homogeneous traffic, and may not be directly applicable to capture the vehicle heterogeneity seen on Indian roads. To address this issue, a non-continuum macroscopic dynamic traffic flow model based on the lumped-parameter approach was developed in this study. The model was developed based on the conservation of vehicles equation and a dynamic speed equation, incorporating an empirically developed traffic stream model, which is an important contribution of this study. Using this model, an estimation scheme has been developed based on the Kalman filtering technique to estimate traffic states in real time. The proposed scheme was implemented and corroborated for the heterogeneous traffic conditions existing in India. The performance of this scheme has been evaluated and the results obtained have been found to be promising.

**Keywords:** dynamic traffic flow modelling; lumped-parameter approach; extended Kalman filter; real-time traffic state estimation

## 1. Introduction

The exponential increase in quantity and complexity of road traffic, coupled with insufficient developments in infrastructure to match, has led to serious traffic problems on the Indian road. A cost-effective way to manage these problems is by utilizing the existing facility more efficiently through operational means. Such an approach hinges on a better understanding of the dynamics of traffic flow through traffic flow modelling. Traffic flow models mathematically study the interactions between vehicles, drivers and infrastructure to understand and predict the dynamic behaviour of traffic and develop optimal transportation patterns and networks. Such dynamic traffic models are particularly critical in the development and implementation of real-time applications such as Intelligent Transportation Systems (ITS).

A popular ITS application is the provision of real-time information about traffic states to travellers. Such a facility requires quantification of traffic congestion in terms of speed, delay, travel time and traffic density on the road. Among these, traffic density and travel time are the most important traffic parameters that can directly be used to determine the level of congestion. However, density and travel time (or speed) are spatial parameters that cannot be easily measured from the field using automated traffic sensors. They are usually estimated from other traffic parameters that can be measured using available sensors.

Accurate mathematical models are required to process the location-based information into spatial information. Model-based estimation schemes using techniques such as Kalman filters (KFs) are being increasingly used for the purpose because deterministic models have been found to be insufficient by themselves for accurate estimation of traffic variables in real time (Jabari and Liu, 2013; Wang, Li, Chen, & Ni, 2011). Such techniques have an additional advantage of being able to account for the uncertainty associated with traffic flow phenomena, which is of more relevance under traffic conditions such as in India where the randomness associated with traffic is high. Another advantage of this approach is that in order to estimate state variables at a given instant of time, one needs only the estimate from the previous instant of time and the measured data collected during that instant of time. Thus, unlike data or pattern-driven methods, the data measured during all the previous instants of time need not be stored, which is advantageous in places where the system is being implemented and hence a database is not available, such as under Indian conditions.

Traffic flow models have been classified into microscopic and macroscopic based on the level of detail used

*Corresponding author. Email: lelitha@iitm.ac.in

to describe traffic. Macroscopic traffic flow models are better suited for real-time applications and hence such an approach is attempted in this study. A majority of the macroscopic models reported in literature treat traffic as a continuum (Hoogerdoorn & Bovy, 2001). However, this continuum approach to model traffic flow has been criticized by several researchers (Papageorgiou, 1998; Tyagi, Darbha, & Rajagopal, 2009). The number of vehicles in a typical section of roadway is limited and cannot be described as a continuum, even in congested condition. Furthermore, continuum models allow the two-way propagation of disturbances, which is unrealistic for the traffic scenario. Because of these issues, alternative non-continuum approaches have been attempted for modelling traffic flow (Daganzo, 1994a, 1994b; Tyagi, Darbha, & Rajagopal, 2008). The present study proposed a non-continuum macroscopic dynamic traffic flow model for the estimation of traffic density using Kalman filtering technique. A few studies in this area of model-based density estimation are listed in the following table.

The modelling studies mentioned above have predominantly dealt with homogeneous traffic conditions. A few models developed for heterogeneous traffic conditions are discussed in this section. Logghe and Immers (2003) developed an extended Lighthill, Whitham and Richard (LWR) model to include different classes of vehicles. A microscopic theory of spatial-temporal congested traffic patterns in heterogeneous traffic flow with a variety of driver behavioural characteristics and vehicle parameters was presented by Kerner and Klenov (2004) based on three-phase traffic theory. A recent addition to heterogeneous traffic modelling is by Tang, Huangb, Zhao, and Shang (2009), where a new dynamic car-following model has been developed by applying the relationship between the microscopic and macroscopic variables. Studies reported from India include a few on the use of macroscopic models (Anand, Vanajakshi, & Subramanian, 2011; Padiath, Vanajakshi, & Subramanian, 2010; Padiath, Vanajakshi, Subramanian, & Manda, 2009; Tiwari, Fazio, Gaurav, & Chatteerjee, 2008). Other reported studies have focused on microscopic models that are not suited for real-time applications (Chakroborty & Kikuchi, 1999; Chakroborty & Maurya, 2008; Gupta, Chakroborty, & Mukherjee, 1998; Mallikarjuna & Rao, 2009; Mathew, Gundaliya, & Dhingra, 2006; Maurya & Chakroborty, 2008; Venkatesan, Gowri, & Sivanandan, 2008).

Thus, although there are a number of reports on macroscopic traffic flow, most of them focused on homogeneous traffic conditions. Reported heterogeneous traffic flow modelling has been mainly of the microscopic type, which is not applicable for real-time applications. Macroscopic models that are computationally tractable are particularly suitable for representing the real-time stream features such as traffic congestion on Indian urban

| No. | Authors (year) | Facility type, data source and type of data | Parameter estimated and technique used |
|---|---|---|---|
| 1. | Gazis and Knapp (1971) | Freeway, location-based, field data | Density, model-based extended Kalman filter (EKF) approach |
| 2. | Nahi and Trivedi (1973) | Freeway, location-based, field data as well as simulated data | Density, model-based KF approach |
| 3. | Gazis and Szeto (1974) | Multilane roadway, location-based, field data | Density and speed, model-based KF approach |
| 4. | Houpt, Kurkjian, Gershwin, and Willsky (1979) | Freeway, location-based, Field data as well as simulated data | Density, model-based scalar KF approach |
| 5. | Willsky et al. (1980) | Freeway, location-based, field data as well as simulated data | Density and speed, model-based KF approach |
| 6. | Kurkjian et al. (1980) | Freeway, location-based, field data as well as simulated data | Density, model-based scalar KF approach |
| 7. | Hoogendoorn and Bovy (2000) | Simulated data | Density and speed, method of moments |
| 8. | Zhongke (2003) | Freeway, location-based, field data as well as simulated data | Density, model-based EKF approach |
| 9. | Sun, Muñoz, and Horowitz (2004) | Highway, location-based and spatial, field data as well as simulated data | Density, mixture KF approach |
| 10. | Wang and Papageorgiou (2005) | Freeway, location-based and field data as well as simulated data | Density and speed, model-based EKF approach |
| 11. | Chu, Oh, and Recker (2005) | Freeway, location-based, field data as well as simulated data | Density and travel time, model-based adaptive Kalman filter approach |
| 12. | Hegyi, Girimonte, Babuska, and DeSchutter (2006) | Freeway, location-based, field data as well as simulated data | Density and speed, METANET Model + EKF and unscented Kalman filter |
| 13. | Wang, Papageorgiou, and Messmer (2007) | Freeway, location-based and field data as well as simulated data | Density and speed, model-based EKF approach |
| 14. | Wang, Papageorgiou, and Messmer (2008) | Freeway, location-based and field data as well as simulated data | Density and speed, model-based EKF approach |

roads. Due to the unsuitability of continuum models, non-continuum models are gaining importance in traffic modelling and this study presents one such model. A lumped-parameter macroscopic dynamic traffic flow model based on a non-continuum approach for the estimation of traffic density using the Kalman filtering technique is proposed here. The model formulation employed the law of conservation of vehicles inside a road section and a dynamic speed equation obtained using a traffic stream model. The chosen stream model was a two-regime steady-state speed–density relation that was developed for the specific traffic condition being analysed. Density and aggregate space mean speed were considered as the state variables of interest in this model. Using this model, an estimation scheme was developed based on the Kalman filtering technique. All traffic variables were quantified without considering traffic lanes in order to account for the lack of lane discipline. The scheme was first designed and implemented without considering the heterogeneity of traffic. In the next stage, the heterogeneity was incorporated into the scheme in two ways. In the first approach, heterogeneity was incorporated by expressing all variables in standard Passenger Car Unit (PCU) equivalent values (IRC, 1990). In the second approach, heterogeneity was incorporated by explicitly considering different categories of vehicles into the modelling and estimation processes. The above estimation schemes were corroborated using field data. The results showed the efficacy of the developed model and the estimation scheme in real-time estimation of traffic state under the heterogeneous traffic conditions existing in India.

## 2. The proposed model

The model proposed in this study is developed in a state-space form appropriate for the design of model-based scheme for estimation of traffic density using the Kalman filer. A dynamic non-continuum lumped-parameter macroscopic model was formulated for describing the flow of traffic. In the lumped-parameter approach, within a small section of roadway, the spatial variation of traffic variables (such as density, speed, etc.) is neglected and it is assumed that the variables depend only on time. The typical section length with automated data collection is less than 1 km and it is reasonable to make this assumption for this distance. The length of the section ($L$) used in this study is also in this range. To apply this procedure to a longer roadway, it must be divided into sections within which it is reasonable to neglect spatial variations in traffic characteristics. A schematic diagram of a typical road section is shown in Figure 1. The lumped-parameter approach results in the governing equations of the model being ordinary differential equations (ODEs) (in the continuous time domain) and ordinary difference equations (in the discrete time domain).

The number of vehicles inside the section per unit length (density) and the average space mean speed of traffic are spatial parameters that are difficult to measure in the field. However, being spatial in nature, they are good indicators of the state of traffic. Hence, they were considered as the macroscopic state variables in this study. The first governing equation of the model was formulated based on the conservation of vehicles inside the section as follows.

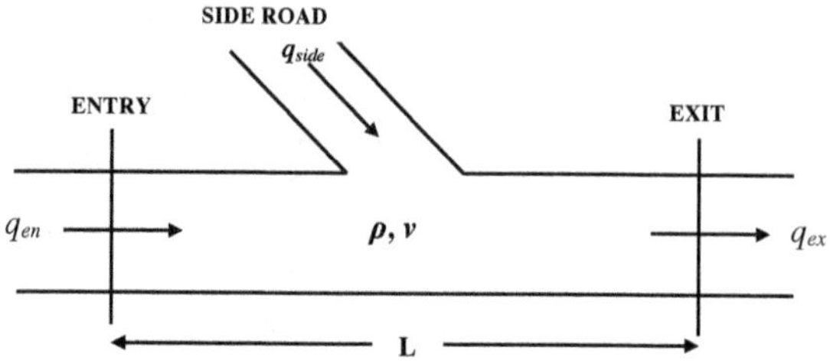


Figure 1. Schematic diagram of a typical road section.

Let $N(k)$ denote the number of vehicles inside the section at the $k$th instant of time. Then, the conservation of vehicles inside the section for a time step of $h$ can be represented as

$$N(k+1) = N(k) + h(q_{\text{en}}(k) - q_{\text{ex}}(k) + q_{\text{side}}(k)), \quad (1)$$

where $q_{\text{en}}(k)$ is the flow rate at which vehicles are entering into the section, $q_{\text{ex}}(k)$ is the flow rate at which vehicles are exiting from the section and $q_{\text{side}}(k)$ is the net flow rate at which the vehicles are entering into the section from the side road in the time interval $(k, k+1)$.

Dividing Equation (1) by the length of the section ($L$) resulted in

$$\rho(k+1) = \rho(k) + \frac{h}{L}(q_{\text{en}}(k) - q_{\text{ex}}(k) + q_{\text{side}}(k)), \quad (2)$$

where $\rho(k+1)$ denotes the density inside the section at the $(k+1)$th instant of time.

The second governing equation of the model is a dynamic speed equation formulated by incorporating the appropriate stream model for the specific traffic under study. A two-regime steady-state speed–density relationship was found to be the best fit for the traffic under study (Ajitha and Vanajakshi, 2012). The second governing equation was then obtained with the motive of minimizing the error ($e$) between the speed values estimated using this steady-state speed–density relation $v(\rho)$ and the observed speed values $v$, that is, $e = v(\rho) - v$. The time evolution of this error was hypothesized to behave as governed by

$$\frac{\mathrm{d}e}{\mathrm{d}t} = -a \cdot e(t), \quad (3)$$

where the parameter $a$ was selected to be positive. This equation is a linear homogeneous ODE and it is well known that its unique solution is $e(t) = e(0)\exp(-at)$ (Coddington, 1989), where $e(0)$ is the initial error (can be either positive or negative). Thus, the error will converge to zero with time. Although there may be other choices for describing the time evolution of the error function, an exponentially decaying error function (an exponential function is a very commonly used function in many phenomenological studies) has been chosen in this study since its performance will be comparably good to any alternate choice. This approach has been applied in other studies involving the dynamical systems approach (Ioannou & Chien, 1993; Swaroop, Hedrick, Chien, & Ioannou, 1994).

Substituting $e = v(\rho) - v$ in Equation (3) and rearranging resulted in

$$\frac{d(v(\rho))}{d\rho} \cdot \frac{d\rho}{dt} - \frac{dv}{dt} = -a \cdot (v(\rho) - v). \quad (4)$$

Discretizing Equation (4) using a time step $h$ resulted in

$$\frac{d(v(\rho))}{d\rho} \frac{(\rho(k+1) - \rho(k))}{h} - \frac{(v(k+1) - v(k))}{h} = -a(v(\rho) - v(k)). \quad (5)$$

By substituting $(\rho(k+1) - \rho(k))/h = (1/L)(q_{\text{en}}(k) - q_{\text{ex}}(k) + q_{\text{side}}(k))$ from Equation (2), the dynamic equation for average space mean speed inside the section (i.e. the equation governing the evolution of $v$) was obtained as

$$v(k+1) = v(k) + ah(v(\rho) - v(k)) + \frac{h}{L}\frac{d(v(\rho))}{d\rho}(q_{\text{en}}(k) - q_{\text{ex}}(k) + q_{\text{side}}(k)), \quad (6)$$

where $v(k + 1)$ denotes the average space mean speed of traffic inside the section at the $(k + 1)$th instant of time.

Thus, the general formulation of the non-continuum lumped-parameter model is represented by Equations (2) and (6). Here, the heterogeneity of traffic was not explicitly considered. The site-specific speed–density relationship $v(\rho)$ was also developed without considering heterogeneity and was incorporated in Equation (6). Brief descriptions of the developed speed–density relation and the details on incorporating it in Equation (6) are provided below.

Based on the field data collected from the study site, the best-fitting speed–density relation was identified empirically (Ajitha & Vanajakshi, 2012). To take into account the lack of lane discipline, the roadway was analysed without considering the traffic lanes and hence flow was expressed in veh/hr and density in veh/km. Using this data, the best-fitting traffic stream model was found to be a two-regime model with a constant speed in the free-flow regime and the speed decreasing nonlinearly in the congested regime up to the point where the jam density is reached. The functional form of this two-regime speed–density relation was obtained as

$$v = \begin{cases} 47 & \text{when } 0 \le \rho \le 156, \\ 20.226\left(\dfrac{519}{\rho} - 1\right) & \text{when } 156 \le \rho \le 519, \end{cases} \quad (7)$$

where the speed $v$ is expressed in kmph and the density $\rho$ in veh/km.

Incorporating this in Equation (6), the dynamic speed equation is reduced to

$$v(k+1) = \begin{cases} v(k) + ah(47 - v(k)) \quad \text{when } 0 \le \rho(k) \le 156, \\ v(k) + ah\left(20.226\left(\dfrac{519}{\rho(k)} - 1\right) - v(k)\right) \\ \quad - \dfrac{10497.29h}{L \cdot (\rho(k))^2}(q_{\text{en}}(k) - q_{\text{ex}}(k) + q_{\text{side}}(k)) \\ \quad \text{when } 156 \le \rho(k) \le 519. \end{cases} \quad (8)$$

Equations (2) and (8) represent the complete model for traffic without considering heterogeneity. This model was represented in the state-space form appropriate for the design of a model-based estimation scheme using the KF (more explanation on the KF is given in Section 3). However, a necessary condition for the KF to work correctly is that the system for which the states are to be determined is observable. A system is said to be observable if the internal states of the system can be determined using only the knowledge about the system out puts. Observability for a discrete time linear system can be tested by checking the rank of the obsevability matrix.

Consider a linear system whose model takes the form:

$$\mathbf{x}_{k+1} = \mathbf{A}\mathbf{x}_k + \mathbf{B}\mathbf{u}_k + \mathbf{w}_k, \quad (9)$$

$$\mathbf{z}_k = \mathbf{H}\mathbf{x}_k + \mathbf{v}_k, \quad (10)$$

where, $\mathbf{x}_k$ is the system state, $\mathbf{z}_k$ is the system output, $\mathbf{u}_k$ is the system input, $\mathbf{w}_k$ is the process disturbance and $\mathbf{v}_k$ is the measurement noise at the $k$th instant of time. The matrix $\mathbf{A}$ relates the state at the $k$th instant of time to the state at $(k + 1)$th instant of time and the matrix $\mathbf{B}$ relates the input to the state.

Then the system governed by Equations (2) and (8) is observable if and only if the observability matrix ($\mathbf{O}$) given by

$$\mathbf{O} = \begin{bmatrix} \mathbf{H} \\ \mathbf{HA} \\ \mathbf{HA}^2 \\ \vdots \\ \mathbf{HA}^{n-1} \end{bmatrix},$$

has rank equal to $n$, the order of the system (Gopal, 1984).

For the system under study represented by Equations (3) and (9), the values of the parameters, $\mathbf{A}$ and $\mathbf{H}$ for the free-flow regime were derived as

$$\mathbf{A} = \begin{bmatrix} 1 & 0 \\ 0 & 1 - ah \end{bmatrix}, \quad \mathbf{H} = [0 \quad 1].$$

The corresponding observability matrix was obtained as

$$\mathbf{O} = \begin{bmatrix} 0 & 1 \\ 0 & 1 - ah \end{bmatrix}.$$

The rank of this observability matrix was found to be 1 and thus the system is found to be unobservable in the free-flow regime.

To make the system observable, the governing equation for density as represented by Equation (2) was modified by expressing the flow passing exit section $q_{ex}$ in terms of the average space mean speed of traffic passing the exit location using the fundamental equation of traffic flow given by

$$q_{ex}(k) = \rho(k)v_{ex}(k), \tag{11}$$

where $\rho$ is the density and $v_{ex}$ is the average space mean speed of vehicles calculated using individual speed measurements at the exit. Substituting Equation (11) in Equation (2) provided:

$$\rho(k+1) = \rho(k) + \frac{h}{L}(q_{en}(k) - \rho(k)v_{ex}(k) + q_{side}(k)). \tag{12}$$

Incorporating this change in the dynamic equation for speed (Equation (7)), the governing equation for speed is reduced to

$$v(k+1) = \begin{cases} v(k) + ah(47 - v(k)) & \text{when } 0 \le \rho(k) \le 156, \\ v(k) + ah\left(20.226\left(\dfrac{519}{\rho(k)} - 1\right) - v(k)\right) & \\ +\dfrac{-10497.29h}{L\cdot(\rho(k))^2}(q_{en}(k) - \rho(k)\cdot v_{ex}(k) & \\ +q_{side}(k)) & \text{when } 156 \le \rho(k) \le 519. \end{cases} \tag{13}$$

The check for observability of the system governed by the modified state equations (Equations (12) and (13)) was again carried out for free-flow and congested regimes as detailed above and the rank for the new observability matrices was found to be equal to 2 and thus the system was observable. Equations (12) and (13) govern the traffic system without considering heterogeneity. This model was then modified to incorporate heterogeneity in two different ways as explained below.

One common approach to consider the mixture of different categories of vehicles in a traffic stream is to convert them into a homogeneous equivalent using standard PCU values (IRC, 1990). This approach has been first adopted in this study to account for the effect of presence of several vehicle types. Thus, in Equations (12) and (13), traffic flow was considered in PCU/hr and traffic density in PCU/km. The site-specific speed–density relationship $v(\rho)$ was developed using PCU converted data (Ajitha & Vanajakshi, 2012) and was incorporated in Equation (13). While developing $v(\rho)$ in this case, the data were measured separately for different categories of vehicles and then converted in to PCU units. The functional form of this speed–density relationship was obtained as

$$v = \begin{cases} 47 & \text{when } 0 \le \rho \le 143, \\ 23.5\left(\dfrac{429}{\rho} - 1\right) & \text{when } 143 \le \rho \le 429, \end{cases} \tag{14}$$

where the density $\rho$ is in PCU/km.

Incorporating this in Equation (13), the dynamic speed equation is reduced to

$$v(k+1) = \begin{cases} v(k) + ah(47 - v(k)) & \text{when } 0 \le \rho(k) \le 143, \\ v(k) + ah\left(23.5\left(\dfrac{429}{\rho(k)} - 1\right) - v(k)\right) & \\ -\dfrac{10081.5h}{L\cdot(\rho(k))^2}(q_{en}(k) - \rho(k)\cdot v_{ex}(k) & \\ +q_{side}(k)) & \text{when } 143 \le \rho(k) \le 429. \end{cases} \tag{15}$$

Thus, the complete model formulation after PCU incorporation was represented by Equations (12) and (15) with flow rates and density expressed in PCU/hr and PCU/km, respectively.

Another way of introducing heterogeneity in traffic flow is to consider different categories separately in the modelling process. The classification considered in this study was the three vehicle groups, namely two wheelers (TWs), three wheelers (ThWs) and four wheelers (FWs). TWs include motorcycles, scooters and mopeds and ThWs include auto-rickshaws and small three-wheeled tempos. The FW category can be further subdivided into classes such as light passenger cars, heavy commercial vehicles, and others. However, due to the difficulty in manual data extraction, this classification was followed in the present study. The procedure can be easily extended for other classes, if data are available. The above developed model was modified by incorporating separate state equations for the three different classes of vehicles considered, namely TWs, ThWs and FWs. The first three equations of the modified model were based on the conservation of the three classes of vehicles inside the section and was obtained as

$$\rho_{TW}(k+1) = \rho_{TW}(k) + \frac{h}{L}(q_{en}^{TW}(k) - \rho_{TW}(k) \cdot v_{ex}^{TW}(k) + q_{side}^{TW}(k)), \tag{16}$$

$$\rho_{ThW}(k+1) = \rho_{ThW}(k) + \frac{h}{L}(q_{en}^{ThW}(k) - \rho_{ThW}(k) \cdot v_{ex}^{ThW}(k) + q_{side}^{ThW}(k)), \tag{17}$$

$$\rho_{FW}(k+1) = \rho_{FW}(k) + \frac{h}{L}(q_{en}^{FW}(k) - \rho_{FW}(k) \cdot v_{ex}^{FW}(k) + q_{side}^{FW}(k)). \tag{18}$$

The remaining three equations were dynamic speed equations formulated using the stream models developed for these three classes. The stream models developed for the three categories of vehicles (Ajitha & Vanajakshi, 2012) were

$$v_{\mathrm{TW}}=\begin{cases}48 & \text{when } 0\le\rho_{\mathrm{TW}}\le 87,\\ 18.316\left(\dfrac{315}{\rho_{\mathrm{TW}}}-1\right) & \text{when } 87\le\rho_{\mathrm{TW}}\le 315,\end{cases}\tag{19}$$

$$v_{\mathrm{ThW}}=\begin{cases}40 & \text{when } 0\le\rho_{\mathrm{ThW}}\le 14,\\ 18.065\left(\dfrac{45}{\rho_{\mathrm{ThW}}}-1\right) & \text{when } 14\le\rho_{\mathrm{ThW}}\le 45,\end{cases}\tag{20}$$

$$v_{\mathrm{FW}}=\begin{cases}50 & \text{when } 0\le\rho_{\mathrm{FW}}\le 57,\\ 23.208\left(\dfrac{180}{\rho_{\mathrm{FW}}}-1\right) & \text{when } 57\le\rho_{\mathrm{FW}}\le 180.\end{cases}\tag{21}$$

Using these equations, the formulated dynamic speed equations for TWs, ThWs and FWs were obtained as

$$v_{\mathrm{TW}}(k+1)=\begin{cases}v_{\mathrm{TW}}(k)+ah(48-v_{\mathrm{TW}}(k))\\ \quad\text{when } 0\le\rho_{\mathrm{TW}}(k)\le 87,\\ v_{\mathrm{TW}}(k)+ah\left(18.316\left(\dfrac{315}{\rho_{\mathrm{TW}}(k)}-1\right)\right.\\ \left.\quad -\,v_{\mathrm{TW}}(k)\right)-\dfrac{5769.54h}{L.(\rho_{\mathrm{TW}}(k))^2}(q_{\mathrm{en}}^{\mathrm{TW}}(k)\\ \quad -\rho_{\mathrm{TW}}(k)\cdot v_{\mathrm{ex}}^{\mathrm{TW}}(k)+q_{\mathrm{side}}^{\mathrm{TW}}(k))\\ \quad\text{when } 87\le\rho_{\mathrm{TW}}(k)\le 315,\end{cases}\tag{22}$$

$$v_{\mathrm{ThW}}(k+1)=\begin{cases}v_{\mathrm{ThW}}(k)+ah(40-v_{\mathrm{ThW}}(k))\\ \quad\text{when } 0\le\rho_{\mathrm{ThW}}(k)\le 14,\\ v_{\mathrm{ThW}}(k)+ah\left(18.065\left(\dfrac{45}{\rho_{\mathrm{ThW}}(k)}-1\right)\right.\\ \left.\quad -\,v_{\mathrm{ThW}}(k)\right)-\dfrac{812.925h}{L\cdot(\rho_{\mathrm{ThW}}(k))^2}(q_{\mathrm{en}}^{\mathrm{ThW}}(k)\\ \quad -\rho_{\mathrm{ThW}}(k)\cdot v_{\mathrm{ex}}^{\mathrm{ThW}}(k)+q_{\mathrm{side}}^{\mathrm{ThW}}(k))\\ \quad\text{when } 14\le\rho_{\mathrm{ThW}}(k)\le 45,\end{cases}\tag{23}$$

$$v_{\mathrm{FW}}(k+1)=\begin{cases}v_{\mathrm{FW}}(k)+ah(50-v_{\mathrm{FW}}(k))\\ \quad\text{when } 0\le\rho_{\mathrm{FW}}(k)\le 57,\\ v_{\mathrm{FW}}(k)+ah\left(23.208\left(\dfrac{180}{\rho_{\mathrm{FW}}(k)}-1\right)\right.\\ \left.\quad -\,v_{\mathrm{FW}}(k)\right)-\dfrac{4177.44h}{L\cdot(\rho_{\mathrm{FW}}(k))^2}(q_{\mathrm{en}}^{\mathrm{FW}}(k)\\ \quad -\rho_{\mathrm{FW}}(k)\cdot v_{\mathrm{ex}}^{\mathrm{FW}}(k)+q_{\mathrm{side}}^{\mathrm{FW}}(k))\\ \quad\text{when } 57\le\rho_{\mathrm{FW}}(k)\le 180.\end{cases}\tag{24}$$

Thus, it can be seen that all the above models have two components – one component that was derived using the conservation of vehicles and the hypothesis regarding the evolution of the error $e$, which will hold for any road segment. The second component was obtained from a traffic stream model. Such a stream model is not available for Indian road traffic conditions and hence is one of the contributions of this study. It is well known that traffic stream models are section-/location-dependent (Gartner, Messer, & Rathi, 2001). Hence, the model is transferable only to sections with similar characteristics. In other cases, the section-specific stream models need to be known or developed.

Next, the estimation scheme was developed using the Kalman filtering technique. The KF is an optimal state estimator applicable for dynamic systems. Although it was originally derived for linear systems, it can be extended to nonlinear systems and the filter so obtained is referred to as the EKF. In the present study, as the models developed are linear in the uncongested or free-flow regime and nonlinear in the congested regime, both the linear and the EKFs were used to estimate the traffic state as detailed below.

## 3. Estimation scheme

The KF (Kalman, 1960) is a popular tool for recursive estimation of variables that characterize a system (these variables are usually referred to as 'state variables'). The KF is a model-based estimation scheme that takes into account the stochastic properties of the process disturbance and the measurement noise. The process disturbance and the measurement noise are assumed to be independent of one another, white and normally distributed with zero mean. The KF works like a predictor corrector algorithm, that is, it first predicts an 'a priori' estimate of the state variables using the system model and the state estimate from the previous time interval, and then corrects the same using measurements to obtain an 'a posteriori' state estimate. The KF has been widely used in many disciplines including the field of transportation (Nanthawichit, Nakatsuji, & Suzuki, 2003; Okutani & Stephanedes, 1984; Wang & Papageorgiou, 2005; Wang, Papageorgiou, & Messmer, 2008). The KF is used for estimation and prediction when the governing equations of the system are linear. When the governing equations are nonlinear, an EKF (Jazwinsky, 1970) is commonly used. The EKF linearizes the governing equations at each time step about the estimate obtained from the previous time step.

Using the traffic flow models presented in the previous section, the estimation scheme was obtained as explained below. In the first case, where heterogeneity was not considered, the density of vehicles ($\rho$) expressed in veh/km and the average space mean speed ($v$) of vehicles inside the section in km/hr were taken as the state variables of interest. The output variable was taken as the measured values of the average space mean speed of vehicles. The rates at which vehicles enter into the section from upstream and from the side road in veh/hr were provided as inputs to the estimation scheme. Similarly, for the second case, where heterogeneity was incorporated by expressing all variables in PCU units, density ($\rho$) expressed in PCU/km and average space mean speed ($v$) in km/hr were considered as state variables, measured values of average space mean speed of vehicles as the output variable and flow passing the upstream and side road in PCU/hr as inputs to the scheme. For the third case, where different classes of vehicles were explicitly considered in modelling, density of TWs ($\rho_{\mathrm{TW}}$), ThWs ($\rho_{\mathrm{ThW}}$) and FWs ($\rho_{\mathrm{FW}}$) and average space mean speed of TWs ($v_{\mathrm{TW}}$), ThWs ($v_{\mathrm{ThW}}$) and FWs ($v_{\mathrm{FW}}$) were taken as the state variables of interest. The output variables were taken as the measured values of space mean speeds of these three classes of vehicles and inputs were taken as flow of these three classes of vehicles from upstream and through side roads.

Since the governing equations of the models in the uncongested regime were linear, the KF was used as below. Consider a linear system whose model takes the form:

$$\mathbf{x}_{k+1} = \mathbf{A}\mathbf{x}_k + \mathbf{B}\mathbf{u}_k + \mathbf{w}_k, \tag{25}$$

$$\mathbf{z}_k = \mathbf{H}\mathbf{x}_k + \mathbf{v}_k, \tag{26}$$

where $\mathbf{x}_k$ is the system state, $\mathbf{z}_k$ is the system output, $\mathbf{u}_k$ is the system input, $\mathbf{w}_k$ is the process disturbance and $\mathbf{v}_k$ is the measurement noise at the $k$th instant of time. The matrix $\mathbf{A}$ relates the state at the $k$th instant of time to the state at $(k+1)$th instant of time and the matrix $\mathbf{B}$ relates the input to the state.

Now the following steps are used recursively for estimation:

(1) The a priori estimate in the $(k+1)^{\text{th}}$ interval of time was obtained through

$$\hat{\mathbf{x}}_{k+1}^- = \mathbf{A}\hat{\mathbf{x}}_k^+ + \mathbf{B}\mathbf{u}_k.$$

(2) The a priori error covariance in the $(k+1)$th interval of time was obtained through

$$\mathbf{P}_{k+1}^- = \mathbf{A}\mathbf{P}_k^+\mathbf{A}^{\mathrm{T}} + \mathbf{Q}.$$

(3) The Kalman gain $\mathbf{K}_{k+1}$ was calculated through

$$\mathbf{K}_{k+1} = \mathbf{P}_{k+1}^-\mathbf{H}^{\mathrm{T}}(\mathbf{H}\mathbf{P}_{k+1}^-\mathbf{H}^{\mathrm{T}} + \mathbf{R})^{-1}.$$

(4) Then, the a posteriori state estimate was calculated through

$$\mathbf{x}_{k+1}^+ = \mathbf{x}_{k+1}^- + \mathbf{K}_{k+1}(\mathbf{z}_{k+1} - \mathbf{H}\mathbf{x}_{k+1}^-).$$

(5) Finally, the a posteriori error covariance was obtained through

$$\mathbf{P}_{k+1}^+ = (1 - \mathbf{K}_{k+1}\mathbf{H})\mathbf{P}_{k+1}^-.$$

Here, $\mathbf{Q}$ is the process disturbance covariance, $\mathbf{R}$ is the measurement noise covariance and $\mathbf{H}$ is the matrix which relates the state to the measurement. Also, $\hat{\mathbf{x}}_k^-$ and $\hat{\mathbf{x}}_k^+$ denote, respectively, the a priori estimate and the a posteriori estimate of the state variables at the $k$th instant of time. Similarly, $\mathbf{P}_k^-$ and $\mathbf{P}_k^+$ denote, respectively, the a priori and the a posteriori error covariance at the $k$th instant of time.

In the present study, since the models in the congested regime were nonlinear, an extended KF was used. The EKF linearizes the governing equations at each time step about the estimate obtained from the previous time step. Consider a nonlinear system whose model is given by

$$\mathbf{x}_{k+1} = \mathbf{f}(\mathbf{x}_k, \mathbf{u}_k, \mathbf{w}_{1k}), \tag{27}$$

$$\mathbf{z}_k = \mathbf{g}(\mathbf{x}_k, \mathbf{v}_{1k}), \tag{28}$$

where $\mathbf{f}$ represents the nonlinear function that relates the state at time step $k$ to the state at time step $k+1$. Similarly, $\mathbf{g}$ is the nonlinear function that relates the state to the measurement. The above equations can be linearized using Taylor's Series expansion to result in

$$\mathbf{x}_{k+1} = \tilde{\mathbf{x}}_{k+1} + \mathbf{A}_1(\mathbf{x}_k - \hat{\mathbf{x}}_k^+) + \mathbf{W}\mathbf{w}_{1k}, \tag{29}$$

$$\mathbf{z}_k = \tilde{\mathbf{z}}_k + \mathbf{H}_1(\mathbf{x}_k - \tilde{\mathbf{x}}_k) + \mathbf{V}\mathbf{v}_{1k}, \tag{30}$$

where $\tilde{\mathbf{x}}$ and $\tilde{\mathbf{z}}$ are the approximate state and measurement variables without considering the process disturbance and measurement noise as indicated by Equations (31) and (32), $\mathbf{A}_1$ is the matrix of the partial derivative of $\mathbf{f}$ with respect to $\mathbf{x}$, $\mathbf{W}$ is the matrix of the partial derivative of $\mathbf{f}$ with respect to $\mathbf{w}_1$, $\mathbf{H}_1$ is the matrix of the partial derivative of $\mathbf{g}$ with respect to $\mathbf{x}$ and $\mathbf{V}$ is the matrix of the partial derivative of $\mathbf{g}$ with respect to $\mathbf{v}_1$. Thus,

$$\tilde{\mathbf{x}}_{k+1} = \mathbf{f}(\hat{\mathbf{x}}_k^+, \mathbf{u}_k, 0), \tag{31}$$

$$\tilde{\mathbf{z}}_k = \mathbf{g}(\tilde{\mathbf{x}}_k, 0). \tag{32}$$

Now the following steps were followed recursively for estimation using EKF:

(1) The a priori estimate in the $(k+1)$th interval of time was obtained through

$$\hat{\mathbf{x}}_{k+1}^- = \mathbf{f}(\hat{\mathbf{x}}_k^+, \mathbf{u}_k).$$

(2) The a priori error covariance in the $(k+1)$th interval of time was obtained through

$$\mathbf{P}_{k+1}^- = \mathbf{A}_1\mathbf{P}_k^+\mathbf{A}_1^{\mathrm{T}} + \mathbf{W}\mathbf{Q}\mathbf{W}^{\mathrm{T}}.$$

(3) The Kalman gain $\mathbf{K}_{k+1}$ was calculated through

$$\mathbf{K}_{K+1} = \mathbf{P}_{k+1}^-\mathbf{H}_1^{\mathrm{T}}[\mathbf{H}_1\mathbf{P}_{k+1}^-\mathbf{H}_1^{\mathrm{T}} + \mathbf{V}\mathbf{R}\mathbf{V}^{\mathrm{T}}]^{-1}.$$

(4) Then, the a posteriori state estimate was calculated through

$$\hat{\mathbf{x}}_{k+1}^+ = \hat{\mathbf{x}}_{k+1}^- + \mathbf{K}_{k+1}(\mathbf{z}_{k+1} - \mathbf{g}(\hat{\mathbf{x}}_{k+1}^-)).$$

(5) Finally, the a posteriori error covariance was obtained through

$$\mathbf{P}^{+}_{k+1} = [\mathbf{I} - \mathbf{K}_{k+1}\mathbf{H}_1]\mathbf{P}^{-}_{k+1}.$$

In the present study, the following parameters were identified in free-flow regime for the first case where heterogeneity was not considered:

$$\mathbf{x}(k) = \begin{bmatrix} \rho(k) \\ v(k) \end{bmatrix}, \quad \mathbf{u} = \begin{bmatrix} q_{\mathrm{en}} \\ q_{\mathrm{side}} \end{bmatrix}, \quad \mathbf{z}(k) = v(k),$$

$$\mathbf{A} = \begin{bmatrix} 1 - \dfrac{h \cdot v_{\mathrm{ex}}(k)}{L} & 0 \\ 0 & 1 - ah \end{bmatrix},$$

$$\mathbf{B} = \begin{bmatrix} \dfrac{h}{L} & \dfrac{h}{L} \\ 0 & 0 \end{bmatrix}, \quad \mathbf{H} = [0 \quad 1].$$

In the congested regime for this case, the parameters were obtained as

$$\mathbf{f} = \begin{bmatrix} \rho(k) + \dfrac{h}{L}(q_{\mathrm{en}}(k) - \rho(k) \cdot v_{\mathrm{ex}}(k) + q_{\mathrm{side}}(k)) \\ v(k) + ah\left(20.226\left(\dfrac{519}{\rho(k)} - 1\right) - v(k)\right) \\ -\dfrac{10497.29h}{L \cdot (\rho(k))^2}(q_{\mathrm{en}}(k) - \rho(k) \cdot v_{\mathrm{ex}}(k) \\ +q_{\mathrm{side}}(k)) \end{bmatrix},$$

$$\mathbf{A}_1 = \begin{bmatrix} 1 - \dfrac{h \cdot v_{\mathrm{ex}}(k)}{L} & 0 \\ \dfrac{-10497.29h}{(\rho(k))^2}\left[a - \dfrac{2}{L \cdot \rho(k)}(q_{\mathrm{en}}(k) + q_{\mathrm{side}}(k)) + \dfrac{v_{\mathrm{ex}}(k)}{L}\right] & 1 - ah \end{bmatrix},$$

$$\mathbf{W} = \begin{bmatrix} h & 0 \\ 0 & h \end{bmatrix}, \quad \mathbf{H}_1 = [0 \quad 1], \quad \mathbf{V} = 1.$$

where density $\rho$ is in veh/km and flow rates $q_{\mathrm{en}}$ and $q_{\mathrm{side}}$ are in veh/hr.

Similarly, for the second case, where all variables were considered in PCU units, the parameters for the free-flow regime were the same as that of the previous case except that the variables such as density and flow rates were considered in PCU units. In the congested regime, the parameters such as $\mathbf{W}$, $\mathbf{H}_1$ and $\mathbf{V}$ remain same as that of the first case and the nonlinear function $\mathbf{f}$ and $\mathbf{A}_1$ was obtained as

$$\mathbf{f} = \begin{bmatrix} \rho(k) + \dfrac{h}{L}(q_{\mathrm{en}}(k) - \rho(k) \cdot v_{\mathrm{ex}}(k) + q_{\mathrm{side}}(k)) \\ v(k) + ah\left(23.5\left(\dfrac{429}{\rho(k)} - 1\right) - v(k)\right) \\ -\dfrac{10081.5h}{L \cdot (\rho(k))^2}(q_{\mathrm{en}}(k) - \rho(k) \cdot v_{\mathrm{ex}}(k) \\ +q_{\mathrm{side}}(k)) \end{bmatrix},$$

$$\mathbf{A}_1 = \begin{bmatrix} 1 - \dfrac{h \cdot v_{\mathrm{ex}}(k)}{L} & 0 \\ \dfrac{-10081.5h}{(\rho(k))^2}\left[a - \dfrac{2}{L \cdot \rho(k)}(q_{\mathrm{en}}(k) + q_{\mathrm{side}}(k)) + \dfrac{v_{\mathrm{ex}}(k)}{L}\right] & 1 - ah \end{bmatrix},$$

where density $\rho$ is in PCU/km and flow rates $q_{\mathrm{en}}$ and $q_{\mathrm{side}}$ are in PCU/hr.

For the third case, where different classes of vehicles were considered separately, the parameters in free-flow regime were obtained as

$$\mathbf{x}(k) = \begin{bmatrix} \rho_{\mathrm{TW}}(k) \\ \rho_{\mathrm{ThW}}(k) \\ \rho_{\mathrm{FW}}(k) \\ v_{\mathrm{TW}}(k) \\ v_{\mathrm{ThW}}(k) \\ v_{\mathrm{FW}}(k) \end{bmatrix}, \quad \mathbf{u} = \begin{bmatrix} q^{\mathrm{TW}}_{\mathrm{en}} \\ q^{\mathrm{TW}}_{\mathrm{side}} \\ q^{\mathrm{ThW}}_{\mathrm{en}} \\ q^{\mathrm{ThW}}_{\mathrm{side}} \\ q^{\mathrm{FW}}_{\mathrm{en}} \\ q^{\mathrm{FW}}_{\mathrm{side}} \end{bmatrix}, \quad \mathbf{z}(k) = \begin{bmatrix} v_{\mathrm{TW}}(k) \\ v_{\mathrm{ThW}}(k) \\ v_{\mathrm{FW}}(k) \end{bmatrix},$$

$$\mathbf{A} = \begin{bmatrix} 1 - \dfrac{h \cdot v^{\mathrm{TW}}_{\mathrm{ex}}}{L} & 0 & 0 & 0 & 0 & 0 \\ 0 & 1 - \dfrac{h \cdot v^{\mathrm{ThW}}_{\mathrm{ex}}}{L} & 0 & 0 & 0 & 0 \\ 0 & 0 & 1 - \dfrac{h \cdot v^{\mathrm{FW}}_{\mathrm{ex}}}{L} & 0 & 0 & 0 \\ 0 & 0 & 0 & 1 - ah & 0 & 0 \\ 0 & 0 & 0 & 0 & 1 - ah & 0 \\ 0 & 0 & 0 & 0 & 0 & 1 - ah \end{bmatrix},$$

$$\mathbf{H} = \begin{bmatrix} 0 & 0 & 0 & 1 & 0 & 0 \\ 0 & 0 & 0 & 0 & 1 & 0 \\ 0 & 0 & 0 & 0 & 0 & 1 \end{bmatrix}$$

and in the congested regime for the same case, the nonlinear function **f** was obtained as

$$f = \begin{bmatrix} \rho_{TW}(k) + \frac{h}{L}(q_{en}^{TW}(k) - \rho_{TW}(k) \\ \cdot v_{ex}^{TW}(k) + q_{side}^{TW}(k)) \\ \rho_{ThW}(k) + \frac{h}{L}(q_{en}^{ThW}(k) - \rho_{ThW}(k) \\ \cdot v_{ex}^{ThW}(k) + q_{side}^{ThW}(k)) \\ \rho_{FW}(k) + \frac{h}{L}(q_{en}^{FW}(k) - \rho_{FW}(k) \\ \cdot v_{ex}^{FW}(k) + q_{side}^{FW}(k)) \\ v_{TW}(k) + ah\left(18.316\left(\frac{315}{\rho_{TW}(k)} - 1\right) - v_{TW}(k)\right) \\ - \frac{5769.54h}{L \cdot (\rho_{TW}(k))^2}(q_{en}^{TW}(k) - \rho_{TW}(k) \\ \cdot v_{ex}^{TW}(k) + q_{side}^{TW}(k)) \\ v_{ThW}(k) + ah\left(18.065\left(\frac{45}{\rho_{ThW}(k)} - 1\right) - v_{ThW}(k)\right) \\ - \frac{812.925h}{L \cdot (\rho_{ThW}(k))^2}(q_{en}^{ThW}(k) - \rho_{ThW}(k) \\ \cdot v_{ex}^{ThW}(k) + q_{side}^{ThW}(k)) \\ v_{FW}(k) + ah\left(23.208\left(\frac{180}{\rho_{FW}(k)} - 1\right) - v_{FW}(k)\right) \\ - \frac{4177.44h}{L \cdot (\rho_{FW}(k))^2}(q_{en}^{FW}(k) - \rho_{FW}(k) \\ \cdot v_{ex}^{FW}(k) + q_{side}^{FW}(k)) \end{bmatrix}.$$

Other parameters such as $\mathbf{A}_1$, $\mathbf{W}_1$, $\mathbf{H}_1$ and $\mathbf{V}$ for this case were derived as before. The initial values of the state variables were assumed in all the three cases and the above estimation schemes were implemented. The results from these estimation schemes were compared with the actual values collected from the field.

## 4. Data collection and extraction

Data collection for the present study was carried out using the video recording technique on a stretch on the Rajiv Gandhi road, Chennai, India. The selected stretch was a six-lane roadway, with three lanes in each direction. For the present study, only one direction of traffic was considered. The section had one side road and the vehicles entering through it were counted manually. Video data were collected at the entry and exit points of the selected section of roadway during one hour each on five week days and two hours and three hours each on two other week days. The collected videos were later analysed in the laboratory to extract the required data. Data extraction was carried out manually due to lack of any reliable automated data extraction methods and was carried out separately for all the three classes of vehicles considered. The flow data passing the entry and exit sections were extracted for every one-minute interval by counting the number of TWs, ThWs and FWs travelling in all the three lanes. The spot speeds of TWs, ThWs and FWs passing the entry and exit sections were determined for every one minute by measuring the time taken to cross a known distance. The space mean speeds were then computed for these three classes by taking the harmonic mean (HM) of spot speeds (May, 1990). The average space mean speed values of vehicles were computed by averaging the HM values at the entry and exit sections. The actual densities of the three classes of vehicles at every one-minute interval required for corroboration of the estimation scheme were determined using input–output analysis (May, 1990). The initial numbers of TWs, ThWs and FWs present inside the section (required for the input output analysis) were measured by taking a still picture of the section at the start of the data collection. For implementing and corroborating the first scheme, data were considered without classifying them into different classes. In the case of the second scheme, PCU converted values were used and for the third scheme, the classified data were used.

## 5. Results

The estimated values of state variables were compared with the actual values obtained from field. The performance of this approach was quantified by calculating the mean absolute percentage error (MAPE) given by

$$\text{MAPE} = \left[\frac{1}{N}\sum_{i=1}^{N}\frac{|x_{est} - x_{obs}|}{x_{obs}}\right]100, \quad (33)$$

where $x_{est}$ and $x_{obs}$ are the estimated and observed values of the state variable, respectively, and $N$ is the total number of observations. The estimated values of average space mean speed values were found to converge with the measured values in all the three cases. The plots of the estimated values of densities using PCU converted data against the actual values for two representative days are shown in Figures 2 and 3, respectively. The MAPE values for traffic density estimates for all days using the three different schemes are tabulated in Table 1.

According to Lewis' interpretation of MAPE results (Kenneth & Ronald, 1982), any forecast with a MAPE value of up to 10% is considered as highly accurate, 11–20% as good, 21–50% as reasonable and greater than 50% as weak and inaccurate. Based on this, it can be seen that the results are good for five out of the seven days and reasonable for the remaining two days for all the schemes.

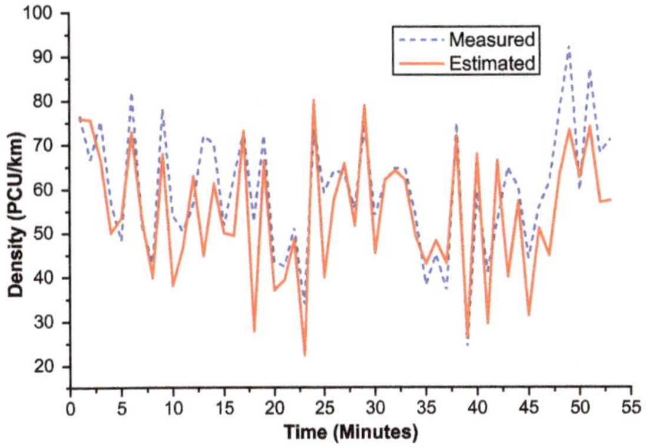


Figure 2. Comparison of the actual and estimated values of density of representative Day 1.

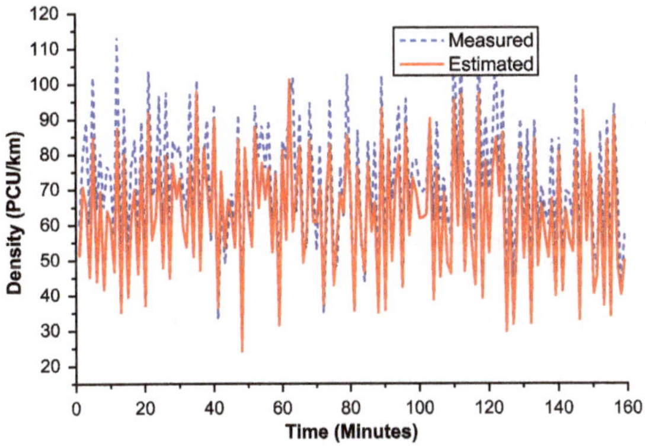


Figure 3. Comparison of the actual and estimated values of density of representative Day 2.

Table 1. MAPE for density estimation.

| | MAPE (%) | | |
|---|---|---|---|
| Day | Without considering heterogeneity | Heterogeneity introduced in terms of PCU | Heterogeneity by including different classes |
| Day 1 | 20.2 | 17.7 | 17.6 |
| Day 2 | 13.1 | 11.6 | 14.0 |
| Day 3 | 13.0 | 13.9 | 12.6 |
| Day 4 | 24.1 | 23.8 | 23.7 |
| Day 5 | 23.3 | 25.6 | 23.8 |
| Day 6 | 15.1 | 14.2 | 14.6 |
| Day 7 | 14.0 | 16.7 | 20.8 |

It can also be seen that introduction of heterogeneity in terms of PCU or considering different classes of vehicles separately improved results for a few days. However, the improvement may not be significant enough compared to the efforts involved in data collection and extraction. Considering the stochastic nature of Indian traffic, the performance of all the estimation schemes is promising.

## 6. Conclusions

A model-based approach was developed for the estimation of traffic states in Indian heterogeneous traffic conditions using the Kalman filtering approach. A non-continuum lumped-parameter macroscopic model was proposed using the law of conservation of vehicles and a dynamic speed equation incorporating an empirically developed traffic stream model. Traffic density and aggregate space mean speed were identified as the state variables of interest. The scheme was implemented initially without considering heterogeneity and then heterogeneity was incorporated in terms of standard PCU units as well as by explicitly considering different categories of vehicles into the modelling and estimation processes. It was observed that converting heterogeneous traffic into a homogeneous equivalent using constant values of PCU was a good representation in the proposed non-continuum macroscopic modelling approach, when compared with the alternative approach of including different categories of vehicles in the model. The model-based scheme with speed as measurement was performing better under all traffic conditions. The estimation scheme was corroborated using data collected from a road stretch in Chennai, India, using the videographic technique. The results obtained were compared with the actual values and were found to be promising. The space mean speed estimates can be used to estimate and predict the travel time of vehicles in a given road stretch and traffic density is useful in the prediction of congestion. An advantage of this scheme is that it can be easily integrated with real-time data, if available, and can be used for real-time implementations. Successful implementation of such systems will help in better management of traffic through real-time ITS applications. The dynamic model developed in this study can also be used to develop control schemes for regulating the flow of traffic on Indian roads.

Overall, the main contributions of this study are:

- This study is one of the first steps towards dynamic modelling of Indian Traffic by treating it as a non-continuum and applying it for real-time traffic state estimation.
- The developed traffic stream model is one of the first stream models for Indian urban road traffic conditions that helped in the development of a dynamic macroscopic model applied to Indian traffic conditions.
- The comprehensive non-continuum macroscopic model developed for describing the flow of traffic will overcome the limitations in modelling traffic as a continuum and will be useful in characterizing Indian traffic.
- The developed model-based scheme for real-time estimation of traffic density using the Kalman filtering technique can take into account the variability

arising from heterogeneity to a large extent, leading to better prediction accuracy.


## Funding

The second author acknowledges the support provided by the Department of Science and Technology, Government of India [grant No. SR/FTP/ETA-55/2007]. The authors also acknowledge the support provided by the Ministry of Urban Development, Government of India [grant No. N-11025/30/2008-UCD].


## Disclosure statement

No potential conflict of interest was reported by the authors.

## References


Ajitha, T., & Vanajakshi, L. (2012, January). *Development of optimized traffic stream models under heterogeneous traffic conditions*. Proc. 91st transportation research board annual meeting, Washington, DC, USA.

Anand, R. A., Vanajakshi, L., & Subramanian, S. C. (2011, June). *Traffic density estimation under heterogeneous traffic conditions using data fusion*. Intelligent vehicles symposium (IV), 2011 IEEE, pp. 31–36.

Chakroborty, P., & Kikuchi, S. (1999). Evaluation of the general motors based car-following models and a proposed fuzzy inference model. *Transportation Research C: Emerging Technologies*, *7*, 209–235.

Chakroborty, P., & Maurya, A. K. (2008). Microscopic analysis of cellular automata based traffic flow models and an improved model. *Transport Reviews*, *28*, 717–734.

Chu, L., Oh, S., & Recker, W. (2005, January). *Adaptive Kalman filter based freeway travel time estimation*. 84th TRB annual meeting, Washington, DC, USA.

Coddington, E. A. (1989). *An introduction to ordinary differential equations*. New York, NY: Dover Publications.

Daganzo, C. F. (1994a). The cell transmission model: A dynamic representation of highway traffic consistent with hydrodynamic theory. *Transportation Research B: Methodological*, *28*(4), 269–287.

Daganzo, C. F. (1994b). The cell transmission model, part II: Network traffic. *Transportation Research B*, *29*(2), 279–293.

Gartner, N., Messer, C. J., & Rathi, A. K. (2001). *Traffic flow theory a state-of-the-art report* (Transportation Research Board Special Report 165). Wasington, DC: Springer-Verlag.

Gazis, D. C., & Knapp, C. H. (1971). On-line estimation of traffic densities from time-series of flow and speed data. *Transportation Science*, *5*(3), 283–301.

Gazis, D. C., & Szeto, M. W. (1974). *Design of density-measuring systems for roadways*. Transportation Research Record 495, Transportation Research Board, pp. 44–52 (No. HS-015 791).

Gopal, M. (1984). *Modern control system theory*. New York, NY: John Wiley and Sons.

Gupta, S., Chakroborty, P., & Mukherjee, A. (1998). *Microscopic simulation of vehicular traffic on congested roads*. Proceedings of the international symposium on Industrial Robotic Systems-98, Bangalore, India.

Hegyi, A., Girimonte, D., Babuska, R., & De Schutter, B. (2006, September). *A comparison of filter configurations for freeway traffic state estimation*. Intelligent transportation systems conference. ITSC'06, Toronto, Canada, IEEE, pp. 1029–1034.

Hoogendoorn, S. P., & Bovy, P. H. L. (2000). Continuum modeling of multiclass traffic flow. *Transportation Research Part B: Methodological*, *34*, 123–146.

Hoogerdoorn, S. P., & Bovy, P. H. L. (2001). State-of-the-art of vehicular traffic flow modelling. Proceedings of Institution of Mechanical Engineers, Part I. *Journal of Systems and Control Engineering*, *215*, 283–304.

Houpt, P., Kurkjian, A., Gershwin, S. B., & Willsky, A. S. (1979, December). *Estimation of roadway traffic density on freeways using presence detector data*. Decision and control including the symposium on adaptive processes, 1979 18th IEEE, 18, pp. 524–525.

Ioannou, P. A., & Chien, C. C. (1993). Autonomous intelligent cruise control. *IEEE Transactions on Vehicular Technology*, *42*(4), 657–672.

IRC: 106. (1990). *Guidelines for capacity of urban roads in plain areas*. New Delhi: IRC.

Jabari, S. E., & Liu, H. X. (2013). A stochastic model of traffic flow: Gaussian approximation and estimation. *Transportation Research B: Methodological*, *47*, 15–41.

Jazwinsky, A. H. (1970). *Stochastic processes and filtering theory*. New York, NY: Academic Press.

Kalman, R. E. (1960). A new approach to linear filtering and prediction problems. *Journal of Basic Engineering*, *82*, 35–45.

Kenneth, D. L., & Ronald, K. K. (1982). *Advances in business and management forecasting*. London: Emerald Books.

Kerner, B. S., & Klenov, S. L. (2004). Spatial–temporal patterns in heterogeneous traffic flow with a variety of driver behavioural characteristics and vehicle parameters. *Journal of Physics A: Mathematical and General*, *37*, 8753–8788.

Kurkjian, A., Gershwin, S. B., Houpt, P. K., Willsky, A. S., Chow, E. Y., & Greene, C. S. (1980). Estimation of roadway traffic density on freeways using presence detector data. *Transportation Science*, *14*(3), 232–261.

Logghe, S., & Immers, L. H. (2003). *Heterogeneous traffic flow modeling with the LWR model using passenger–car equivalents*. Proc.10th world congress on ITS, Madrid.

Mallikarjuna, C., & Rao, K. R. (2009). Cellular automata model for heterogeneous traffic. *Journal of Advanced Transportation*, *43*, 321–345.

Mathew, T. V., Gundaliya, P., & Dhingra, S. L. (2006, August). *Heterogeneous traffic flow modeling and simulation using cellular automata*. Proceedings of 9th international conference on Applications of Advanced Technology in Transportation, Chicago.

Maurya, A. K., & Chakroborty, P. (2008, September). *Microscopic model for simulation of uninterrupted mixed traffic streams without lane discipline*. Proceedings of the international conference on Best Practices to Relieve Congestion on Mixed-Traffic Urban Streets in Developing Countries, Chennai, India.

May, A. D. (1990). *Traffic flow fundamentals*. Englewood Cliffs, NJ: Prentice Hall.

Nahi, N. E., & Trivedi, A. N. (1973). Recursive estimation of traffic variables: Section density and average speed. *Transportation Science*, *7*(3), 269–286.

Nanthawichit, C., Nakatsuji, T., & Suzuki, H. (2003, January). *Application of probe vehicle data for real time traffic state estimation and short term travel time prediction on a freeway*. 82nd transportation research board annual meeting, Washington, DC, USA.

Okutani, I., & Stephanedes, Y. J. (1984). Dynamic prediction of traffic volume through Kalman filtering theory. *Transportation Research B: Methodological*, *18*, 1–11.

Padiath, A., Vanajakshi, L., Subramanian, S. C., & Manda, H. (2009, October). *Prediction of traffic density for congestion analysis under Indian traffic conditions*. Intelligent transportation systems, 2009. ITSC'09. 12th International IEEE, pp. 1–6.

Padiath, A. S., Vanajakshi, L. D., & Subramanian, S. C. (2010, January). *Estimation of traffic density under Indian traffic conditions*. Transportation research board 89th annual meeting (No. 10–2877), Washington, DC, USA.

Papageorgiou, M. (1998). Some remarks on macroscopic traffic flow modeling. *Transportation Research A*, *32*, 323–329.

Swaroop, D., Hedrick, J. K., Chien, C. C., & Ioannou, P. (1994). A comparison of spacing and headway control laws for automatically controlled vehicles. *Vehicle System Dynamics*, *23*, 597–625.

Sun, X., Muñoz, L., & Horowitz, R. (2004, June). *Mixture Kalman filter based highway congestion mode and vehicle density estimator and its application*. American control conference, Boston, USA. IEEE, 3, pp. 2098–2103.

Tang, Q., Huangb, H. J., Zhao, S. G., & Shang, H. Y. (2009). A new dynamic model for heterogeneous traffic flow. *Physics Letters A*, *373*, 2461–2466.

Tiwari, G., Fazio, J., Gaurav, S., & Chatteerjee, N. (2008). Continuity equation validation for nonhomogeneous traffic. *Journal of Transportation Engineering ASCE*, *134*, 118–127.

Tyagi, V., Darbha, S., & Rajagopal, K. R. (2008). A dynamical systems approach based on averaging to model the macroscopic flow of freeway traffic. *Nonlinear Analysis: Hybrid Systems*, *2*, 590–612.

Tyagi, V., Darbha, S., & Rajagopal, K. R. (2009). A review of the mathematical models for traffic flow. *International Journal of Advances in Engineering Sciences and Applied Mathematics*, *1*, 53–68.

Venkatesan, K., Gowri, A., & Sivanandan, R. (2008). Development of microscopic simulation model for heterogeneous traffic using object oriented approach. *Transportmetrica*, *4*(3), 227–247.

Wang, H., Li, J., Chen, Q. Y., & Ni, D. (2011). Logistic modeling of the equilibrium speed-density relationship. *Transportation Research Part B*, *45*, 554–566.

Wang, Y., & Papageorgiou, M. (2005). Real-time freeway traffic state estimation based on extended Kalman filter: A general approach. *Transportation Research Part B: Methodological*, *39*(2), 141–167.

Wang, Y., Papageorgiou, M., & Messmer, A. (2007). Real-time freeway traffic state estimation based on extended Kalman filter: A case study. *Transportation Science*, *41*(2), 167–181.

Wang, Y., Papageorgiou, M., & Messmer, A. (2008). Real-time freeway traffic state estimation based on extended Kalman filter: Adaptive capabilities and real data testing. *Transportation Research Part A: Policy and Practice*, *42*(10), 1340–1358.

Willsky, A. S., Chow, E., Gershwin, S., Greene, C., Houpt, P., & Kurkjian, A. (1980). Dynamic model-based techniques for the detection of incidents on freeways. *IEEE Transactions on Automatic Control*, *25*(3), 347–360.

Zhongke, S. (2003, October). *A new estimation method for multi-section traffic states of freeway*. Intelligent transportation systems, 2003. IEEE, 1, pp. 618–623.

# The effect of torque feedback exerted to driver's hands on vehicle handling – a hardware-in-the-loop approach

S. Samiee[a,b]*, A. Nahvi[a], S. Azadi[a], R. Kazemi[a], A.R. Hatamian Haghighi[a] and M.R. Ashouri[a]

[a]*Faculty of Mechanical Engineering, K.N. Toosi University of Technology, Tehran, Iran;* [b]*Institute of Automotive Engineering, Graz University of Technology, Graz, Austria*

In this paper, road forces on tire are exerted on driver's hands via an equivalent torque applied to the steering wheel using a hardware-in-the-loop method to analyze the effect of steering torque feedback in vehicle handling. An electrical torque-feedback steering system is used for experimental validation. A 14-degree-of-freedom vehicle dynamic model, including engine, tire, and steering system mechanism are simulated. The required inputs such as throttle angle, brake demand, and steering wheel angular position are transmitted to the computer via an I/O interface card. Tire forces and steering gear torque are solved. This torque is then sent via an I/O interface card to a DC motor connected to the steering shaft. All equations are solved in real time. To investigate the influence of torque feedback on vehicle handling, several experiments are executed on 25 users. For this purpose, an experimental protocol is defined. In the experiments, the users had to drive along a specific path with constant speed using the designed electrical torque-feedback steering system. During the tests, the driving pattern of each user was recorded and the simulator's instantaneous position was compared with its desirable value. The results show that the torque feedback improved the driver's perception from the surrounding environment and enables her/him to handle the vehicle satisfactorily.

**Keywords:** vehicle handling; electrical steering; hardware-in-the-Loop

## 1. Introduction

Virtual reality systems expose humans to simulated environments and through interaction with the five senses they can simulate physical presence of the real world for the users (Burdea & Coiffet, 2003). Generally, driving simulators are divided into three categories including high-cost, medium-cost and low-cost (Gregersen, Falkmer, Dol, & Pardo, 2001). About 81% of driving simulators are used in research projects and the remaining 19% serve as training simulations for novice drivers (Straus, 2005). The importance of steering system with torque feedback in driving simulators makes it preferable even in low-cost simulations. Such a system increases the environmental interaction between the user and the simulator, and intensifies the user's sense of immersion and enhances the process of driver training. It is not only essential for training of novice drivers, but also can be utilized for evaluation and validation of the electrical steer control designs in brand-new vehicles in the car industry. Car manufacturers can use this system to evaluate and modify their conceptual designs without bearing the cost of initial modeling.

Replacement of mechanical and hydraulic systems by electronic ones has led to the by-wire technology. Enhancement of efficiency, increased safety and reliability, and reduction of manufacturing costs are the main advantages of electronic steering systems. If the mechanical steering system is fully replaced by a steer-by-wire system, the following benefits are achieved:

(1) Removal of steering shaft simplifies the internal design of the vehicle. Removal of various parts allows more space around the engine. Hence, there will be fewer difficulties for installation of the combustion engine and the steering system can be designed and installed as small units.
(2) There is no direct physical connection between tires and steering wheel and in the case of accidents, fewer impacts are transferred to the driver via steering wheel.
(3) The specifications of the steering system can be easily changed. Hence, desirable steer response and steering feel can be achieved based on the specified needs.
(4) Safety can be improved by providing computer-controlled intervention of vehicle controls with systems such as Electronic Stability Control, Adaptive Cruise Control, and Lane Keeping System.

*Corresponding author. Email: S.samiee@tugraz.at

A considerable number of research works have already been carried out on steer-by-wire systems. Kim and Song (2002) worked on the electronic power steering (EPS) control system. They focused on designing a control logic, which was able to reduce the torque applied to the driver, produce a different steering feel, and improve return-to-center performance. They implemented their proposed approach on a hardware-in-the-loop system to validate it. A robust control technique was used by Chen and Ulsoy (2002) to control the EPS system and provide the driver with desirable driving experiences. They conducted experiments in both time and frequency domains. The obtained results revealed that the robust control technique can reduce a vehicle's lateral position deviation and provide slight improvement to the marginal stability of the vehicle. This controller was implemented on the steering system of a human-in-the-loop driving simulator. Park, Han, and Lee (2005) studied steer-by-wire system control. They intended to improve the driving experience and enhance a vehicle stability and investigated the advantages of the removal of the steering shaft in the steer-by-wire system and creation of more space around the engine. A state feedback controller was developed by Yih and Gerdes (2004) with lateral slip as the state variable to improve handling of the vehicle. To measure lateral slip, an observer was designed using steering torque, steering angle, and yaw angle rate. The state feedback controller was implemented on a vehicle with the steer-by-wire system. They estimated the steering torque using the steer system's motor current and calculated the yaw angle rate based on a linear model of the vehicle and the steering system. Segawa, Nakano, Nishihara, and Kumamoto (2001). designed a controller to improve the stability of the vehicles with a steer-by-wire system and implemented it on a driving simulator. They demonstrated that the proposed controller is able to reduce the effect of the lateral winds on the vehicle's stability. Moreover, based on some experiments, they showed that the controller improves vehicle's stability in the case of driving on low-friction roads where behavior of the vehicle is different from the driver's input command. Yao (2006) transformed various driving scenarios, such as experiencing different steering feel, return-to-center performance, and fast and accurate wheel angle tracking based on driver's input into the control problems. Modern control techniques, system parameter identification, and state variable estimation were used to execute these scenarios. Finally, Yao implemented different control loops on seven vehicles with steer-by-wire systems and eight hardware-in-the-loop systems.

This paper focuses on the effect of torque feedback – exerted to driver's hands – on vehicle handling. An electrical torque-feedback steering system is designed and implemented in a driving simulator for experimental tests. The developed system is a segment of a by-wire control system and hence can be used for the construction and control of the steer-by-wire systems in domestic vehicles. Other applications of the proposed system include teleoperation of vehicles and driver behavior investigations such as works done in Friedrichs and Yang (2010) and Samiee, Azadi, Kazemi, Nahvi, and Eichberger (2014).

In this paper, forces applied to the driver's hand via the steering wheel will be identified and reproduced for implementation on the hardware. Vehicle's dynamics is modeled using equations of motion. Dynamic equations of tires are also solved instantaneously to obtain friction forces. Furthermore, geometry of the steering mechanism and tires are simulated to transform the generated forces on the tires to an equivalent torque on the steering wheel. Then, the equivalent torque is sent to the motor driver using an interface card to produce the required torque by providing the equivalent current. The generated torque is measured by a sensor, installed on the output of the gearbox, and is sent back to the computer using another interface card. Hence, the generated torque is compared with the reference (desirable) value and the error signal is fed into the controller.

All of the aforementioned calculations and the data transfers between computer and hardware must be carried out in real time. The used hardware and software are able to perform real-time calculations, generate graphical pictures, and communicate with interface cards simultaneously. The vehicle is modeled by a 14-degree-of-freedom (DOF) model. The dynamic model of tires is of a Fiala type. Other dynamics of the system are ignored and only kinematics is modeled for the generation of torques on the steering wheel based on the generated tire forces. A classical proportional-integrator-derivative (PID) control is used in this paper which demonstrates favorable performance according to the experiments results.

## 2. Dynamic modeling of vehicle in ADAMS

In this paper, the virtual vehicle's dynamic equations, tires dynamic equations, and geometry of the front wheels are validated by a full car model with 251-DOF, implemented in ADAMS/Car, which had been previously validated using experimental results of a real car. The characteristics of the subsystems of the vehicle modeled in ADAMS are as follows:

- Front wheel's suspension system is of McPherson type with linear spring.
- The compound suspension system with nonlinear spring is considered for rear wheels.
- The steering system is of rack-pinion type.
- Disk-type braking system is modeled for each of the four wheels.
- The power generation and transmission units are modeled by a series of mathematical equations, with throttle, brake, and gear as the inputs and the torque on the wheels as the output.

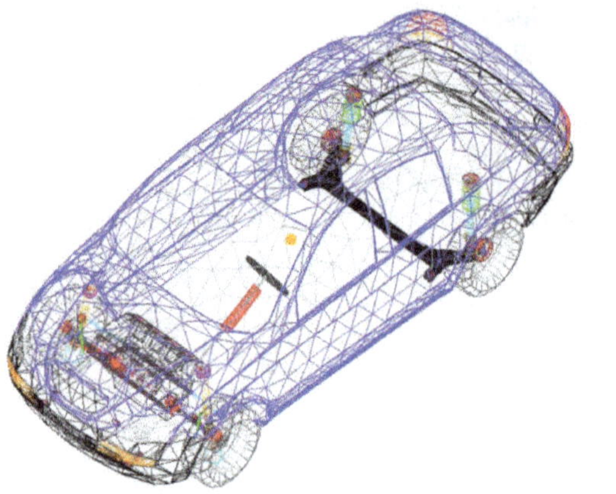

Figure 1. The vehicle's 251-DOF model implemented in ADAMS.

- The tires are represented by the Fiala model. The vehicle's body is modeled as a concentrated mass on the center of gravity of the vehicle.

By proper interconnection of the vehicle subsystems in ADAMS/Car, the full dynamic model of the vehicle is obtained. Figure 1 illustrates the vehicle model.

## 3. Simulation of vehicle's dynamics

A 14-DOF model is used to represent the vehicle's dynamics. The main advantage of the 14-DOF model against other simpler models (e.g. three- and seven-DOF models) is that it is composed of full six-DOF models for vehicle's body and suspension system dynamics. Figure 2 shows the suspension system of the 14-DOF model. It is assumed that each wheel has its own independent suspension system. For each wheel, there is a spring and damper, located between the body and suspension arm.

The parameters in Figure 2 are defined as follows:

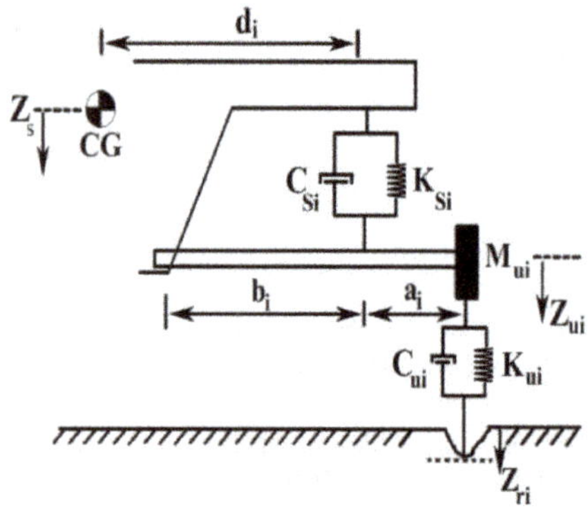


Figure 2. Suspension system of the 14-DOF model.

- The distance from wheel $i$ to the connection point of spring and damper to the suspension arm is indicated by $a_{\mathrm{i}}$.
- The distance between the connection point of the spring and damper to the suspension arm of wheel $i$ and the connection point of the suspension arm to the vehicle's body is denoted by $b_{\mathrm{i}}$.
- The distance from the spring and damper of wheel $i$ to the longitudinal axis is shown by $d_{\mathrm{i}}$.
- Parameters $M_{\mathrm{u}i}$ and $Z_{\mathrm{u}i}$ represent the $i$th unsprung mass and its height from the ground, respectively, and $Z_{\mathrm{r}i}$ indicates the height of the road beneath the $i$th wheel.
- Parameters $K_{\mathrm{u}i}$ and $C_{\mathrm{u}i}$ are the spring stiffness and equivalent damping coefficient of the $i$th tire, respectively.
- The spring stiffness and equivalent damping coefficient of the $i$th wheel are designated by $K_{\mathrm{s}i}$ and $C_{\mathrm{s}i}$, respectively.
- And $Z_{\mathrm{s}}$ shows the distance from the center of gravity to the ground.

After deriving the equations of linear and angular motion for the vehicle and some algebraic manipulations, the 14-DOF equations of motion can be expressed as follows (Bastow, Howard, & Whitehead, 2004).

### 3.1. *Longitudinal equations of vehicle*

Using Newton's second law, the longitudinal dynamic equation of the vehicle can be presented as Equation (1),

$$M_{\mathrm{t}}(\dot{u} + qw - rv) = X_1 + X_2 + X_3 + X_4 - F_{\mathrm{ax}}, \quad (1)$$

where $M_{\mathrm{t}}$ is the total vehicle mass, $\dot{u}$ is the longitudinal acceleration of the vehicle's center of gravity, $q$ is the pitch rate, $w$ is the vertical velocity of the vehicle's center of gravity, $r$ is the yaw rate, $v$ is the lateral velocity of the vehicle's center of gravity, and $F_{\mathrm{ax}}$ is longitudinal air resistance force. In addition, $X_i$ is found from Equation (2),

$$X_i = F_{xi}\cos(\delta_i) - F_{yi}\sin(\delta_i) \quad (2)$$

where $\delta_1 = \delta_{\mathrm{fr}}$ is the steering angle of the front right wheel, $\delta_2 = \delta_{\mathrm{fl}}$ is the steering angle of the front left wheel, and $\delta_3 = \delta_4 = 0$ are the steering angles of rear wheels, which are always equal to zero. Also, $F_{xi}$ and $F_{yi}$ represent the applied longitudinal and lateral forces on the $i$th tire, respectively.

### 3.2. *Lateral equations of vehicle*

Using Newton's second law, the lateral dynamic equation of the vehicle can be presented as Equation (3),

$$M_{\mathrm{t}}(\dot{v} + ru - pw) = Y_1 + Y_2 + Y_3 + Y_4 - F_{\mathrm{ay}}, \quad (3)$$

where $p$ indicates the roll rate and $F_{\mathrm{ay}}$ shows the lateral air resistance force. Furthermore, $Y_i$ can be stated as

Equation (4)

$$Y_i = F_{xi}\sin(\delta_i) + F_{yi}\cos(\delta_i). \quad (4)$$

### 3.3. Vertical equations of vehicle

The vertical dynamic equation can be shown by

$$M_s(\dot{w} + pv - qu) = \sum_{i=1}^{4} \frac{F_{si}}{R_{ri}}, \quad (5)$$

where $M_s$ represents the sprung mass and $F_{si}$ is the suspension force of wheel $i$, expressed by Equation (6),

$$\begin{aligned} F_{s1} &= K_{s1}(Z_{u1} - Z_s + L_f\sin(\theta) + d_1\sin(\varphi)) + C_{s1}(w_{u1} - \\ &\quad w + L_f q\cos(\theta) + d_1 p\cos(\varphi)), \\ F_{s2} &= K_{s2}(Z_{u2} - Z_s + L_f\sin(\theta) - d_2\sin(\varphi)) + C_{s2}(w_{u2} \\ &\quad - w + L_f q\cos(\theta) - d_2 p\cos(\varphi)), \\ F_{s3} &= K_{s3}(Z_{u3} - Z_s - L_r\sin(\theta) + d_3\sin(\varphi)) + C_{s3}(w_{u3} \\ &\quad - w - L_r q\cos(\theta) + d_3 p\cos(\varphi)), \\ F_{s4} &= K_{s4}(Z_{u4} - Z_s - L_r\sin(\theta) - d_4\sin(\varphi)) + C_{s4}(w_{u4} \\ &\quad - w - L_r q\cos(\theta) - d_4 p\cos(\varphi)), \end{aligned} \quad (6)$$

where $\theta$ and $\varphi$ are pitch and roll angles of vehicle's body, respectively, and $L_f$ and $L_r$ indicate the distance of the vehicle's center of gravity from the front and rear axles, respectively. $R_{ri}$ is a coefficient related to the geometry of the suspension system and is found from Equation (7),

$$R_{ri} = \frac{a_i + b_i}{b_i}. \quad (7)$$

### 3.4. Roll angle

Angular equations of vehicle around longitudinal axle represent Equation (8).

$$I_{xs}\dot{P} + (I_{zs} - I_{ys})\mathrm{qr} = \sum_{i=1}^{4} \overline{R_{ri}} F_{si} - h_{cg}\sum_{i=1}^{4} Y_i, \quad (8)$$

where $h_{cg}$ is the distance of vehicle's center of gravity from ground in standstill and $I_{xs}$, $I_{ys}$, and $I_{zs}$ indicate the moment of inertia of the sprung mass around longitudinal, lateral, and vertical axles, respectively. In addition, $\overline{R_{ri}}$ is a coefficient related to the geometry of the suspension and,

$$\begin{aligned} \overline{R_{r1}} &= -d_1 + (d_1 - b_1)\frac{a_1}{a_1 + b_1}, \\ \overline{R_{r2}} &= d_2 - (d_2 - b_2)\frac{a_2}{a_2 + b_2}, \\ \overline{R_{r3}} &= -d_3 + (d_3 - b_3)\frac{a_3}{a_3 + b_3}, \\ \overline{R_{r4}} &= d_4 - (d_4 - b_4)\frac{a_4}{a_4 + b_4}. \end{aligned} \quad (9)$$

### 3.5. Pitch angle

Vehicle's equations around lateral axle represent Equation (10)

$$I_{ys}\dot{q} + (I_{xs} - I_{zs})pr = -L_f\left(\frac{F_{s1}}{R_{r1}} + \frac{F_{s2}}{R_{r2}}\right) + L_r\left(\frac{F_{s3}}{R_{r3}} + \frac{F_{s4}}{R_{r4}}\right) + h_{cg}\sum_{i=1}^{4} X_i\sqrt{2}. \quad (10)$$

### 3.6. Yaw angle

Vehicle's equations around vertical axle result in Equation (11)

$$I_z\dot{r} + (I_y - I_x)pq = L_f(Y_1 + Y_2) - L_r(Y_3 + Y_4) + \frac{T_f}{2}(X_1 - X_2) + \frac{T_r}{2}(X_3 - X_4) + \sum_{i=1}^{4} M_{zi}, \quad (11)$$

where $T_f$ is the distance of front wheels from each other, $T_r$ is the distance between rear wheels, and $I_x$, $I_y$, and $I_z$ represent the moment of inertia around longitudinal, lateral, and vertical axles, respectively.

### 3.7. Unsprung masses (four equations)

Equations of motion for unsprung masses lead to Equation (12).

$$M_{ui}\dot{w}_{ui} = K_{ui}(Z_{ri} - Z_{ui}) + C_{ui}(w_{ri} - w_{ui}) - \frac{F_{si}}{R_{ri}}, \quad (12)$$

where $w_{ri}$ indicates the rate of change of the road input and $w_{ui}$ is the rate of change of the motion, which is perpendicular to the unsprung mass.

### 3.8. Wheels dynamics (four equations)

Considering one DOF for each tire results in Equation (13)

$$I_{wi}\dot{\omega}_i = T_{wi} - R_w F_{xi} - T_{Ri} - T_b \quad (13)$$

where $R_w$ is the effective radius of the tires, and $T_b$ and $T_{Ri}$ represent the braking and resistant torques, respectively. The dynamic parameters of the vehicle are presented in Table 1. The 14-DOF model is obtained after performing some algebraic manipulations on the linear and angular equations of motion. To validate the 14-DOF model, a full vehicle model in ADAMS/Car was used.

Two different maneuvers were applied to both the models. In the first maneuver, the vehicle was traveling with a constant velocity of 70 km/h and then a sinusoidal displacement with the frequency of 0.2 Hz and amplitude of 70 degrees are applied to the steering wheel. In the second maneuver, a step input of 70 degrees was applied to

Table 1. Dynamic parameters of the vehicle.

| | |
|---|---|
| Distance of vehicle's center of gravity from front axle | $l_f = 1$ (m) |
| Distance of vehicle's center of gravity from rear axle | $l_r = 2.6$ (m) |
| Distance between front wheels | $T_f = 1.401$ (m) |
| Distance between rear wheels | $T_r = 1.408$ (m) |
| Distance of vehicle's center of gravity from ground in standstill | $h_{cg} = 0.497$ (m) |
| Mass of vehicle | $M_t = 924$ (kg) |
| Coefficient of front spring | $K_{s1} = 17,650$ (N/m) |
| Coefficient of rear spring | $K_{s2} = 22,600$ (N/m) |
| Coefficient of front damper | $C_{s1} = 1600$ (N s/m) |
| Coefficient of rear damper | $C_{s2} = 3730$ (N s/m) |
| Steer ratio | $N = 1 : 17.5$ |

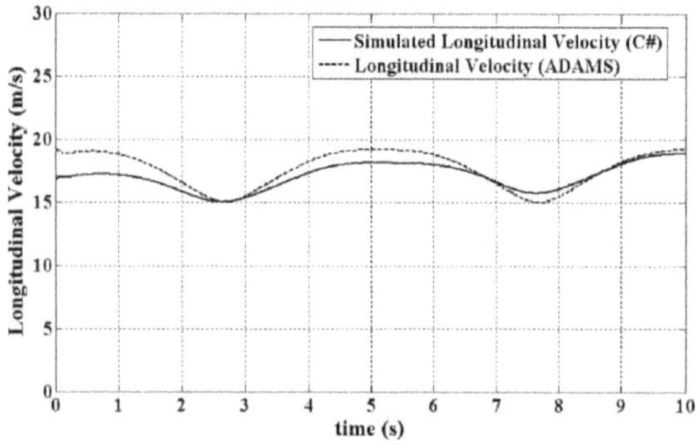


Figure 3. Longitudinal velocity of 14-DOF and ADAMS models (sinusoidal input of steering wheel).

the steering wheel of the vehicle, traveling with a constant velocity of 70 km/h.

Figures 3–6 show the results of the comparison between the dynamic parameters of both the models, including the longitudinal and lateral velocities and the rate of yaw angle. These figures demonstrate a close match between the outputs of both the models. The maximum longitudinal and lateral velocity errors for both the models are 3 m/s. Moreover, the 14-DOF model has accurately generated the rate of change of the yaw angle. The maximum error for the 14-DOF model in the case of step input maneuvering is 0.05 rad/s.

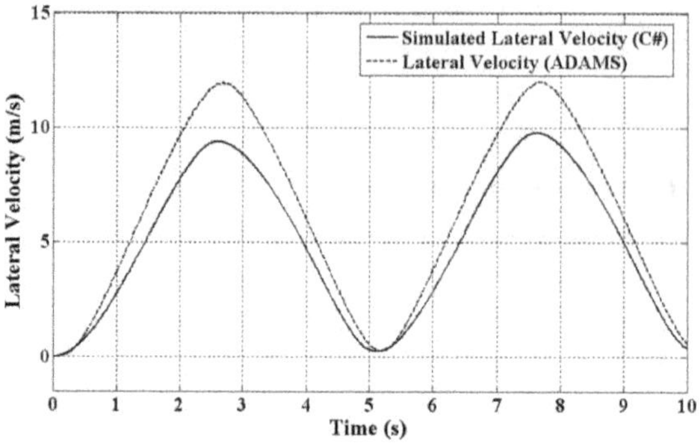


Figure 4. Lateral velocity of 14-DOF and ADAMS models (sinusoidal input of steering wheel).

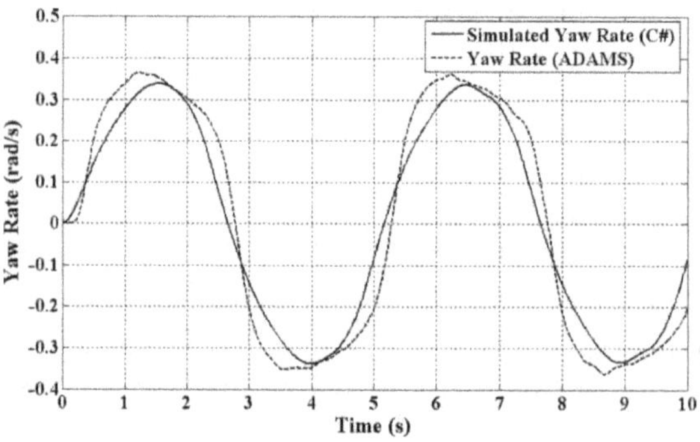


Figure 5. Yaw rate of 14-DOF and ADAMS models (sinusoidal input of steering wheel).

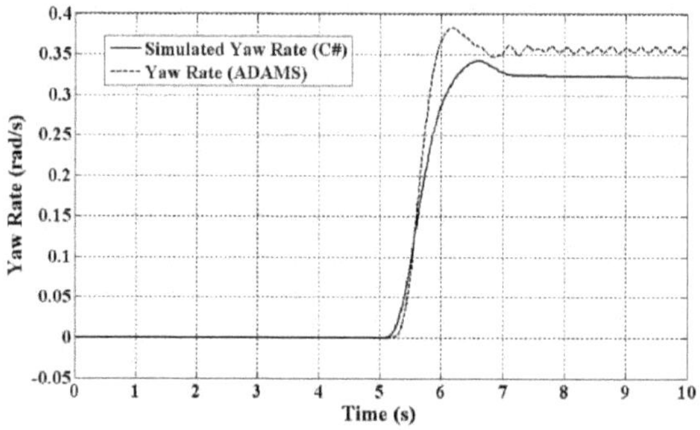


Figure 6. Yaw rate of 14-DOF and ADAMS models (step input of steering wheel).

## 4. Dynamic simulation of tire

Correct modeling of tires significantly influences the process of the vehicle dynamic simulation. The applied forces are transferred to the vehicle through the tires. Therefore, the accurate calculation of the tire forces is the very first step toward the simulation of the vehicle. Different tire models have been proposed, which produce dependable estimations for tire forces and torques if the range of applied forces is not broad and tire deflection is slight. This paper uses the Fiala model to simulate the tire. In this model, it is assumed that tire contact to the road is rectangular and camber angle does not affect forces. The Fiala model is described as follows.

### *4.1. Calculation of longitudinal slip angle of tire*

Longitudinal slip happens due to the difference between the velocity of the tire center and the contact patch center. According to Figure 7, theoretical calculation of longitudinal slip is as follows:

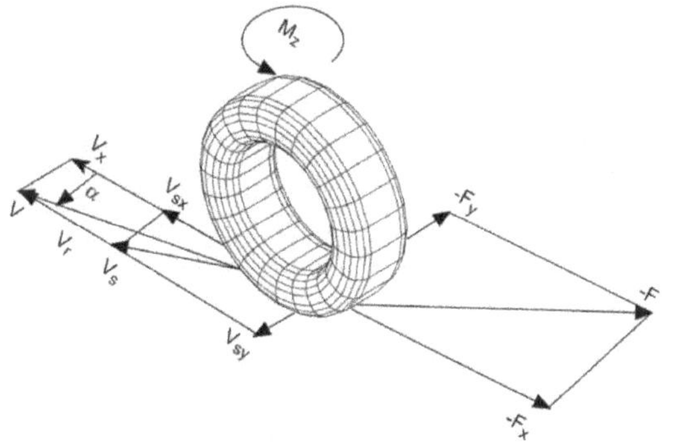


Figure 7. Velocity and force vectors directions (MSC Software Corporation, 2003).

The Fiala model uses the following equations for stable estimation of the slip.

*4.1.1. Acceleration slip*

$$S_s = \frac{V_r - V_x}{V_r} = \frac{(\omega_{\text{Actual}} - \omega_{\text{Free}})}{\omega_{\text{Actual}}}. \tag{14}$$

*4.1.2. Braking slip*

$$S_s = \frac{V_r - V_x}{V_x} = \frac{(\omega_{\text{Actual}} - \omega_{\text{Free}})}{\omega_{\text{Free}}}, \tag{15}$$

where $V_r$ is the linear velocity of the contact point between tire and the road, obtained from Equation (16) and $V_x$ indicates the image of the velocity of wheel center along the longitudinal axle of the wheel and is found from Equations (17) and (18)

$$V_r = \omega R_w, \tag{16}$$

$$V_{x\text{fr}} = V_x - 0.5\text{r} \cdot T_f \cos\delta_{\text{fr}} + (V_y + r \cdot L_f) \sin \delta_{\text{fr}},$$
$$V_{x\text{fl}} = V_x + 0.5\text{r} \cdot T_f \cos\delta_{\text{fl}} + (V_y + r \cdot L_f) \sin \delta_{\text{fl}}, \tag{17}$$
$$V_{x\text{rr}} = (V_x - 0.5r \cdot T_r),$$
$$V_{x\text{rl}} = (V_x + 0.5r \cdot T_r). \tag{18}$$

In these equations, $V_{x\text{fr}}$, $V_{x\text{fl}}$, $V_{x\text{rr}}$, and $V_{x\text{lr}}$ represent the image of the velocity of wheel center along the longitudinal axle of the front right, front left, rear right, and rear left wheels, respectively, and $R_w$ is the radius of the unloaded wheel.

## 4.2. *Calculation of lateral slip angle of tire*

The lateral slip angle of the tires is found from the difference between the longitudinal axle of tire and the axis of velocity vector on the tire plane ($\alpha$ in Figure 8).

The sign of the angle is positive in the upward direction and negative in the downward direction. The slip can be stated based on the state variables such as longitudinal,

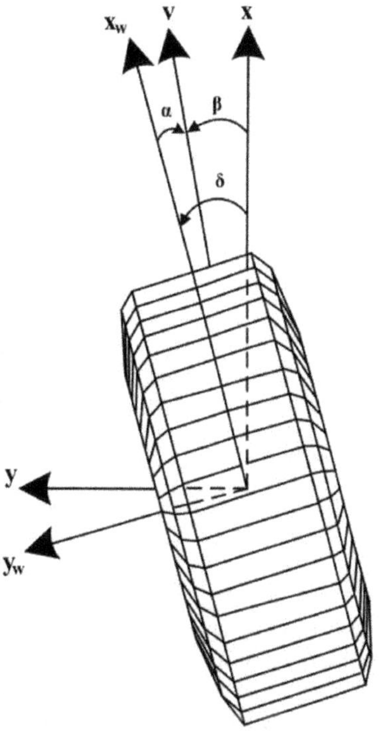


Figure 8. Slip angle of the front tire.

lateral, and yaw velocities. The lateral slip is expressed by Equation (19)

$$\begin{aligned}
\alpha_{\text{fr}} &= \delta_{\text{fr}} - a\tan\left(\frac{(V_y + r \cdot L_f)}{(V_x - 0.5r \cdot T_f)}\right),\\
\alpha_{\text{fl}} &= \delta_{\text{fl}} - a\tan\left(\frac{(V_y + r \cdot L_f)}{(V_x + 0.5r \cdot T_f)}\right),\\
\alpha_{\text{rr}} &= -a\tan\left(\frac{(V_y - r \cdot L_r)}{(V_x - 0.5r \cdot T_r)}\right),\\
\alpha_{\text{rl}} &= -a\tan\left(\frac{(V_y - r \cdot L_f)}{(V_x + 0.5r \cdot T_r)}\right).
\end{aligned} \tag{19}$$

## 4.3. *Calculation of vertical force*

In the Society of Automotive Engineers coordinate system, vector $+Z$ faces downwards (towards the road). Hence, the vertical force of the road on the tire in the contact point is always negative. The vertical force is expressed by Equation (20),

$$F_z = \min(0.0, (F_{ZK} + F_{ZC})), \tag{20}$$

where $F_{ZK}$ is the vertical force due to the vertical stiffness of the tire and is found from Equation (21):

$$F_{ZK} = -\text{VS} \times \text{TD}, \tag{21}$$

where VS and TD indicate vertical stiffness and tire deflection, respectively. Moreover, $F_{ZC}$ is the vertical force due to the tire damping, expressed by Equation (22):

$$F_{zc} = -\text{VD} \times \frac{\text{d}}{\text{d}t}(\text{TD}), \tag{22}$$

where VD represents vertical damping of the tire.

### 4.4. *Calculation of longitudinal force*

The longitudinal force on the tires in any instance depends on the vertical force, $F_Z$, instantaneous friction coefficient, $U$, the ratio of the longitudinal slip, $S_s$, and the lateral slip angle $\alpha$ in that instance. Also, total slip ($S_{s\alpha}$), instantaneous friction coefficient ($U$), and critical longitudinal slip ($S_{\text{Critical}}$) must be calculated prior to calculation of the longitudinal force. The total slip is expressed by Equation (23),

$$S_{s\alpha} = \sqrt{S_s^2 + \tan^2(\alpha)}. \tag{23}$$

The instantaneous friction coefficient depends on the static friction coefficient ($U_{\max}$), dynamic friction coefficient ($U_{\min}$), and the total slip. The $U_{\max}$ represents the friction coefficient between the tire and the road at zero slip, which does not happen in practice as there always exists non-zero slip. On the other hand, $U_{\min}$ indicates friction coefficient between the tire and the road at 100% slip

$$U = U_{\min} + (U_{\max} - U_{\min}) \times S_{s\alpha}. \tag{24}$$

The critical slip ($S_{\text{Cr}}$) can be expressed by Equation (25)

$$S_{\text{Cr}} = \left| \frac{U \times F_Z}{2 \times C_{\text{Sl}}} \right|, \tag{25}$$

where $C_{\text{Sl}}$ is the longitudinal slip constant found by the limit of changes in $F_x$ with respect to $S_s$ while slip tends to go to zero. For the critical mode, that is, when the slip is below its critical value ($|S_S| < S_{\text{Cr}}$), the longitudinal force is calculated by Equation (26):

$$F_X = -C_{\text{Sl}} \times S_S. \tag{26}$$

For the full slip mode when slip is higher than its critical value ($|S_S| > S_{\text{Cr}}$), the longitudinal force is given by

$$F_X = -(F_{X1} - F_{X2}) \times \text{sign}(S_S), \tag{27}$$

where $F_{X1}$ and $F_{X2}$ are stated as,

$$F_{X1} = U \times F_Z, \tag{28}$$

$$F_{X2} = \left| \frac{(U \times F_Z)^2}{4 \times |S_S| \times C_{\text{Slip}}} \right|. \tag{29}$$

### 4.5. *Calculation of lateral force*

The lateral force on a tire is a function of the vertical force and the friction coefficient. Similar to the calculation of the longitudinal force, the Fiala model uses a critical lateral slip ($\alpha_{\text{Cr}}$), to obtain the lateral force,

$$\alpha_{\text{Cr}} = a\tan\left(\frac{3 \times U \times |F_Z|}{C_\alpha}\right), \tag{30}$$

where $C_\alpha$ is the lateral stiffness, found by the limit of changes in $F_y$ with respect to $\alpha$ when $\alpha$ is close to zero. When the slip angle $\alpha$ reaches its critical value ($\alpha_{\text{Cr}}$), the lateral force takes its maximum value ($U \times |F_Z|$). Furthermore, if $|\alpha| \leq \alpha_{\text{Critical}}$ then $F_y$ is given by

$$F_y = -U \times |F_z| \times (1 - H^3) \times \text{sign}(\alpha), \tag{31}$$

where H is expressed by

$$H = 1 - \frac{C_\alpha \times |\tan(\alpha)|}{3 \times U \times |F_Z|}. \tag{32}$$

For the full slip mode, that is, $|\alpha| > \alpha_{\text{Cl}}$, $F_y$ is given by

$$F_y = -U \times |F_Z| \times \text{sign}(\alpha). \tag{33}$$

Table 2 summarizes some dynamic parameters of the tire.

Dynamic equations of the tire were coded in C# and then two aforementioned maneuvers were used for validation. The forces applied to the tire from the road, derived from dynamic equations, were compared with those obtained through the full car model in ADAMS.

Table 2. Dynamic parameters of tire.

| | |
|---|---|
| Unload tire radius | $R_\omega = 0.283$ (m) |
| Mass of tire | $M_{\text{tire}} = 14.9$ (kg) |
| Friction coefficient at full slip | $U_{\min} = 0.7$ |
| Friction coefficient at zero slip | $U_{\max} = 0.9$ |
| Vertical damping factor of tire | VD = 153,000 (N/m) |
| Vertical stiffness factor of tire | VS = 100 (N s/m) |

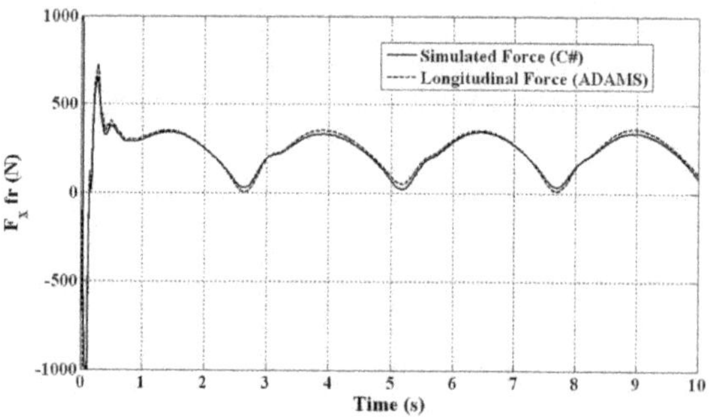


Figure 9. Longitudinal force on the tire of ADAMS and C# models (sinusoidal input of the steering wheel).

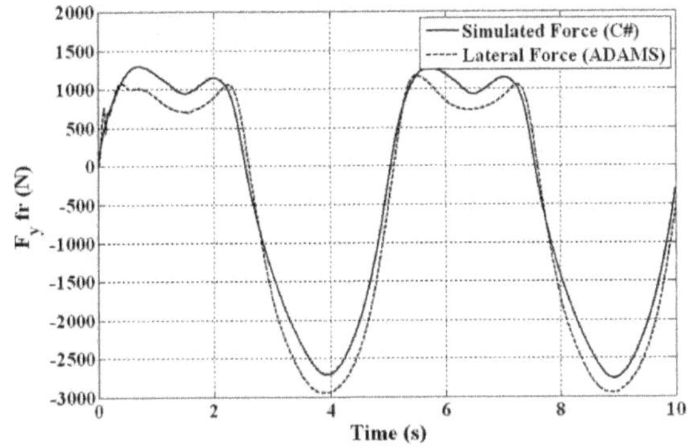


Figure 10. Lateral force on the tire of ADAMS and C# models (sinusoidal input of the steering wheel).

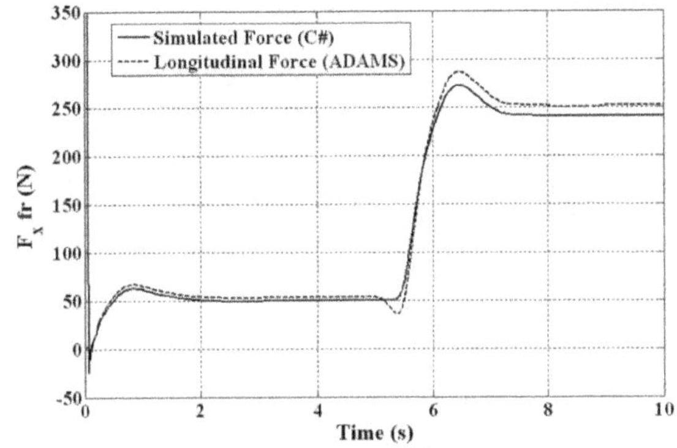


Figure 11. Longitudinal force on the tire of ADAMS and C# models (step input of the steering wheel).

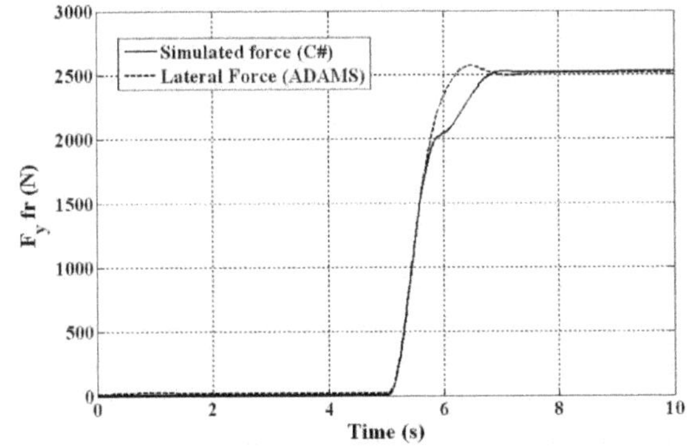


Figure 12. Lateral force on the tire of ADAMS and C# models (step input of the steering wheel).

Figures 9–12 illustrate the results of comparison of the tire longitudinal and lateral forces.

It is clear that the simulated model has accurately produced longitudinal and lateral forces. The maximum error for the lateral force is 200 N, obtained in the sinusoidal input maneuvering. The average error over one cycle is less than 100 N, indicating an error level below 5%. This value for longitudinal force is below 50 N.

## 5. Simulation of front wheels kinematics

The effect of tire forces and torques on the steering wheel must be calculated. In real vehicles, these forces and torques are transferred to the steering wheel through linkage of the front wheels. In the simulated steering system, the linkage connecting the steering box to the steering wheel is incorporated into the control system in a hardware-in-the-loop form and hence its dynamic modeling is not required. However, the linkage from the tire to the steering box does not exist and therefore must be simulated. To perform this simulation, some simplifications are considered to reduce computational burden and allow real-time implementation with insignificant computational error. These simplifications are as follows:

- Wheel angle $\delta$ is calculated by dividing the steering angle by the pinion ratio. Hence, this angle is identical for both wheels. For the real car, the average value of the left and the right wheel's angle is considered.
- To calculate the torques on the steering wheel, only the influence of the longitudinal and lateral forces, applied from the road to the tire, and the steering angle are taken into account and the effect of other forces and torques is ignored.
- The contact point of the kingpin axle to the ground is considered as the center of rotation.
- The tie-rod shaft is modeled as a bi-force element and its deflection during movement is ignored. Hence, the force applied on the tie-rod is equivalent to the rack force.
- The ground-tire contact point is always constant.

After simplifications, the diagram of the applied forces on the tire is developed as illustrated in Figure 13.

Based on this figure, rack forces are obtained by vector summation of the forces from wheels on the tie-rod.

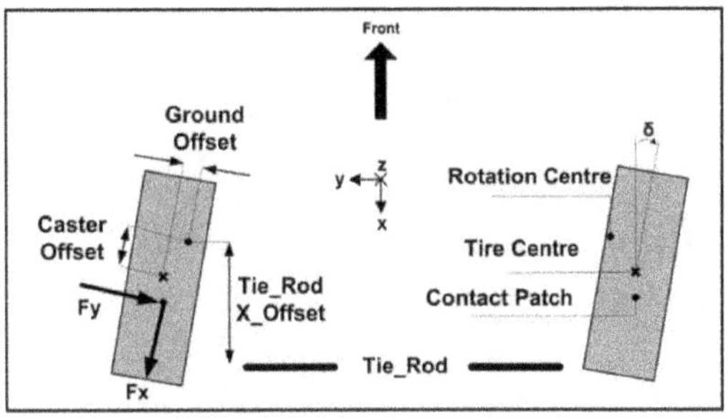


Figure 13. Diagram of the applied forces on the tire.

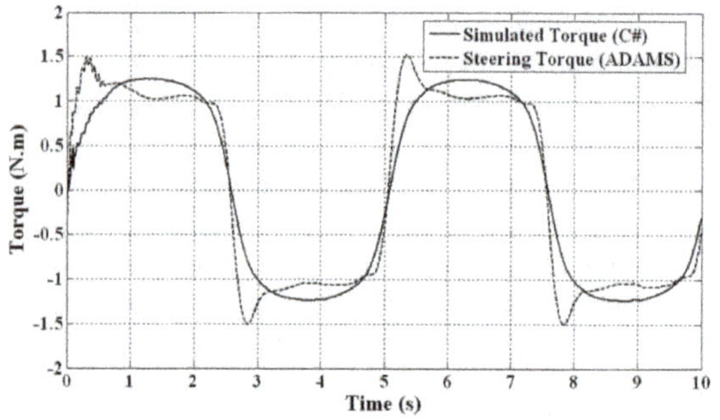


Figure 14. Torques on steering wheel of ADAMS and C# models (sinusoidal input of steering wheel).

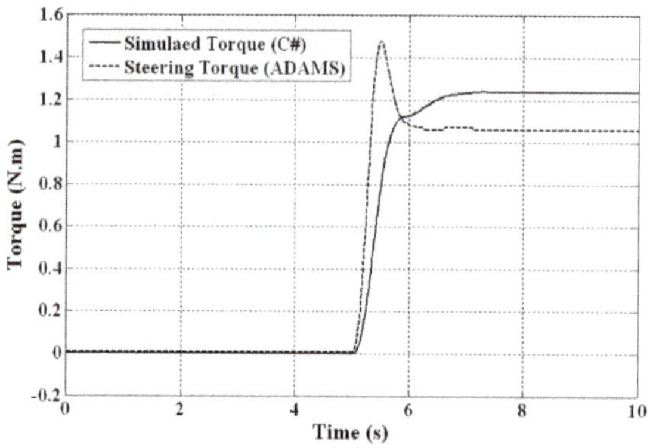


Figure 15. Torques on steering wheel of ADAMS and C# models (step input of steering wheel).

The forces on the tie-rod are calculated by equations in Equation (34),

$$\begin{aligned} A &= \left(\frac{1}{\mathrm{TO}_x + \mathrm{CO}}\right), \\ B &= (F_y + (F_x \cdot \sin\delta))(\mathrm{CO_x} + \mathrm{Caster\_Offset}), \\ C &= (F_x \cdot \cos\delta \cdot \mathrm{Ground\_Offset}). \end{aligned} \tag{34}$$

where $\mathrm{TO}_x$ is the vertical distance of the tie-rod from the longitudinal rotation center of the tire and $\mathrm{CO}_x$ indicates the vertical distance of the tire–road contact center from the longitudinal rotation center of the tire.

### *5.1. Simulation for the left wheel*

$$(F_{\mathrm{Tie\_Rod}})_{\mathrm{l}} = A(B + C). \tag{35}$$

### *5.2. Simulation for the right wheel*

$$(F_{\mathrm{Tie\_Rod}})_{\mathrm{r}} = A(B - C). \tag{36}$$

The geometry of the front wheels was validated using the maneuvers described in Section 3. The developed torques on the pinion of the steering box were compared with those obtained by the ADAMS model during the maneuvers. Figures 14 and 15 illustrate this comparison.

It is seen from these figures that the maximum error for both the maneuvers is about 0.2 Nm. This error can be justified by the fact that only the effect of the longitudinal and lateral forces on development of torques on the steer is considered in the simulated model and the influence of other forces and torques, such as vertical force and rolling resistance torque, is ignored.

## 6. Hardware-in-the-loop implementation

In this paper, dynamic equations of the vehicle, friction force on tires and the resulting torques on the steering box are simulated in the computer, while elements of the steering system such as torque actuator, sensors, steering shaft, and steering wheel are physically realized in the hardware-in-the-loop system. An illustration of the hardware-in-the-loop system is shown in Figure 16.

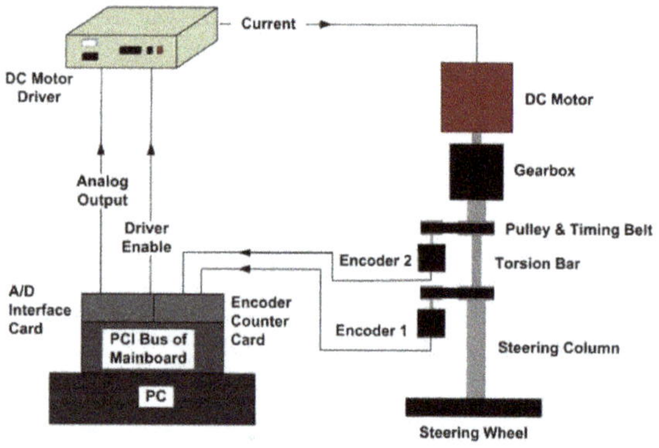


Figure 16. Hardware-in-the-loop system.

This system uses a brushed DC motor, which is directly connected to the end of the steering shaft and produces the torques developed by the friction forces between the tire and the road. The used DC motor has the rated power of 400 W and produced 3.1 Nm torque at the maximum current of 16 A.

The simulated dynamic systems required the instantaneous angular position of the steering wheel to calculate the steering torques. Moreover, the control system needs the motor torque to properly control the applied torque. These requirements are fulfilled by the use of a torque sensor, which is shown in Figure 17.

In this scheme, the angular positions of both ends of the torsion bar are estimated by two 12-bit encoders. The changes in the angles of both ends of the torsion bar are transferred by pulleys and timing belts to the encoders. The pulleys have a 1:2 diameter ratio and are connected to high-resolution encoders (3600 pulse/revolution). The torque sensor resolution is 0.0375 Nm. The sensor works based on a simple principle that the applied torque on a rod is proportional to the difference in rotational angles of the

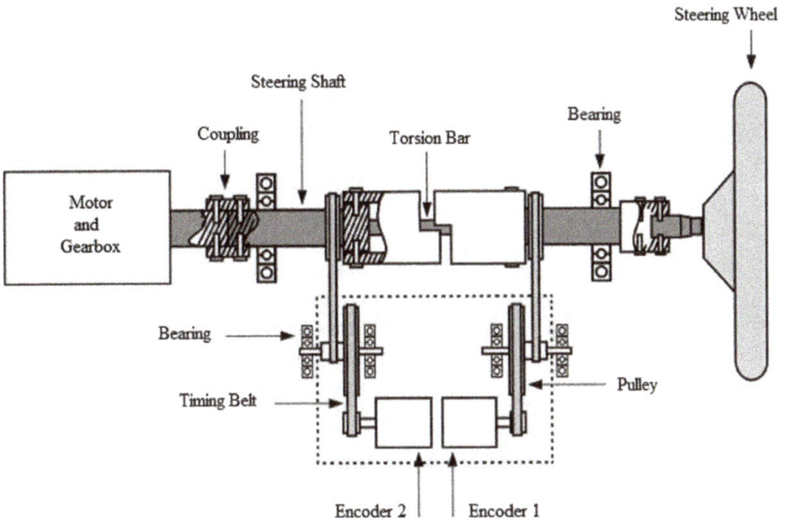


Figure 17. Torque sensor placement in the system.

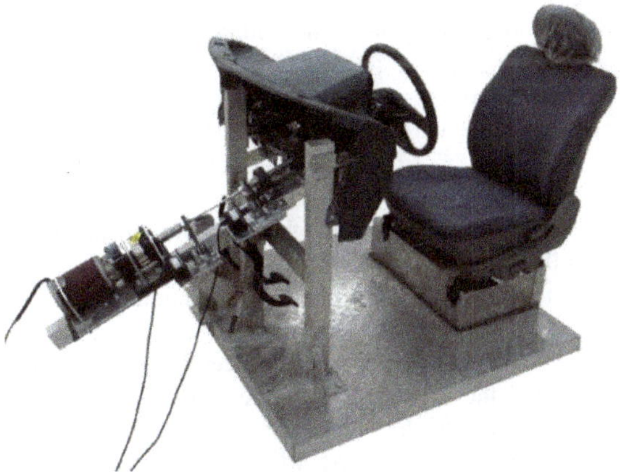

Figure 18. The hardware system made at the laboratory.

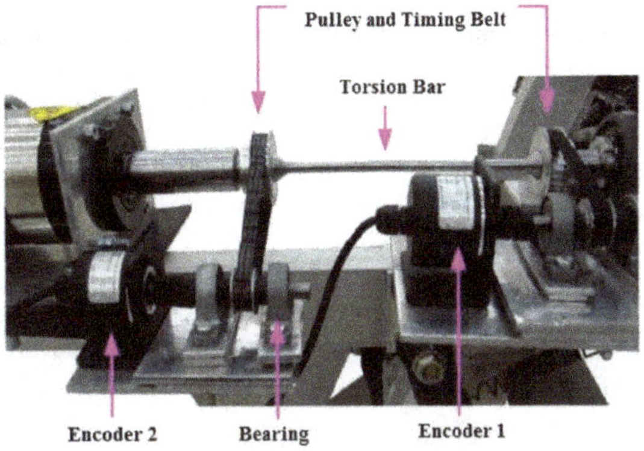


Figure 19. Implemented torque sensor and its components.

rod's two ends (Beer & Johnston, 2008),

$$T_{gb} = K_{tb} \cdot \Delta\theta = K_{tb} \cdot (\theta_{sw} - \theta_{gb}), \qquad (37)$$

where $T_{gb}$ is the output torque of the gearbox, $K_{tb}$ indicates stiffness coefficient of the torsion bar, which is about 186.6 N m./rad, and $\theta_{sw}$ and $\theta_{gb}$ are the rotation angles of the steering wheel and the gearbox, respectively. Hence, the motor torque is calculated and sent as a feedback signal to the computer. The picture of the implemented hardware system is shown in Figure 18 whereas a detailed overview of the torque sensor is illustrated in Figure 19.

## 7. Simulation of control system

Before hardware implementation of the control system, the controller was simulated in the computer to investigate its functionality and performance. Hence, all elements of the hardware system were modeled by equivalent dynamic equations as shown in Figure 20.

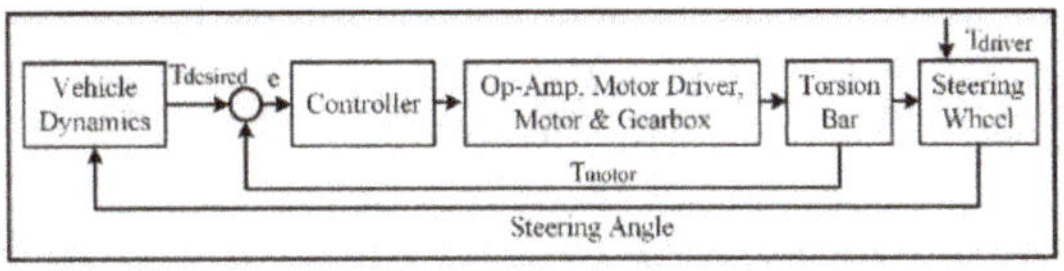


Figure 20. Control system block diagram.

In Figure 20, $T_{desired}$ is the desired torque which is calculated by the vehicle dynamic model and is compared with the motor torque ($T_{motor}$) and the generated error is fed to the controller. The controller's output signal is amplified by an operational amplifier and then sent to the driver

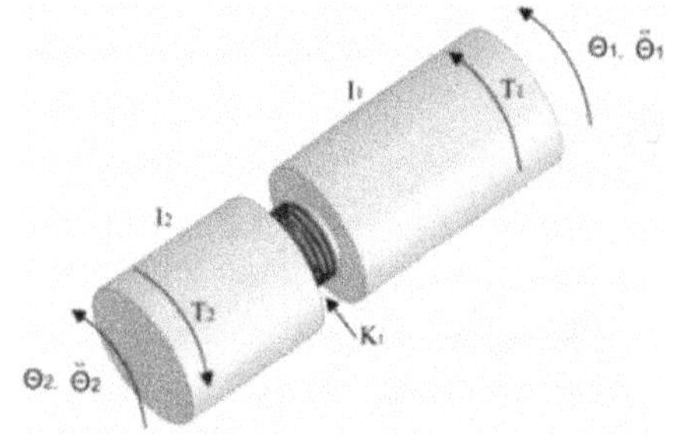


Figure 21. Simplified model of the steering system from the steering wheel to the pinion.

of the DC motor to produce the suitable current for generation of the desired torque. The motor torque is passed through the gearbox and then is measured by the torque sensor. The resultant of this torque and the driver torque, $T_{\text{driver}}$, are simultaneously applied to the steering wheel. The rotation angle of the steering wheel is the output of the system, which is re-used to solve the dynamic equations of the vehicle.

To derive the dynamic model of the steering system from the steering wheel to the connection point of motor and gearbox, the overall system is considered by two masses with rotational inertia. Since the mass and inertia of the torsion bar are negligible compared with other system's components, it is modeled by a torsional spring alone. Hence, the overall system is modeled by two rotational inertias interconnected by a torsion spring as shown in Figure 21.

In this figure, $T_1$ is equivalent to $T_{\text{driver}}$, applied by the driver to the steering wheel and $T_2$ is the motor torque after the gearbox and is expressed by

$$T_2 = T_m \cdot N_1 = T_{\text{gb}}. \tag{38}$$

Moreover, $I_1$ is the sum of the steering wheel inertia ($I_{\text{sw}}$) and the steering shaft inertia ($I_{\text{ss}}$). In addition, $I_2$ indicates the sum of the motor inertia ($I_{\text{m}}$), the gearbox inertia, $I_{\text{gb}}$, and the coupling inertia, $I_{\text{coupling}}$, between the gearbox shaft and the torsion bar (Craig, 2004),

$$I_1 = I_{sw} + I_{ss}, \tag{39}$$

$$I_2 = I_{\text{m}} \cdot N_1^2 + I_{\text{gb}} + I_{\text{coupling}}. \tag{40}$$

The equation governing DC motor is expressed by Equation (41) as stated in Dorf & Bishop (2004),

$$V_{\text{a}} = V_{\text{b}} + R_{\text{a}} I_{\text{a}} + L_{\text{a}} \frac{dI_{\text{a}}}{dt}, \tag{41}$$

where $V_{\text{a}}$, $I_{\text{a}}$, and $R_{\text{a}}$ are armature voltage, current, and resistance, respectively, and $L_{\text{a}}$ is the armature self-inductance. $V_{\text{b}}$ indicates the induced voltage, which is proportional to motor speed. The relationship between the armature current and the output torque is expressed by

$$T_{\text{m}} = K_{\text{t}} I_{\text{a}},$$

where $K_{\text{t}}$ is the motor constant. The classical PID controller is used in this paper. To analyze the system response, a step input is applied. To find the controller coefficients, the Zeigler–Nichols approach is used. Then the coefficients are tuned for a more accurate response. Since the input of the actual system changes frequently, the transient response of the system is very important. To investigate the transient response, a sinusoidal input with 1 Hz frequency and unit amplitude is applied to the system. The system response reveals that the designed controller is able to desirably follow variable inputs.

## 8. Experimental results

After investigating the controller performance by simulation inputs, the control was implemented on hardware.

Before doing the main tests, and to evaluate systems performance in a virtual driving condition, a user was asked to use the system and follow the desired route, specified by graphical illustrations.

The desired steering shaft torque, obtained by dynamic equations, along with the real torque generated by the steering system of the driving simulator and measured by the torque sensor are shown in Figure 22.

The generated torque by the dynamic equations and controller is transferred to driver's hands through the motor connected to the steering shaft. The maximum delay for this case is below 0.05 s.

For more investigation of the steering system performance and the influence of torque feedback on driver's proper steering of the vehicle, various experiments were executed on 25 users in the driving simulator. To do so, first an experimental protocol and a driving scenario were defined. Then, the users had to drive on the route of Figure 23, using the steering system, as described in this paper.

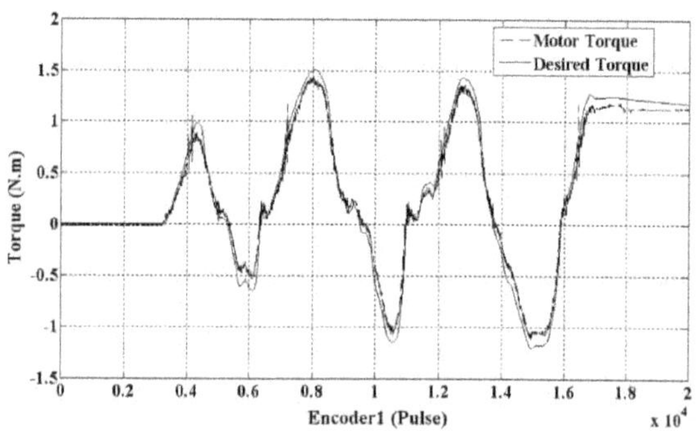


Figure 22. Desired torque and torque generated with the steering system's motor in a driving scenario.

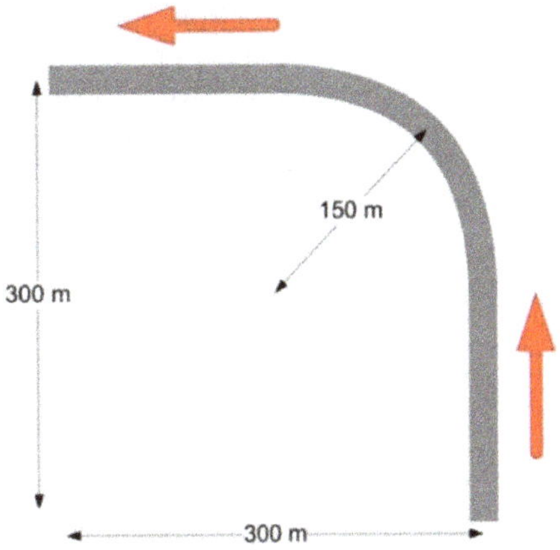


Figure 23. Schematic image of the path used in a driving scenario.

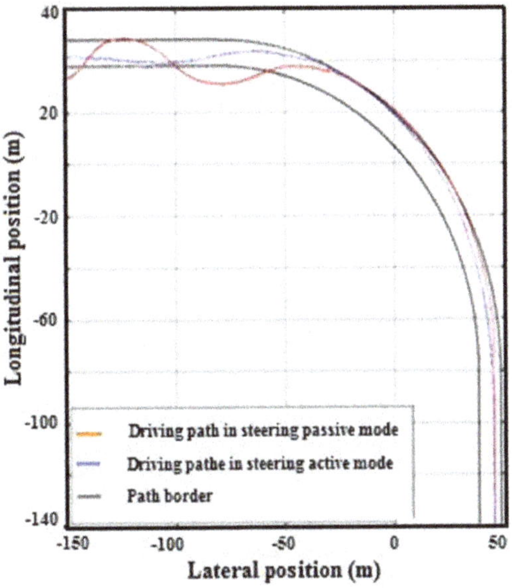


Figure 24. Different paths traveled by a chosen user while driving with two different steering modes.

Before doing the main tests and to familiarize each participant with the path, each subject was asked to drive in the specified scenario once when there is torque feedback on the steering system (active mode) and another time when there is no torque feedback on the steering system (passive mode). For the main test, the users had to drive along the path with constant speed three times in the active steering mode and three times in the passive steering mode. During the experiment, the driving pattern of each user was recorded and the simulator's instantaneous position was compared with its desirable value (keeping within the road limits). Figure 24 shows the path passed by one of the users during the driving scenario for both active and inactive steering modes.

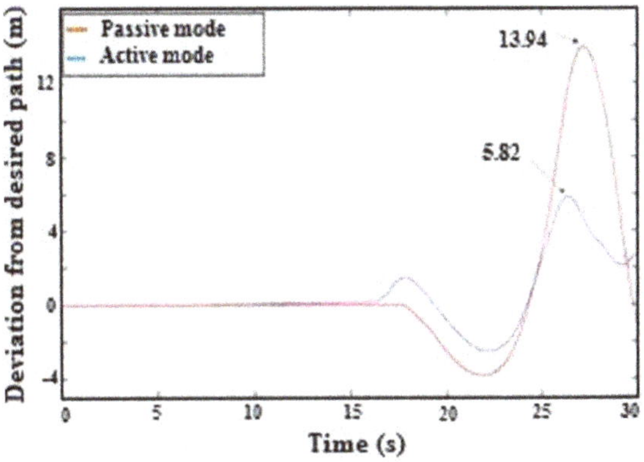


Figure 25. Deviation from desired position in two different steering modes (selected user result).

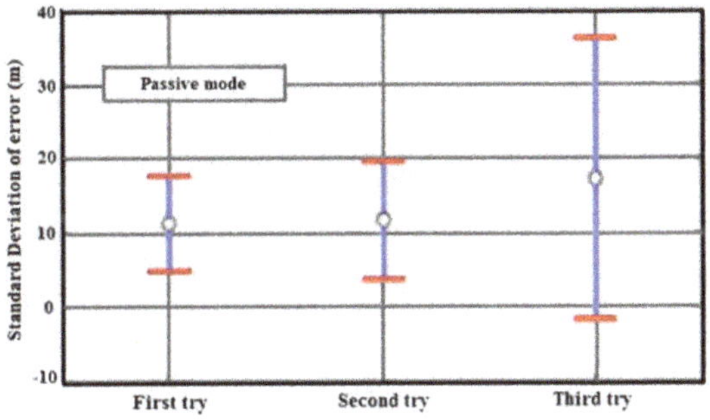


Figure 26. Maximum and standard deviation of error from the desired path (passive steering mode).

The deviation of the user from the desired position is shown in Figure 25. Obviously, it can be seen that there is a small deviation when the steering system with torque feedback is used.

To have a broader analysis, the results obtained for all users are considered. The average error of maximum deviations from the road for all 25 users and the standard deviation of this error for three experiments, in the passive mode, are shown in Figure 26.

In this figure, the white circles indicated the root mean square error of all the users in every experiment and the blue vertical lines show the standard deviation of the error.

It is seen that, on average, the maximum deviation of the users for the first, second, and third stage is 11.5, 12 and 17.2 m, respectively. Clearly, increasing the number of experiments has led to higher deviations due to the driver fatigue. The standard deviation of error has also increased with the number of experiments, which is not desirable.

The average of maximum errors during the three experiments in the active mode (with torque feedback) is depicted in Figure 27.

Comparing Figures 26 and 27 confirms the favorable influence of the torque feedback on error reduction.

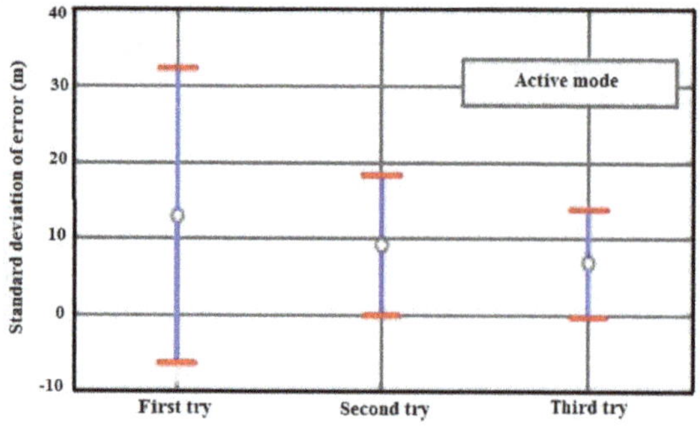


Figure 27. Maximum and standard deviation of error in comparison with the desired path (active steering mode).

Moreover, the torque feedback has improved users' performance during the experiments. For instance, the average of maximum errors in the first round of experiment is 13 m, whereas this value in the third round of experiments had been reduced to 7 m.

## 9. Conclusion

This paper presented the design, construction, and control of the electrical steering system for a driving simulator. The simulator can be used in driving simulation to have a deeper sense of immersion by the users as well as for construction of steer-by-wire systems. The system was then used to analyze the effect of torque feedback exerted to driver's hands on vehicle handling.

To produce the steering torques, the 14-DOF equations of the vehicle, the engine model, the dynamic equations of the wheels, and the steer geometry were simulated. The validity of all the models was investigated by comparison with a 251-DOF car model, developed and validated in ADAMS.

After construction of simulator, it was used by a user to travel along a specified path. The torque calculated by the dynamic equations was compared with those generated by the system actuator and it was seen that the system followed the generated torques. On the other hand, though the PID controller fulfilled the control requirements of the system, it was sensitive to variations of system parameters. Hence, using adaptive control schemes may improve system response.

Various experiments were conducted on several users to investigate the influence of the torque feedback on the steering wheel on proper steering of the vehicle. By comparing the results obtained with and without torque feedback during these experiments, it was shown that the torque feedback improved the driver's perception from the surrounding environment and enabled her/him to handle the vehicle satisfactorily.

### Disclosure statement

No potential conflict of interest was reported by the authors.

### References

Bastow, D., Howard, G., & Whitehead, J. P. (2004). *Car suspension and handling*. Warrendale, PA: SAE International.

Beer, F. P., & Johnston, E. R. (2008). *Mechanics of materials*. New York: McGraw Hill.

Burdea, G. C., & Coiffet, P. (2003). *Virtual reality technology*. Hoboken, NJ: John Wiley & Sons.

Chen, L. K., & Ulsoy, A. G. (2002). *Experimental validation of a robust steering assist controller on a driving simulator* (Vol. 3, pp. 2528–2533). Proceedings of the 2002 American Control Conference, Anchorage, AK, 2002.

Craig, J. J. (2004). *Introduction to robotics: Mechanics and control*. Upper Saddle River, NJ: Addison-Wesley.

Dorf, R. C., & Bishop, R. H. (2004). *Modern control systems*. Upper Saddle River, NJ: Prentice Hall.

Friedrichs, F., & Yang, B. (2010). *Drowsiness monitoring by steering and lane data based features under real driving conditions* (pp. 209–213). Paper presented at the 18th European Signal Processing Conference (EUSIPCO-2010), Denmark.

Gregersen, N. P., Falkmer, T., Dol, J., & Pardo, J. (2001). *Driving simulator scenarios and requirements*. EU; Competitive and Sustainable Growth Programme.

Kim, J. H., & Song, J. B. (2002). Control logic for an electric power steering system using assist motor. *Mechatronics*, *12*, 447–459.

MSC Software Corporation. (2003). MSC ADAMS 2003. *Help, Tire*.

Park, T., Han, C., & Lee, S. (2005). Development of the electronic control unit for the rack-actuating steer-by-wire using the hardware-in-the-loop simulation system. *Mechatronics*, *15*, 899–918.

Samiee, S., Azadi, S., Kazemi, R., Nahvi, A., & Eichberger, A. (2014). Data fusion to develop a driver drowsiness detection system with robustness to signal loss. *Sensors*, *14*, 17832–17847.

Segawa, M., Nakano, S., Nishihara, O., & Kumamoto, H. (2001). Vehicle stability control strategy for steer by wire system. *JSAE Review*, *22*, 383–388.

Straus, S. H. (2005). *New, improved, comprehensive, and automated driver's license test and vision screening system* (Report No. FHWA-AZ-04-559(1)). USA: U.S. Department of Transportation, Federal Highway Administration.

Yao, Y. (2006). *Vehicle steer-by-wire system control* (SAE Technical Paper). Detroit, MI: SAE.

Yih, P., & Gerdes, J. C. (2004). *Steer-by-wire for vehicle state estimation and control* (pp. 785–790). Paper presented at the Seventh International Symposium on Advanced Vehicle Control, The Netherlands.

# Sensorless control for the brushless DC motor: an unscented Kalman filter algorithm

Haidong Lv, Guoliang Wei*, Zhugang Ding and Xueming Ding

*Department of Control Science and Engineering, University of Shanghai for Science and Technology, Shanghai 200093, People's Republic of China*

In this paper, a new mathematical model is built according to the characteristics of the brushless DC (BLDC) motor and a new filtering algorithm is proposed for the sensorless BLDC motor based on the unscented Kalman filter (UKF). The proposed UKF algorithm is employed to estimate the speed and rotor position of the BLDC motor only using the measurements of terminal voltages and three-phase currents. In order to observe the drive performance, two simulation examples are given and the feasibility and effectiveness of the UKF algorithm are verified through the simulation results, and the accurate estimate performance is shown in simulation figures.

**Keywords:** BLDC; sensorless control; UKF algorithm; nonlinear wave function; trigonometricseries

## 1. Introduction

With the rapid development of modern industries, the brushless DC (BLDC) motor is increasingly being used in computer peripherals, office automation and machine tool industries because of its high efficiency, easy control, compact form, etc. (Pillay and Krishnan, 1991), and the BLDC motor also has the characteristics of trapezoidal electromotive force (emf) and quasi-rectangular current waveforms. In order to obtain the appropriate commutation signals every 60 electrical degrees, rotor position sensors must be used, such as hall sensors and photoelectric encoders. However, speed or position sensors require additional mounting space, they increase the cost and the complexity of the system as well as reduce the reliability of the system. In recent years, the sensorless control technology has received wide attention. The sensorless technology improves the system reliability and it is of great significance to further expand the application fields of the BLDC motor.

At present, there are so many papers that have reported the sensorless control technology. Among them, the most popular and widely used method is the back electromotive force (back-emf) method (Iizuka, Uzuhashi, Kano, Endo, and Mohri, 1985). However, this method generally has a drawback that the back-emf cannot be detected exactly under low-speed conditions. In addition, some other methods have been discussed, such as the estimation flux method (Ji and Li, 2008) and the freewheeling diodes detection method. Nevertheless, these methods cannot provide continual rotor position information when a high accuracy of the rotor speed and position are required. The extended Kalman filter (EKF) is an optimal recursive estimation algorithm for nonlinear systems and has been applied for estimating the state variable of the BLDC motor (Lenine, Reddy, and Kumar, 2007; Terzic and Jadric, 2001). However, there are also some limitations for the EKF algorithm, such as the complex computation of the Jacobian matrices, only first-order accuracy, etc.

In this paper, a new method has been introduced to overcome the above drawbacks by means of the unscented Kalman filter (UKF) algorithm. The UKF algorithm uses a deterministic sampling approach to capture the mean and covariance estimates with a minimal set of sample points. The related applications of the UKF have been reported in many papers, such as permanent magnet synchronous motor drive (Chan and Borsje, 2009) and induction motor drive (Rigatos and Siano, 2012). The simulation results by MATLAB/Simulink indicate that the speed could be tracked and adjusted precisely, and the dynamical property of systems is evidently improved.

## 2. Mathematical model of the BLDC

In this paper, the three-phase BLDC motor with star connection is considered, and the voltage equation of the phase A of the stator is as follows (Pillay and Krishnan, 1989):

$$u_a = Ri_a + (L - M)\frac{di_a}{dt} + e_a + u_n, \tag{1}$$

*Corresponding author. Email: guoliang.wei1973@gmail.com

where $u_a$ and $u_n$ are the terminal voltage of phase A and neutral point voltage, respectively, $R$, $i_a$, $L$ and $e_a$ are the phase resistance, the phase current, the phase inductance and the phase back-emf of the phase A, respectively, and $M$ is the mutual inductance. Especially, we have similar voltage equations for phases B and C.

According to the operation principle of the BLDC motor, it is known that only two phases conduct in three-phase stator winding at each time point. Thus, the neutral point voltage can be deduced as follows:

$$u_n = \tfrac{1}{3}[(u_a + u_b + u_c) - (e_a + e_b + e_c)], \qquad (2)$$

where $u_b$ and $e_b$ are the voltage and the back-emf of the phase B, respectively, $u_c$ and $e_c$ are the voltage and the back-emf of the phase C, respectively.

By substituting Equation (2) into the voltage equation of phase A, we can obtain the following equation:

$$\frac{di_a}{dt} = \frac{u_{ab} - u_{ca}}{3(L-M)} - \frac{R}{L-M} i_a + \frac{e_b + e_c - 2e_a}{3(L-M)}, \qquad (3)$$

where $u_{ab} = u_a - u_b$ and $u_{ca} = u_c - u_a$.

According to the structure of the BLDC motor, the back-emf of phase A can be written as (Chen, Huang, Wang, and Wu, 2011)

$$e_a = \omega \varphi_m f_a(\theta), \qquad (4)$$

where $\omega$ is the motor angular velocity, $\varphi_m$ is the magnet flux linkage of the stator winding, $f_a(\theta)$ changeing along with the rotor position is the wave function of the back-emf of phase A and its maximum value and minimum value are 1 and $-1$, respectively.

The nonlinear function $f_a(\theta)$ can be described as follows:

$$f_a(\theta) = \begin{cases} \frac{6}{\pi} \cdot \theta, & 2k\pi \le \theta \le \frac{\pi}{6} + 2k\pi, \\ 1, & \frac{\pi}{6} + 2k\pi \le \theta \le \frac{5\pi}{6} + 2k\pi, \\ -\frac{6}{\pi} \cdot (\theta - \pi), & \frac{5\pi}{6} + 2k\pi \le \theta \le \frac{7\pi}{6} + 2k\pi, \\ -1, & \frac{7\pi}{6} + 2k\pi \le \theta \le \frac{11\pi}{6} + 2k\pi, \\ \frac{6}{\pi} \cdot (\theta - 2\pi), & \frac{11\pi}{6} + 2k\pi \le \theta \le 2\pi + 2k\pi \end{cases} \qquad (5)$$

and represented by the trigonometric series as follows:

$$f_a(\theta) = 1.21 \sin\theta + 0.27 \sin 3\theta + 0.05 \sin 5\theta - 0.02 \sin 7\theta - 0.03 \sin 9\theta \ldots. \qquad (6)$$

For the symmetry structure of the BLDC motor, we have $f_b(\theta) = f_a(\theta - 2\pi/3)$ and $f_c(\theta) = f_a(\theta + 2\pi/3)$.

From Equations (3) and (4), it can be derived that

$$\frac{di_a}{dt} = \frac{u_{ab} - u_{ca}}{3(L-M)} - \frac{R}{L-M} i_a + \frac{\omega\varphi_m[f_b(\theta) + f_c(\theta) - 2f_a(\theta)]}{3(L-M)}. \qquad (7)$$

In the motor control community, because of the extensive application of the digital control, the discrete-time system is widely used to describe the motor motions, and hence we consider the corresponding discrete-time equation of Equation (7) in this paper:

$$i_a(k+1) = \frac{T[u_{ab}(k) - u_{ca}(k)]}{3(L-M)} + \left(1 - \frac{TR}{L-M}\right) i_a(k) + \frac{T\omega(k)\varphi_m[f_b(\theta_k) + f_c(\theta_k) - 2f_a(\theta_k)]}{3(L-M)}, \qquad (8)$$

where $T$ is the sampling time.

Similarly, we have

$$i_b(k+1) = \frac{T[u_{bc}(k) - u_{ab}(k)]}{3(L-M)} + \left(1 - \frac{TR}{L-M}\right) i_b(k) + \frac{T\omega(k)\varphi_m[f_a(\theta_k) + f_c(\theta_k) - 2f_b(\theta_k)]}{3(L-M)}, \qquad (9)$$

$$i_c(k+1) = \frac{T[u_{ca}(k) - u_{bc}(k)]}{3(L-M)} + \left(1 - \frac{TR}{L-M}\right) i_c(k) + \frac{T\omega(k)\varphi_m[f_a(\theta_k) + f_b(\theta_k) - 2f_c(\theta_k)]}{3(L-M)}, \qquad (10)$$

where $i_b$ and $i_c$ are the phase current of phases B and C, respectively.

According to Equation (4), we can obtain the following torque equation of the BLDC motor:

$$T_e = \frac{e_a i_a + e_b i_b + e_c i_c}{\Omega} = \frac{\omega\varphi_m[f_a(\theta)i_a + f_b(\theta)i_b + f_c(\theta)i_c]}{\omega/p} = p\varphi_m[f_a(\theta)i_a + f_b(\theta)i_b + f_c(\theta)i_c], \qquad (11)$$

where $T_e$ is the electromagnetic torque, $\Omega$ is the mechanical angular velocity and $p$ is the number of pole pairs of the BLDC motor.

Consider the motion equation of the BLDC motor:

$$T_e - T_L = J\frac{d\Omega}{dt} + B_v\Omega, \qquad (12)$$

where $T_L$ is the load torque of motor, $J$ is the rotational inertia and $B_v$ is the viscous friction coefficient.

From Equations (11) and (12) and by converting the mechanical angular velocity into the electrical angular velocity, we have

$$\omega(k+1) = \frac{p^2\varphi_m T}{J}[f_a(\theta)i_a + f_b(\theta)i_b + f_c(\theta)i_c] - \frac{pT}{J}T_L + \left(1 - \frac{B_v T}{J}\right)\omega(k). \quad (13)$$

According to Newton's law of motion, we can have

$$\theta(k+1) = T\omega(k) + \theta(k). \quad (14)$$

By combining the relevant equations (7), (9), (10), (13) and (14), the nonlinear state equations can be expressed in the following form:

$$\begin{aligned} x_{k+1} &= F_k(x_k)x_k + G_k u_k, \\ y_k &= Hx_k, \end{aligned} \quad (15)$$

where $x_k = [i_a\ i_b\ i_c\ \omega\ \theta]_k^T$; $y_k = [i_a\ i_b\ i_c]_k^T$; $u_k = [u_{ab} - u_{ca}\ u_{bc} - u_{ab}\ u_{ca} - u_{bc}\ T_L]_k^T$

$$F_k(x_k) = \begin{bmatrix} 1-\frac{RT}{L-M} & 0 & 0 & F14 & 0 \\ 0 & 1-\frac{RT}{L-M} & 0 & F24 & 0 \\ 0 & 0 & 1-\frac{RT}{L-M} & F34 & 0 \\ F41 & F42 & F43 & 1-\frac{B_v\cdot T}{J} & 0 \\ 0 & 0 & 0 & T & 1 \end{bmatrix},$$

$$F14 = \frac{T\omega(k)\varphi_m[f_b(\theta_k) + f_c(\theta_k) - 2f_a(\theta_k)]}{3(L-M)},$$

$$F24 = \frac{T\omega(k)\varphi_m[f_a(\theta_k) + f_c(\theta_k) - 2f_b(\theta_k)]}{3(L-M)},$$

$$F34 = \frac{T\omega(k)\varphi_m[f_a(\theta_k) + f_b(\theta_k) - 2f_c(\theta_k)]}{3(L-M)},$$

$$F41 = \frac{p^2\varphi_m T}{J}f_a(\theta_k),$$

$$F42 = \frac{p^2\varphi_m T}{J}f_b(\theta_k),$$

$$F43 = \frac{p^2\varphi_m T}{J}f_c(\theta_k),$$

$$G_k = \begin{bmatrix} \frac{T}{3(L-M)} & 0 & 0 & 0 \\ 0 & \frac{T}{3(L-M)} & 0 & 0 \\ 0 & 0 & \frac{T}{3(L-M)} & 0 \\ 0 & 0 & 0 & -\frac{T\cdot p}{J} \\ 0 & 0 & 0 & 0 \end{bmatrix},$$

$$H = \begin{bmatrix} 1 & 0 & 0 & 0 & 0 \\ 0 & 1 & 0 & 0 & 0 \\ 0 & 0 & 1 & 0 & 0 \end{bmatrix}.$$

## 3. UKF algorithm

For the Kalman filtering problem of a nonlinear system, although the EKF algorithm maintains the efficient recursive update form of the Kalman filter, it suffers a number of serious limitations. For instance, the calculation of the Jacobian matrices may be a difficult and error-prone process. For the purpose of overcoming the limitations of the EKF algorithm, the unscented transformation (UT) was proposed by Julier and Uhlman and a new Kalman filter algorithm (UKF) was presented in Julier and Uhlmann (2004). In this algorithm, a set of sample points are used to parameterize the mean and covariance of the probability distribution of the state variables.

In this section, the basic UT is reviewed first and the UKF algorithm is then introduced in detail.

### 3.1. *Basic UT*

Consider the following nonlinear function:

$$y = h(x), \quad (16)$$

where $x$ is the $n$-dimensional variable with the mean $\bar{x}$ and covariance matrix $P_1$, $y$ is the $m$-dimensional variable with the mean $\bar{y}$ and covariance matrix $P_2$.

For the variable $x$ in Equation (16), a set of $2n$ sigma points are selected as follows (Simon, 2006):

$$\begin{aligned} x^{(i)} &= \bar{x} + \tilde{x}^{(i)}, \quad i = 1, \ldots, 2n, \\ \tilde{x}^{(i)} &= (\sqrt{nP_1})_i^T, \quad i = 1, \ldots, n, \\ \tilde{x}^{(n+i)} &= -(\sqrt{nP_1})_i^T, \quad i = 1, \ldots, n, \end{aligned} \quad (17)$$

where $\sqrt{nP_1}$ is the matrix square root of $nP_1$ with $\sqrt{nP_1}^T \cdot \sqrt{nP_1} = nP_1$, $(\sqrt{nP_1})_i$ is the $i$th row of $\sqrt{nP_1}$. Since $nP_1$ is a positive-definite symmetric matrix, the square root can be calculated using the Cholesky factorization which can simplify the calculation procedure.

Each sample point $x^{(i)}$ is propagated through the nonlinear function to yield corresponding transformed sigma points $y^{(i)}$, that is

$$y^{(i)} = h(x^{(i)}) \quad i = 1, \ldots, 2n.$$

The mean $\bar{y}$ and covariance $P_2$ are approximated by the average mean and covariance of the transformed sigma points.

$$\bar{y} = \frac{1}{2n}\sum_{i=1}^{2n} y^{(i)},$$

$$P_2 = \frac{1}{2n}\sum_{i=1}^{2n}(y^{(i)} - y)(y^{(i)} - y)^T,$$

where $1/2n$ is the weight being used to calculate the mean and covariance.

### 3.2. *UKF algorithm*

Consider the following BLDC motor model:

$$\begin{aligned} x_{k+1} &= F_k(x_k)x_k + G_k u_k + w_k, \\ y_k &= H x_k + v_k, \end{aligned} \tag{18}$$

where $w_k$ and $v_k$ are, respectively, the process noise and the measurement noise, which is assumed as Gaussian white noise with covariance matrices $Q_k$ and $R_k$.

In the actual motor systems, the noises are unavoidable, such as the unmodeled noise, the detecting noise and so on. And these noises are likely to affect the normal operation of the motor. In the state-space model (18), all kinds of noises are described by the random variables $w_k$ and $v_k$.

Based on Equation (18), the UKF algorithm can be summed up as follows:

(1) Compute the set of sigma points $x_{k-1}^{(i)}$ based on the current optimal state estimation $\hat{x}_{k-1}^{+}$ and the covariance estimation $P_{k-1}^{+}$ according to Equation (17), that is,

$$\begin{aligned} \hat{x}_{k-1}^{(i)} &= \hat{x}_{k-1}^{+} + \tilde{x}^{(i)}, \quad i = 1, \ldots, 2n, \\ \tilde{x}^{(i)} &= (\sqrt{nP_{k-1}^{+}})_i^{\mathrm{T}}, \quad i = 1, \ldots, n, \\ \tilde{x}^{(n+i)} &= -(\sqrt{nP_{k-1}^{+}})_i^{\mathrm{T}}, \quad i = 1, \ldots, n. \end{aligned}$$

(2) Propagate the sigma points $\hat{x}_{k-1}^{(i)}$ to $\hat{x}_k^{(i)}$ through the nonlinear systems of the BLDC motor

$$\hat{x}_k^{(i)} = F_{k-1}(\hat{x}_{k-1}^{(i)})\hat{x}_{k-1}^{(i)} + G_k u_k, \quad i = 1, \ldots, 2n.$$

(3) Obtain the predicted mean $\hat{x}_k^-$

$$\hat{x}_k^- = \frac{1}{2n}\sum_{i=1}^{2n} \hat{x}_k^{(i)}.$$

(4) Compute the predicted covariance $P_k^-$

$$P_k^- = \frac{1}{2n}\sum_{i=1}^{2n} (\hat{x}_k^{(i)} - \hat{x}_k^-)(\hat{x}_k^{(i)} - \hat{x}_k^-)^{\mathrm{T}} + Q_{k-1}.$$

(5) Select the sigma points $\hat{x}_k^{(i)}$ based on the predicted mean and covariance

$$\begin{aligned} \hat{x}_k^{(i)} &= \hat{x}_k^{+} + \tilde{x}^{(i)}, \quad i = 1, \ldots, 2n, \\ \tilde{x}^{(i)} &= (\sqrt{nP_k^-})_i^{\mathrm{T}}, \quad i = 1, \ldots, n, \\ \tilde{x}^{(n+i)} &= -(\sqrt{nP_k^-})_i^{\mathrm{T}}, \quad i = 1, \ldots, n. \end{aligned}$$

(6) Transform the new sigma points through the measurement model

$$\hat{y}_k^{(i)} = H\hat{x}_k^{(i)}, \quad i = 1, \ldots, 2n.$$

(7) Calculate the predicted mean of the observation $\hat{y}_k$

$$\hat{y}_k = \frac{1}{2n}\sum_{i=1}^{2n} \hat{y}_k^{(i)}.$$

(8) Compute the predicted covariance matrices of the observation $P_y$

$$P_y = \frac{1}{2n}\sum_{i=1}^{2n} (\hat{y}_k^{(i)} - \hat{y}_k)(\hat{y}_k^{(i)} - \hat{y}_k)^{\mathrm{T}} + R_k.$$

(9) Compute the cross covariance matrices

$$P_{xy} = \frac{1}{2n}\sum_{i=1}^{2n} (\hat{x}_k^{(i)} - \hat{x}_k^-)(\hat{y}_k^{(i)} - \hat{y}_k)^{\mathrm{T}}.$$

(10) Update the estimation using the Kalman filter algorithm

$$\begin{aligned} K_k &= P_{xy} \cdot P_y^{-1}, \\ \hat{x}_k^+ &= \hat{x}_k^- + K_k(y_k - \hat{y}_k), \\ P_k^+ &= P_k^- - K_k P_y K_k^{\mathrm{T}}. \end{aligned}$$

## 4. Simulation results

In this section, under the environment of MATLAB/Simulink, the control system model of sensorless BLDC motor is built in Figure 1 and simulation examples are presented to illustrate the effectiveness of the UKF filter design method developed in this paper.

The parameters of the BLDC motor are given as follows: the stator resistance $R = 0.62\,\Omega$, the equivalent inductance of the stator $L - M = 1.0 \times 10^{-3}$ H, the maximum of each phase winding permanent magnet flux $\varphi_m = 0.066$ Wb, the inertia $J = 0.362 \times 10^{-3}$ kg m$^2$, the viscous friction coefficient $B_v = 9.444 \times 10^{-5}$ N m s, poles of the permanent magnet $p = 4$ and simulation step length $T = 5 \times 10^{-7}$ s, $x_0 = [0\ 0\ 0\ 0\ 0]^{\mathrm{T}}$.

According to the UKF algorithm, the results will be more accurate by using the appropriate noise covariance matrices $Q_k$ and $R_k$. After repeated experiments, they are chosen as follows: $Q_k = \mathrm{diag}(0.01\ \ 0.01\ \ 0.01\ \ 0.001\ \ 0)$, $R_k = \mathrm{diag}(0.1\ 0.1\ 0.1)$.

In order to verify the estimate performance of our algorithm, two different experiments are given to check the speed tracking performance by the effect of the constant load and the changing load.

### 4.1. *Constant load*

In this example, the constant load torque is assumed to be 2 N m, and the reference speed changes from 2000 to 2500 rpm at time $t = 0.2$ s. Then, the experiment results (i.e. the performance curves) are obtained and presented in Figure 2.

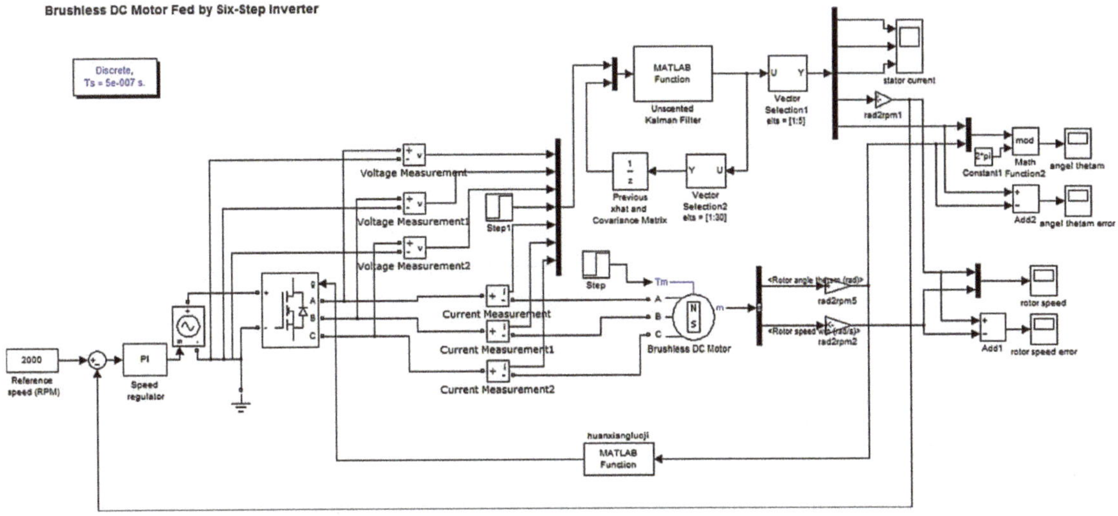


Figure 1. The system model.

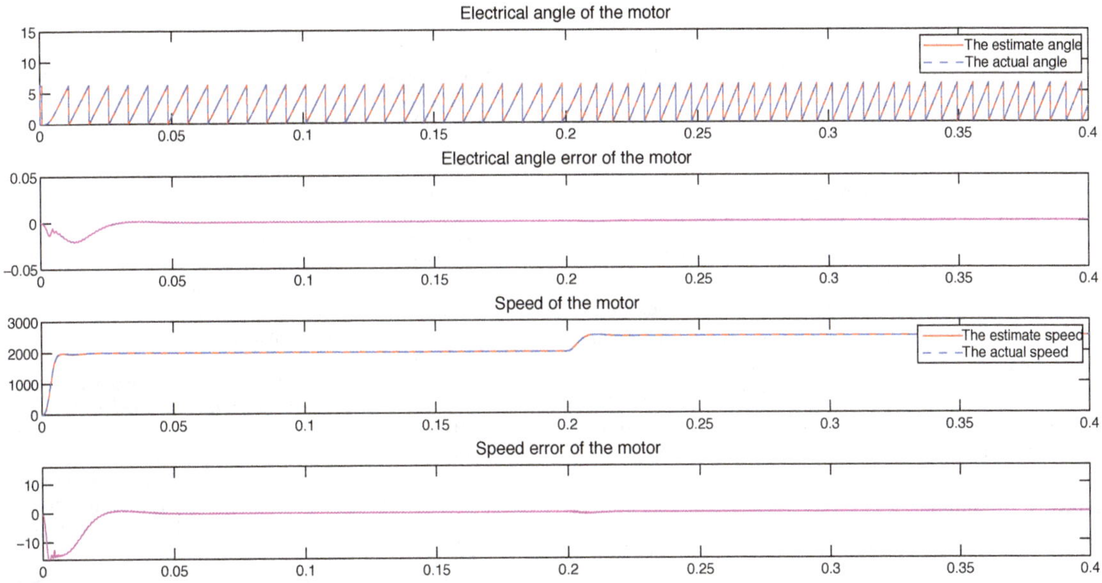


Figure 2. Performance curves of speed change.

From Figure 2, we can see that the estimated speed can track the actual speed accurately when the reference speed changes. The error between the estimated speed and the actual speed is about 0.2 rpm. From the first performance curve, we can see that the estimated position and the actual position of the rotor are almost the same. And the error is about $1 \times 10^{-3}$ electrical angle, which confirms that the motor can operate normally with little torque ripple. The experiment results illustrate the effectiveness of our filtering algorithm.

### *4.2. Changing load*

In this simulation experiment, the speed changes from 0 to reference speed 2000 rpm. Then, the load torque $T_L = 2$ N m is added to this motor at time $t = 0.1$ s and removed at $t = 0.3$ s. The performance curves are presented in Figure 3.

From Figure 3, we can see that the estimated and the actual speed are almost the same when the load torque suddenly changes. Moreover, the error curve changes slightly.

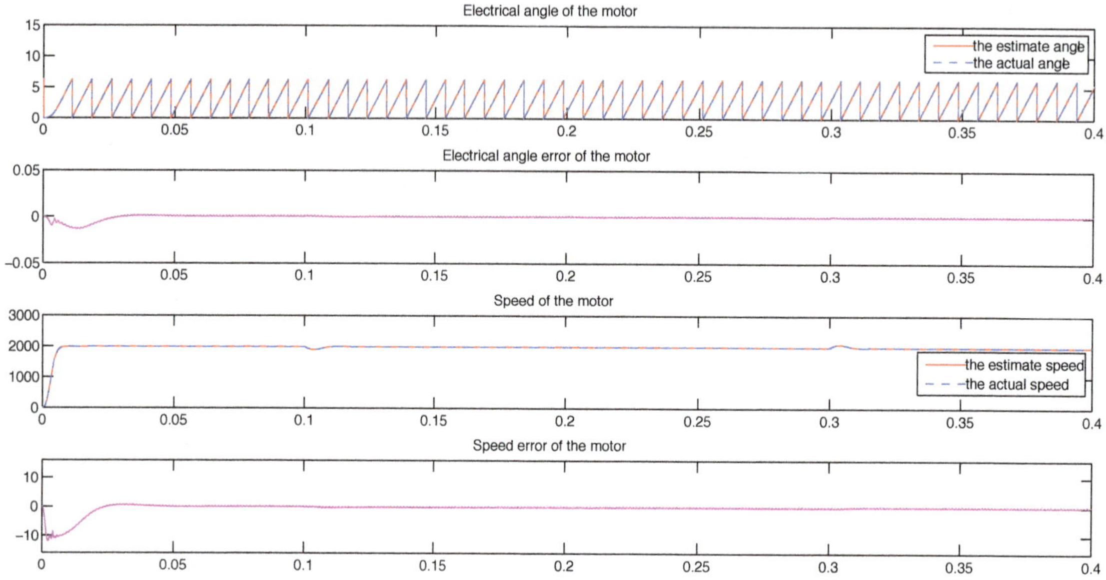


Figure 3. Performance curves of load torque change.

It can be concluded that the designed UKF algorithm is very effective when the torque suddenly changes.

## 5. Conclusion

In this paper, a mathematical model of a sensorless BLDC motor system has been built and a new filtering problem has been considered for this model based on the UKF algorithm. In order to evaluate the estimate performance, simulation experiments are presented in the paper. It is obvious to see that, from the simulation results, the accurate estimation performance can be obtained and the effectiveness of our designed algorithm can be demonstrated. Moreover, the sensorless BLDC motor can be controlled precisely according to the designed UKF algorithm.

## Disclosure statement

No potential conflict of interest was reported by the author(s).

## Funding

This work was supported in part by the National Natural Science Foundation of China [grant number 61374039]; the Program for Professor of Special Appointment (Eastern Scholar) at Shanghai Institutions of Higher Learning; the Program for New Century Excellent Talents in University [grant number NCET-11-1051] and Shanghai Pujiang Program under [grant number 13PJ1406300].

## References

Chan T., & Borsje P. (2009). *Application of unscented Kalman filter to sensorless permanent-magnet synchronous motor drive*. Paper presented at IEEE international conference on electric machines and drives conference, Miami, Florida, USA (pp. 631–638).

Chen Y., Huang S., Wang S., & Wu F. (2011). *Dynamic equations algorithm of the sensorless BLDC motor drive control with Back-EMF filtering based on dsPIC30F2010*. Paper presented at the 30th Chinese control conference (CCC), Yantai, China (pp. 3626–3630).

Iizuka K., Uzuhashi H., Kano M., Endo T., & Mohri K. (1985). Microcomputer control for sensorless brushless motor. *IEEE Transactions on Industry Applications*, *IA-21*(3), 595–601.

Ji H., & Li Z. (2008). *A new position detecting method for brushless DC motor*. Paper presented at IEEE international conference on automation and logistics, Qingdao, China (pp. 1110–1114).

Julier S. J., & Uhlmann J. K. (2004). Unscented filtering and nonlinear estimation. *Proceedings of the IEEE*, 401–422.

Lenine D., Reddy B.R., & Kumar S.V. (2007). *Estimation of speed and rotor position of BLDC motor using extended Kalman filter*. Paper presented at IET-UK international conference on information and communication technology in electrical sciences (ICTES 2007), Tamil Nadu, India (pp. 433–440).

Pillay P., & Krishnan R. (1989). Modeling, simulation, and analysis of permanent-magnet motor drives. II. The brushless DC motor drive. *IEEE Transactions on Industry Electronics*, *25*(2), 274–279.

Pillay P., & Krishnan R. (1991). Application characteristics of permanent magnet synchronous and brushless DC motors for servo drives. *IEEE Transactions on Industry Applications*, *27*(5), 986–996.

Rigatos G., & Siano P. (2012). *Sensorless nonlinear control of induction motors using unscented Kalman filtering*. Paper presented at the 38th annual conference on IEEE industrial electronics society, Montreal, Canada (pp. 4654–4659).

Simon D. (2006). *Optimal state estimation: Kalman, $H_\infty$, and nonlinear approaches*. Hoboken, NJ: Wiley.

Terzic B., & Jadric M. (2001). Design and implementation of the extended Kalman filter for the speed and rotor position estimation of brushless DC motor. *IEEE Transactions on Industry Electronics*, *48*(6), 1065–1073.

# 14

## Analytic and linear prognostic model for a vehicle suspension system subject to fatigue

Abdo Abou Jaoude*

*Department of Mathematics and Statistics, Faculty of Natural and Applied Sciences, Notre Dame University-Louaize, Lebanon*

Recent developments in system design technology, such as in aerospace, defense, petro-chemistry and automobiles, are represented earlier in the literature by simulated models during the conception step and this is to ensure the high availability of the industrial systems. Given that the integration of diagnostic–prognostic models in these industrial systems is facilitated by these developments. In fact, the monitoring of the degradation indicators is used indirectly in failure prognostic models and is just a measurement of an unwanted situation. Therefore, the diagnostic is not only a failure detection procedure but it also indicates the actual state and the history of the system. Hence, a predictive maintenance is done by the subsequent prognostic model. Consequently, from a predefined threshold of degradation, the remaining useful lifetime is estimated. Based on a physical dynamic vehicle suspension system, this research paper elaborates a procedure to create a failure prognostic model. I will adopt here analytic laws of degradation such as the Paris–Erdogan law for fatigue degradation and the Palmgren–Miner law for cumulative damage instead of applying degradation abaci largely used in prognostic studies.

**Keywords:** analytic laws; diagnostic; fatigue; Palmgren–Miner law; Paris–Erdogan law; degradation; prognostic; remaining useful lifetime

### 1. Introduction

Predicting the remaining useful lifetime (RUL) of industrial systems becomes currently an important aim for industrialists given that the failure which can occur suddenly is generally very expensive at the level of reparation, of production interruption, and is bad for reputation. The classical strategies of maintenance (Vachtsevanos, Lewis, Roemer, Hess, & Wu, 2006, chaps. 5–7) are no more efficient and practical because they do not take into consideration the instantaneous evolving product state, so it is important to understand the product in real time in order to prevent a failure during operation. In fact, we introduce a prognostic approach that seeks to provide an intelligent maintenance.

In the specialized literature, several studies on the prognostic procedure are presented, among them we mention the model-based, statistic-based and data-based models. The works based on abaci of degradation as in the work of Peysson et al. (Chelidze & Cusumano, 2004; Peysson et al., 2009) are useful at this level. As the latter is related to the three influent components: process, mission and environment, it is a non-analytic-based model founded on expert knowledge and on a large database.

A proposed analytic prognostic methodology based on some laws of damage in fracture mechanics is developed here. The damages are generally: crack propagation, corrosions, chloride attack, creep, excessive deformation and deflection, and damage accumulation. Whenever their analytic laws are available, the advantage of a prognostic approach based on a known damage law for a mechanical system is that it is adaptable to new situations and useful in determining the RUL of the system.

The procedure proposed in this work belongs to the model-based prognosis approach related to the physical model. It is focused on developing and implementing effective diagnostic and prognostic technologies with the ability to detect faults in the early stages of degradation. Early detection and analysis may lead to a better prediction and end-of-life estimations by tracking and modeling the degradation process. The idea is to use these estimations to make accurate and precise prediction of the time to failure of components. Early detection also helps avoid catastrophic failures.

Any prognostic methodology must lie on a type of damage. In mechanical systems, the damage can take many shapes. In this research paper, the case of fatigue degradation has been chosen due to the fact that it can be mathematically formulated by available analytic laws such as Paris–Erdogan's and Palmgren–Miner's laws.

This approach seems to be important in ensuring the high availability of industrial systems, such as in aerospace, defense, petro-chemistry and automobiles.

*Email: abdoaj@idm.net.lb

Among these systems, the petrochemical industries can be cited as an important example of prognostic models due to its favorable economic and availability consequences on their exploitation cost (El-Tawil, Abou Jaoude, Kadry, Noura, & Ouladsine, 2010).

In the automobile industry, for example the suspension component, this approach also shows its importance for the same earlier reasons and it will be explained and elaborated later in this paper. In vehicle suspension study, the results of model simulations are done for three cases of road profile excitations.

An analytic linear prognostic model is developed in this research paper that permits one to predict the RUL of a dynamic suspension system. This model considers the fatigue as a damage parameter, and hence, it is based on existing and well-known damage laws in fracture mechanics, such as the crack propagation law of Paris–Erdogan besides the linear damage accumulation law of Palmgren–Miner. An index of degradation that varies from zero to one will be derived. In fact, my proposed model is based on the link between this damage index $D$ and the crack length $a$, given that failure is produced when $a$ reaches a critical length $a_C$. Hence, my model is given by a simple function relating the instantaneous degradation to actual crack length as a measurement of actual damage.

This work is organized as follows: first the mechanical model of fatigue is presented in the linear cumulative damage case, then the prognostic model of fatigue failure is developed, and finally a case study of vehicle suspensions is illustrated (Abou Jaoude, 2012).

## 2. Proposed prognostic model

The purpose of this paper is to construct a process of prognostic models capable of predicting the degradation trajectories of a complex system for a given mission under a given environment and starting from an initial known damage. The complex system is decomposable into subsystems where each one can comprise a damage function.

The fatigue failure is one of the famous damage phenomena in mechanical systems such as in aircraft where the wings are subject to the fluctuation of air pressure between a maximal value ($\sigma_{max}$) and a minimal value ($\sigma_{min}$) (Figure 1; Lemaitre & Chaboche, 1990). This type of loadings leads to a crack propagation that can accelerate rapidly. Usually, micro-cracks exist originally in the materials due to the fabrication process where stresses remain after manufacture. These micro-cracks are detected and measured and denoted by $a_0$.

The advantage of the choice of fatigue damage for the developed prognostic methodology is that it is a failure mechanism very well studied in the literature and described under many known analytic laws. This mechanism has relatively the simplest formulation in comparison to the other damage phenomena. The fatigue characterizes the main failure cause of industrial equipments.

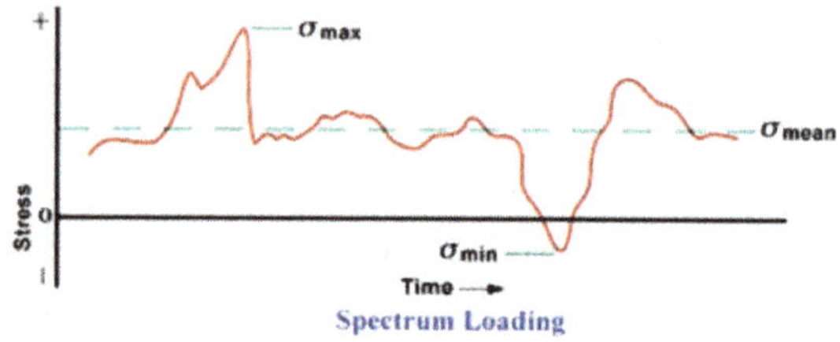


Figure 1. Load fluctuation.

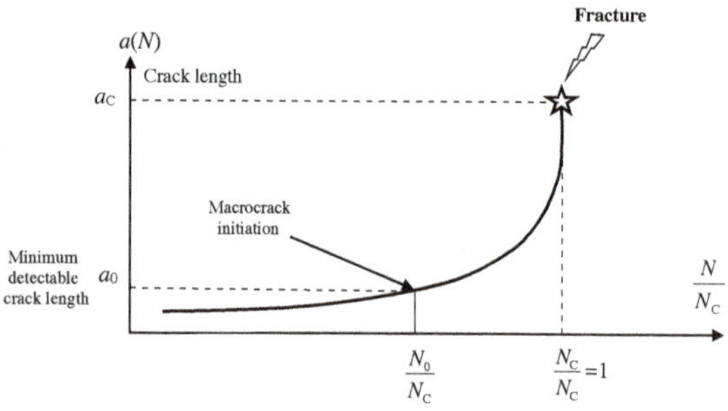


Figure 2. Pre-crack fatigue damage.

### 2.1. *Damage evolution law*

The fatigue of materials under cyclic loading creates micro-cracks. Starting from an initial length $a_0$ corresponding to an initial cycle number $N_0$, the macro-cracks become detectable and unstable. These macro-cracks will grow under loading cycles $N$ to a critical length $a_C$ reached at $N_C$ cycles and creating, thus, fractures that lead to failure. This evolution is represented in Figure 2 in terms of the normalized number of cycles $N/N_C$ for the simplicity of reading.

It can be assumed that $a_C = e/8$, where $e$ and $\ell$ are, respectively, the device dimension in the crack direction and the perpendicular dimension to the crack direction (Figure 3). $\Delta a_N$ is the crack length increment due to a loading cycle dN. $t_N$ is the instant corresponding to a cycle $N$ and to a crack length $a_N$.

### 2.2. *Paris–Erdogan's law*

Paris–Erdogan's law (Paris & Erdogan, 1963) permits one to determine the propagation rate of a crack length $a$ after its detection. The law of damage growth is given by

$$\frac{da(N)}{dN} = C \cdot (\Delta K)^m, \quad (1)$$

where $C$ and $m$ are the material and environment parameters: $(0 < C \ll 1)$; $(2 \leq m \leq 4)$, respectively, $a$ is the crack length, $N$ is the number of cycles (where the RUL is derived directly) and $\Delta K$ is the stress intensity factor.

It can be distinguished (Figure 4) that:

- The long cracks obey Paris–Erdogan's law.

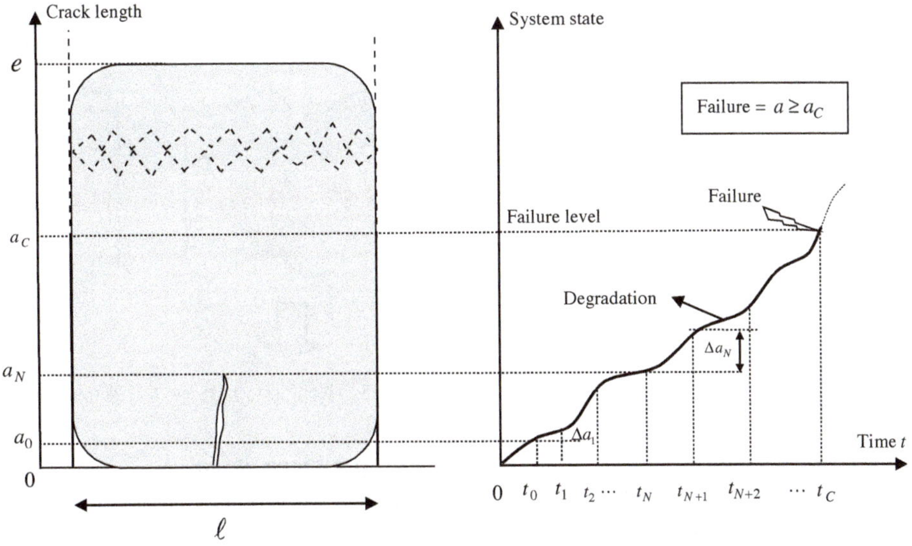


Figure 3. Crack length evolution.

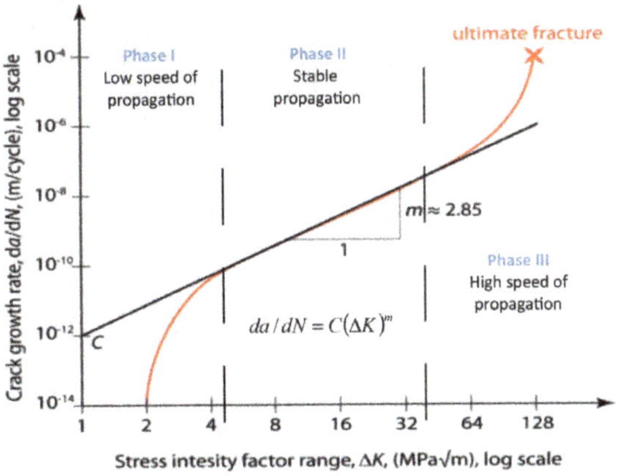


Figure 4. The three phases of crack growth, Paris–Erdogan's law.

- The short cracks serve to decrease the speed of propagation.
- The short physical cracks serve to increase the speed of propagation.

The law can be written also as follows:

$$\log\left(\frac{\mathrm{d}a}{\mathrm{d}N}\right) = \log C + m\,\log(\Delta K). \tag{2}$$

From the general form of Paris–Erdogan's law, McEvily and Ritchie (1988) have proven the following form:

$$\frac{\mathrm{d}a}{\mathrm{d}N} = C \cdot (\Delta K_{\mathrm{eff}})^m \cdot (K_{\max})^m, \tag{3}$$

where $\Delta K_{\mathrm{eff}} = K_{\max} - K_{\mathrm{op}}$, $K_{\max}$ is the maximum stress intensity factor and $K_{\mathrm{op}}$ is the stress intensity factor required to open the fatigue crack.

So, the decoupled form where two different functions of crack length $a$ and of load $P$ can be deduced is given as follows:

$$\frac{\mathrm{d}a}{\mathrm{d}N} = C \cdot \phi_1(a) \cdot \phi_2(P), \tag{4}$$

where the function $\phi_1(a) = \left(Y(a) \cdot \sqrt{\pi}\, a\right)^m$ and the load function $\phi_2(P) = (P)^m$, where $P = K_{\max}$, with $Y(a)$ being the geometric factor function of the body dimensions and $P$ being the load parameter.

Palmgren–Miner's rule can be used now to count the damages (Miner, 1945).

### 2.3. *Palmgren–Miner's rule*

Palmgren–Miner's rule (Miner, 1945) serves to compute the cumulative damage $d_i$ of different stress levels $\sigma_i$ ($i = 1$, $i = 2$, …, $i = k$) applied for $n_i$ cycles. Given that $N_i$ is the total cycle's number of stresses $\sigma_i$ to be applied and that lead to failure. The linear cumulative damage relative to applied stresses ($i = 1$ to $k$) is given by (Figure 5):

$$D_k = \sum_{i=1}^{k} d_i = \sum_{i=1}^{k} \frac{n_i}{N_i}. \tag{5}$$

### 2.4. *Wöhler's curve*

In material fatigue, it is important to know the critical level of applied stresses. When repeated stresses $\sigma(t)$ are applied along time under a cyclic model, they are limited between

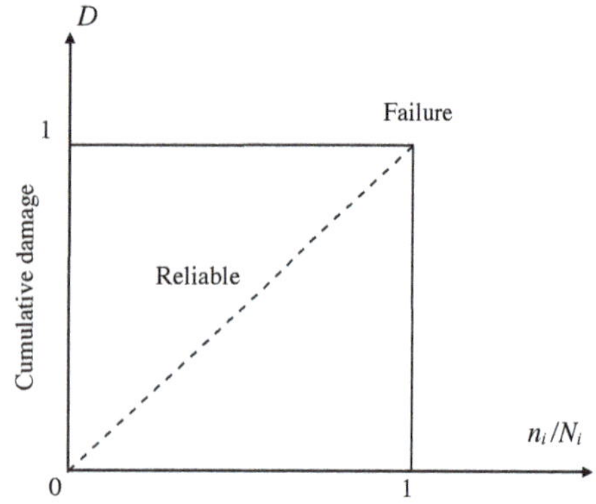


Figure 5. Palmgren–Miner's linear rule of damage.

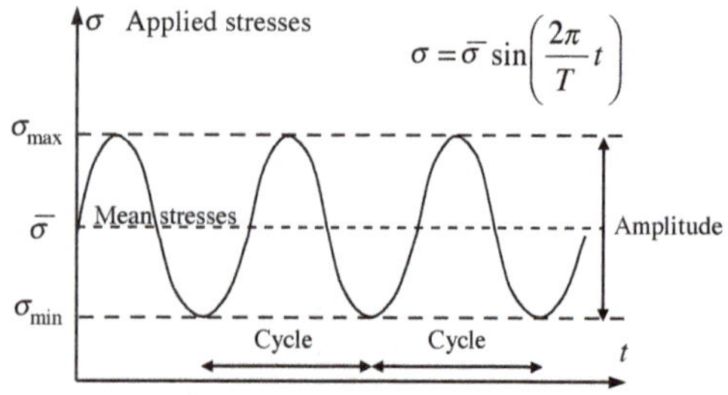


Figure 6. Cyclic applied stresses.

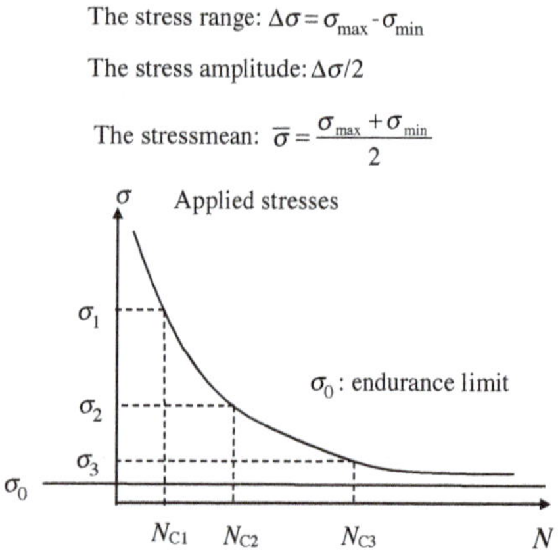


Figure 7. Wöhler's curve of fatigue.

two extreme values $\sigma_{max}$ and $\sigma_{min}$. Wöhler's curve governs the relation between the applied stress levels $\sigma$ and the critical number of cycles $N_C$ during the fatigue process of the material (Figures 6 and 7). For example, if the equipment is loaded by a stress level $\sigma_1$, then the critical cycle number is $N_{C1}$. Each stress level has its own critical cycle number.

The stress range: $\Delta\sigma = \sigma_{max} - \sigma_{min}$
The stress amplitude: $\Delta\sigma/2$
The stress mean: $\bar{\sigma} = ((\sigma_{max} + \sigma_{min})/2)$.

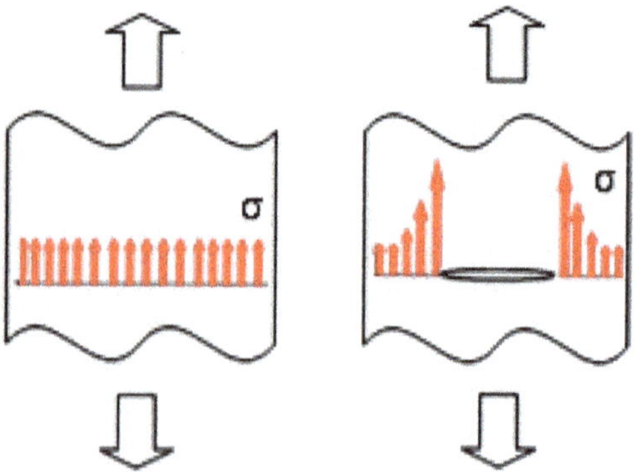


Figure 8. Non-uniform stress distribution near crack.

### 2.5. *Stress intensity factor*

The stress intensity factor is an important term in Paris's law expression; it represents the effect of stress concentration in the presence of a flat crack. When a flat crack occurs in the system body, the internal stresses in this section change from a uniform to a non-uniform distribution around the crack. This factor represents an amplification of the stress field near the crack (Figure 8). This change is expressed by a factor $K_I$ called the stress intensity factor (Langon, 1999; Lemaitre & Desmorat, 2005, chap. 6) for mode-I crack opening (mode I: the crack opening is in the same direction of applied stresses), and is given by

$$(K_I)^m = \left(Y(a) \cdot \sqrt{\pi a}\right)^m \cdot (\sigma_{max})^m = \phi_1(a) \cdot \phi_2(P_j). \quad (6)$$

Note that $K_I$ must verify the inequality:

$$K_I \leq K_{IC} = \sqrt{\frac{J_{IC} \cdot E}{1 - (\nu)^2}},$$

where $Y(a)$ is the geometric factor function of the equipment geometric parameters $(a, e)$ and $K_{IC}$ is the tenacity of material (or critical stress intensity factor) and is given by

$J_{IC}$ is the resistant crack force of the material,
$E$ is Young's modulus,
$\nu$ is the Poisson ratio.

At failure, the factor $K_I$ is equal to the value of $K_{IC}$, then we can deduce the critical length $a_C$ from $K_I(a) = K_{IC} \Rightarrow a = a_C$; but since it is difficult to deduce it from the last relation, I preferred to adopt the simple form: $a_C = e/8$.

### 2.6. *Additivity rule in Palmgren–Miner's rule*

The case where damage is caused by fatigue is an important application of the additivity rule (Todinov, 2001, 2005). In this case, the measurement of damage is the length of the fatigue crack. The additivity rule in Palmgren–Miner's rule (Miner, 1945) has been proposed as an empirical rule in the case of damage due to fatigue controlled by the crack propagation. The rule states that in a fatigue test at a constant stress amplitude $\Delta\sigma_i$, damage could be considered to accumulate linearly with the number of cycles. Accordingly, if at stress amplitude $\Delta\sigma_1$ the component has $N_1$ cycles of life (where total life corresponds to an amount of damage $a_C$), then after $\Delta n_1$ cycles the amount of damage will be $(\Delta n_1/N_1)a_C$. After $\Delta n_2$ stress cycles spent at stress amplitude $\Delta\sigma_2$, characterized by a life of $N_2$ cycles, the amount of damage will be $(\Delta n_2/N_2)a_C$.

Failure occurs when, at a certain amplitude $\Delta\sigma_M$, the sum of the partial amounts of damage attains the amount $a_C$, that is, when

$$\frac{\Delta n_1}{N_1}a_C + \frac{\Delta n_2}{N_2}a_C + \cdots + \frac{\Delta n_M}{N_M}a_C = a_C \quad (7)$$

is fulfilled.

As a result, the analytical expression of Palmgren–Miner's rule becomes

$$\sum_{i=1}^{M} \frac{\Delta n_i}{N_i} = 1, \quad (8)$$

where $N_i$ is the number of cycles needed to reach the specified amount of damage $a_C$ at constant stress amplitude $\Delta\sigma_i$

Palmgren–Miner's rule is central to reliability calculations, yet no comments are made whether it is compatible with the damage development laws characterizing the different stages of fatigue crack growth. The necessary and sufficient condition for validity of the empirical Palmgren–Miner's rule is the possibility of factorizing the rate of damage as a function of the amount of accumulated damage $a$ and the stress or strain amplitude $\Delta p$:

$$\frac{\mathrm{d}a(N)}{\mathrm{d}N} = F(a) \cdot G(\Delta p). \quad (9)$$

The theoretical derivation of Palmgren–Miner's rule can be found in Todinov (2001).

A widely used fatigue crack growth model is Paris–Erdogan's power law given by

$$\frac{\mathrm{d}a(N)}{\mathrm{d}N} = C \cdot (\Delta K)^m, \quad (1)$$

where $\Delta K = Y(a) \cdot \Delta\sigma \cdot \sqrt{\pi\, a}$ is the stress intensity factor range, $C$ and $m$ are material constants and $Y(a)$ is a parameter which can be expressed as a function of the amount of damage $a$.

Clearly, Paris–Erdogan's fatigue crack growth law can be factorized as in the previously stated equation and, therefore, it is compatible with Palmgren–Miner's rule. In the cases where this factorization is impossible, Palmgren–Miner's rule does not hold. Such as, for example, the fatigue crack's growth law, given by

$$\frac{\mathrm{d}a(N)}{\mathrm{d}N} = B \cdot \Delta\gamma \cdot a^{\beta} - D, \quad (10)$$

discussed by Miller (Todinov, 2001), and which characterizes physically small cracks.

In Equation (10), $B$ and $\beta$ represent the material constants, $\Delta\gamma$ represents the applied shear strain range, $a$ represents the crack length at cycle $N$ and $D$ represents a threshold value.

Thus, following what has been said, the proposed model can use the additivity characteristic of Paris's law.

## 2.7. *Maintenance and diagnostic/prognostic*

It is proved that the schedule-based inspection/maintenance NDI (non-destructive inspection) is less beneficial than the on-demand (or continuous) inspection with permanently installed sensors/condition-based maintenance SHM (structural health monitoring) for many reasons such as the increased availability, the quick assessment of potential/actual damage events, the increasing safety and the performance of advanced materials.

But the major technical challenges for SHM reside in the sensors. The monitoring should be directed to the detection of the cracks and corrosion, the multiple damage modes, the pre-crack fatigue damage and the amount of residual stress.

We can say that the NDI leads to prognostics based on the following:

- NDI performed at the time of fabrication and as in-service inspections.
- Condition-based maintenance-active component monitoring.
- Move from diagnosis to the prediction of the remaining life and the structural health monitoring/management.
- Prognostics (for machinery) are the prediction of a remaining safe or service life, based on an analysis of the system or the material condition, stressors and degradation phenomena.

For example, bearing crack faults may be prognosed by examining and predicting their vibration signals.

The relation between maintenance and prognostic is summarized in Figure 9.

### 2.7.1. *Flow chart of the various components of the diagnostic/prognostic/maintenance process*

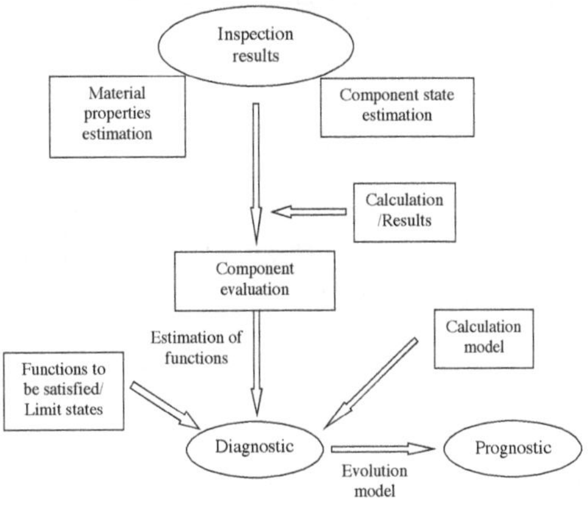

### *2.7.2. Cycle of prognostic/diagnostic/maintenance*

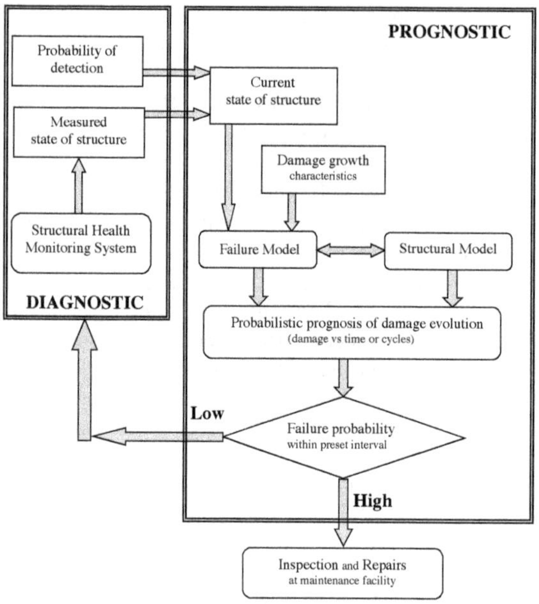


### *2.8. Accumulation of fatigue damage*

In fatigue damage, in order to study the prognosis of a degraded component, my idea is to predict and estimate the end of life of an equipment component subject to fatigue by tracking and modeling the corresponding degradation function. To facilitate the analysis, it is convenient to adopt a normalized damage measurement $D \in [0, 1]$ by exploiting the advantage of the cumulative damage law of Palmgren–Miner (Figure 5). In fact, this law helps to estimate the lifetime of components subject to load cycles, and it considers that the damage fraction $d_i$ at stress level $\sigma_i$ is the ratio of $n_i$ over the total cycle number $N_i$ producing failure.

For a body of equipment of thickness $e$, we take the initial crack length as $a_0$ ($a_0 \le a \le a_C$). Given that $1.01 \le e/a \le 10$ and $e/a_C = 8$, then from Equation (1) a recurrent form of crack length growth $a$ can be deduced as (Abou Jaoude, 2012)

$$da = a_N - a_{N-1} = C \cdot \phi_1(a_{N-1} - a_0) \cdot \phi_2(P_j),$$

$P_j$ is a realization of $P$

$$\Rightarrow a_N = a_{N-1} + C \cdot \phi_1(a_{N-1}) \cdot \phi_2(P_j)$$

where ($a_0 \approx 0$).

For the other sequences,

$$\begin{aligned} a_1 &= a_0 + C\phi_1(a_0)\phi_2(P_j) \\ a_2 &= a_1 + C\phi_1(a_1)\phi_2(P_j) \\ \vdots \quad & \quad \vdots \\ a_N &= a_{N-1} \\ &\quad + C\phi_1(a_{N-1})\phi_2(P_j) \end{aligned} \qquad \begin{cases} \phi_1(a_N) = \left(Y(a_N) \cdot \sqrt{\pi a_N}\right)^m \\ \phi_2(P_j) = (\sigma_{\max})^m \end{cases}$$

$$\Rightarrow \mathrm{d}a = a_N - a_{N-1} = C \cdot (K_I)^m \times \mathrm{d}N$$

where $(K_{\mathrm{I}}(a_N))^m = \phi_1(a_N) \cdot \phi_2(P_j)$ is a function of the crack length $a$.

For each cycle, we have $\mathrm{d}N = 1$; therefore, $a_N = a_{N-1} + C \cdot (K_{\mathrm{I}})^m$.

$$\text{As } \mathrm{D}_k = \sum_{i=1}^{k} d_i = \sum_{i=1}^{k} \frac{n_i}{N_i}$$

(Miner's law; with $i$ in Miner's law $= N$ in my model)

and based on the additivity characteristic of Paris's law, the addition of damages gives the total crack growth at failure point ($a_C - a_0$) realized at the total number of cycle $N_C$:

$$a_C - a_0 = \sum_{N=1}^{N_C} \mathrm{d}a_N = \text{total damage}.$$

At each $n_i$, the crack grows to length $\mathrm{d}a_i = \mathrm{d}a_N$, therefore, Miner's damage fraction, for any stress level (Figure 10), is given in terms of the crack length by

$$d_i = \frac{n_i}{N_i} = \frac{\mathrm{d}a_N}{(a_C - a_0)}, \tag{11}$$

where $n_i$ is the damage increment due to stress number $i$, $N_i$ is the total damage for stress number $i$.Then, the cumulated total damage at cycle $N$ is given by

$$D_N = \sum_{i=1}^{N} d_i = \sum_{i=1}^{N} \frac{\mathrm{d}a_i}{a_C - a_0} = \frac{\sum_{i=1}^{N} \mathrm{d}_i}{a_C - a_0}. \tag{12}$$

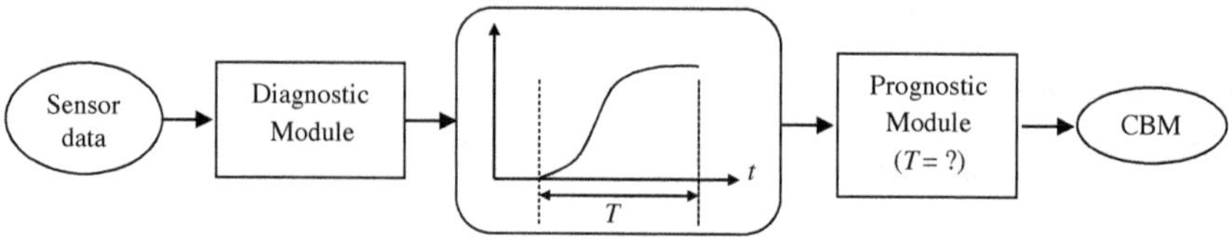


Figure 9. Diagnostic–prognostic–maintenance. Note: CBM, condition-based maintenance.

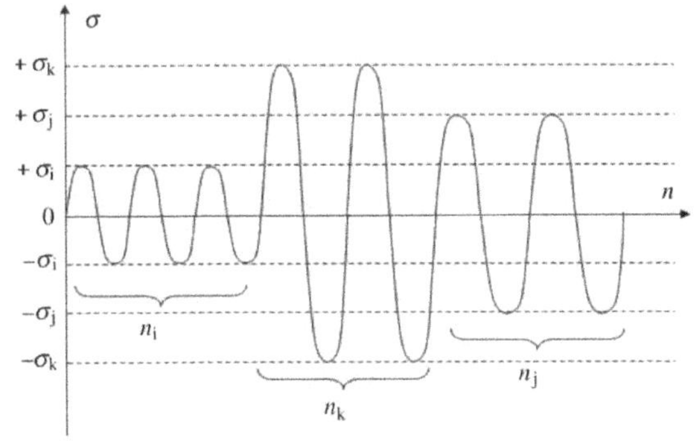

Figure 10. Cumulative stress levels.

where

$$\mathrm{d}a_N = a_N - a_{N-1},$$

$$\sum_{i=1}^{N} \mathrm{d}a_i = \mathrm{d}a_1 + \mathrm{d}a_2 + \cdots + \mathrm{d}a_N$$
$$= (a_1 - a_0) + (a_2 - a_1) + \cdots + (a_N - a_{N-1})$$
$$= a_N - a_0.$$

The value of the critical crack length due to fatigue effects is taken equal to a conservative value $e/8 - a_0$, given that $e$ is the equipment thickness measured in the direction of the crack.

Nevertheless, the subtraction of the initial crack length $a_0$ from $a_C$ in the denominator has a negligible effect on the degradation curve and it is also in the conservative direction (i.e. larger value of degradation): $a_C - a_0 \approx a_C$ ($a_0 \ll a_C$).

It can be easily proved that:

$$D_C = \sum_{i=1}^{N_C} d_i = \frac{\sum_{i=1}^{N_C} \mathrm{d}a_i}{a_C - a_0} = \frac{\mathrm{d}a_1 + \mathrm{d}a_2 + \cdots + \mathrm{d}a_{N_C}}{a_C - a_0}$$
$$= \frac{a_C - a_0}{a_C - a_0} = 1.$$

$$\text{As}: a_0 \le a \le a_C \text{ then}: \quad D_0 \le D \le D_C = 1$$

$$\text{and } D_0 = \frac{a_0}{a_C - a_0} \Rightarrow a_0 = \frac{D_0 a_C}{1 + D_0}$$

The other sequences are $(0 \le N \le N_C)$

$$D_0 = \frac{a_0}{a_C - a_0}, \quad D_1 = \frac{a_1}{a_C - a_0}, \quad D_2 = \frac{a_2}{a_C - a_0}, \ldots,$$
$$D_N = \frac{a_N}{a_C - a_0}.$$

Consequently, a recurrent form of degradation can be deduced as follows:

$$\begin{aligned}
D_N &= \frac{a_N}{a_C - a_0} \\
&= \frac{a_{N-1} + C \cdot (K_I)^m}{a_C - a_0} \\
&= \frac{a_{N-1}}{a_C - a_0} + \frac{C \cdot (K_I)^m}{a_C - a_0} \\
&= \frac{a_{N-1}}{a_C - a_0} + \frac{C \cdot \left(Y(a_{N-1}) \cdot \sqrt{\pi \cdot a_{N-1}} \cdot \sigma_j\right)^m}{a_C - a_0} \\
&= \frac{a_{N-1}}{a_C - a_0} + \frac{C \cdot Y(a_{N-1})^m \cdot \left(\sqrt{\pi \cdot a_{N-1}}\right)^m \cdot \sigma_j^m}{a_C - a_0} \\
&= \frac{a_{N-1}}{a_C - a_0} + \frac{C}{a_C - a_0} \cdot Y(a_{N-1})^m \cdot \left(\sqrt{\pi \cdot a_{N-1}}\right)^m \cdot \sigma_j^m \\
&= D_{N-1} + \eta \cdot \phi_1(D_{N-1}) \cdot \phi_2(P_j),
\end{aligned} \tag{13}$$

where

$$\begin{aligned}
D_{N-1} &= \frac{a_{N-1}}{a_C - a_0}, \\
\eta &= \frac{C}{a_C - a_0}, \\
\phi_1(D_{N-1}) &= Y(a_{N-1})^m \cdot \left(\sqrt{\pi \cdot a_{N-1}}\right)^m, \\
\phi_2(P_j) &= \sigma_j^m.
\end{aligned}$$

Hence, the new prognostic analytic model is expressed by the general function given by

$$D_N = D(N) = P_{\text{rog}}(a_N) = \frac{a_{N-1}}{a_C - a_0} + \frac{C}{a_C - a_0} \cdot Y(a_{N-1})^m \cdot \left(\sqrt{\pi \cdot a_{N-1}}\right)^m \cdot \sigma_j^m. \tag{14}$$

And therefore, the degradation trajectories $D(N)$ along the total number of loading cycles $N$ can be drawn (Abou Jaoude, El-Tawil, Kadry, Noura, & Ouladsine, 2010).

### 2.9. *Flow chart of the prognostic model*

The following flow chart summarizes all the procedures of the proposed model (Abou Jaoude, Kadry, El-Tawil, Noura, & Ouladsine, 2011):

### 2.10. *Environment effects in the proposed prognostic model*

The environment effects are taken into account by the two parameters $C$ and $m$. These parameters are related to the material interacting with its environment.

Large values of $m$ ($m > 40$) correspond to the case of brittle materials (brittle failure) and small values of $m$ ($m \to 2$) correspond to the case of ductile materials

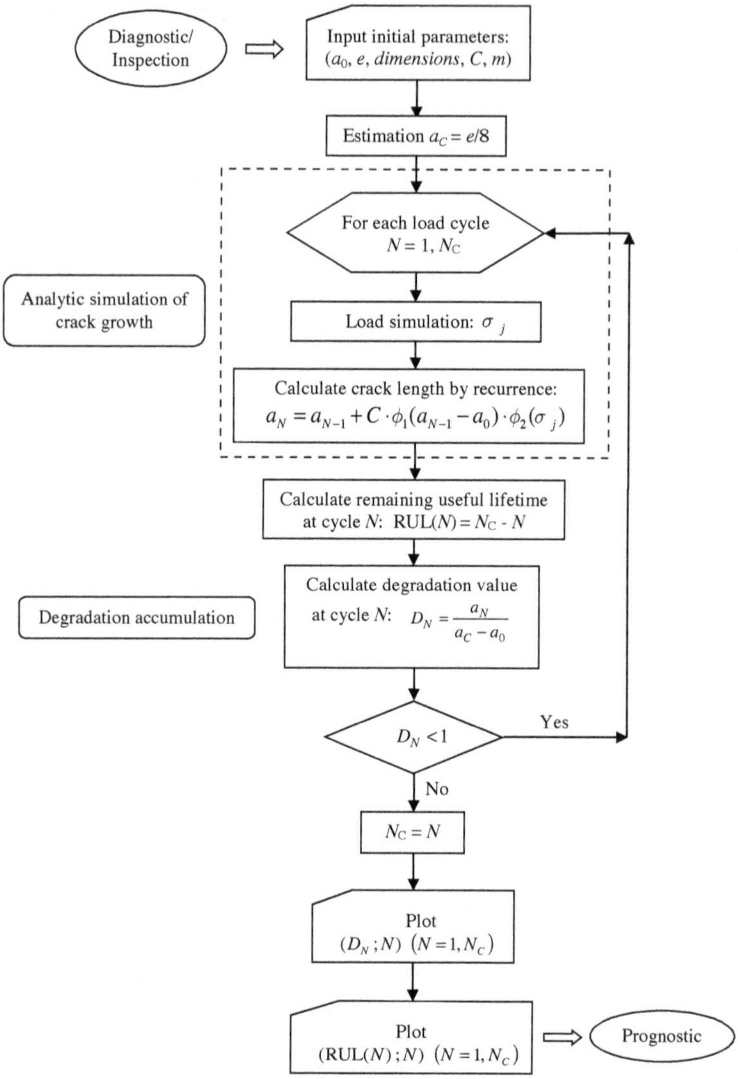


($m$ = 2: fully plastic). Otherwise for fatigue failure, the range value of $m$ is $2 \leq m \leq 3$. The parameter $m$ depends mainly on the specimen length. For lower toughness, steels $m$ is greater than or equal to 3 (Sukumar, Chopp, & Moran, 2003).

The coefficient $C$ is affected by the edges and consequently its value depends on whether it is the case of a plane stress or a plane strain. However, for the case of an infinite equipment body and far from the edge effects, the coefficient $C$ takes a constant value (Newman & Raju, 1981).

Moreover, $C$ and $m$ depend on the testing conditions such as the loading ratio $\sigma_{min}/\sigma_{max}$, on the geometry and the size of the specimen, and on the initial crack length.

These two parameters govern the behavior of the material during the fatigue process through the crack propagation. The environment influencing parameters in this process, such as temperature, humidity, geometry dimensions, material nature, water action, soil action, applied load location and body shape, are also represented by these two parameters $C$ and $m$.

These two parameters are evaluated by the mean of experiments in true conditions.

Examples (Newman & Raju, 1981; Sukumar et al., 2003):

$C = 5.2 \times 10^{-13}$ (free air)
$C = 1.3 \times 10^{-14}$ (under soil)
$C = 2 \times 10^{-11}$ (offshore)
and $m = 3$ (metal).

## 3. Application of the prognostic method to industrial systems

To illustrate the proposed new analytic approach, in this section an important mechanical system will be applied which is the suspension in the automotive industry, given that its corresponding prognostic study is essential for economic reasons.

### 3.1. *Vehicle suspension fatigue life*

Fatigue analysis of a vehicle suspension (Figure 11) by finite element models was done in many works (Colquhoun & Draper, 2000) as a consequence of experimental results. It permits one to define the location of potential fatigue cracks. The major aspect of local strain fatigue is determined by the crack initiation and propagation. The original theories that were developed for uniaxial stress conditions were elaborated and improved later to eliminate the errors due to simplified uniaxial conditions.

It was proposed in the literature (Colquhoun & Draper, 2000; Frost & Dugdale, 1957) that for high-cycle fatigue, successful life estimates for biaxial stress conditions could be made using combinations of axial and shear stresses.

There is a lot of experimental evidence from fatigue testing carried out in the middle of the last century showing that stress gradients have an important effect on the total fatigue life of a component. Stress gradients have also been used in an attempt to explain the effect of notch sensitivity.

Moreover, finite element analysis provides surface strains on the model, but for real engineering components it is very difficult to determine the stress concentration factor at a notch (Figure 12).

The stress concentration factor is the same as the stress intensity factor which was explained in Section 2.5.

The endurance limit stress is the stress level for which the critical number of loading cycles tends to infinity (refer to Section 2.4).

Referring to Figure 13, $\sigma_e$ is the smooth specimen endurance limit stress, $S_{th}$ is the threshold stress for non-propagation cracks, that is, below $S_{th}$, fatigue is not influent and $S_e = S_{th}$, Kt is the stress concentration factor, which is

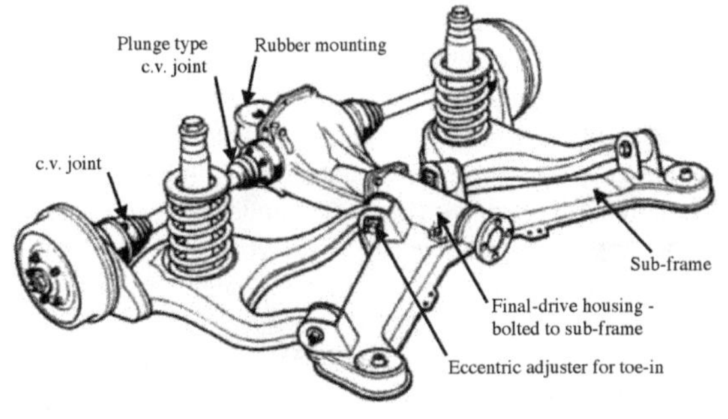


Figure 11. Vehicle suspension system (Newman & Raju, 1981).

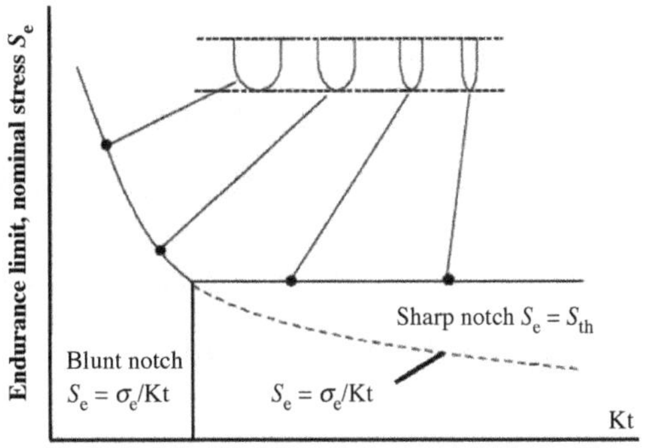


Figure 12. Relationship between endurance limit stress $\sigma_e$ and the stress concentration factor Kt (Colquhoun & Draper, 2000).

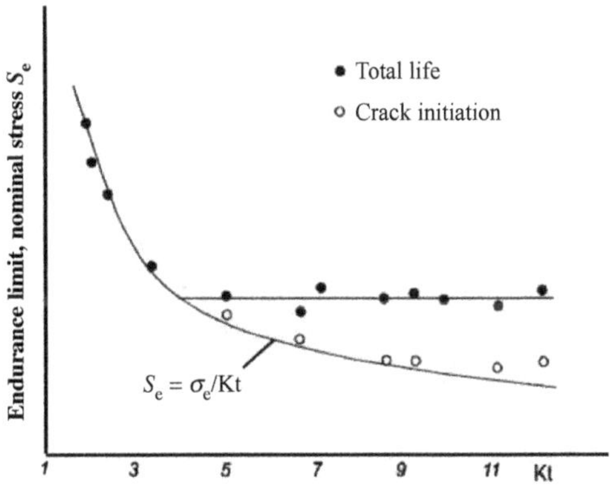


Figure 13. Relationship between endurance limit stress $\sigma_e$ and the stress concentration factor Kt for crack initiation and total life (Colquhoun & Draper, 2000).

given as

$$\text{Kt} = \frac{\text{Endurance limit of a notch} - \text{free specimen}}{\text{Endurance limit of a notched specimen}}.$$

The endurance limits (Frost & Dugdale, 1957) are obtained from standard rotating beam experiments carried out under certain specific conditions. They are given by $S_e = \sigma_e/\text{Kt}$.

As the stress concentration factor increases, where this is the case for many ductile metals, a minimum value of fatigue limit stress that is expressed by $S_{th}$ occurs. Hence, increasing the stress concentration factor by sharpening the notch produces no further reduction in fatigue strength (Figure 13).

The parts forming the vehicle suspension are indicated in Figure 14 where the damper's element can be seen.

Using test data on plate and round bar, specimens in aluminum alloy and steel materials have shown that if fatigue life to the first crack initiation is considered, then the fatigue strength reduces with increasing stress concentration with no limiting value (Figure 15).

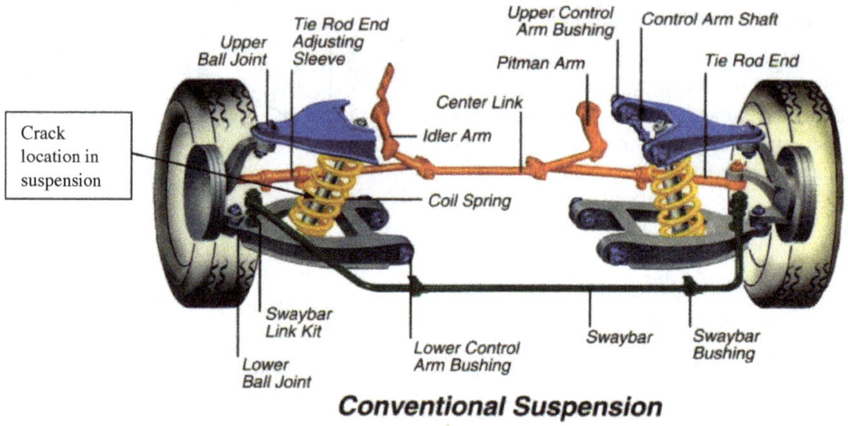


Figure 14. Vehicle suspension components and crack possible location (Frost, 1960).

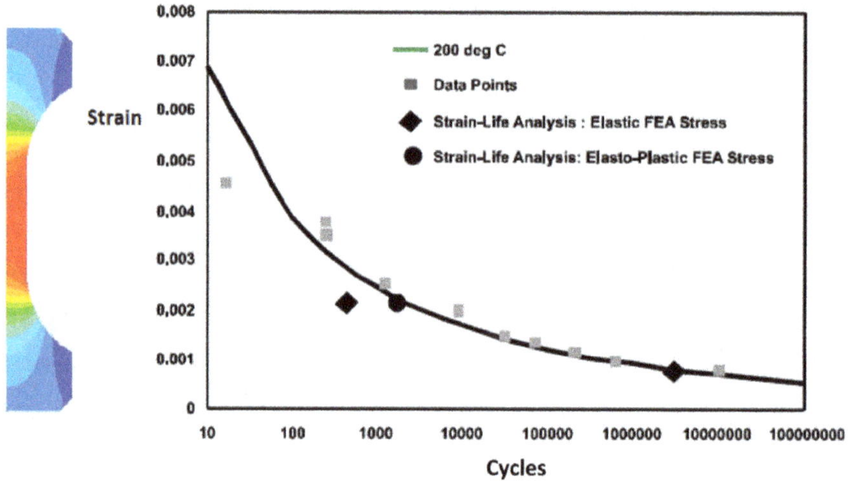


Figure 15. Comparison of test data with calculated lives from elastic and elastic–plastic FE analysis (Frost, 1960).

Many works (Conle & Topper, 1980; Duquesnay, Pompetzki, & Topper, 2002; Frost, 1960) have shown that the constant amplitude endurance limit does not apply to the analysis of real service loading if some cycles in the loading exceed the constant amplitude endurance limit stress amplitude. For a finite life design, the larger cycles in the loading cause the endurance limit stress to be reduced significantly, with the result that small cycles contribute to the fatigue damage process.

Figure 15 (Conle & Topper, 1980) shows the results of strain-controlled constant amplitude tests on an aluminum alloy at high temperature. The finite element calculation made by the software SAFE (FE-SAFE) from an elastic finite element analysis (FEA) shows excellent correlation for high-cycle fatigue. For low-cycle fatigue, at 1000 cycles the calculated fatigue life is conservative by a factor of 3. This is a commonly observed phenomenon at such low fatigue lives in components where yielding occurs across the entire section. For comparison, an elastic–plastic FEA analysis of the model was used as an input into the FE-SAFE analysis, and the correlation with the test result was then excellent.

This component was analyzed in FE-SAFE and compared with the results of fatigue testing. A scale factor was applied to the test loading to produce a failure. The correlation between the calculated life of 1631 repeats of the load history and the test life of 1650 repeats is extremely good.

The steel component was analyzed (Duquesnay et al., 2002) with a load–time history in one direction (Figure 16). A scale factor was applied to produce a failure. The analysis used stresses from an elastic FEA; fatigue lives were calculated for each node on the model, using averaged nodal stresses. Experience has shown that this is much more accurate than using stresses at integration points or at the element centroid.

In designing engine crank shafts (Figure 16), the finite elements analysis is used to generate stress solutions. The FEA analysis shows that the principal stresses change their

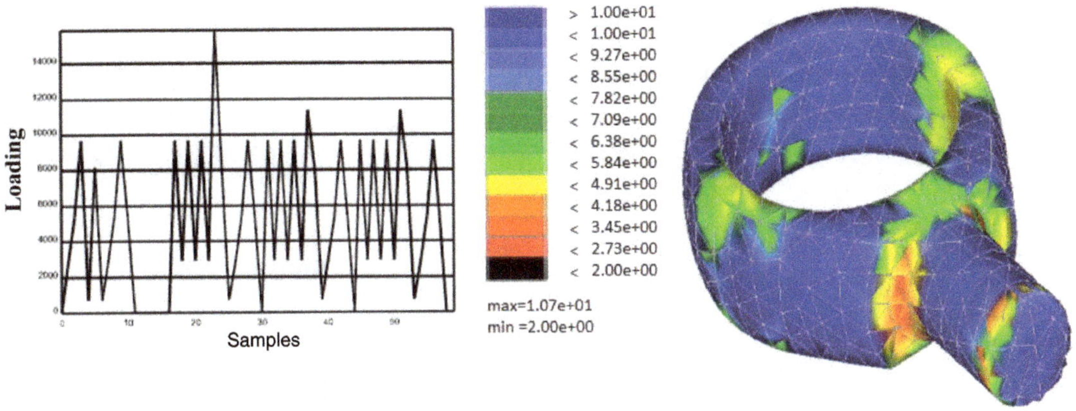


Figure 16. Loading history for accelerated testing (left) and fatigue life contours (right). Test life: 1650 repeats of loading. Calculated life: 1631 repeats of loading (Conle & Topper, 1980).

orientation and magnitude during the load cycle applied to the crank shaft.

FE-SAFE uses the sequence of FEA analysis results to calculate the fatigue life at each node. FE-SAFE correctly identified the critical location in the crank shaft, using a Brown–Miller fatigue analysis, and correlated well with test results.

A common theme from these validation exercises is that a uniaxial strain-life using the maximum principal stress can fail to identify the critical location, for components where biaxial stresses (Von-Mises) and particularly non-proportional stresses are present at the critical locations.

In the computer-based fatigue analysis of the finite element model, the type of loading depends very much on the customer's requirements. Some companies (Duquesnay et al., 2002) specify a validation using simple sinusoidal loading, whereas other companies, such a Ford, require the application of measured time histories of vertical, braking and cornering forces on the tire contact patch or wheel center (Figure 17). At present, the test procedure uses a single actuator to apply the forces at the tire contact patch, angled to produce a specific relationship between the three forces. FE-SAFE allows for different time histories to be applied in each direction, up to 4096 load histories of unlimited length.

#### 3.1.1. *Types of mechanical effects, their mechanisms and the possible consequences*

The following flow chart describes the relationship between the sources, the mechanical effects and the consequences of various loading stresses (Lemaitre & Chaboche, 1990).

#### 3.1.2. *Automatic diagnostic of a bad suspension bushing*

Automobile suspension bushings come in a variety of shapes, sizes and thicknesses, according to their application. Bushings may be made from several materials, including rubber, polyurethane, urethane and graphite composites. Bushings prevent wear to expensive suspension components by absorbing vertical and lateral forces produced by the vehicle over different terrains. They cushion and absorb shock on the chassis to keep the shock from entering the passenger compartment. While absorbing these vibrations, they still allow limited movement and flex in the suspension joints, keeping the wheels firmly grounded and on track during turning maneuvers. A vehicle's owner may check all its suspension bushings for proper shape and condition.

#### 3.1.3. *Prognostic study for vehicle suspension systems*

Let us consider a half-vehicle suspension system (Figure 18) subject to non-regular road surface excitations (Lee, 2004). It is composed of a front part and a rear part. To study the prognostic of this system, it is important to define the mechanical model in order to deduce the output response from the input excitation road. The system has four degrees of freedom that can be reduced to two degrees of freedom by considering the front suspension alone.

The dynamic equations of the system are given by

$$
\begin{aligned}
& m\ddot{x} + (f_{ca} + f_{ka}) + (f_{cb} + f_{kb}) = 0, \\
& I\ddot{\theta} + l_a(f_{ca} + f_{ka}) - l_b(f_{cb} + f_{kb}) = 0, \\
& m_{2a}\ddot{x}_{2a} - (f_{ca} + f_{ka}) + k_{2a}(x_{2a} - w_a) = 0, \\
& m_{2b}\ddot{x}_{2b} - (f_{cb} + f_{kb}) + k_{2b}(x_{2b} - w_b) = 0, \\
& x = (l_b x_{1a} + l_a x_{1b})/l, \quad \tan\theta \approx \theta = \frac{(x_{1a} - x_{1b})}{l}, \\
& l = l_a + l_b, \\
& f_{ci} = c_i(\dot{x}_{1i} - \dot{x}_{2i}), \quad i = a, b, \\
& f_{ki} = k_{1i}(x_{1i} - x_{2i}), \quad i = a, b,
\end{aligned}
$$

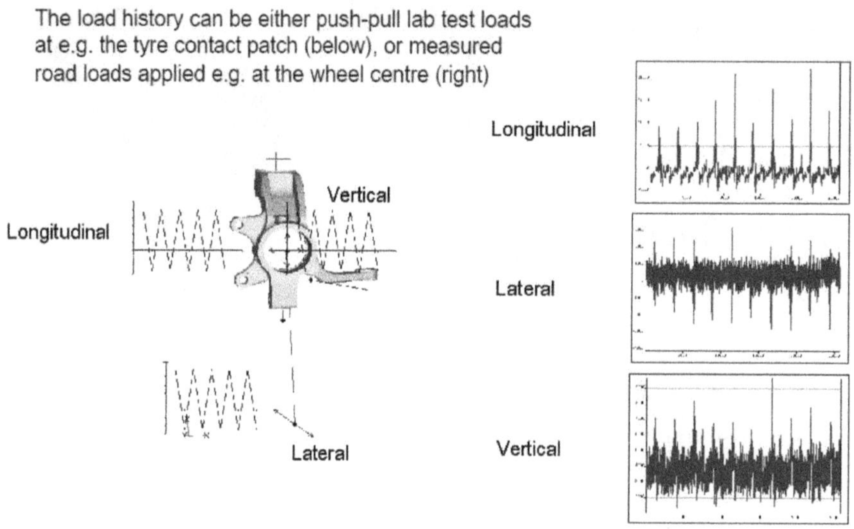


Figure 17. Application of force time histories (Duquesnay et al., 2002).

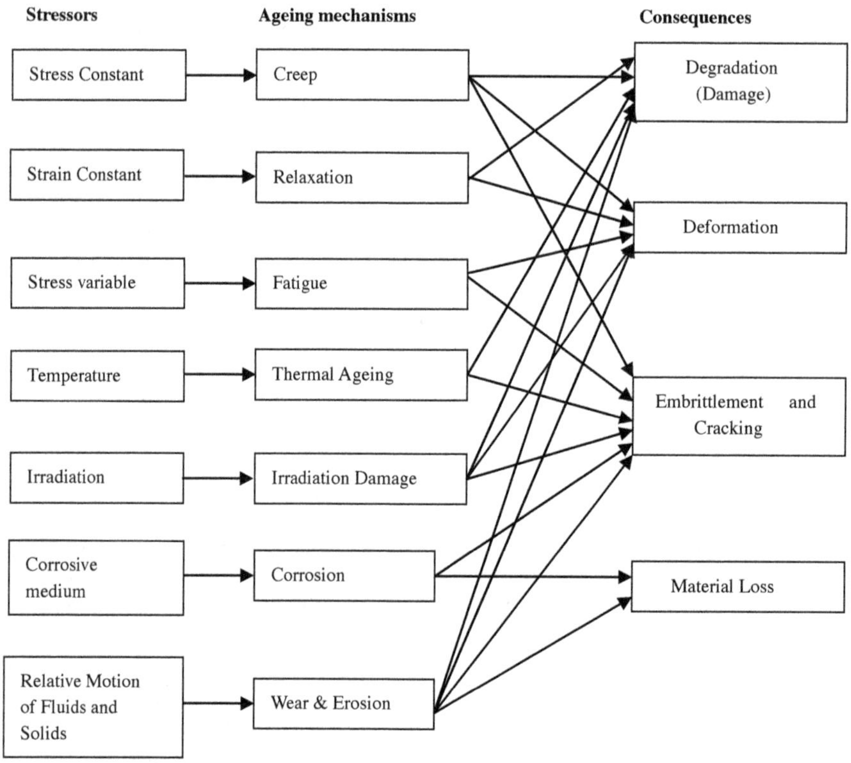


where $m$ is the vehicle mass; $I$ is the moment of inertia; $m_{2a}$ is the mass of front wheel; $m_{2b}$ is the mass of rear wheel; $\theta$ is the rotary angle of vehicle; $x$ is the vertical displacement; $c_i$ is the friction coefficient of dumping $(i = a, b)$; $f_{ca}$, $f_{cb}$ are the dumping forces of the front/rear wheel; $f_{ka}, f_{kb}$ are the restoring forces of the front/rear wheel; $k_{1a}$, $k_{1b}$ are the spring constants of the front/rear suspension; $k_{2a}$, $k_{2b}$ are the spring constants of the front/rear wheel; $x_{2a}, x_{2b}$ are the vertical displacements of the front/rear wheel; $x_{1a}, x_{1b}$ are the displacements of the vehicle body at front/rear wheel; $l_a, l_b$ are the distances of the front/rear suspension to center; and $w_a, w_b$ are the irregular excitations from the road surface (see Figure 19).

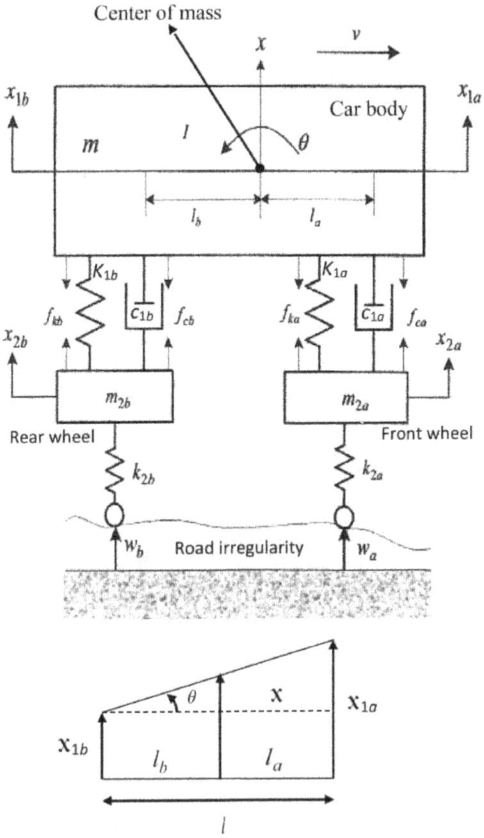


Figure 18. Vehicle suspension model.

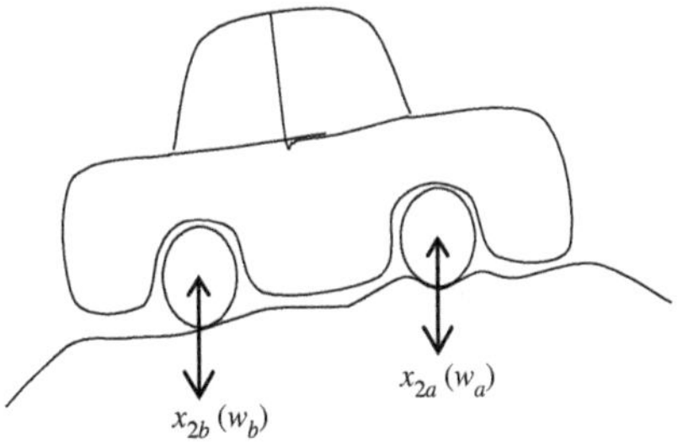

Figure 19. Road profile excitation.

### 3.1.4. System identification

The model parameters are given by the following numerical data (Lee, 2004):

$m = 1200$ kg, $I = 2100$ kg m$^2$,
$m_{2a} = 30$ kg, $m_{2b} = 25$ kg,
$c_b = 4000$ N/m/s, $c_a = 5000$ N/m/s,
$k_{1a} = 56000$ N/m, $k_{1b} = 42000$ N/m,
$k_{2a} = k_{2b} = 152$ kN/m, $l_a = 0.9$ m, $l_b = 1.2$ m.

The matrix form of the previous equations is given by

$$M\ddot{z} + N\dot{z} + Kz = Eu, \tag{15}$$

where $M$ is the mass matrix, $N$ is the damper coefficients matrix and $K$ is the stiffness matrix.

The input excitation vector is $u = [\ w_a \quad w_b\ ]^{\mathrm{T}}$.

The output damper displacement vector is $z = [\ x_{1a} \quad x_{2a} \quad x_{1b} \quad x_{2b}\ ]^{\mathrm{T}}$.

The vertical accelerations $\ddot{x}_{1a}, \ddot{x}_{1b}, \ddot{x}_{2a}, \ddot{x}_{2b}$ are measured variables. The matrices $M, N, K$ and $E$, respectively, are given by

$$M = \begin{bmatrix} l_b m/l & 0 & l_a m/l & 0 \\ I/l & 0 & -I/l & 0 \\ 0 & m_{2a} & 0 & 0 \\ 0 & 0 & 0 & m_{2b} \end{bmatrix},$$

$$N = \begin{bmatrix} c_a & -c_a & c_b & -c_b \\ l_a c_a & -l_a c_a & -l_b c_b & l_b c_b \\ -c_a & c_a & 0 & 0 \\ 0 & 0 & -c_b & c_b \end{bmatrix},$$

$$K = \begin{bmatrix} k_{1a} & -k_{1a} & k_{1b} & -k_{1b} \\ l_a k_{1a} & -l_a k_{1a} & -l_b k_{1b} & l_b k_{1b} \\ -k_{1a} & k_{1a} + k_{2a} & 0 & 0 \\ 0 & 0 & -k_{1b} & k_{1b} + k_{2b} \end{bmatrix},$$

$$E = \begin{bmatrix} 0 & 0 \\ 0 & 0 \\ k_{2a} & 0 \\ 0 & k_{2b} \end{bmatrix}.$$

The state vectors (damper displacements and velocity) are

$$x = \begin{bmatrix} z(t) \\ \dot{z}(t) \end{bmatrix}, \quad \dot{x} = \begin{bmatrix} \dot{z}(t) \\ \ddot{z}(t) \end{bmatrix}.$$

The state vectors are obtained from the solution of the dynamic equation (15) by a convenient method applied to an ordinary differential equation. As this type of solution is not in the scope of this paper, a simplification was done later (Section 3.1.6) to deduce the damper displacements.

### 3.1.5. Stress intensity factor of a suspension

The modeling of the suspension damage begins by determining the stress intensity factor composed of the multiplication of two functions:

$$(K_{\mathrm{I}})^m = \left(Y(a) \cdot \sqrt{\pi a}\right)^m \cdot (\sigma_{\max})^m = \phi_1(a) \cdot \phi_2(P_j), \tag{6}$$

where $\phi_1(a)$ is the crack length function determined in terms of a geometric function $Y(a)$ and $\phi_2(P_j)$ is the loading function.

*Determination of the first function* $\phi_1(a)$:

Assume that the front suspension of the system has a crack length $a$ perpendicular to the exterior load (Figure 20).

Let $m = 2$ be the material constant, then $\phi_1(a) = \left(Y(a) \cdot \sqrt{\pi a}\right)^2$.

Experimental and empirical results of the validation of the crack propagation models permit one to define some

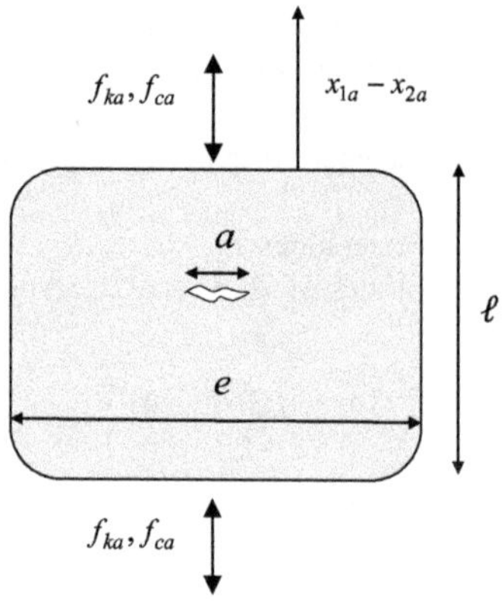


Figure 20. Suspension fatigue crack modeling.

Table 1. Characteristics of each mode of roads.

| Road mode | Mean of $\Delta x_j$ ($\overline{\Delta x_j}$ in mm) | Coefficient of variation of $\Delta x_j$ in % |
|---|---|---|
| Severe (mode 1) | 100 | 15 |
| Fair (mode 2) | 50 | 10 |
| Good (mode 3) | 25 | 5 |

geometric functions $Y(a)$ that depend on the body of the studied equipment (El-Tawil, 2004). For the case of suspensions, $Y(a)$ is defined with a sufficient precision by a development limited to order 4 (Equation (16)). Therefore, the first function can be considered as given by

$$\phi_1(a_N) = (\pi\, a_N)\left[1.122 - 1.4\left(\frac{a_N}{e}\right) + 7.33\left(\frac{a_N}{e}\right)^2 - 13.08\left(\frac{a_N}{e}\right)^3 + 14\left(\frac{a_N}{e}\right)^4\right]^2, \tag{16}$$

where

$$Y(a) = 1.122 - 1.4\left(\frac{a_N}{e}\right) + 7.33\left(\frac{a_N}{e}\right)^2 - 13.08\left(\frac{a_N}{e}\right)^3 + 14\left(\frac{a_N}{e}\right)^4$$

and $a_N$ is the crack length at cycle $N$, $e$ is the width of the mechanical component of the suspension.

*Determination of the second function $\phi_2(P_j)$:*

Given that $P_j$ is the load parameter, and we have $\phi_2(P_j) = P_j^m = P_j^2$, we will simulate the degradation model by generating the load $P_j$ of road profile $[\ w_a \quad w_b\ ]^{\mathrm{T}}$ (Chelidze & Cusumano, 2004) under the Gaussian Normal law for the three modes of roads (Table 1).

From the system of equations (15), the solution of this system of matrices gives the output vector $z$.

Then, the range of the suspension displacement for the front wheel is given by

$$\Delta x_j = x_{1a}^j - x_{2a}^j. \tag{17}$$

If we take as mean value $\bar{x}_j$ and as standard variation$\sigma_{x_j}$, then we obtain a set of $\{x\}^r$ for each road mode$(r = 1, 2, 3)$, given that the load parameter is always $P_j$. Hence

$$\phi_2(P_j) = P_j^m = P_j^2 = \sigma_j^2.$$

The amplitude of the stresses developed in the suspension due to $\Delta x_j$ is simplified by

$$\Delta\sigma_j = E \times \frac{\Delta x_j}{\ell}, \tag{18}$$

where $\ell$ is the length of the suspension device ($\ell =$ 500 mm), $\Delta x_j$ is the variation of this length (dilation) under road profile excitation and $E$ is Young's modulus of the suspension material ($E = 200$ GPa).

### 3.1.6. *Fatigue damage modeling of the suspensions*

Assume that the maximum of $a_N$is $a_C = e/8$ (Lemaitre & Chaboche, 1990).

We define

$$D_N = \frac{a_N}{a_C - a_0} \approx \frac{a_N}{a_C} = \frac{8a_N}{e} \quad (\text{as } a_0 \ll a_C).$$

Then, we replace $\phi_1(a_N = eD_N/8)$ in Equation (13) and we get

$$D_N = D_{N-1} + \eta \cdot \phi_1(D_{N-1}) \cdot \phi_2(P_j). \tag{19}$$

Moreover, $\eta$ is a material constant and we have $\eta = 8 \times 10^{-6}$ (El-Tawil, 2004).

Thus, we have the recursive formula (20) in terms of the crack length:

$$a_N = a_{N-1} + C \cdot \phi_1(a_{N-1}) \cdot \phi_2(P_j). \tag{20}$$

With (El-Tawil & Kadry, 2010) $m = 2$, $C = \eta \cdot (a_C - a_0)$, $\eta = 8 \times 10^{-6}$

$$\phi_1(a_{N-1}) = (\pi a_{N-1})\left[1.122 - 1.4\left(\frac{a_{N-1}}{e}\right) + 7.33\left(\frac{a_{N-1}}{e}\right)^2 - 13.08\left(\frac{a_{N-1}}{e}\right)^3 + 14\left(\frac{a_{N-1}}{e}\right)^4\right]^2 \tag{21}$$

and

$$\phi_2(P_j) = P_j^m = P_j^2 = \sigma_j^2.$$

Therefore, the recursive expression of the crack length for the suspension model is given by

$$a_N = a_{N-1} + C \times (\pi a_{N-1}) \Big[ 1.122 - 1.4 \left(\frac{a_{N-1}}{e}\right) + 7.33 \left(\frac{a_{N-1}}{e}\right)^2 - 13.08 \left(\frac{a_{N-1}}{e}\right)^3 + 14 \left(\frac{a_{N-1}}{e}\right)^4 \Big]^2 \times \sigma_j^2. \quad (22)$$

From the equation $D_N = (a_N / a_C - a)$, the recursive expression of the degradation index for the suspension model becomes

$$D(N) = \frac{a_{N-1}}{a_C - a_0} + \frac{C}{a_C - a_0} \times (\pi a_{N-1}) \times \Big[ 1.122 - 1.4 \left(\frac{a_{N-1}}{e}\right) + 7.33 \left(\frac{a_{N-1}}{e}\right)^2 - 13.08 \left(\frac{a_{N-1}}{e}\right)^3 + 14 \left(\frac{a_{N-1}}{e}\right)^4 \Big]^2 \times \sigma_j^2.$$

### 3.1.7. *Simulation of three road profiles*

To take into account various states of roads, we consider three different types of roads which are severe, fair and good. In the following table, we indicate the model characteristics of each type of roads.

The parabolic road profile of a vehicle circulation time with $T = 2$ s as a recurrent interval is considered. Given that this interval is repeated as needed until reaching the failure level ($D_C = 1$). Figure 21 illustrates the road profile.

Each interval shows that the road profile contains a symmetric curve of width $T/8 = 0.25$ s with a peak value followed by a horizontal run of zero amplitude.

Moreover, the period of road profile $T = 2$ s must be compared to the proper period of the suspension system to verify if there is a risk of dynamic amplification (i.e. mechanical resonance).

### 3.1.8. *Simulation results*

The prognostic study of a suspension is realized by using the degradation simulation (Equation (23)). The methodology is composed of two parts:

- In the first part, the simulation of the road profile for the three modes (severe, fair and good) (Table 1) is done and from which $\Delta x$ and $\Delta\sigma$ are deduced.
- In the second part, the crack length $a_N$ is cumulated at each cycle $N$ (Equation (22)).

The resulting curves of $D(N)$ are represented in Figures 22–24.

In mode 1 case (severe), it is noted that (Figure 22) for $N = 6{,}836{,}000$ cycles, the degradation $D_N$ reaches the critical value $D_C = 1$. The deduced lifetime of the suspension is 6,836,000 cycles of road excitation in mode 1. Moreover, the first sign of damage appears at about 2,500,000 cycles. Starting from 6,000,000 cycles, the slope of the degradation curve becomes very acute; hence, damage is increasing very fast.

In mode 2 case (fair), it is noted that (Figure 23) for $N = 10{,}850{,}000$ cycles, the degradation $D_N$ reaches the critical value $D_C = 1$. The deduced lifetime of the suspension is 10,850,000 cycles of road excitation in mode 2. Moreover, the first sign of damage appears at about 4,000,000 cycles. Starting from 10,000,000 cycles, the slope of the degradation curve becomes very steep; hence, damage is increasing very fast.

In mode 3 case (good), it is noted that (Figure 24) for $N = 17{,}222{,}000$ cycles, the degradation $D_N$ reaches the critical value $D_C = 1$. The deduced lifetime of the suspension is 17,222,000 cycles of road excitation in mode

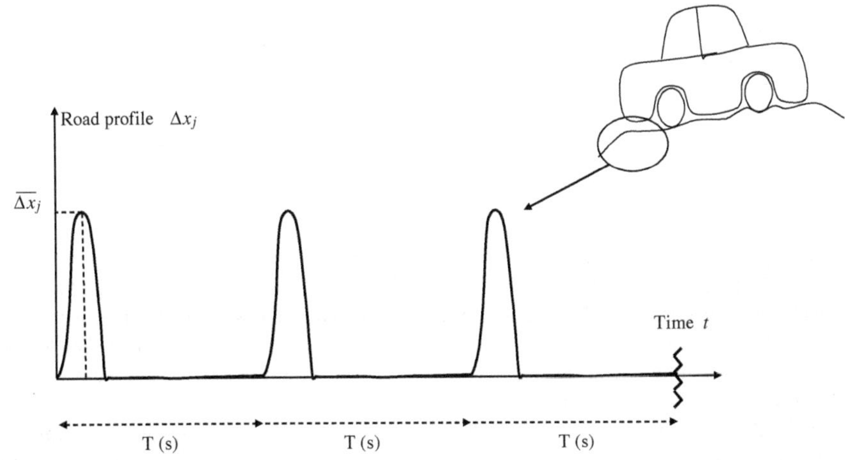


Figure 21. Simulated road profile.

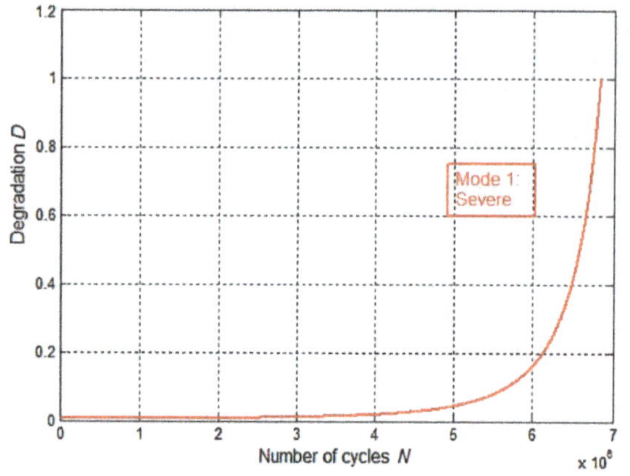


Figure 22. Degradation trajectory for the road with mode 1 profile.

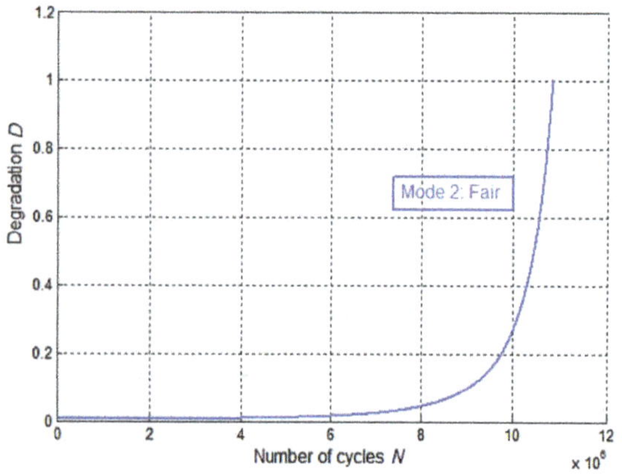


Figure 23. Degradation trajectory for the road with mode 2 profile.

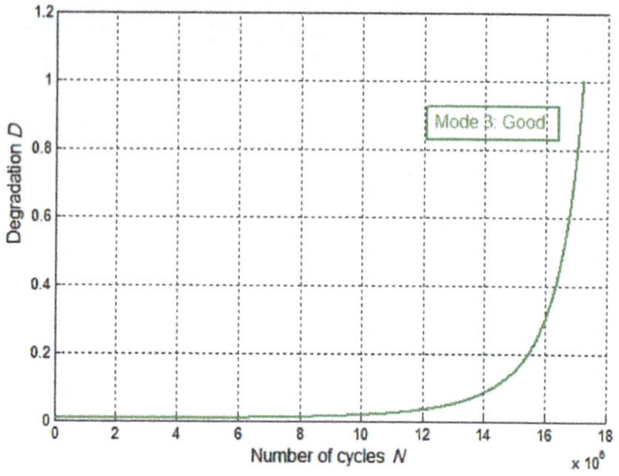


Figure 24. Degradation trajectory for the road with mode 3 profile.

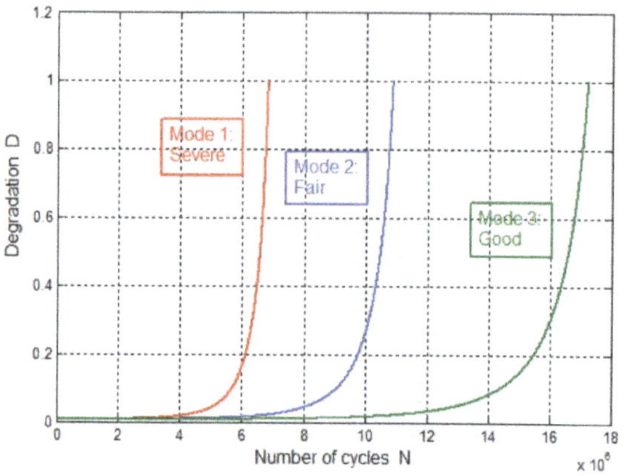


Figure 25. Degradation trajectory for the three modes of road profiles.

3. Moreover, the first sign of damage appears at about 6,200,000 cycles. Starting from 16,000,000 cycles, the slope of the degradation curve becomes very acute; hence, damage is increasing very fast.

In addition, Figure 25 recapitulates the three previous figures.

### *3.1.9. Analysis of the simulation results*

The expectation of the lifetime for mode 1 is nearly 63% of that of mode 2 and the expectation of the lifetime for mode 2 is nearly 63% of mode 3 (Figure 25). It can be noticed from the obtained results that the increase in the suspension lifetime relative to the road of mode 3 is as follows: mode (1)/mode (3) ≈ 152 and mode (2)/mode (3)≈ 59.

From the results above, the three expected lifetimes are as follows: $N_{C1} = 6{,}836{,}000$ cycles; $N_{C2} = 10{,}850{,}000$ cycles; $N_{C3} = 17{,}222{,}000$ cycles. Then, our prognostic procedure yields the RUL for the three modes (Figure 26) that can now be easily deduced from these three curves at any instant or any active cycle $N$ as follows:

$$\text{For mode 1}: \text{RUL}_1(N) = N_{C1} - N,$$
$$\text{For mode 2}: \text{RUL}_2(N) = N_{C2} - N,$$
$$\text{For mode 3}: \text{RUL}_3(N) = N_{C3} - N.$$

### *3.1.10. Conversion of RUL into years and km*

Given that each cycle duration is 2 s (refer to Figure 21), we convert the suspension lifetime into years units by using: $\text{RUL(s)} = 2 \times \text{RUL}(N)$. We assume that the suspension time usage is 10% of a day, which corresponds to 2.4 hours/day.

The conversions from cycles to km and to years, for a vehicle running with 50 km/hour, are given by the following literal expressions:

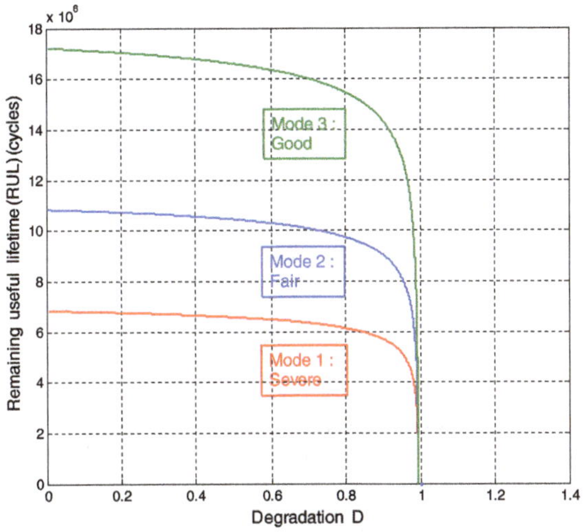


Figure 26. RULs estimated by the prognostic model.

*From cycles to km*:

$$\mathrm{RUL(km)} = \frac{\mathrm{RUL(cycles)} \times 2(\mathrm{s/cycle}) \times 50(\mathrm{km/hour})}{60(\mathrm{s/min}) \times 60(\mathrm{min/hour})} = \frac{\mathrm{RUL(cycles)}}{36(\mathrm{cycles/km})}.$$

*From km to years*:

$$\mathrm{RUL(years)} = \frac{\mathrm{RUL(km)}}{2.4(\mathrm{hours/day}) \times 50(\mathrm{km/hour}) \times 365(\mathrm{days/year})} = \frac{\mathrm{RUL(km)}}{43{,}800(\mathrm{km/year})}.$$

Therefore, the RUL results can be expressed by the following units: cycles, or km, or years.

Thus, the expected lifetimes' durations are:

$$\text{For mode 1}: \frac{6{,}836{,}000(\text{cycles}) \times 2(\text{s})}{60(\text{s}) \times 60(\text{min}) \times 2.4(\text{hours}) \times 365(\text{days})} = 4.34\,\text{years} \Rightarrow 190{,}092\,\text{km},$$

$$\text{For mode 2}: \frac{10{,}850{,}000(\text{cycles}) \times 2(\text{s})}{60(\text{s}) \times 60(\text{min}) \times 2.4(\text{hours}) \times 365(\text{days})} = 6.88\,\text{years} \Rightarrow 301{,}344\,\text{km},$$

$$\text{For mode 3}: \frac{17{,}222{,}000(\text{cycles}) \times 2(\text{s})}{60(\text{s}) \times 60(\text{min}) \times 2.4(\text{hours}) \times 365(\text{days})} = 10.92\text{years} \Rightarrow 478{,}296\,\text{km}.$$

Moreover, the validation of these results can be found in the work of Vakili-Tahami, Zehsaz, & Alidadi (2009) on the fatigue life of suspensions. An average life of 200,000 km is deduced under severe conditions and which corresponds to 4.57 years for a vehicle running with 50 km/hour and for 2.4 hours/day. The results obtained are thus realistic.

## 4. Conclusion

An analytic linear prognostic model is developed in this research paper that permits one to predict the RUL of a dynamic suspension system. This model considers the fatigue as a damage parameter, and hence it is based on well-known laws of damage such as Paris's and Miner's laws. An index of degradation that varies from zero to one is derived. My proposed model is based on the link between this index $D$ and the crack length $a$. Given that failure is produced when $a$ reaches a critical length $a_C$. Hence, my model is given by a simple function relating the instantaneous degradation to actual crack length as a measurement of actual damage.

My aim is to evaluate the evolution of the system lifetime at each instant. For this purpose, the degradation trajectories have been used in terms of cycles' numbers or the time of operation. From these degradation trajectories, the RUL variations were deduced. The prognostics of a complex system can be deduced from the prognostic of its sub-systems when their damage laws are available.

To demonstrate the effectiveness of my model, an industrial example has been considered in the simulation in this paper. This example is the vehicle suspension system where three modes of road profiles are simulated and examined.

In such an industrial system, this model proves that it is very convenient and it provides a useful tool for a prognostic analysis. Moreover, it is less expensive than other models that need a large number of data and measurements.

## Disclosure statement

No potential conflict of interest was reported by the author.

## References

Abou Jaoude, A. (2012). *Advanced analytical model for the prognostic of industrial systems subject to Fatigue*. (Ph.D. thesis). Aix-Marseille Université and the Lebanese University, pp. 55–91, defended on December 7, 2012.

Abou Jaoude, A., El-Tawil, K., Kadry, S., Noura, H., & Ouladsine, M. (2010). Analytic prognostic model for a dynamic system. *International Review of Automatic Control*, *3*(6), 568–577.

Abou Jaoude, A., Kadry, S., El-Tawil, K., Noura, H., & Ouladsine, M. (2011). Analytic prognostic for petrochemical pipelines. *Journal of Mechanical Engineering Research*, *3*(3), 64–74.

Chelidze, D., & Cusumano, J. P. (2004). A dynamical systems approach to failure prognosis. *Journal of Vibrations and Acoustics*, *126*, 2–8.

Colquhoun, C., & Draper, J. (2000). *Fatigue analysis of an FEA model of a suspension component and comparison with experimental data*. Retrieved from http://www.slideshare.net/

IndranilBhattacharyya/fatigue-analysis-of-an-fea-model-of-a-suspension-component

Conle, F. A., & Topper, T. H. (1980). Overstrain effects during variable amplitude service history testing. *International Journal of Fatigue*, *2*(3), 130–136.

Duquesnay, D. L., Pompetzki, M. A., & Topper, T. H. (2002). *Fatigue life prediction for variable amplitude strain histories*, SAE Paper 930400, Society of Automotive Engineers.

El-Tawil, K. (2004). *Mécanique Aléatoire et Fiabilité*, Cours de Master 2R Mécanique, EDST, Université Libanaise.

El-Tawil, K., Abou Jaoude, A., Kadry, S., Noura, H., & Ouladsine, M. (2010, November). *Prognostic based on analytic laws applied to petrochemical pipelines*. Proceedings of the international conference on computer-aided manufacturing and design, CMD 2010, China (pp. 105–110).

El-Tawil, K., & Kadry, S. (2010). *Fatigue Stochastique des Systèmes Mécaniques Basée sur la Technique de Transformation Probabiliste* (Internal report). Beirut: Lebanese University.

Frost, N. E. (1960). Notch effects and the critical alternating stress required to propagate a crack in an aluminum alloy subject to fatigue loading. *Journal of Mechanical Engineering Science*, *2*, 109–119.

Frost, N. E., & Dugdale, D. S. (1957). Fatigue tests on notched mild steel plates with measurements of fatigue cracks. *Journal of the Mechanics and Physics of Solids*, *5*, 182–192.

Langon, M. (1999). *Introduction à la fatigue et mécanique de la rupture*. Toulouse: Centre d'essais aéronautique de Toulouse, ENSICA.

Lee, J. (2004). *Smart products and service systems for e-business transformation*. 3e Conférence Francophone de Modélisation et Simulation « Conception, Analyse et Gestion des Systèmes Industriels » MOSIM'01, du 25 au 27 avril, Troyes, France.

Lemaitre, J., & Chaboche, J. (1990). *Mechanics of solid materials*. New York: Cambridge University Press.

Lemaitre, J., & Desmorat, R. (2005). *Engineering damage mechanics*. New York: Springer.

McEvily, A. J., & Ritchie, R. O. (1988). Crack closure and the fatigue-crack propagation threshold as a function of load ratio. *Fatigue and Fracture of Engineering Matererial & Structures*, *21(7)*, 847–855.

Miner, M. A. (1945). Cumulative damage in fatigue. *Journal of Applied Mechanics*, *12*, A159–A164.

Newman Jr., J. C., & Raju, I. S. (1981). An empirical stress-intensity factor equation for the surface crack. *Engineering Fracture Mechanics*, *15*(1/2), 185–192.

Paris, P., & Erdogan, F. (1963). A critical analysis of crack propagation laws. *Journal of Basic Engineering, Transactions of the American Society of Mechanical Engineers*, *85*(4), 528–534.

Peysson, F., Ouladsine, M., Outbib, R., Leger, J-B., Myx, O., & Allemand, C. (2009). A generic prognostic methodology using damage trajectory models. *IEEE Transactions on Reliability*, *58*(2), 277–285.

Sukumar, N., Chopp, D. L., & Moran, B. (2003). Extended finite element method and fast marching method for three-dimensional fatigue crack propagation. *Engineering Fracture Mechanics*, *70*, 29–48.

Todinov, M. (2001). Necessary and sufficient condition for additivity in the sense of Palmgren–Miner rule. *Computational Materials Science*, *21*(1), 101–110.

Todinov, M. (2005). *Reliability and risk models setting reliability requirements*. Cranfield, UK: Cranfield University, Wiley.

Vachtsevanos, G., Lewis, F., Roemer, M., Hess, A., & Wu, B. (2006). *Intelligent fault diagnosis and prognosis for engineering systems*. Hoboken, NJ: Wiley.

Vakili-Tahami, F., Zehsaz, M., & Alidadi, M. R. (2009). Fatigue analysis of the weldments of the suspension-system-support for an off-road vehicle under the dynamic loads due to the road profiles. *Asian Journal of Applied Sciences*, *2*(1), 1–21.

# Dynamic modelling and simulation of IGCC process with Texaco gasifier using different coal

Yue Wang[a], Jihong Wang[a*], Xing Luo[a], Shen Guo[a], Junfu Lv[b] and Qirui Gao[b]

[a]*School of Engineering, The University of Warwick, Coventry CV4 7AL, UK;* [b]*Department of Thermal Engineering, Tsinghua University, Beijing 100084, People's Republic of China*

Integrated gasification combined cycle (IGCC) is considered as a viable option for low emission power generation and carbon dioxide sequestration. As a part of the process of IGCC plant design and development, modelling and simulation study of the whole IGCC process is important for thermodynamic performance evaluation, study of carbon capture readiness and economic analysis. The work presented in the paper is to develop such a whole system model and simulation platform. A simplified dynamic model for the IGCC process is developed, in which Texaco gasifier is chosen to give the basic representation for the IGCC process. The chemical equilibriums principle is used to predict the syngas contents in the modelling procedure. The influences of key parameters to regulate the input such as oxygen/coal ratio and water/coal ratio to syngas generation are studied. The simulation results are validated by comparing with the industry data provided by the Lu-nan fertilizer factory. Water-shift reactor, gas turbine and heat recovery steam-generation modules are modelled to study the dynamic performance with respect to the variation from the input of syngas stream. The simulation results reveal the dynamic changes in the plant outputs, including gas temperature, power output and mole percentages of hydrogen and carbon dioxide in the syngas. The process dynamic responses with three types of coal inputs are studied in the paper and their dynamic variation trends are presented via the simulation results.

**Keywords:** chemical equilibrium; dynamic performance; IGCC; syngas; Texaco gasifier

## Nomenclature

| | |
|---|---|
| $A$ | carbon conversion rate (–) |
| $C_p$ | specific heat capacity (kJ/(kmol·K)) |
| $N$ | molar flow rate (kmol/s) |
| $D$ | derivative |
| $H$ | enthalpy (kJ/kmol) |
| $\Delta H_f$ | enthalpy of formation (kJ/kmol) |
| $T$ | Kelvin temperature (K) |
| $X$ | mole faction |
| $S$ | simulation |
| $R$ | reference |
| $P$ | pressure |
| $Ar$ | Argon |
| C | carbon |
| $CH_4$ | methane |
| CO | carbon monoxide |
| $CO_2$ | carbon dioxide |
| COS | cabonyl sulphide |
| $g$ | gas |
| H | hydrogen |
| $H_2$ | hydrogen element |
| $H_2O$ | water, vapour |
| $H_2S$ | hydrogen sulphide |
| $i$ | gas species |
| $N_2$ | nitrogen |
| $O_2$ | oxygen |
| S | sulphur element |
| $SO_2$ | sulphur dioxide |
| 0 | original value |

## Abbreviation

| | |
|---|---|
| CC | combined cycle |
| HRSG | heat recovery steam generator |
| WGS | water gas shift reactor |

## 1. Introduction

Integrated gasification combined cycle (IGCC) offers the benefits over conventional coal-fired power plants, especially with regard to the environment and feedstock flexibility (Casella & Colonna, 2012; Yang et al., 2011). The gasification process of solid fuel, such as coal and biomass, generates much less pollutants than the direct burning process in traditional coal-fired power plants (Maurstad, 2005; Sun, Liu, Chen, Zhou, & Su 2011). IGCC shows its own

*Corresponding author. Email: jihong.wang@warwick.ac.uk

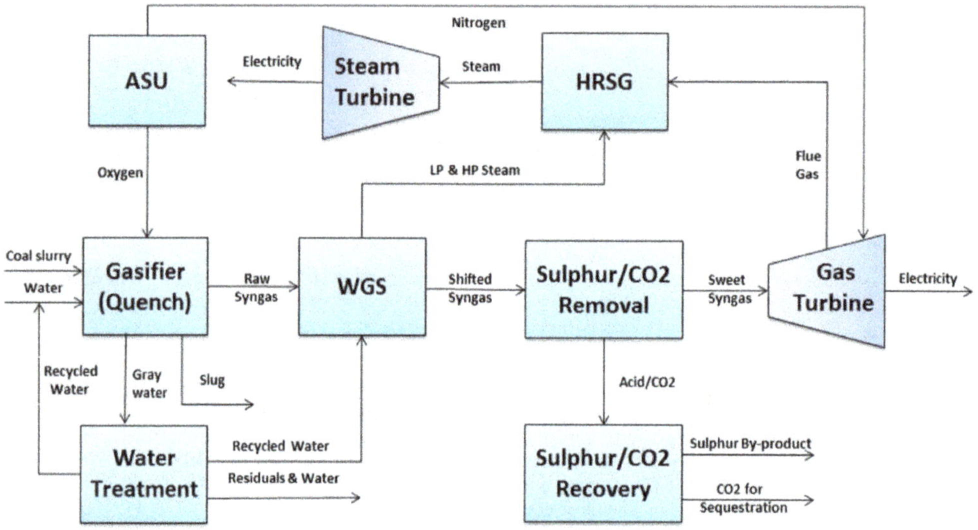


Figure 1. Simplified schematic diagram of an IGCC system (Casella & Colonna, 2012).

merits for integration with carbon capture and storage units. Figure 1 shows a schematic diagram of the whole IGCC process which is, no doubt, a complicated energy conversion process formed by many interconnected subsystem modules. In the IGCC process, the coal slurry and oxygen react in the gasifier and generate syngas primarily made of $H_2$ and CO. The WGS raises the $H_2$ and $CO_2$ concentration while reducing the CO content. The sweet syngas combustion in the gas turbine will generate power and hot flue gas, which is used in HRSG to generate power as well.

The work on IGCC process modelling can be dated back to the 1970s. Researchers in the chemical engineering field studied the coal gasification process and developed models based on mass and energy balances (Beér, 2000; Brown, Smoot, & Hedman, 1986; Buskies, 1996; Chen et al., 2004; Govind & Shah, 1984; Ni & Williams, 1995; Smoot & Smith, 1979; Ubhayakar, Stickler, & Gannon, 1977; Watkinson, Lucas, & Lim, 1991; Wen & Chaung, 1979). Researchers in thermal engineering developed gas turbine and steam power plant models (Colonna & van Putten, 2007; Lu & Hogg, 2000; van Putten & Colonna, 2007). However, dynamic modelling of the whole process is still not mature and requires further study.

Among all the sub-modules of the IGCC process, the most important and complicated one is the one for the gasifier. The challenge of modelling the gasification process will need to deal with complicated chemical reactions involved in the process. The earliest report on modelling this process was published in the 1970s (Ubhayakar et al., 1977), in which a one-dimensional model was reported with consideration of fluid mix in the axial direction. Smoot and Smith (1979) provided an approach to evaluate different chemical kinetics data and to estimate the input parameters; this method has laid the foundation of many subsequent works. Wen and Chung (1979) built a model of Texaco gasifier which divided the furnace into three zones to describe the processes from pyrolytic cracking to gasification; mass balance and energy balance equations are built for each zone. Govind and Shah (1984) introduced momentum conservation to the former work and calculated the temperature, concentration and fluid field in the axial direction.

Most of the models reported in the published literature are based on experimental data using a data-driven approach, which limits the suitability of a model for industry use as its working conditions vary in a wide range. To provide good prediction for syngas output, a generic gasifier model is developed based on the process engineering operation principles discussed in this paper and its steady-state prediction is validated. The syngas output stream from the gasifier varies depending on the reactions in the gasifier and the coal slurry feeding speed. It increases from its initial rate of 0.1 mol/s to settle at 100 mol/s in 100 s.

The auxiliary modules including shift reactor, gas turbine, and HRSG are built with Matlab and a Simulink-based toolbox – Thermolib. The syngas generated by the gasifier will first enter a water quencher and will then be further cooled by a syngas cooler. After hydrolysis reaction and desulphuration reaction, the COS and $H_2S$ contents in the syngas will be removed. The sweet syngas will then pass the shift reactor where the CO contents will be shifted to hydrogen and $CO_2$, which not only can enhance the fuel gas quality but also raise the $CO_2$ concentration. After the shift reaction, the shifted syngas will be compressed and heated again and injected to the gas turbine to generate electricity; the heat carried by the exhaust gas will be utilized to generate superheated steam in HRSG and produce more electricity, thus improving the overall system efficiency.

## 2. Description of Texaco gasifier and mathematical model

Texaco gasification technology, also known as coal slurry gasification technology, is developed by Texaco Company, initially for heavy oil gasification applications. Texaco gasifier structure is shown in Figure 2. It gasifies coal slurry, which is mixed by pulverized coal particles and water, as raw material, and uses oxygen as the gasification agent. Coal slurry is injected into the gasifier furnace through nozzles; the moisture content of coal slurry droplets will evaporate rapidly and the pulverized coal particles will devolatilize and yield coal tar, gaseous hydrocarbons and oil. The gaseous components and volatiles will be consumed rapidly with steam and oxygen. The combustion of carbon char will react with oxygen, carbon dioxide and hydrogen while the reaction products react with each other as well. The whole gasification process involves complicated physical and chemical reactions. The gasifier generates wet syngas composed of CO, $CO_2$, $H_2$ and steam. Syngas will leave the gasification zone with slag and enter the water quench zone where the slag will be deposited in the slag tank. Raw syngas will be cooled and cleaned after the quenching process.

The mathematic model for this process is developed by following the work of Watkinson et al. (1991). To simplify the modelling procedure, the following assumptions are made:

(1) The flow in the gasifier furnace is uniformed laminar flow, and the differences in temperature, concentration, pressure and material exchanges in the radial direction are not taken into consideration. Actually, laminar flow only exists in the lower part of the gasifier; the flow between nozzles and the lower part should be jet flow surrounded by a strong back flow zone. The eddy turbulent is not considered in the modelling process as the syngas content will not be affected by this flow type. Thus, it is acceptable for the global laminar assumption.

(2) Preheating of slurry droplets, moisture evaporation and coal devolatilization will be complete as soon as the coal slurry is injected into the gasifier. The nozzles are surrounded by high-pressure high-temperature gas flow.

(3) The released volatile combustion and carbon pyrolytic and char combustion reactions reach chemical equilibrium as soon as the slurry enters the furnace. The chemical equilibrium constants of homogeneous reactions inside the gasifier are used to describe the reactions.

(4) Nitrogen and argon are assumed to be steady and will not participate in any chemical reaction. It is assumed that all oxygen is consumed, and the carbon conversion is 99.5% in the entire gasifier.

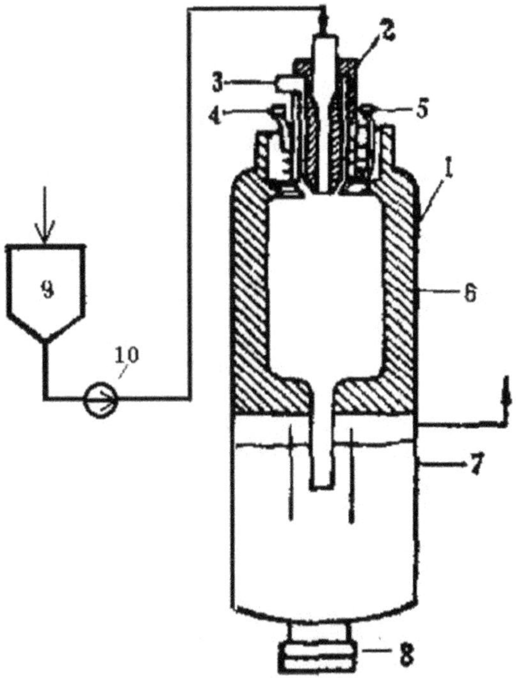


Figure 2. Structure of Texaco gaisifer. 1 – gasifier, 2 – Nozzle, 3 – oxygen input, 4 – cooling water input, 5 – cooling water output, 6 – refractory bricks liner, 7 – quenching water input, 8 – slag output, 9 – coal slurry tank, 10 – coal slurry pump.

Chemical reactions considered in this paper are as follows:

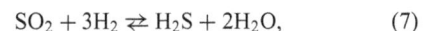

$$C + O_2 \rightarrow CO_2, \tag{1}$$

$$C + \tfrac{1}{2}O_2 \rightarrow CO, \tag{2}$$

$$C + H_2O \rightleftarrows H_2 + CO, \tag{3}$$

$$C + CO_2 \rightleftarrows 2CO, \tag{4}$$

$$CO + H_2O \rightleftarrows H_2 + CO_2, \tag{5}$$

$$CO + 3H_2 \rightleftarrows CH_4 + H_2O, \tag{6}$$

$$SO_2 + 3H_2 \rightleftarrows H_2S + 2H_2O, \tag{7}$$

$$COS + H_2O \rightleftarrows H_2S + CO_2. \tag{8}$$

For the aforementioned chemical reactions, mass balance equations of carbon, oxygen, hydrogen, nitrogen and sulphur can be derived as follows:

$$N_{C,0}A = N_g(X_{CO} + X_{CO_2} + X_{CH} + X_{COS}), \tag{9}$$

$$N_{O_2,0} = N_g(0.5X_{CO} + X_{CO_2} + X_{SO_2} + 0.5X_{COS} + 0.5X_{H_2O}),^{y} \tag{10}$$

$$N_{H_2,0} = N_g(2X_{CH_4} + X_{H_2S} + X_{H_2} + X_{H_2O}), \tag{11}$$

$$N_{N_2,0} = N_g(X_{N_2}), \quad (12)$$

$$N_{S,0} = N_g(X_{SO_2} + X_{H_2S} + X_{COS}), \quad (13)$$

$$N_{Ar,0} = N_g(X_{Ar}). \quad (14)$$

According to Dalton's law (Watkinson et al., 1991), we have

$$X_{CO} + X_{CO_2} + X_{CH_4} + X_{H_2} + X_{H_2O} + X_{H_2S} + X_{SO_2} + X_{COS} + X_{N_2} + X_{Ar} = 1. \quad (15)$$

Equations (9)–(14) are derived based on the mass conservation of C, O, H, N, S and Ar. From Dalton's law (Watkinson et al., 1991), Equation (15) means that the sum of all the syngas contents equals 1.

Based on chemical equilibriums:

$$\frac{X_{H_2}X_{CO_2}}{X_{CO}X_{H_2O}} = 0.0265\, e^{3956/T_g}, \quad (16)$$

$$\frac{X_{CH_4}X_{H_2O}}{X_{CO}X_{H_2}^3 P^2} = 6.7125 \times 10^{-14}\, e^{27020/T_g}, \quad (17)$$

$$\frac{X_{H_2S}X_{H_2O}^2}{X_{SO_2}X_{H_2}^3 P} = 4.3554 \times 10^{-4}\, e^{26281/T_g}, \quad (18)$$

$$\frac{X_{H_2S}X_{CO_2}}{X_{COS}X_{H_2O}P} = 0.75314\, e^{4083/T_g}. \quad (19)$$

Equations (16)–(19) are derived based on the chemical equilibrium of reactions (5)–(8) listed earlier; the reaction temperature and pressure will affect the equilibrium and change the concentration of syngas contents. There are 11 variables in Equations (9)–(19); this nonlinear equation system can be solved using the Newton–Raphson method (Wang, Wang, Guo, Lv, & Gao, 2013). Energy balance can be expressed as {Total enthalpy input} − {Total enthalpy output} = {Heat loss to environment}; the enthalpy input includes raw material enthalpy, the chemical reaction formation enthalpy and enthalpy carried by recycled gas. The enthalpy output includes enthalpy carried by output syngas, tar and char. Gaseous enthalpies are calculated as

$$H_{gi} = N_{gi}\left[\sum_i X_i\left(\Delta H_{f\,i} + \int_{298}^{T_g} C_{p\,i}\, dT\right)\right]. \quad (20)$$

## 3. Governing equations

In the Thermolib toolbox, the syngas stream is organized as a vector formed by the data flow of mole flow, contents concentration, temperature, pressure, enthalpy flow, entropy flow, Gibbs energy rate, heat capacity rate and vapour fraction of all compounds, which is described as combined flow bus. The gas phase contents are considered as the real gas form, and their enthalpy, entropy and heat capacity rate will be calculated by using the Peng–Robinson real gas equations of state (Peng & Robinson, 1976).

The most important governing equations are formed of mass balance and energy balance; the mass balance for a normal (standard) block is as follows:

Mass balance:

$$\frac{dM_i}{dt} = \sum Y_{i,\text{in}}\dot{M}_{\text{in}} - \sum Y_{i,\text{out}}\dot{M}_{\text{out}} + \sum R_i, \quad (21)$$

where $Y_{i,\text{in}}$ denotes the mass concentration of content $i$ in inlet flow, $\dot{M}_{\text{in}}$ denotes the inlet mass flow rate, $Y_{i,\text{out}}$ denotes the mass concentration of content $i$ in outlet flow, $Y_{i,\text{out}}$ denotes the mass concentration of content $i$ in outlet flow, $\dot{M}_{\text{out}}$ denotes the outlet mass flow rate and $R_i$ denotes the net production rate of $i$ by chemical reactions.

When no chemical reaction occurs in the simulated block, the factor $R_i$ will equal zero. In addition, mass accumulation is not considered in all the blocks simulated in this paper.

The governing equation of energy balance is derived by following the first law of thermodynamics:

$$\frac{dU}{dt} = \sum \dot{H}_{i,\text{in}} - \sum \dot{H}_{j,\text{out}} + \sum \dot{Q}_k + \sum P_m, \quad (22)$$

where $U$ denotes the internal energy in block, $\dot{H}_{i,\text{in}}$ denotes the enthalpy flow rate of content $i$ in the inlet flow, $\dot{H}_{j,\text{out}}$ denotes the enthalpy flow rate of content $j$ in the outlet flow, $\dot{Q}_k$ denotes the heat flow and $P_m$ denotes the mechanical power.

All the auxiliary blocks built with Thermolib in this paper will follow Equations (21) and (22). The individual blocks will be explained one by one in the following.

## 4. Description of the WGS reactor

The syngas generated by the gasifier will be cooled in the quench water pool. The wet syngas will then enter the WGS and the exothermic water gas shift reaction will take place:

$$CO + H_2O \rightleftarrows H_2 + CO_2. \quad (23)$$

In the industry, a shift catalyst is used in the reactor; the catalyst can convert most of the CO to $CO_2$ with the evolution of heat. There are a number of specific advantages from incorporating a shift reactor into the coal gasification flow scheme in real world; it will improve the $H_2$ extraction while decreasing the CO concentration in the syngas stream. The shift reactor with heat recovery is able to increase the power output for the same gas turbine investment as well. Moreover, the WGS is important for the preparation of $CO_2$ sequestration because the $CO_2$ concentration rises as well, which is ideal for PSA (pressure swing adsorption) adsorption capture (Karmarkar, 2005).

The shift reactor in this study is used to prepare high $H_2$ and $CO_2$ content-shifted syngas stream for the hydrogen gas turbine and future carbon capture module. The complicated catalysis or equilibrium process is hard to be developed and validated for this simulation; thus, a reaction rate-controlled reactor block developed using the Thermolib toolbox is adopted to simulate the WGS reaction process, and the heat exchange with the environment is considered.

In this mode, two shift reactors are connected in series and the reaction rates are defined, respectively, as 0.8 and 0.9 (Karmarkar, 2005); meanwhile, the pressure loss and heat exchange with the environment are considered. By defining the reaction rate, the conversion of CO can be controlled; thus, the partial pressure of $CO_2$ in the shifted syngas will satisfy the demand for further PSA carbon capture simulation. The development of the carbon capture model and its integration with the current IGCC model will be studied in the future.

Based on the energy balance, the reactor model in the WGS process can be described by

$$\sum_{\text{reac}} \dot{m}_{\text{in}} h_{\text{in}} = \sum_{\text{prod}} \dot{m}_{\text{out}} h_{\text{out}} + \dot{Q}. \tag{24}$$

The syngas temperature, pressure, mass flow rate and contents dynamic change are obtained. In this paper, the contents of the syngas entering the WGS are simplified as a mixture of CO, $CO_2$, $H_2$ and $H_2O$ only. This simplification will not cause big error because the total portion of $CH_4$, $N_2$, $H_2S$, $SO_2$, COS and Ar is less than 2% in the syngas, and they will not affect the model to represent the process correctly in terms of the dynamic features concerned in the analysis.

## 5. Description of the gas turbine

The gas turbine-centred power-generation process is based on Brayton cycle (Lichty, 1967). It has a rotating compressor coupled to the shaft of the turbine, and a combustion chamber in between. The shifted syngas is compressed, mixed with compressed air in the mixer and then ignited in the combustion chamber. The combustion of $H_2$ under a constant pressure will generate the hot flue gas. The flue gas expands through the turbine to perform work.

An isentropic compressor developed in Thermolib is adopted to model the compression process; this block can increase the pressure of an incoming flow to a given outlet pressure. To achieve more accurate simulation results, a certain isentropic efficiency is given in this process modelling, which can decrease the error between simulation results and practical working conditions. The isentropic efficiency is defined by Equation (25):

$$\eta_s = \frac{h_{\text{out},s} - h_{\text{in}}}{h_{\text{out}} - h_{\text{in}}}, \tag{25}$$

where $h_{\text{out},s}$ is the enthalpy in ideal isentropic state, while $h_{\text{out}}$ is the enthalpy in actual state. Thus, the mechanical power consumption can be calculated as

$$W_{mch} = \frac{\dot{m}(h_{\text{out},s} - h_{\text{in}})}{\eta_s}. \tag{26}$$

The compressed syngas flow is then mixed with compressed air in a mixer. It is assumed that there is no mass loss in the mixer. According to the energy balance, the following equation can be obtained:

$$\sum_{\text{input}} \dot{m}_{\text{in}} h_{\text{in}} = \dot{m}_{\text{out}} h_{\text{out}}. \tag{27}$$

The mixed syngas and air will then be ignited in the combustion chamber, where the fuel gas combustion will take place, and the chemical reactions considered include

$$2H_2 + O_2 \rightleftarrows 2H_2O, \tag{28}$$

$$2CO + O_2 \rightleftarrows 2CO_2. \tag{29}$$

The dominant reaction in the chamber is hydrogen combustion. The remaining carbon monoxide combustion is also considered. After passing the WGS, the CO mole content in shifted syngas will drop to less than 2.47% (Karmarkar, 2005); thus, the influence of CO combustion is tenuous.

The heat generation of the combustion will be calculated by hydrogen and carbon monoxide lower heating value (LHV) as 120 and 10.112 MJ/kg. In this model, a reactor block like WGS reactor is adopted to simulate the combustion, and the reaction rate is defined as 1, which means complete combustion happens inside the chamber.

The high-temperature and high-pressure flue gas is then injected into the turbine block. The turbine can decrease the pressure of the incoming flow to a given outlet pressure. It determines the thermodynamic state of the outgoing flow along with the produced mechanical power at a given isentropic efficiency, which is similar to that of the isentropic compressor. The mechanical power generated by the turbine can be calculated by the following equations:

$$\eta_s = \frac{h_{\text{in}} - h_{\text{out}}}{h_{\text{in}} - h_{\text{out},s}}, \tag{30}$$

$$W_{mch} = \dot{m}(h_{\text{in}} - h_{\text{out},s})\eta_s. \tag{31}$$

The parameters with subscript $s$ are isentropic state change. This turbine block is defined as passive, which means the mass flow rate remains unchanged during the calculation. The isentropic efficiency used in this block is retrieved from a lookup table as a function of the mass flow. It is defined as the ratio of actual enthalpy difference to enthalpy difference for isentropic change of state with the same pressure drop. In this study, the value of isentropic efficiency is set as 0.8.

The thermodynamic characteristics of the syngas flow and flue gas flow and the mechanical power-generation results are obtained. The thermal efficiency is calculated based on the fuel heat value and gas turbine net power generation.

## 6. The HRSG description

The HRSG unit is based on the Rankine cycle (Wong, 2012). It has a heat exchanger to pick up the waste heat of the flue gas to heat the feed water and generate steam. Then, the vapour will drive the turbine and then the generator to produce electricity. Using the HRSG, the thermal efficiency of the power cycle can be improved.

A pump module is adopted to provide feed water to the heat exchanger. The pump increases the pressure of the incoming water flow to target pressure. It determines the thermodynamic state of the outgoing flow along with the required mechanical power consumption. The energy balance of the compression process in the pump is calculated using the following equation (32):

$$\dot{m}\left(h_{\text{in}} + \frac{1}{2}v_{\text{in}}^2\right) = \dot{m}\left(h_{\text{out}} + \frac{1}{2}v_{\text{out}}^2\right) + \dot{W}_{\text{mch}}, \tag{32}$$

while

$$\dot{m}(h_{\text{out}} - h_{\text{in}}) = \frac{P_{\text{out}} - P_{\text{in}}}{\rho}, \tag{33}$$

where $P_{\text{out}}$ and $P_{\text{in}}$ are the pressure of the pump outlet and inlet. $\rho$ is density of the working flow. The power consumption is calculated by

$$W_{\text{mch}} = \frac{\dot{m}\Delta P}{\rho\eta_{\text{pump}}}, \tag{34}$$

where $\eta_{\text{pump}}$ is the pump efficiency and is set as 0.8. The power consumption of the pump is considered in the total HRSG output calculation. With the mechanical power, the enthalpy of the water outlet is equal to the sum of input enthalpy and electric power.

$$\dot{H}_{\text{out}} = \dot{H}_{\text{in}} + W_{\text{mch}}. \tag{35}$$

The pumped feed water is then transported to the heat exchanger, where its temperature will rise to the target by adsorbing heat from the flue gas generated by the gas turbine. A heat exchanger module based on Thermolib is used for calculating the heat transfer between the flue gas and the water stream. It is important to notice that the heat transfer between these two streams is indirect and they are treated as counter flow. The two media state dynamic change will be simulated by using the number of transfer units method (Incropera & DeWitt, 1985). The actual heat transfer rate can be determined by calculating the effectiveness $\varepsilon$, which is the actual heat transfer divided by the maximum possible heat transfer.

$$\varepsilon = \left(\frac{\dot{Q}}{\dot{Q}_{\max}}\right). \tag{36}$$

For counter flow:

$$\varepsilon = \left(\frac{1 - \exp(-M(1 - C))}{1 - C \times \exp(-M(1 - C))}\right), \tag{37}$$

where

$$M = \frac{UA}{\dot{C}_{\min}}, \tag{38}$$

$$C = \frac{\dot{C}_{\min}}{\dot{C}_{\max}}, \tag{39}$$

$$\dot{C}_{\min} = \min(\dot{m}_1 c_{p1}, \dot{m}_2 c_{p2}), \tag{40}$$

$$\dot{C}_{\max} = \max(\dot{m}_1 c_{p1}, \dot{m}_2 c_{p2}), \tag{41}$$

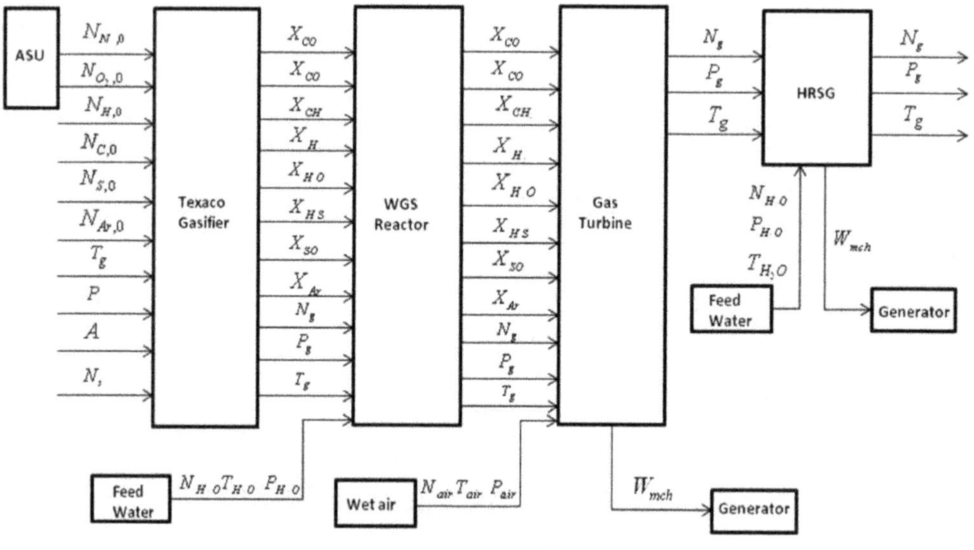


Figure 3. Schematic diagram of process simulation.

where $U$ is the overall heat transfer coefficient, $A$ is the surface area available for the heat transfer and thus $UA$ is the overall heat transfer rate. It represents the heat transfer between the flow and wall as well as the heat conduction in the wall. This is the product of heat transfer coefficient and effective heat exchange area between the flows.

Using the aforementioned equations (36–41), we can have

$$Q_{\max} = \dot{C}_{\min}(T_{\mathrm{hi}} - T_{\mathrm{ci}}), \quad (42)$$

where $T_{\mathrm{hi}}$ is the temperature of the hot fluid input to the heat exchanger and $T_{\mathrm{ci}}$ is the cold fluid input temperature. In the calculation of this heat exchanger block, the heat exchanges of both the flow streams with environment and pressure loss of the flows are considered.

The heated water steam is then injected to the steam turbine block, where the isentropic expansion process happens. The same block as described in the gas turbine system introduction is adopted for the steam turbine. The net power output of the whole HRSG system will be the difference between steam turbine power generation and pump power consumption. The dynamic change in net power of HRSG and the gas turbine system is obtained and analysed as the power output of the combined cycle.

## 7. Air separation unit

The air separation unit (ASU) is an important part in the IGCC power plants using oxygen-feed gasifiers (Wang et al., 2013). Air is liquefied and separated into oxygen and nitrogen along with some by-products such as helium. The generated oxygen is then compressed to 40 bar and injected into the gasifier for raw syngas production. Some plants integrate ASU with gas turbine, using nitrogen for combustion diluent aiming to reduce $NO_x$ emission (Karmarkar, 2005). In this paper, only hydrogen and carbon monoxide combustion are considered in the gas turbine module, the integration of ASU and gas turbine will not be included.

The working principle of ASU in this study is based on the Linde–Hampson cycle (Timmerhaus & Reed, 2007). Air is isothermally compressed first and then passes the main heat exchanger; the isobaric heat transfer between the air phase and liquid phase results in huge temperature drop. After the isenthalpic expansion process in the throttle valve, the air will be liquefied and transported to the distillation tower for further separation. In the separation process, liquid nitrogen will reach boiling point first and be separated, and pure nitrogen and oxygen will be generated for the IGCC process.

A polytropic compressor in Thermolib is adopted to simulate the isothermal process. The power consumption is calculated by using the following equation:

$$W_{\mathrm{mch}} = \frac{n(p_{\mathrm{out}} v_{\mathrm{out}} - p_{\mathrm{in}} v_{\mathrm{in}})}{1 - n} \quad (43)$$

Table 1. Model input.

| Input | Unit | Illinois No. 6 | Australia | Fluid coke |
|---|---|---|---|---|
| Slurry flow rate | kg/s | 1 | 1 | 1 |
| Slurry concentration | kg coal/kg slurry | 0.665 | 0.621 | 0.606 |
| $O_2$ purity | Vol% | 98 | 99.6 | 100 |
| $O_2$/coal | kg $O_2$/kg dry coal (no ash) | 0.86 | 0.87 | 1.03 |
| Ar/$O_2$ | kg Ar/kg $O_2$ | 0 | 0 | 0 |
| Gasifier pressure | MPa | 4.083 | 4.083 | 4.083 |
| Temp. | °C | 1141 | 1044 | 1060 |
| Heat loss | H.H.V.% | 2 | 2 | 2 |
| Ultimate analysis(dry) | Mass % | | | |
| C | % | 69.6 | 66.8 | 86 |
| H | % | 5.3 | 5.0 | 2.0 |
| O | % | 10 | 7.3 | 2.3 |
| N | % | 1.3 | 1.7 | 1.0 |
| S | % | 3.9 | 4.2 | 8.3 |
| Ash | % | 10 | 15 | 0.5 |

Table 2. Comparison of the simulation results and reference data.

| | Illinois No. 6 | | Australia | | Fluid coke | |
|---|---|---|---|---|---|---|
| | $R$ (mol%) | $S$ (mol%) | $R$ (mol%) | $S$ (mol%) | $R$ (mol%) | $S$ (mol%) |
| CO | 41.0 | 41.0 | 35.2 | 35.4 | 47.1 | 47.2 |
| $H_2$ | 29.80 | 30.1 | 29.9 | 29.6 | 24.3 | 23.7 |
| $CO_2$ | 10.2 | 10.0 | 12.8 | 12.8 | 13.2 | 13.3 |
| $H_2O$ | 17.1 | 16.8 | 20.3 | 20.0 | 12.7 | 13.0 |
| $CH_4$ | 0.3 | 0.15 | 0.02 | 0.22 | 0.09 | 0.33 |
| $N_2$ | 0.80 | 0.9 | 0.63 | 0.63 | 0.4 | 0.3 |
| $H_2S$ | 1.1 | 1.01 | 1.14 | 1.10 | 2.2 | 2.07 |
| COS | | 0.04 | | 0.25 | | 0.10 |
| Error | | 0.26 | | 0.14 | | 0.21 |

Table 3. Model input.

| Input | Unit | Data |
|---|---|---|
| Slurry rate | t/d | 650 |
| Slurry concentration | kg coal/kg slurry | 0.66 |
| Oxygen purity | Vol% | 98 |
| Oxygen/coal | kg $O_2$/kg dry coal (no ash) | 0.96 |
| Pressure | MPa | 4.0 |
| Temperature | °C | 1350 |
| Heat loss | H.H.V.% | 2 |
| Ultimate analysis | % | |
| C | % | 71.5 |
| H | % | 4.97 |
| O | % | 11.15 |
| N | % | 1.07 |
| S | % | 2.16 |
| Ash | % | 9.15 |

Table 4. Comparison of dry syngas output content and industry data.

| | CO | $H_2$ | $CO_2$ | $CH_4$ + Sulphide + $N_2$ + Ar |
|---|---|---|---|---|
| Industry | 48.82 | 36.58 | 14.41 | 0.19 |
| Simulation | 49.54 | 35.69 | 12.79 | 1.98 |

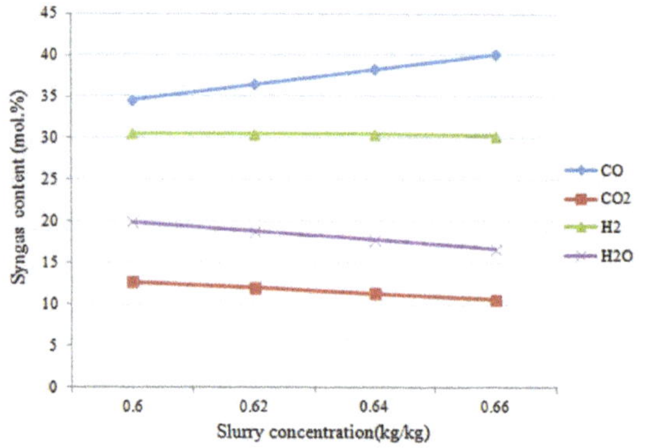


Figure 4. Syngas content change with coal slurry concentration unit (kg/kg).

where $p$, $v$ are the pressure and volume value of different states, and $n$ is the coefficient in the polytrophic process. The value of $n$ in this study is set as 0.997 to stabilize the air temperature, which is required for isothermal condition.

In the isobaric heat exchanger, the gas phase air temperature is reduced to the target, and the difference in enthalpy between the input and output states is the transferred heat.

The throttle valve module is based on the Joule–Thompson effect (Perry & Green, 1984), and liquefaction of the air which has been cooled enough consists of an isenthalpic expansion. The pressure loss over the valve is calculated by the Thermolib valve block. This isenthalpic process will cause full liquefaction of the air. The valve block is based on the following equations:

$$\dot{H}_{\text{out}} = \dot{H}_{\text{in}}, \tag{44}$$

$$\dot{m}_{\text{out}} = \dot{m}_{\text{in}}, \tag{45}$$

$$p_{\text{out}} = p_{\text{in}} - k(\text{pos})\dot{m}, \tag{46}$$

where $k(\text{pos})$ is the function of valve position. The value is calculated based on a well-validated lookup table in the Thermolib toolbox.

In this study, the liquefied air temperature is 70 K (Karmarkar, 2005). The liquid phase air is than fractionally distilled to generate pure nitrogen and oxygen. The media flow in the model is set as a vector carrying the state parameters, which include flow rate, temperature, pressure, enthalpy, entropy, contents concentration and gas phase fraction.

## 8. Results and discussion

The whole process model implemented in Simulink and Thermolib is shown in Figure 3. Three types of coal are applied to test the working process of the model. The model input variables and model parameters are listed in Table 1. The simulation results of the final steady state and their associated reference data (Watkinson et al., 1991) are given in Table 2. For error analysis, the mean absolute error between the simulation results and reference data is given in Table 2.

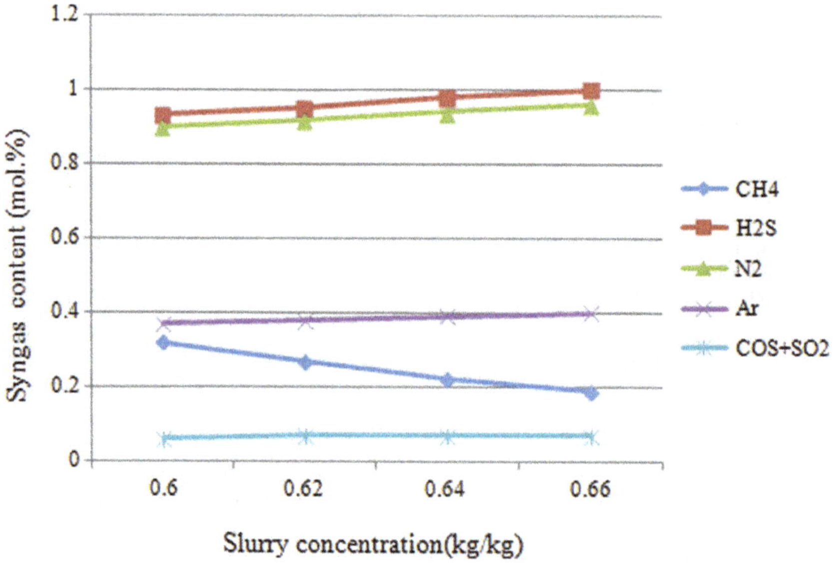


Figure 5. Syngas content change with coal slurry concentration unit (kg/kg).

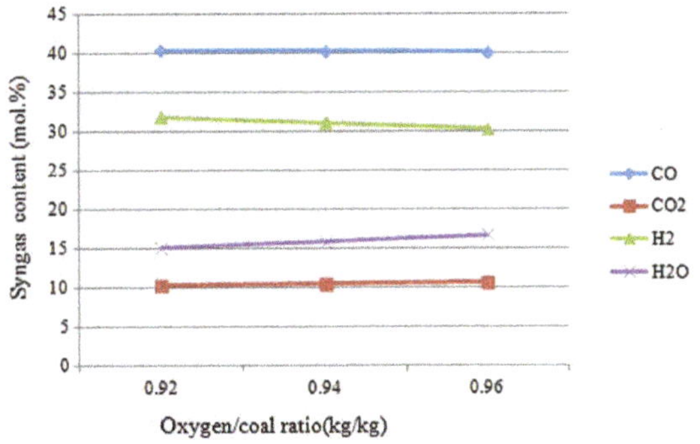


Figure 6. Syngas content change with oxygen/coal ratio unit (kg/kg).

Column R denotes the reference data of syngas contents concentration in mole percentage, the column S presents the simulation results of three coal types. The comparison in Table 2 shows that the simulation results match well with the reference data (Watkinson et al., 1991).

The gasifier simulation results are then compared with the reference data from the Lu-nan fertilizer factory. The model input data are given in Table 3. Comparison of the predicted output dry syngas content results and industry data is given in Table 4.

Table 4 shows that the simulation results can match well with the industry data. This means that the assumptions and the mathematic model used in the modelling work are reasonable.

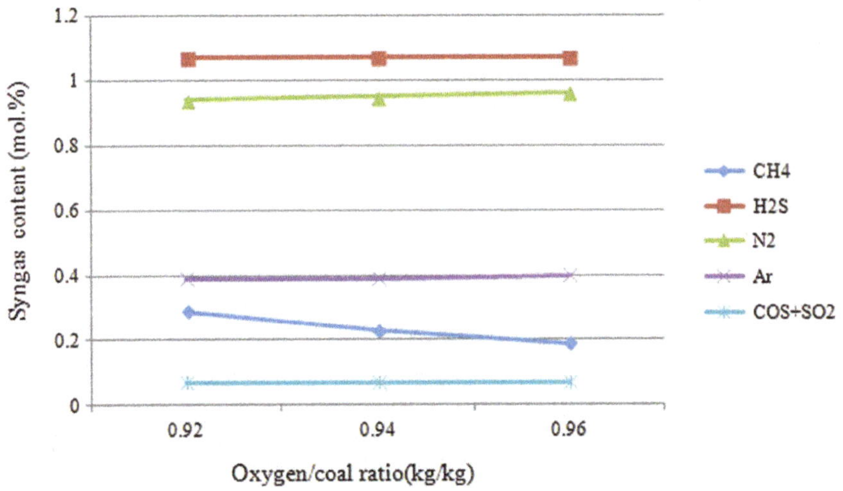


Figure 7. Syngas content change with oxygen/coal ratio unit (kg/kg).

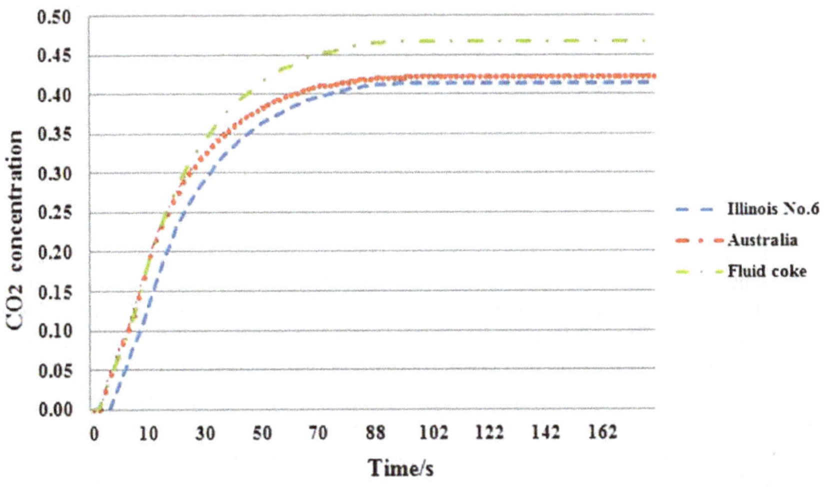


Figure 8. Shifted syngas $CO_2$ concentration dynamic change.

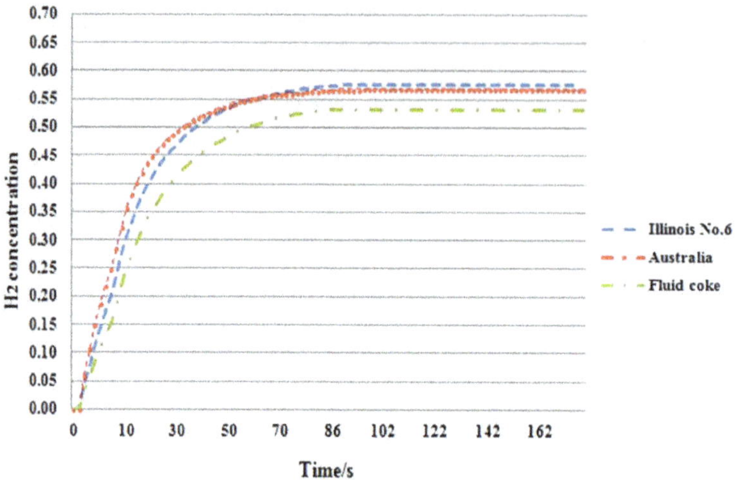


Figure 9. Shifted syngas $H_2$ concentration dynamic change.

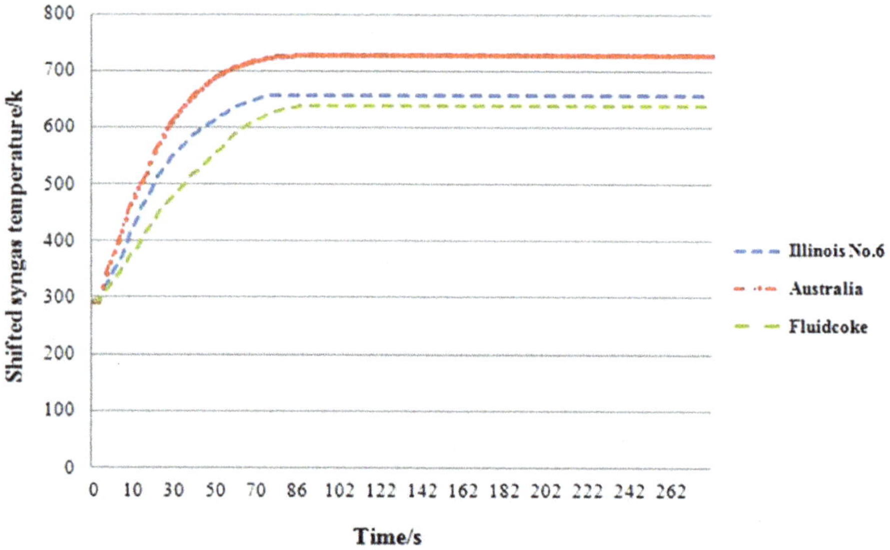


Figure 10. Shifted syngas temperature dynamic change.

To further test the model, the change in oxygen/coal ratio and slurry concentration's effects on the syngas content are studied. Figures 4 and 5 show the slurry concentration's effect on syngas contents. When the slurry concentration increases from 60% to 66%, the CO content increases; while $H_2$ remains stable, $CO_2$ and $CH_4$ decrease. When slurry concentration increases, more coal feed enters the gasifier; but when the oxygen/coal ratio remains stable, it means the oxygen input will rise as well, thus enhancing the gasification process and inducing content change. The results well match with Azuhata's experiment (Azuhata, Hedman, & Smoot, 1986).

Figures 6 and 7 show the syngas content change with oxygen/coal ratio. The rise in this ratio means increase in the oxygen supply, which will enhance the combustion and raise the gasifier temperature, thus enhancing the gasification process. But it will also consume more CO and $H_2$ released from the volatile, resulting in the decrease in CO and $H_2$ content and the increase in $CO_2$ content. The results match with Azuhata's experiment (Azuhata et al., 1986) and Vamvuka's simulation results (Vamvuka, Woodburn, & Senior, 1995a, 1995b).

Three different types of coal are applied to the simulation to test the process dynamic response. The coal feeding

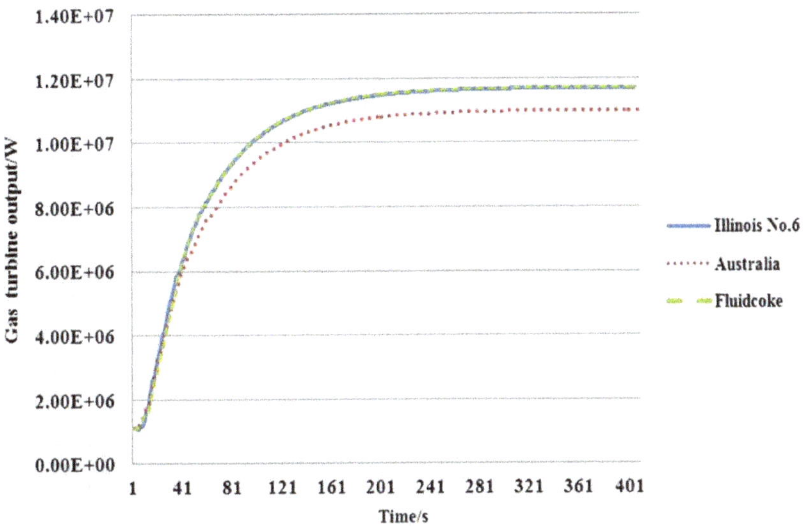


Figure 11. Gas turbine power output dynamic change.

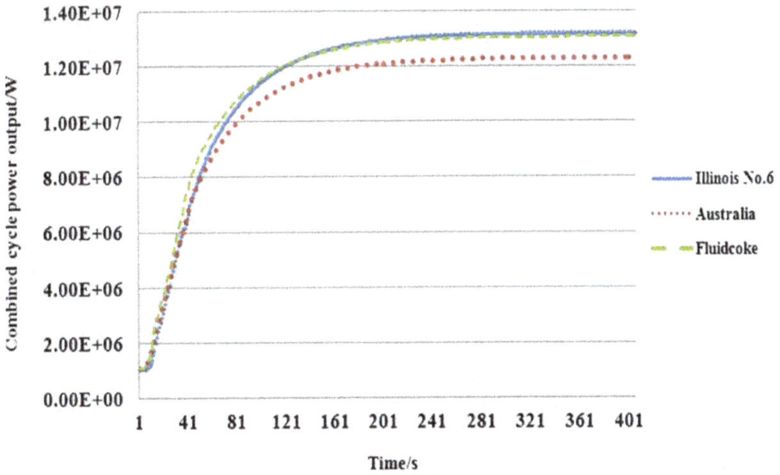


Figure 12. Combined cycle power output dynamic change.

is the input of the whole system. In this test, the coal flow rate will be inputted to the gasifier model and simulated to generate syngas output. The syngas flow rises from 0.1 to 100 mol/s within 100 s and it will first enter the WGS module for shift reaction.

The concentration of $CO_2$ and $H_2$ and the temperature of the shifted syngas are shown in Figures 8–10.

For the three types of coal, a similar trend is observed for the dynamic change in $CO_2$, $H_2$ concentration and syngas temperature. The simulation results prove that the WGS module can enhance the $H_2$ and $CO_2$ extraction; the carbon sequestration and hydrogen combustion process will thus benefit from it. Meanwhile, the shift reaction is exothermic, and the temperature of the shifted syngas rises in the WGS. In a real power plant, the heat can be used to raise HP steam to HRSG, but in this study, the heat recovery process is not considered.

The dynamic behaviours of the gas turbine power output are shown in Figure 11, while the net power output of the combined cycle is shown in Figure 12. The power consumptions of the compressors coupled with turbines are considered.

The power outputs of the gas turbine and combined cycle show similar trends for the three types of coal. The thermal efficiency of the gas turbine cycle is 30–33%. Integrated with HRSG, the combined cycle thermal efficiency is 51–55%. Hence, the HRSG is important in improving the power generation of the IGCC process.

For the ASU module, the liquefied air flow rate is set as 150 mol/s, which can satisfy the demand for the Texaco

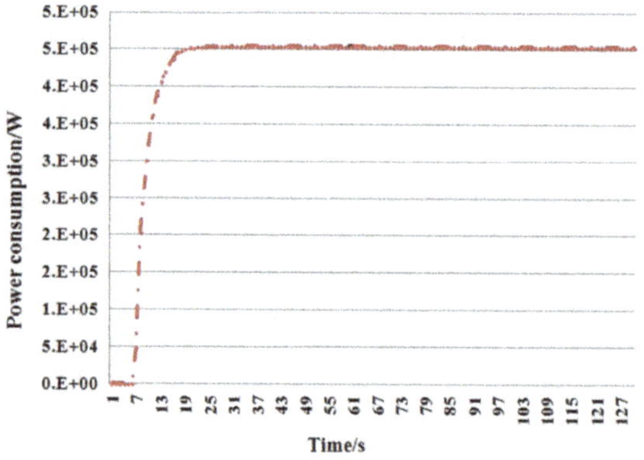


Figure 13. ASU compressor power consumption dynamic change.

gasifier use based on the model input oxygen/dry coal ratio. The compressor consumption dynamic change is shown in Figure 13. In this paper, the power consumption of air liquefaction process is mainly caused by compressor usage.

In this model, pure oxygen and wet air are used as the oxidant in the gasifier and the gas turbine, respectively. In the real industry, the ASU will not only provide oxygen to the gasifier but also integrate with the gas turbine; but in this paper, since $NO_x$ is not considered as the combustion product, integration of the ASU and the gas turbine is not included.

## 9. Conclusion

In this paper, a mathematical model for an IGCC process using Taxaco gasifier is developed which can predict the syngas contents. The model is derived by applying the engineering principle, the chemical equilibrium, mass balance and energy balance. The simulation results of dry syngas match well with the industry data. The influence of coal slurry concentration and oxygen/coal ratio on the syngas contents is studied as CO, $CO_2$ and $H_2$ contents can reveal the syngas quality. To enhance the gasification process and generate raw syngas with high quality, higher slurry concentration and oxygen supply is shown to be one of the suitable solutions. The dynamic behaviour of WGS shows its importance in improving syngas quality potential to heat steam. To improve the syngas quality and prepare for further carbon capture, WGS plays a key role. HRSG improves the thermal efficiency of the combined cycle and the overall power-generation process. ASU can lead to efficiency loss of the whole power plant since the liquefaction and transportation processes will both consume a large amount of energy.

## Disclosure statement

No potential conflict of interest was reported by the authors.

## Funding

This work was supported by Engineering and Physical Sciences Research Council (EPSRC, EP/I010955/1). The validation data used in the gasification model are provided by Tsinghua University. The author Yue Wang is partially supported by Chinese Scholarship Council (CSC).

## References

Azuhata, S., Hedman, P. O., & Smoot, L. D. (1986). Carbon conversion in an atmospheric-pressure entrained coal gasifier. *Fuel*, *65*, 212–217.

Beér, J. M. (2000). Combustion technology developments in power generation in response to environmental challenges. *Progress in Energy and Combustion Science*, *26*, 301–327.

Brown, B. W., Smoot, L. D., & Hedman, P. O. (1986). Effect of coal type on entrained gasification. *Fuel*, *65*, 673–678.

Buskies, U. (1996). The efficiency of coal-fired combined-cycle powerplants. *Applied Thermal Engineering*, *16*, 959–974.

Casella, F., & Colonna, P. (2012). Dynamic modeling of IGCC power plants. *Applied Thermal Engineering*, *35*, 91–111.

Chen, X., He, M. Y., Spitsberg, I., Fleck, N. A., Hutchinson, J. W., & Evans, A. G. (2004). Mechanisms governing the high temperature erosion of thermal barrier coatings. *Wear*, *256*, 735–746.

Colonna, P., & van Putten, H. (2007). Dynamic modeling of steam power cycles. Part I – modeling paradigm and validation. *Applied Thermal Engineering*, *27*, 467–480.

Govind, R., Shah, J. (1984). Modeling and simulation of an entrained flow coal gasifier. *AIChE*, *30*, 79–91.

Incropera, F. P., & DeWitt, D. P. (1985). *Fundamentals of heat and mass transfer*. New York, NY: Wiley.

Karmarkar, M. (2005). *Jacob Consultancy UK, University of Nottingham, E.ON UK, Power Asset Modelling, Watergrid, Mitsui Babcock and EPRI*. Impact of $CO_2$ removal on coal gasificaiton based fuel plants final Report, 57–58.

Lichty, L. C. (1967). *Combustion engine processes*. New York, NY: McGraw-Hill.

Lu, S., & Hogg, B. W. (2000). Dynamic nonlinear modelling of power plant by physical principles and neural networks. *International Journal of Electrical Power & Energy Systems*, *22*, 67–78.

Maurstad, O. (2005). *An overview of coal based integrated gasification combined cycle (IGCC) technology*. Cambridge, MA: Massachusetts Institute of Technology, 8 Publication no. LFEE 2005–002 WP.

Ni, Q., & Williams, A. (1995). A simulation study on the performance of an entrained-flow coal gasifier. *Fuel*, *74*, 102–110.

Peng, D. Y., & Robinson, D. B. (1976). A new two-constant equation of state. *Industrial & Engineering Chemistry Fundamentals*, *15*, 59–64, 1976/02/01.

Perry, R. H.Green, D. W., et al. (1984). *Perry's chemical engineers' handbook*. New York, NY: McGraw-Hill.

van Putten, H., & Colonna, P. (2007). Dynamic modeling of steam power cycles: Part II – simulation of a small simple Rankine cycle system. *Applied Thermal Engineering*, *27*, 2566–2582.

Smoot, L. D., & Smith, P. J. (1979). *Pulverized-coal combustion and gasification*. New York, NY: Plenum Press. Ch.13.

Sun, B., Liu, Y., Chen, X., Zhou, Q., & Su, M. (2011). Dynamic modeling and simulation of shell gasifier in IGCC. *Fuel Processing Technology*, *92*, 1418–1425.

Timmerhaus, K. D., & R. P. Reed. (2007). *Cryogenic engineering: Fifty years of progress*. New York, NY: Springer.

Ubhayakar, S. K., Stickler, D. B., & Gannon, R. E. (1977). Modelling of entrained-bed pulverized coal gasifiers. *Fuel, 56*, 281–291.

Vamvuka, D., Woodburn, E. T., & Senior, P. R. (1995a). Modelling of an entrained flow coal gasifier. 1. Development of the model and general predictions. *Fuel, 74*, 1452–1460.

Vamvuka, D., Woodburn, E. T., & Senior, P. R. (1995b). Modelling of an entrained flow coal gasifier. 1. Effect of operating conditions on reactor performance. *Fuel, 74*, 1461–1465.

Wang, Y., Wang, J., Guo, S., Lv, J., & Gao, Q. (13–14 September, 2013). *Dynamic modelling and simulation study of Texaco gasifier in an IGCC process*. 19th International Conference on Automation and Computing, London, UK. pp. 1, 6.

Watkinson, A. P., Lucas, J. P., & Lim, C. J. (1991). A prediction of performance of commercial coal gasifiers. *Fuel, 70*, 519–527.

Wen, C. Y., & Chaung, T. Z. (1979). Entrainment coal gasification modeling. *Industrial Engineering, Chemistry, Process Design and Development, 18*, 684–694.

Wong, K. (2012). *Thermodynamics for engineers*. Boca Raton, FL: CRC Press.

Yang, Z., Wang, Z., Wu, Y., Wang, J., Lu, J., Li, Z., & Ni, W. (2011). Dynamic model for an oxygen-staged slagging entrained flow gasifier. *Energy and Fuels, 25*, 3646–3656.

# 16

# Legendre spectral collocation method for approximating the solution of shortest path problems

Emran Tohidi* and Omid Reza Navid Samadi

*Department of Mathematics, Islamic Azad University, Zahedan Branch, Zahedan, Iran*

In the current article, we propose an accurate spectral approximation for solving the shortest path problems with boundary and interior barriers. For this goal, the shortest path problems are modelled as variational problems (VPs). Then, the Legendre polynomials are used as a basis for approximating the solution of these problems, and by using the Chebyshev–Gauss–Lobatto collocation points together with the Legendre–Gauss quadrature rule, the VPs will be changed into nonlinear programming problems (NLPs). The resulting NLPs are solved by the *NLPSolve* command in MAPLE software. Three numerical examples are provided for showing the robustness of the proposed method.

**Keywords:** shortest path problems; modelling; Legendre polynomials; Gauss–Lobatto collocation points; Gauss quadrature rules; nonlinear programming problem

## 1. Introduction

The shortest path problems (SPPs) have received considerable attention in engineering especially in robot industrial and recently in surgery planning. For instance, Latombe (1991) has gathered novel methods for path planning in the presence of obstacles and some extensions of them. Moreover, the problem of optimal path planning in Wang, Lane, and Falconer (2000) is considered as a semi-infinite constrained optimization problem. An important application of SPPs is in the Military industry for rocket range optimization. The task of planning trajectories for a rocket is one of the most applicable research subjects in the military research literature. Most of research works assume that the rocket has just boundary barriers in the model of its environment (see, for instance, Zamirian, Farahi, & Nazemi, 2007). However, less attention has been paid to the problem of interior together with boundary barriers in a known environment.

The optimal algorithms should be employed to find the lowest cost path from the robot (or rocket) start state to the goal state. Cost can be defined to be distance travelled, energy expended, time exposed to danger, etc. However, in this paper cost is defined to be distance travelled. Existing approaches such as Borzabadi, Kamyad, Farahi, and Mehne (2005) and Zamirian et al. (2007) planed traditional and classical methods such as measure theoretical approaches. These techniques usually need to transform the basic problem to an optimal control problem (OCP). These schemes have two disadvantages such as an increase in the dimension of associated algebraic problem and also the weakness of the approximate solution. For instance, in Borzabadi et al. (2005) the authors solve a collection of SPPs using a method based on measure theory. They considered the SPPs as optimization problems and then they convert these problems into OCPs by defining some artificial control functions. Using properties of some kind of measures, they obtained a linear programming problem (LPP) that their solutions give rise to constructing approximate optimal trajectory of the original problem. Defining the artificial control functions usually gives rise to the increase in the associated algebraic problem dimension. Moreover, the methods by Borzabadi et al. (2005) and Zamirian et al. (2007) deal with the local approximations (such as finite-difference methods), meanwhile global approximations have more accuracy with respect to the local approximations.

Because of the aforementioned reasons, we propose a global approximate method in which the dimension of the algebraic problem is very low. These are our motivations for presenting our basic idea. Therefore, in this paper by using Legendre polynomials as a basis for approximating the solutions we convert the SPPs into the associated nonlinear programming problems (NLPs). In other words, the infinite-dimensional SPPs will be transformed into the associated finite-dimensional NLPs in which the cost functionals are approximated by the Legendre–Gauss quadrature rule and the constraints of SPPs are collocated at the Chebyshev–Gauss–Lobatto (CGL) points. For clarity of

*Corresponding author. Email: emrantohidi@gmail.com

presentation, we assume the computational interval to be $[-1, 1]$. Other computational intervals can be transformed into $[-1, 1]$ by simple change of variables.

A general class of SPPs with boundary barriers $\psi_1(s)$ and $\psi_2(s)$ and interior barriers in the shape of circles with the centers $(\alpha_i, \beta_i)$, $i = 1, 2, \ldots, k$, can be modelled by the following variational problem (VP):

$$\begin{aligned} \text{Min}\quad & J = \int_{-1}^{1} \sqrt{1 + x'^2(s)}\, \mathrm{d}s \\ \text{s.t.}\quad & \psi_1(s) \le x(s) \le \psi_2(s), \quad \forall s \in [-1, 1], \qquad (1) \\ & (s-\alpha_i)^2 + (x(s)-\beta_i)^2 \ge D_i^2, \quad i = 1, 2, \ldots, k, \\ & x(-1) = \alpha, \quad x(1) = \beta, \end{aligned}$$

where the start state and the final state are $\alpha$ and $\beta$, respectively.

The rest of this paper is organized as follows. In the next section, some preliminaries about the spectral approximations especially Legendre polynomials will be provided. The basic idea of this paper, in which the VP (1) is changed into a NLP, is stated in Section 3. Three numerical examples are given in Section 4 for illustrating the efficiency and applicability of the proposed method. In Section 5, conclusions of the proposed method are provided. Finally, the MAPLE codes of the first example are given in the appendix.

## 2. Preliminaries

Spectral methods have proven to be powerful tools that are frequently employed in many fields of numerical analysis. They have a higher order of accuracy with respect to the finite-difference methods (FDMs) and finite element methods (FEMs) in the case of smooth solutions of any considered problem. For instance in Samadi and Tohidi (2012), spectral methods have shown their robustness with regard to the Bessel collocation scheme (Sahin, Yuzbasi, & Gulsu, 2011), homotopy perturbation method (Biazar & Ghazvini, 2009) and Block-by-Block technique (Katani & Shahmorad, 2010) for solving the system of Volterra integral equations. Also, in Bhrawy, Assas, Tohidi, and Alghamdi (2013) and Toutounian, Tohidi, and Kilicman (2013), Legendre and Fourier methods depict their spectral accuracy with regard to the reproducing kernel methods and other methods for solving Pantograph delay differential equations. In addition, spectral methods show their higher order accuracy in Doha and Bhrawy (2008, 2012), Doha, Bhrawy, and Abd-Elhameed (2009) and Doha, Bhrawy, and Ezz-Eldien (2011). Moreover, the efficiency of spectral methods has been proved for OCPs governed by Volterra integral equations in Tohidi and Samadi (2012). Among the spectral approximations, the Gegenbauer polynomials are used in many research works (Abd-Elhameed & Youssri, 2014). The simple type of the Gegenbauer polynomials is the Legendre polynomials which are orthogonal in the interval $[-1, 1]$ and satisfy the following recurrence relation (Tohidi & Samadi, 2012):

$$P_{i+1}(s) = \frac{2i+1}{i+1}\, sP_i(s) - \frac{i}{i+1} P_{i-1}(s), \quad i \ge 1,$$

where $P_0(s) = 1$ and $P_1(s) = s$.

The orthogonal property of Legendre polynomials is given by

$$\int_{-1}^{1} P_i(s)P_j(s)\,\mathrm{d}s = \begin{cases} 0, & i \ne j, \\ \dfrac{2}{2i+1}, & i = j. \end{cases}$$

A function $f(s)$, which is absolutely integrable within $-1 \le s \le 1$, may be expressed in terms of a Legendre series as

$$f(s) = \sum_{i=0}^{\infty} f_i P_i(s), \qquad (2)$$

where

$$f_i = \frac{2i+1}{2} \int_{-1}^{1} f(s)P_i(s)\,\mathrm{d}s.$$

PROPOSITION 1 *If we assume that the derivative of $f(s)$ in Equation (2) is described by*

$$f'(s) = \sum_{i=0}^{\infty} g_i P_i(s), \qquad (3)$$

*the relationship between the coefficients $f_i$ in Equation (2) and $g_i$ in Equation (3) can be obtained as follows*:

$$(2i+3)g_{i-1} - (2i-1)g_{i+1} - (2i-1)(2i+3)f_i = 0, \quad i = 1, 2, \ldots. \qquad (4)$$

*Proof* The proof has been provided in Canuto, Hussaini, Quarteroni, and Zang (1987). ∎

The Chebyshev–Gauss (CG), Chebyshev–Gauss–Rad au (CGR), and CGL collocation points lie on the open interval $s \in (-1, 1)$, the half open interval $s \in [-1, 1)$ or $s \in (-1, 1]$, and the closed interval $s \in [-1, 1]$, respectively. A depiction of these three sets of collocation points is shown in Figure 1 where it is seen that the CG points contain neither $-1$ nor 1, the CGR points contain only one of the points $-1$ or 1 (in this case, the point $-1$), and the CGL points contain both $-1$ and 1. In the procedure of collocating the constraints of VP (1), we use the CGL points for clarity of presentation and also for this reason that they cover the hole parts of the computational interval $[-1,1]$. CGL points can be demonstrated in the following form:

$$s_i = -\cos\left(\frac{i\pi}{N}\right), \quad i = 0, 1, \ldots, N. \qquad (5)$$

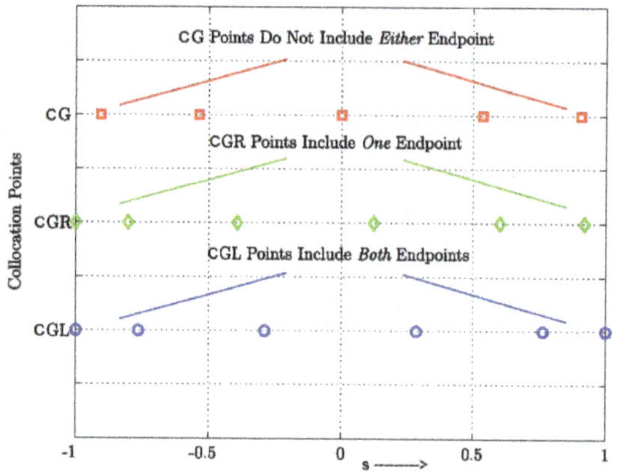


Figure 1. Schematic showing the differences between CGL, CGR, and CG collocation points.

The Legendre–Gauss quadrature rule can be defined as follows (Krylov, 1962):

$$\int_{-1}^{1} h(s)\,\mathrm{d}s \approx \sum_{i=0}^{N} w_i h(t_i), \tag{6}$$

where $t_i$ for $i = 0, 1, \dots, N$ are the distinct roots of the $(N+1)$th Legendre polynomial $P_{N+1}(s)$ and $w_i = 2/(1 - t_i^2)P'_{N+1}(t_i)$.

## 3. Basic idea

We again consider the VP (1). Our aim is to discretize Equation (1) and then optimize itself. For this purpose, a discretization of the interval $-1 = s_0 < s_1 < \cdots < s_N = 1$ is considered, where $s_i$'s are defined in Equation (5). The following expansions for approximating both $x(s)$ and $x'(s)$ are assumed:

$$\begin{aligned} x(s) &\approx x^N(s) = \sum_{i=0}^{N} a_i P_i(s), \\ x'(s) &\approx x'^N(s) = \sum_{i=0}^{N} c_i P_i(s), \end{aligned} \tag{7}$$

where $P_i(s)$ is the $i$th Legendre polynomial. The relationship between $a_i$'s and $c_i$'s is stated in Proposition 1. Now, we approximate the cost functional of VP (1) by the Legendre–Gauss quadrature rule (6). Therefore,

$$\int_{-1}^{1} \sqrt{1 + (x'(s))^2}\,\mathrm{d}s \approx \int_{-1}^{1} \sqrt{1 + \left(\sum_{i=0}^{N} c_i P_i(s)\right)^2}\,\mathrm{d}s$$

$$= \int_{-1}^{1} F(c_0, \dots, c_N, s)\,\mathrm{d}s \approx \sum_{i=0}^{N} w_i F(c_0, \dots, c_N, t_i), \tag{8}$$

where $F(c_0, \dots, c_N, s) := \sqrt{1 + (\sum_{i=0}^{N} c_i P_i(s))^2}$ is a nonlinear function in terms of its variables $c_0, c_1, \dots, c_N$ and $s$ and all of the $t_i$'s and $w_i$'s are defined in Equation (6). For the sake of clarity, we assume that

$$G(c_0, c_1, \dots, c_N) := \sum_{i=0}^{N} w_i F(c_0, \dots, c_N, t_i).$$

After approximating the cost functional, we use the CGL collocation points for discretizing the inequality state constraints, the VP (1) is changed into the following NLP:

$$\begin{aligned} \text{Min}\quad & G(c_0, c_1, \dots, c_N) \\ \text{s.t.}\quad & \psi_1(s_j) \le x(s_j) \le \psi_2(s_j), \quad j = 0, 1, \dots, N, \\ & (s_j - \alpha_i)^2 + (x(s_j) - \beta_i)^2 \ge D_i^2, \\ & \quad j = 0, 1, \dots, N, \ i = 1, 2, \dots, k, \\ & x^N(-1) = \sum_{i=0}^{N} a_i P_i(-1) = \alpha, \\ & x^N(1) = \sum_{i=0}^{N} a_i P_i(1) = \beta, \\ & [a_1, a_2, \dots, a_N]^{\mathrm{T}} = \mathbf{M}[c_0, c_1, \dots, c_N]^{\mathrm{T}}, \end{aligned} \tag{9}$$

where $G(c_0, c_1, \dots, c_N)$ is a nonlinear objective function, $\hat{P}_i(0) = \hat{P}_i(s_0) = (-1)^i$ and $\hat{P}_i(h) = \hat{P}_i(s_N) = 1$.

Note that the last constraints of Equation (9) arise from the following relations:

$$a_i = \left[\frac{c_{i-1}}{2i-1} - \frac{c_{i+1}}{2i+3}\right], \quad i = 1, 2, \dots, N,$$

where $c_{N+1} = c_N = 0$.

Before presenting numerical examples, it should be noted that the convergence analysis of the proposed scheme can be done using some similar ideas provided in Doha, Abd-Elhameed, and Youssri (2014) and Tohidi and Samadi (2012). Since this subject needs some theoretical sections, we just provide the mentioned references.

## 4. Numerical examples

In this section, we conduct two numerical examples to illustrate the effectiveness of the proposed method. We use the method stated in the previous section to transform the main problem (1) into the equivalent NLP (9). All the problems are programmed in MAPLE 13 and run on a Laptop PC with 1.8 GHz and 2 GB RAM. It should be noted that, for solving the NLPs in MAPLE (i.e. the *NLPSolve* command), there are some options in which the NLPs are solved. If the NLP is univariate and unconstrained except for finite bounds, the quadratic interpolation method may be used. If the problem is unconstrained and the gradient of the objective function is available, the preconditioned conjugate gradient (PCG) method may be used. Otherwise, the sequential quadratic programming (SQP) method

Table 1. Numerical results of Example 1 for objective functionals.

| $N$ | $J_N^*$ |
|---|---|
| 4 | 3.26019 |
| 8 | 3.26492 |
| 18 | 3.26920 |
| 32 | 3.26911 |

can be used. According to the structure of our NLP in Equation (9), the SQP method is used.

*Example 1* As the first example, we consider the following SPP with one lower boundary barrier (Hestenes, 1996):

$$\text{Min} \quad J = \int_{-5/4}^{5/4} \sqrt{1 + x'^2(\tau)}\, d\tau$$

$$\text{s.t.} \quad x(\tau) \geq 1 - \tau^2, \quad \forall \tau \in \left[-\frac{5}{4}, \frac{5}{4}\right],$$

$$x\left(-\frac{5}{4}\right) = 0, \quad x\left(\frac{5}{4}\right) = 0.$$

The optimal value of the objective functional is $J^* = 3.26911$ and the exact solution of this SPP is as follows:

$$x^*(\tau) = \begin{cases} \tau + \frac{5}{4}, & -\frac{5}{4} \leq \tau \leq -\frac{1}{2}, \\ 1 - \tau^2, & -\frac{1}{2} \leq \tau \leq \frac{1}{2}, \\ -\tau + \frac{5}{4}, & \frac{1}{2} \leq \tau \leq \frac{5}{4}. \end{cases}$$

Since the computational interval of this problem is $[-\frac{5}{4}, \frac{5}{4}]$, we should transform this interval into $[-1, 1]$ by changing variable $\tau = \frac{5}{4}s$. The readers can see this subject in the appendix using MAPLE codes. By assuming different values of $N$ such as 4, 8, 18 and 32, we solve this problem by the proposed idea. In Table 1, the approximated objective functionals $J_N^*$ are given for these values of $N$. It is seen from this table that we reach the exact objective functional in the case of $N = 32$. Also, a comparison between the numerical solution $x^N(s)$ $(N = 32)$ together with the exact solution $x^*(s)$ at a uniform mesh is given in Table 2. From this table, one can conclude that our method has high order of accuracy and fits with the exact solution. Moreover, the history of the numerical solution $x^N(s)$ $(N = 32)$ is depicted in Figure 2.

*Example 2* As the second example, we consider the following SPP with two lower and upper boundary barriers (Zamirian et al., 2007):

$$\text{Min} \quad J = \int_3^6 \sqrt{1 + x'^2(\tau)}\, d\tau$$

$$\text{s.t.} \quad \sin(\tau - 0.25) \leq x(\tau) \leq 1 + \sin(\tau),$$

$$x(3) = 0.75, \quad x(6) = 0.5.$$

Table 2. Comparison of the exact solution $x^*(s)$ and $x^N(s)$ at the selected points for $N = 32$.

| $s_i$ | $x^*(s_i)$ | $x^N(s_i)$ |
|---|---|---|
| −1 | 0.0 | 0.0 |
| −0.8 | 0.25 | 0.2503 |
| −0.6 | 0.50 | 0.5007 |
| −0.4 | 0.75 | 0.7505 |
| −0.2 | 0.9375 | 0.9375 |
| 0.0 | 1.0 | 1.0 |
| 0.2 | 0.9375 | 0.9375 |
| 0.4 | 0.75 | 0.7505 |
| 0.6 | 0.5 | 0.5007 |
| 0.8 | 0.25 | 0.2503 |
| 1.0 | 0.0 | 0.0 |

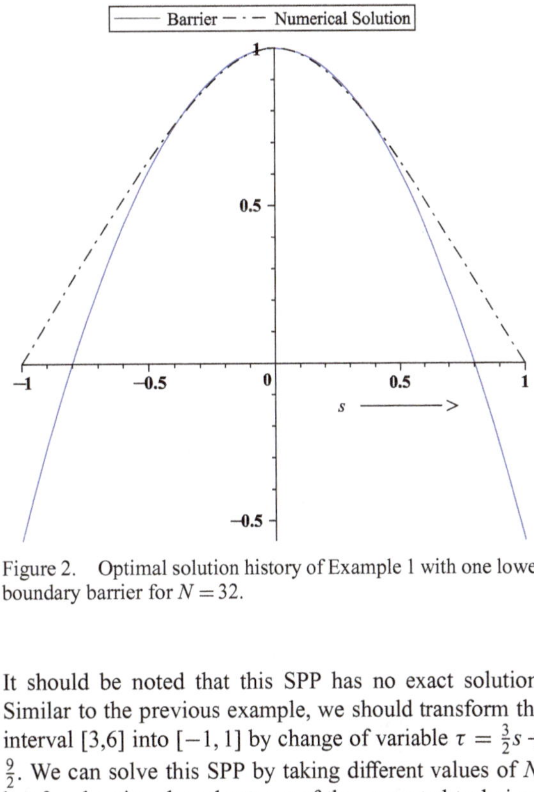

Figure 2. Optimal solution history of Example 1 with one lower boundary barrier for $N = 32$.

It should be noted that this SPP has no exact solution. Similar to the previous example, we should transform the interval [3,6] into $[-1, 1]$ by change of variable $\tau = \frac{3}{2}s + \frac{9}{2}$. We can solve this SPP by taking different values of $N$, but for showing the robustness of the presented technique with regard to the method of Zamirian et al. (2007), the considered problem has been solved by taking $N = 25$ and we reach $J_N^* = 3.2772$ $(N = 25)$, meanwhile the optimal $J^*$ of Zamirian et al. (2007) is 3.4191. This confirms that our idea reaches the shorter path with respect to the method of Zamirian et al. (2007). The history of this optimal path $x^N(s)$ $(N = 25)$ is depicted in Figure 3.

*Example 3* As our final numerical example, we consider the following SPP with two boundary barriers and three

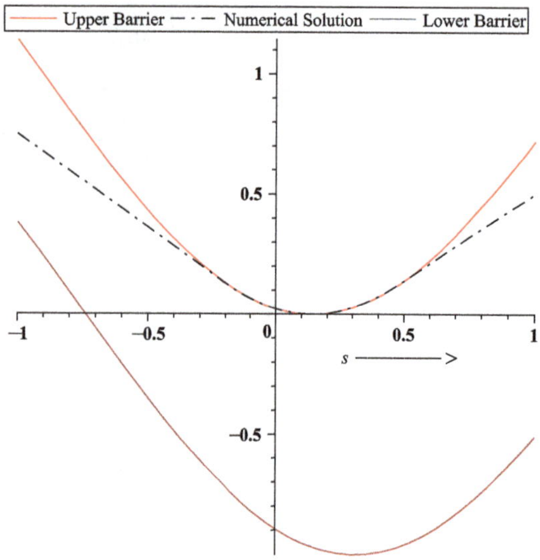


Figure 3. Optimal solution history of Example 2 with two boundary barriers for $N = 25$.

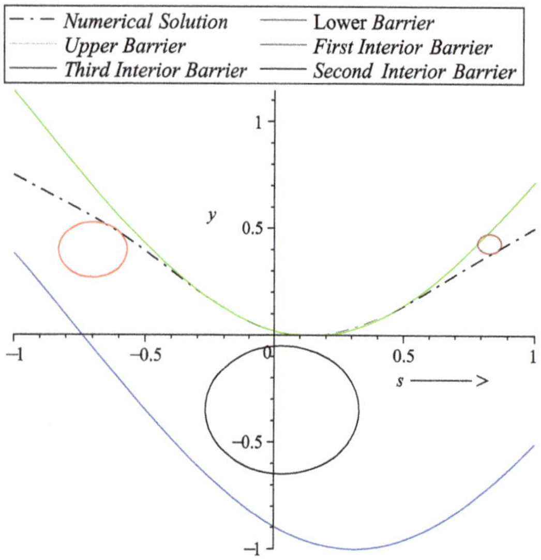


Figure 4. Optimal solution history of Example 3 with two boundary and three interior barriers for $N = 11$.

interior barriers:

$$\begin{aligned}
\text{Min} \quad & J = \int_3^6 \sqrt{1 + x'^2(\tau)}\, d\tau \\
\text{s.t.} \quad & \sin(\tau - 0.25) \le x(\tau) \le 1 + \sin(\tau), \\
& (\tau + 0.7)^2 + (x(\tau) - 0.4)^2 \ge (0.13)^2, \\
& (\tau - 0.83)^2 + (x(\tau) - 0.43)^2 \ge (0.045)^2, \\
& (\tau - 0.03)^2 + (x(\tau) + 0.35)^2 \ge (0.3)^2, \\
& x(3) = 0.75, \quad x(6) = 0.5.
\end{aligned}$$

Similar to the previous problem, this SPP has no exact solution. Moreover, a transformation of the interval [3,6] into $[-1, 1]$ by change of variable $\tau = \frac{3}{2}s + \frac{9}{2}$ should be done. After doing this, we solve this SPP using our basic technique and reach the optimal objective functional $J_N^* = 3.279366169$ ($N = 11$). The history of this optimal path $x^N(s)$ ($N = 11$) is depicted in Figure 4. From this figure, one can see that the numerical optimal path $x^N(s)$ ($N = 11$) does not meet the interior barriers. This confirms the efficiency and applicability of the presented idea.

## 5. Conclusion

The aim of this paper is to determine the optimal solution of SPPs with boundary and interior barriers by a direct method of solution based upon truncated Legendre series expansions together with the CGL points as collocation nodes. The method is based upon reducing a nonlinear SPP to an NLP. The unity of the weight function of orthogonality for the truncated Legendre series and the simplicity of the discretization are merits that make our idea very attractive. Moreover, only a small number of truncated Legendre series are needed to obtain a very satisfactory solution. The given numerical examples support this claim.

## Acknowledgments

The authors would like to thank the referees for the valuable comments and nice suggestions which led to the final version of the manuscript.

## Disclosure statement

The authors declare that they do not have any conflict of interest in their submitted manuscript.

## References

Abd-Elhameed, W. M., & Youssri, Y. H. (2014). New ultraspherical wavelets spectral solutions for fractional Riccati differential equations. *Abstract and Applied Analysis*, *2014*, 1–8.

Bhrawy, A. H., Assas, L. M., Tohidi, E., & Alghamdi, M. A. (2013). A Legendre–Gauss collocation method for neutral functional–differential equations with proportional delay. *Advances in Difference Equations*, *2013*. doi:10.1186/1687-1847-2013-63

Biazar, J., & Ghazvini, H. (2009). He's homotopy perturbation method for solving system of Volterra integral equations of the second kind. *Chaos Solitons Fractals*, *39*, 770–777.

Borzabadi, A. H., Kamyad, A. V., Farahi, M. H., & Mehne, H.H. (2005). Solving some optimal path planning problems using an approach based on measure theory. *Applied Mathematics and Computation*, *170*, 1418–1435.

Canuto, C., Hussaini, Y., Quarteroni, A., & Zang, T. A. (1987). *Spectral Methods in Fluid Dynamics (Scientific Computation)*. New York: Springer.

Doha, E. H., Abd-Elhameed, W. M., & Youssri, Y. H. (2014). New algorithms for solving third and fifth-order two point boundary value problems based on nonsymmetric generalized Jacobi Petrov–Galerkin method. *Journal of Advanced Research*, *2014*. doi:10.1016/j.jare.2014.03.003

Doha, E. H., & Bhrawy, A. H. (2008). Efficient spectral-Galerkin algorithms for direct solution of fourth-order differential equations using Jacobi polynomials. *Applied Numerical Mathematics*, *58*, 1224–1244.

Doha, E. H., & Bhrawy, A. H. (2012). An efficient direct solver for multidimensional elliptic Robin boundary value problems using a Legendre spectral-Galerkin method. *Computers & Mathematics with Applications*, *64*, 558–571.

Doha, E. H., Bhrawy, A. H., & Abd-Elhameed, W. M. (2009). Jacobi spectral Galerkin method for elliptic Neumann problems. *Numerical Algorithms*, *50*, 67–91.

Doha, E. H., Bhrawy, A. H., & Ezz-Eldien, S. S. (2011). Efficient Chebyshev spectral methods for solving multi-term fractional orders differential equations. *Applied Mathematical Modelling*, *35*, 5662–5672.

Hestenes, M. R. (1996). *Calculus of variational and optimal control theory*. New York: Wiley.

Katani, R., & Shahmorad, S. (2010). Block by block method for the systems of nonlinear Volterra integral equations. *Applied Mathematical Modelling*, *34*, 400–406.

Krylov, V. I. (1962). *Approximate calculation of integrals*. Mineola, New York: Dover.

Latombe, J. C. (1991). *Robot motion planning*. Boston: Kluwer.

Sahin, N., Yuzbasi, S., & Gulsu, M. (2011). A collocation approach for solving systems of linear Volterra integral equations with variable coefficients. *Computers & Mathematics with Applications*, *62*, 755–769.

Samadi, O. R. N., & Tohidi, E. (2012). The spectral method for solving systems of Volterra integral equations. *Journal of Applied Mathematics and Computing*, *40*, 477–497.

Tohidi, E., & Samadi, O. R. N. (2012). Optimal control of nonlinear Volterra integral equations via Legendre polynomials. *IMA Journal of Mathematical Control and Information*, *30*, 67–83.

Toutounian, F., Tohidi, E., & Kilicman, A. (2013). Fourier operational matrices of differentiation and transmission: Introduction and applications. *Abstract and Applied Analysis*, *2013*, 1–11.

Wang, Y., Lane, D. M., & Falconer, G. J. (2000). Two novel approaches for unmanned underwater vehicle path planning: Constrained optimisation and semi-infinite constrained optimisation. *Robotica*, *18*, 123–142.

Zamirian, M., Farahi, M. H., & Nazemi, A. R. (2007). An applicable method for solving the shortest path problems. *Applied Mathematics and Computation*, *190*, 1479–1486.

**Appendix. MAPLE codes of the first example**

```
restart;
with(orthopoly) :with(Optimization) :with(Student[Calculus1]) :
r := 33; a := -5/4; b := 5/4;   #r = N + 1
f1 := (t)-> (1 - t^2) :
Digits := 16 :
t := (1/2) * (b - a) * s + (b + a) * (1/2) :
F1 := f1(t) :
F1 := unapply(F1, s) :
x0 := 0; x1 := 0 :
c[r] := 0; c[r + 1] := 0
for j from 1 to r - 1 do
A[j + 1] := c[j]/(2 * j - 1) - c[j + 2]/(2 * j + 3)
end do :
for j from 1 to r do
s[j] := -cos((j - 1) * Pi/(r - 1))
end do :
X := sum(A[i] * P(i - 1, s), i = 1..r);
X := unapply(X, s);
X1 := sum(c[i] * P(i - 1, s), i = 1..r);
S1 := {seq(X(s[i]) >= F1(s[i]), i = 1..r)};
S2 := {X(-1) - x0 = 0};
S3 := {X(1) - x1 = 0};
S := S1 union S2 union S3 :
x := fsolve(P(r, s));
PP := diff(P(r, s), s);
PP := unapply(PP, s);
for j from 1 to r do
w[j] := 2/((1 - x[j]^2) * PP(x[j])^2)
end do :
g := ((b - a) * (1/2)) * (1 + (2 * X1/(b - a))^2)^(1/2);
g := unapply(g, s);
z := sum(w[i] * g(x[i]), i = 1..r);
sol := NLPSolve(z, S) :
assign(sol[2]);
X(s) :
plot(1 - (25/16) * s^2, X(s), s = -1..1, color = [red, black], style = [line, line], font = [5, BOLD]);
```

describing the remainder of the structure (or *numerical substructure*). Although the two coupled passive subsystems resulting from this process are stable (Anderson & Vongpanitlerd, 2006; Desoer & Vidyasagar, 1975; Lozano et al., 2000; Ortega, Praly, & Landau, 1985; Ortega et al., 2001), this stability can be destroyed by the digital implementation of the numerical substructure and the corresponding actuator and sensor dynamics that couple the substructures (Gawthrop, Wallace, Neild, & Wagg, 2007). Fortunately, this problem of coupling the numerical and experimental substructures using real-time digital implementation has been solved and a suite of techniques for robust *numerical-experimental substructuring* is now available (Blakeborough, Williams, Darby, & Williams, 2001; Gawthrop, Wallace, & Wagg, 2005; Wagg, Neild, & Gawthrop, 2008). Furthermore, the substructuring approach is particularly suitable for the type of resonant systems that are the focus of this paper (Gawthrop et al., 2007).

As discussed by Gawthrop et al. (2005), the bond-graph approach (Borutzky, 2011; Gawthrop & Bevan, 2007; Gawthrop & Smith, 1996; Karnopp, Margolis, & Rosenberg, 2012; Mukherjee, Karmaker, & Samantaray, 2006) gives a natural and convenient formulation of substructuring and control (Gawthrop, 2004; Gawthrop et al., 2005; Vink, Ballance, & Gawthrop, 2006) and the concept of actuator/sensor collocation has a clear bond graph interpretation. For these reasons, the bond-graph approach is adopted in this paper.

As discussed by Den Hartog (1985), choosing the structure of a vibration absorber for a single degree of freedom system is straightforward. However, choosing the structure for multi-degree of freedom systems such as those arising from modal decomposition is considerably more involved (Moheimani & Behrens, 2004). This complexity motivates the novel approach of this paper based on the concept of of a *dynamically dual*[1] system.

As discussed in more detail in Section 3, a dynamically dual mechanical system is obtained by interchanging the rôles of velocity and force. This concept of duality has been used for analysis of dynamical systems (Cellier, 1991; Karnopp, 1966; Samanta & Mukherjee, 1985, 1990; Shearer, Murphy, & Richardson, 1971), and this paper uses the concept to design *dynamically dual vibration absorbers* (DDVA).

Although the DDVA method originated an extension of the physically based design of Den Hartog (1985), it will be shown that the method also includes the well-established *acceleration feedback* approach (Preumont, 2002).

In summary, placing both the traditional Den Hartog mechanical vibration absorber and acceleration feedback into the wider context of the DDVA of this paper has two advantages: the method immediately extends to multi degree of freedom systems and the implementation and theoretical results (including robustness to sensor/actuator dynamics) from substructuring can be directly applied.

Section 2 reviews the substructuring approach to provide a framework for the paper. Section 3 gives the foundations of the DDVA approach; Section 3.2 focuses on the Den Hartog (1985) absorber version and Section 3.3 focuses on the acceleration feedback controller (Preumont, 2002). Section 4 discusses a number of multi-mode examples. Section 5 gives illustrative experimental results; Sections 5.1 and 5.2 experimentally verify the approach when applied to a rigid beam with an flexible joint and a flexible cantilever beam, respectively. Section 6 concludes the paper.

## 2. The substructuring formulation

Substructuring is a novel dynamic testing technique that allows the experimental testing of a component within the context of a larger system. This is achieved through the coupling of the physical component with a controller that numerically simulates the dynamics of the remainder of the system. Note that as the controller dynamics are designed to simulate part of a real system, the dynamics are physically realisable.

Figure 1 summarises the basic substructuring formulation Gawthrop et al. (2005) and Gawthrop, Wagg, & Neild (2009). For simplicity, Figure 1 will be assumed to represent a system with scalar quantities, although this can readily be extended to vectors. The three key parts are shown in Figure 1:

(1) **Phy** representing the physical component, with transfer function $p(s)$, to be controlled,
(2) **Num** representing the controller, with physically realisable dynamics, which is implemented numerically and has a transfer function $n(s)$,
(3) **Se**:$\mathbf{F_0}$ representing a disturbing external force, $F_0$,

where $s$ is the Laplace domain independent variable. Firstly the velocity feedback case is shown, Figure 1(a) as a bond graph[2] and in Figure 1(b) as a block diagram. Here, the physical component **Phy** has a force input, $F_0 - F$, and a measured velocity output, $v$. In addition, the parameters in the physical system are represented by a vector, $\theta_p$, Similarly, $\theta_n$, represents the vector of parameters in the numerical system.

An advantage of the bond graph representation is that it emphasises the fact that the physical system **Phy** and the controller **Num** are connected by *power bonds* and thus the control system is collocated – meaning that the actuator and sensor are located at the same point. In Figure 1(a) the parts are joined by a common flow (velocity) junction denoted as **1**. The bond graph also indicates causality and **Phy** and **Num** are represented by the positive real transfer functions $p(s, \theta_p)$ and $n(s, \theta_n)$, respectively. The transfer

# Dynamically dual vibration absorbers: a bond graph approach to vibration control

Peter Gawthrop[a]*, S.A. Neild[b] and D.J. Wagg[c]

[a]*Systems Biology Laboratory, Melbourne School of Engineering, University of Melbourne, Victoria 3010, Australia;* [b]*Department of Mechanical Engineering, Queens Building, University of Bristol, Bristol BS8 1TR., UK;* [c]*Department of Mechanical Engineering, Sir Frederick Mappin Building, University of Sheffield, Mappin Street Sheffield S1 3JD, UK*

This paper investigates the use of an actuator and sensor pair coupled via a control system to damp out oscillations in resonant mechanical systems. Specifically the designs emulate passive control strategies, resulting in controller dynamics that resemble a physical system. Here, the use of the novel dynamically dual approach is proposed to design the vibration absorbers to be implemented as the controller dynamics; this gives rise to the dynamically dual vibration absorber (DDVA). It is shown that the method is a natural generalisation of the classical single-degree of freedom mass–spring–damper vibration absorber and also of the popular acceleration feedback controller. This generalisation is applicable to the vibration control of arbitrarily complex resonant dynamical systems. It is further shown that the DDVA approach is analogous to the hybrid numerical-experimental testing technique known as substructuring. This analogy enables methods and results, such as robustness to sensor/actuator dynamics, to be applied to dynamically dual vibration absorbers. Illustrative experiments using both a hinged rigid beam and a flexible cantilever beam are presented.

**Keywords:** vibration absorber; bond graph; acceleration feedback control; dynamic dual

## 1. Introduction

The use of a secondary resonant mechanical systems to damp out oscillations in a resonant mechanical system by absorbing and dissipating energy has a long history and early work is summarised in the classical textbook by Den Hartog (1985). An alternative method for damping unwanted oscillations is to use some form of active vibration control. To achieve this some type of actuator and sensor system needs to be used. For example, vibrations can be damped from a mechanical system using a piezo-electric transducer and an associated electrical circuit (Hagood & von Flotow, 1991). This can have considerable advantages, although, as discussed by Moheimani & Behrens (2004) multi-modal resonant structures require sophisticated circuit synthesis.

The adjective 'passive' applied to 'system' has two different but related meanings: a physical system not containing a power source and a mathematical expression imposing the corresponding property on the input and output variables of a set of equations (Hogan, 1985; Sharon, Hogan, & Hardt, 1991; Slotine & Li, 1991). In general, this means that passive mechanical (or electrical) vibration absorbers can be replaced by a computer and associated sensor-actuator pairs which emulate the physical passivity in the equivalent mathematical sense. The algorithm implemented in the computer does not have to represent a physical system and can be designed using conventional control-theoretic methods (Balas, 1978; Fleming & Moheimani, 2005; Hong & Bernstein, 1998; Hogsberg & Krenk, 2006; Moheimani & Fleming, 2006), optimisation (Krenk & Hogsberg, 2009) or via system inversion (Ali & Padhi, 2009).

However, it can be argued that there are advantages in implementing *physical* systems within the digital computer (Gawthrop, 1995; Gawthrop, Bhikkaji, & Moheimani, 2010; Hogan, 1985; Lozano, Brogliato, Egelund, & Maschke, 2000; Ortega, Loria, Nicklasson, & Sira-Ramirez, 1998; Ortega, van der Schaft, Mareels, & Maschke, 2001; Sharon et al., 1991; Slotine & Li, 1991); this is the approach explored in this paper. In particular, the well-known relationship between dissipativity, passivity and physical systems (Lozano et al., 2000; Ortega et al., 1998, 2001; Willems, 1972) is exploited. Such energy based concepts rely on the properties of physical connections. In particular, the concept of *collocation* is a key system property in the context of active vibration control (Gawthrop et al., 2010; Preumont, 2002).

Replacing a mechanical vibration absorber by a digital computer is analogous to the well-known hybrid numerical-experimental testing technique where the structure under consideration is split into an experimental test piece (or *physical substructure*) and a numerical model

*Corresponding author. Email: peter.gawthrop@unimelb.edu.au

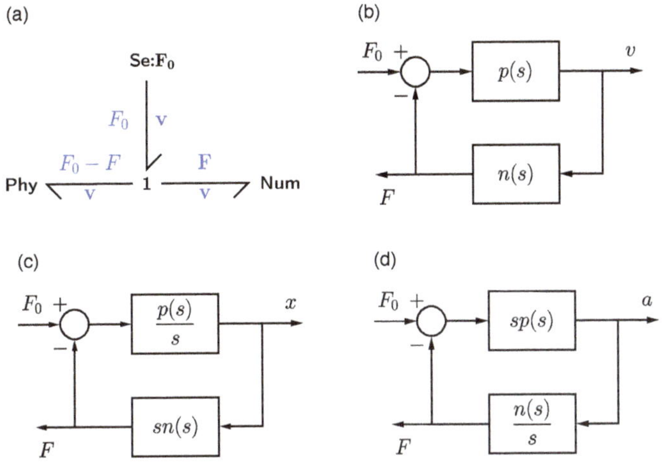


Figure 1. The substructuring formulation. The mathematically equivalent formulations (b)–(d) allow for a choice of sensors. (a) Bond graph, (b) block-diagram: velocity formulation, (c) block-diagram: displacement formulation and (d) block-diagram: acceleration formulation.

functions are related by the following relationships:

$$v = p(s, \theta_p)(F_0 - F), \tag{1}$$

$$F = n(s, \theta_n)v. \tag{2}$$

Although it is natural to work in terms of velocity $v$ rather than displacement $x$, Figure 1(b) can be easily rewritten in terms of displacement as Figure 1(c) where $dx/dt = v$ or in terms of acceleration as Figure 1(d) where $a = dv/dt$. The choice of formulation (displacement, velocity or acceleration) does not change the theoretical closed-loop stability properties defined by the loop-gain $L(s) = n(s, \theta_n)p(s, \theta_p)$, but allows flexibility in the choice of sensor and actuator. As well as providing a conceptual basis for this paper, the substructuring approach links to classical control system concepts useful for stability and robustness analysis. Details can be found elsewhere (Gawthrop et al., 2007, 2009; Wagg et al., 2008).

The substructuring formulation of Figure 1(a) assumes an inertia-like physical component driven by a force; as discussed by Gawthrop et al. (2005), compliance-like physical components can be treated by the formulation of Figure 2(a) where the external force $F_0$ is replaced by an external velocity $v_0$ and the three components are now connected by a common force, or **0**, junction. To distinguish this velocity-driven formulation from the force-driven formulation in Figure 1(a) an over-bar is used:

$$F = \bar{p}(s, \theta_p)(v_0 - v), \tag{3}$$

$$v = \bar{n}(s, \theta_n)F. \tag{4}$$

Using the definitions of Equations (3) and (4), the block diagram equivalent of Figure 2(a) is Figure 2(b). Once again, displacement and acceleration formulations are given by Figure 2(c) and 2(d), respectively.

## 3. Dynamically dual design

As already discussed, in this paper a vibration absorber attached to a system is considered. This vibration absorber, while based on a physical component thus ensuring that the system is passive, is implemented as a controller. This setup can be considered within the substructuring framework, with the vibration absorber forming **Num** and the system which the vibration absorber is attached being **Phy**. One possible absorber is the Den Hartog resonant vibration absorber, which is usually represented by a conventional mass–spring–damper schematic. However, as pointed out by Den Hartog (1985), and discussed in greater depth by Shearer et al. (1971), can equally well be described by an equivalent electrical circuit analogue.

Here, the use of DDVAs is proposed as a method for generating suitable **Num** dynamics. As discussed in Section 3.2, the resonant vibration absorber of Den Hartog (1985) is an example of a DDVA and provides the motivation for this approach. In formulating the DDVA approach the following features of the Den Hartog resonant vibration absorber are abstracted and generalised:

(1) it is a one-port[3] passive[4] physical system,
(2) it is causally compatible with the system, in that, the output velocity of the system provides the input to the absorber and the force output of the absorber provides the input to the system,
(3) there is a variable coupling parameter,
(4) the absorber has the *same* resonant frequency as the system, and
(5) the damping ratio of the absorber is *greater* than that of system.

The DDVA design approach is to set **Num** to be a dynamic-dual of the key mode or modes of the system that

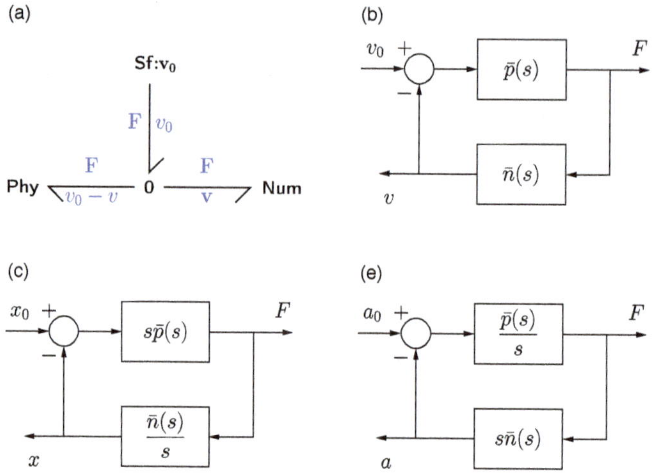


Figure 2. The velocity-driven substructuring formulation. (a) Bond graph, (b) block-diagram: velocity formulation, (c) block-diagram: displacement formulation and (d) block-diagram: acceleration formulation.

the absorber is attached to (which is contained in **Phy**). The method of obtaining a dynamic dual is now discussed. This is followed by discussions of two common absorber strategies which are DDVA; the Den Hartog absorber and the acceleration feedback method proposed by Preumont (2002).

### 3.1. *A dynamic-dual*

A dynamic-dual of a system is obtained by interchanging the rôles of velocity and force, is defined in Shearer et al. (1971). An extended version of this concept, the *scaled dual* is used here and, in the context of mechanical systems is defined as follows:

(1) Each force $F_i$, and each velocity $v_i$, in the original system has a scaled dual $v_i^D$, and $F_i^D$ in the dual system given by:

$$v_i^D = \frac{1}{g}F_i, \tag{5}$$

$$F_i^D = g v_i, \tag{6}$$

where $g$ is the scaling factor and $g = 1$ corresponds to the unscaled dual.

(2) Each mass component with mass $m_i$ is replaced in the dual system with a spring component of stiffness $K_i$, each spring component with stiffness $k_i$ is replaced in the dual system with a mass component of mass $M_i$, and each damper component with damping coefficient $r_i$ is replaced in the dual system with a damper component with damping coefficient $R_i$ where:

$$K_i = \frac{g^2}{m_i}, \tag{7}$$

$$M_i = \frac{g^2}{k_i}, \tag{8}$$

$$R_i = \frac{g^2}{r_i}. \tag{9}$$

(3) Common force connections become common velocity connections and common velocity connections become common force connections in the dual system.

(4) If the system transfer function $h(s)$ has force $F$ as input and velocity $v$ as output, then the dual transfer function $H(s)$ has velocity $v^D$ as input and force $F^D$ as output and is given by:

$$F^D = gv, \tag{10}$$

$$v^D = \frac{1}{g}F, \tag{11}$$

$$H(s) = \frac{F^D}{v^D} = \frac{gv}{(1/g)F} = g^2 h(s). \tag{12}$$

Equations (10) and (11) are *power conserving* in the sense that

$$F^D v^D = Fv. \tag{13}$$

As these Equations (10) and (11) also interchange the roles of force and velocity, they correspond to the bond graph *gyrator* (**GY**) component of Figure 3.

Equations (7) and (8) ensure that the scaled dual retains the same natural frequencies as the system; in the sequel, the value $R_i$ given using Equation (9) is not used, instead it

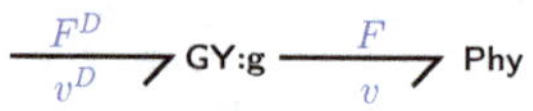


Figure 3. Gyrator interpretation of dynamic-dual.

is replaced by the user-selected value $R'_i$. This allows the damping of the modified scaled dual system, which forms the controller implemented in **Num**, to be adjusted.

Although not essential to the approach of this paper, the bond graph formulation provides a clear exposition of the notion of a scaled dual. In particular, the scaled dual system can be described in two different but equivalent ways as:

(1) the bond graph dual where the component moduli are given by Equations (7)–(9) or
(2) following Equation (13), the system obtained by appending a gyrator of modulus $g$ to the system **Phy** port as in Figure 3. This point is also discussed by Gawthrop et al. (2010).

### 3.2. *Den Hartog absorber*

In his classical text book (Den Hartog, 1985, Section 3.3), Den Hartog considers the design of a damped vibration absorber for an undamped mass–spring system which is subject to a force disturbance. The specifications

(1) 'the main mass is 20 times greater than the damper mass',
(2) 'the frequency of the damper is equal to the frequency of the main system',
(3) The damping ratio of the damper is $\zeta = 0.1$.

were considered.

In the terminology used in this paper, the physical system requiring vibration suppression (**Phy**) is the undamped mass–spring oscillator. In Den Hartog (1985) the vibration absorber was considered to be a physical mechanical device but here it is considered to be a controller (with sensor and actuator) with the same dynamics as the absorber and forms **Num**. The disturbance force acts on the undamped mass–spring oscillator **Phy**, as does a force due to the presence of the absorber **Num**, therefore the system can be represented by the block diagram given in Figure 1(b).

Figure 4(b) and 4(a) gives the schematic diagram of the damped vibration absorber **Num** and the undamped oscillator **Phy**, respectively. A damper with $r = \infty$ is included in the subsystem **Phy** of Figure 4(a) to allow for the corresponding component in the subsystem **Num**. Using standard manipulations, the transfer function of the physical system, **Phy**, of Figure 4(a) is:

$$p(s,\theta_p) = \frac{s(ms+r)}{mrs^2 + kms + kr}. \tag{14}$$

Letting $r \to \infty$ gives:

$$p(s,\theta_p) = \frac{s}{ms^2 + k}. \tag{15}$$

Similarly, from Figure 4(b), the transfer function for **Num**, which represents the Den Hartog absorber, is

$$n(s,\theta_n) = \frac{Ms(Rs+K)}{Ms^2 + Rs + K}. \tag{16}$$

It can be shown that this absorber is a scaled dynamic-dual of the system, **Phy**, it is applied to. Considering the system **Phy**, given in Equation (14), and applying the dual transforms, given in Equations (7)–(9), the parameters $m$, $k$ and $r$ can be rewritten in terms of $M$,$K$ and $D$ to give:

$$\begin{aligned} p(s,\theta_p) &\\ &= \frac{s((g^2/k)s + (g^2/R))}{(g^2/k)(g^2/R)s^2 + (g^2/M)(g^2/K)s + (g^2/M)(g^2/R)} \\ &= \left(\frac{1}{g^2}\right)\frac{Ms(Rs+K)}{Ms^2+Rs+Kr}. \end{aligned} \tag{17}$$

Applying the scaling given in Equation (12), the scaled dual of **Phy** is

$$P(s,\Theta_p) = g^2 p(s,\theta_p) = \frac{Ms(Rs+K)}{Ms^2+Rs+Kr}. \tag{18}$$

Thus, by comparing this to Equation (16), it can be seen that the Den Hartog absorber in **Num** corresponds to the scaled dual of **Phy**:

$$n(s,\theta_n) = P(s,\Theta_p). \tag{19}$$

The first part of the Den Hartog specifications is achieved by setting:

$$M = \alpha m \quad \text{where } \alpha = \frac{1}{20}. \tag{20}$$

The second part of the specification is achieved by setting

$$\frac{K}{M} = \frac{k}{m}. \tag{21}$$

Equations (20) and (21) imply that

$$K = \alpha k. \tag{22}$$

Moreover, using Equations (7) and (22), the scaling gain $g$ is given by

$$g^2 = Km = \alpha mk. \tag{23}$$

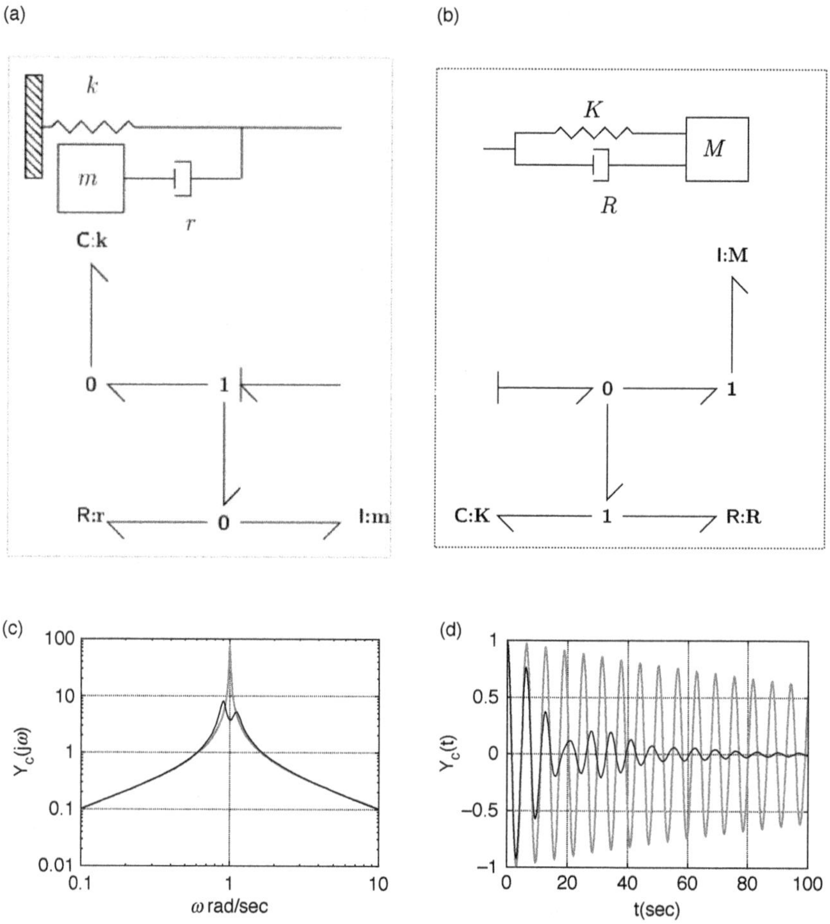


Figure 4. Den Hartog absorber. The physical system (a) and its vibration absorber (and dynamic-dual) (b) are first-order mass–spring—damper systems. (c) The closed-loop frequency response (black line) has a lower resonant peak than the open-loop response (grey line). (d) The closed-loop impulse response (black line) exhibits more damping than the open-loop impulse response (grey line). (a) **Phy**: the physical system, (b) **Num**: the dual physical system, (c) frequency response $H(j\omega)$ and (d) impulse response $h(t)$.

Finally, the third part of the specification is achieved by replacing the damping coefficient $R$ of $n(s, \theta_n)$ by

$$R' = 2\zeta\sqrt{MK}. \tag{24}$$

To illustrate the properties of this particular vibration absorber, the unit system with $m = k = 1$ was used. Using the specification described above, this gives the numerical system parameters $M = K = 0.05$ and $R' = 0.01$, and so the DDVA is given by:

$$n(s, \theta_n') = \frac{Ms(R's + K)}{Ms^2 + R's + K} = \frac{0.01s^2 + 0.05s}{s^2 + 0.2s + 1}. \tag{25}$$

The corresponding closed-loop frequency response appears in Figure 4(c); this shows the 'split peak' phenomenon described by Den Hartog (1985). The corresponding closed-loop impulse response appears in Figure 4(d); this decays exponentially over the time scales determined by the specified damping ratio.

### 3.3. *Acceleration feedback*

The acceleration feedback method has been proposed by Preumont (2002). This section rederives the algorithm from the DDVA point of view. In particular, the undamped physical system of Figure 4(a) (with $1/r = 0$) can equally well be represented in Figure 5(a) with $r = 0$. This system has a different modified dual and thus gives a different form of control; this turns out to be a form of acceleration feedback.

As with the last example the vibration absorber is acting on an undamped mass–spring oscillator. The undamped oscillator forms **Phy** as shown in Figure 5(a). Note that a damper with $r = 0$ is included to allow a dynamic-dual to be formulated. Using standard manipulations, the transfer function of the physical system **Phy** of Figure 5(a) is

$$p(s, \theta_p) = \frac{s}{ms^2 + rs + k}. \tag{26}$$

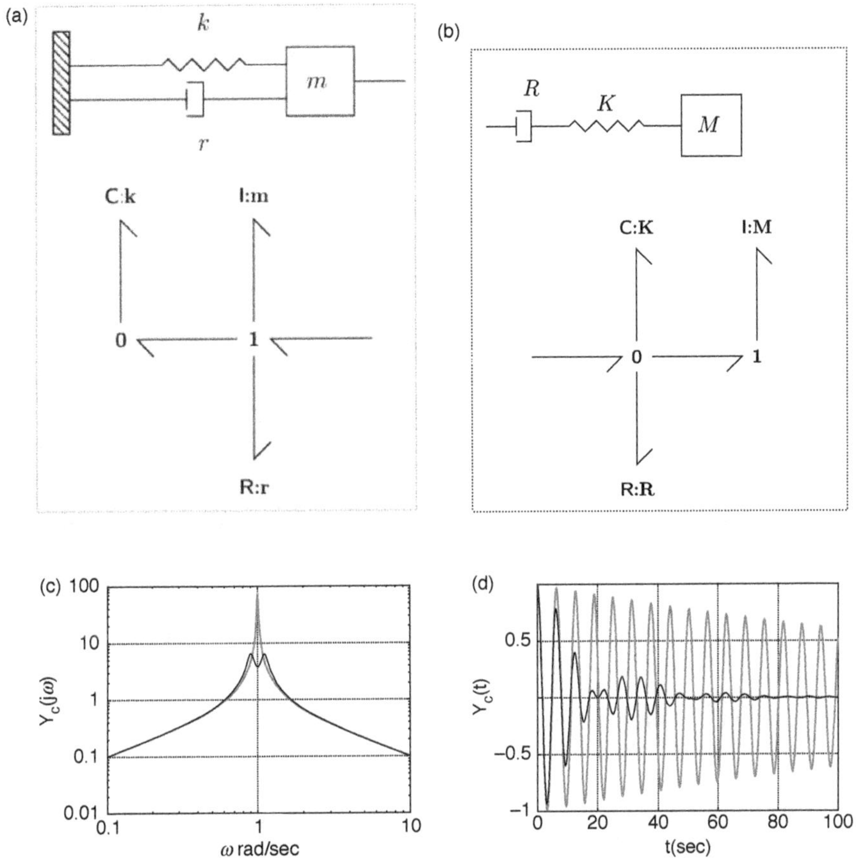


Figure 5. Acceleration feedback. Like the Den Hartog absorber of Figure 4, the physical system and the dual are mass–spring–damper systems, but the configuration is different. The closed and open-loop responses are similar to those of the Den Hartog absorber of Figure 4, but the split peaks are more symmetrical. (a) **Phy**: the physical system, (b) **Num**: the dual physical system, (c) frequency response $H(j\,\omega)$ and (d) impulse response $h(t)$.

Setting $r = 0$ gives

$$p(s,\theta_p) = \frac{s}{ms^2 + k}. \tag{27}$$

The controller transfer function, forming **Num**, for the acceleration feedback method (Preumont, 2002) is given by

$$n(s,\theta_n) = g^2 p(s,\theta_p) = \frac{g^2 s}{ms^2 + rs + k}. \tag{28}$$

Using Equation (12), it can be seen that the numerical system **Num** is the scaled dynamic-dual of **Phy**. Figure 5(b) gives a physical representation of the acceleration feedback controller, where the component values $R$, $K$ and $M$ can be calculated using Equations (7)–(9) but are not required.

To give a direct comparison with Section 3.2, the same system and design considerations are used to give the DDVA:

$$n(s,\theta_n') = \frac{0.05s}{s^2 + 0.2s + 1}. \tag{29}$$

This is similar to the DDVA of Equation (25) except that the numerator $s^2$ term is not present.

The corresponding closed-loop frequency response appears in Figure 5(c); this is similar to Figure 4(c) except that the peaks have a more similar amplitude. The corresponding closed-loop step response appears in Figure 5(d); again, this decays exponentially over the time scales determined by the specified damping ratio.

## 4. Systems with multiple modes of vibration

The examples discussed in the previous section demonstrate that the DDVA approach gives the same type of vibration absorber for the well known cases associated with mass–spring–damper systems. The real advantage of the DDVA approach is when using it to reduce vibrations in systems with multiple modes of vibration. The steps involved are the same as above: (i) define a physical model of the system **Phy**, (ii) set **Num** as the modified scaled dual of **Phy**, and (iii) connect the systems via a single (one-port) connection. As discussed in Section 3.1, step (ii) can either be accomplished directly or indirectly using the **GY** approach of Figure 3. In this paper, attention is focused

Table 1. Modal system: resonant frequencies.

| $n$ | $\omega_n$ (rad/s) |
|---|---|
| 1 | 1.0 |
| 2 | 2.0 |

on linear models thus giving rise to transfer-function representations.

Two examples are considered here. The first is a two degree-of-freedom lumped mass system which is shown schematically in Figure 6(a) and 6(b). Figure 6(a) and 6(b) are similar to Figure 5(a) and 5(b) except that there are *two* coupled mass–spring damper systems involved. Thus **Phy** can be regarded as the modal decomposition of a 2DOF system and **Num** the corresponding vibration absorber. For the purposes of illustration, each subsystem of **Num** has the same parameters as in the example of Figure 5 of Section 3.3 (Table 1).

Figure 6(c) shows the open (without the vibration absorber) and closed-loop (with the vibration absorber) frequency response magnitudes. The magnitude of the closed-loop response (black line) is clearly smaller than the corresponding open-loop response (grey line) at the two resonant frequencies. Figure 6(d) shows the equivalent impulse response; as predicted by the frequency responses, the closed-loop impulse response decays more rapidly than the equivalent open-loop response.

As a second example, a uniform Euler-Bernoulli[5] cantilever beam (with one end fixed and the other free) modelled using a 10 element, finite-element bond graph model (Karnopp, Margolis, & Rosenberg, 2000; Margolis, 1985) is considered. Such beam models are undamped; but the DDVA approach needs to include damping in **Num**. For the purposes of this example, Rayleigh damping is assumed; in particular, each compliant element in the lumped model has an associated damping term represented by a damper connected across the ends of the compliant elements.

Following Balas (1978, Section V) who considers a 'unit beam', the cantilever beam is normalised to have unit mass per unit length and unit compliance per unit length. **Phy** is assumed to have a small (but non-zero) damping of $10^{-6}$ per unit length. The 10 modal frequencies appear in Table 2. For the purposes of illustration, the vibration absorber was applied to the beam using a collocated point Force/Velocity actuator/sensor halfway along the beam. As discussed in the sequel, this point corresponds to a nodal point of the third-resonance and thus this mode cannot be controlled with this choice. The choice of actuator/sensor location is an interesting topic not considered in this paper.

Two versions of DDVA were used. The first DDVA was obtained by considering the complete dynamic-dual of **Phy**. The scaled dual **Num** was obtained using the **GY** approach of Figure 3. A feature of this approach is that

Table 2. Cantilever beam model modal frequencies.

| $n$ | $\omega_n$ (rad/s) |
|---|---|
| 1 | 2.919 |
| 2 | 18.28 |
| 3 | 50.37 |
| 4 | 95.59 |
| 5 | 150.4 |
| 6 | 210.0 |
| 7 | 269.1 |
| 8 | 322.0 |
| 9 | 363.9 |
| 10 | 390.8 |

Note: Not all frequencies appear in Figures 7 and 8 due to coincident zeros.

there are only two control parameters. These were chosen as the gyrator gain $g^2 = 0.05$ and the damping of the cantilever beam model in **Num** as 2 per unit length.

Figure 7(a) shows the open (without the vibration absorber) and closed-loop (with the vibration absorber) frequency response magnitudes. This figure has been expanded to show the frequency responses close to the first–fourth resonances in Figure 7(c)–7(f), respectively. Near the first two resonances (Figure 7(c) and 7(d)), the magnitude of the closed-loop response (black line) is clearly smaller than the corresponding open-loop response (grey line) at the two resonant frequencies. The third resonance corresponds to a node at the sensor/actuator and the fourth is well damped anyway. Thus this controller design naturally applies control authority at the important resonances. Figure 7(b) shows the equivalent impulse response; as predicted by the frequency responses, the closed-loop impulse response decays more rapidly than the equivalent open-loop response. As noted above, the third resonance is not controlled using this approach. However, it could be controlled either by moving the sensor/actuator away from the node or by having a second sensor/actuator away from the node.

The second approach is to use the scaled dynamic-dual of a two mode modal model (as in Figure 6(b)), capturing the dynamics of the first two modes of the cantilever. This is then connected to the same cantilever beam. The parameters of **Num** are the same as in the example of Figure 6 and those of **Phy** the same as those of the example of Figure 7.

Figure 8(a) shows the open (without the vibration absorber) and closed-loop (with the vibration absorber) frequency response magnitudes. This figure has been expanded to show the frequency responses close to the first–fourth resonances in Figure 8(c)–8(f), respectively. Near the first two resonances (Figure 8(c) and 8(d)), the magnitude of the closed-loop response (black line) is clearly smaller than the corresponding open-loop response (grey line) at the two resonant frequencies; these Figures are not the same as Figure 7(c) and 7(d) because

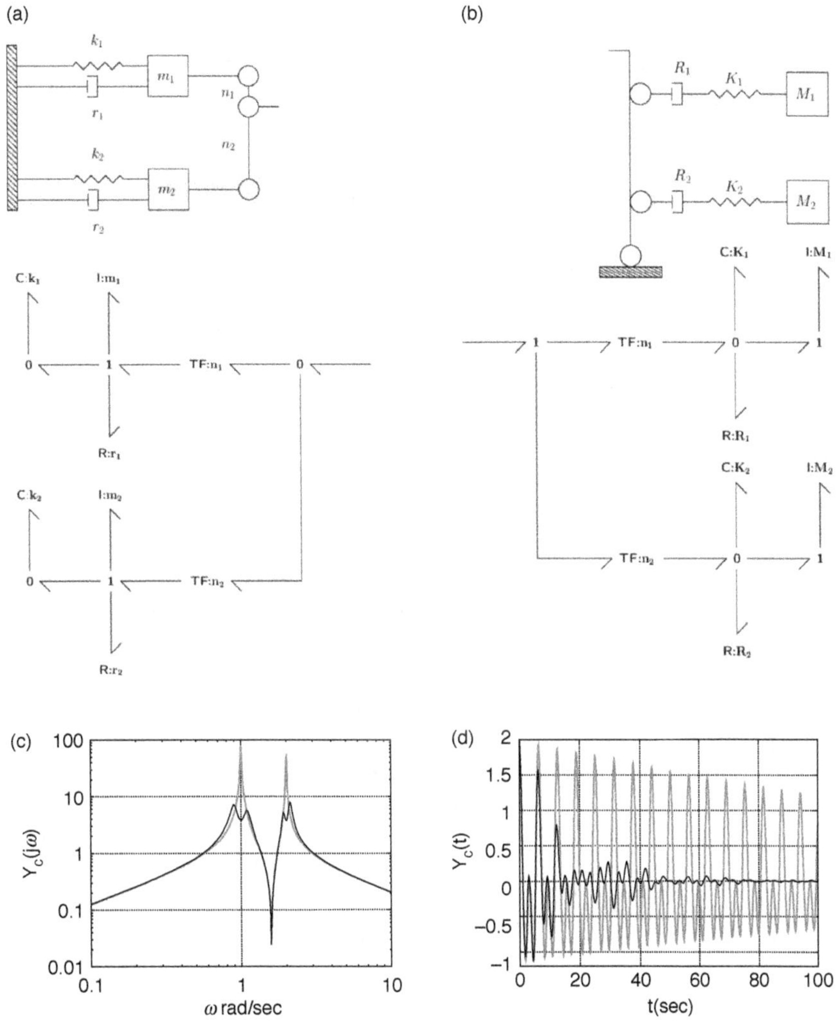


Figure 6. Modal system. Both the physical system and its dual are coupled mass–spring–damper systems and are thus fourth-order. (c) The closed-loop frequency response (black line) has both resonant peaks lower than the open-loop response (grey line). (d) Again, the closed-loop impulse response (black line) exhibits more damping than the open-loop impulse response (grey line). (a) **Phy**: the physical system, (b) **Num**: the dual physical system, (c) frequency response $H(j\omega)$ and (d) impulse response $h(t)$.

the controller is different; but the effect is similar. The third and fourth resonances are explicitly not controlled with this method; but, in this case, the effect is the same as that of the controller of the example of Figure 7. In particular Figure 8(b) shows that the closed-loop impulse response decays more rapidly than the equivalent open-loop response in a similar fashion to that of Figure 7(b). In this particular example, the performance of the two controllers is quite similar.

## 5. Experimental results

As indicated in Figure 9, the experiments were based on the Quanser (Apkarian, 1995) SRV02 rotational servo-motor and associated UPM-15-03-240 power and instrumentation module. The SRV02 was firmly clamped to a rigid bench and interfaced to a Intel CoreTM 2 Duo Processor (2.66 GHz) based computer via a National Instruments PCI-8024E analogue-digital conversion card and cable and the corresponding Quanser interface board.

In the experiment described here, the computer used the real-time Linux operating system RTAI together with the control-orientated software RTAI-Lab (Bucher & Balemi, 2006) running at a sampling frequency of 500 Hz. Using this software, the SRV02 rotational servo motor, rotational position sensor and associated power supply were controlled to give high-gain position control using a proportional and derivative (PD) controller. The servo angle was measured using a potentiometer and scaled within the computer to measure angular position in radians.

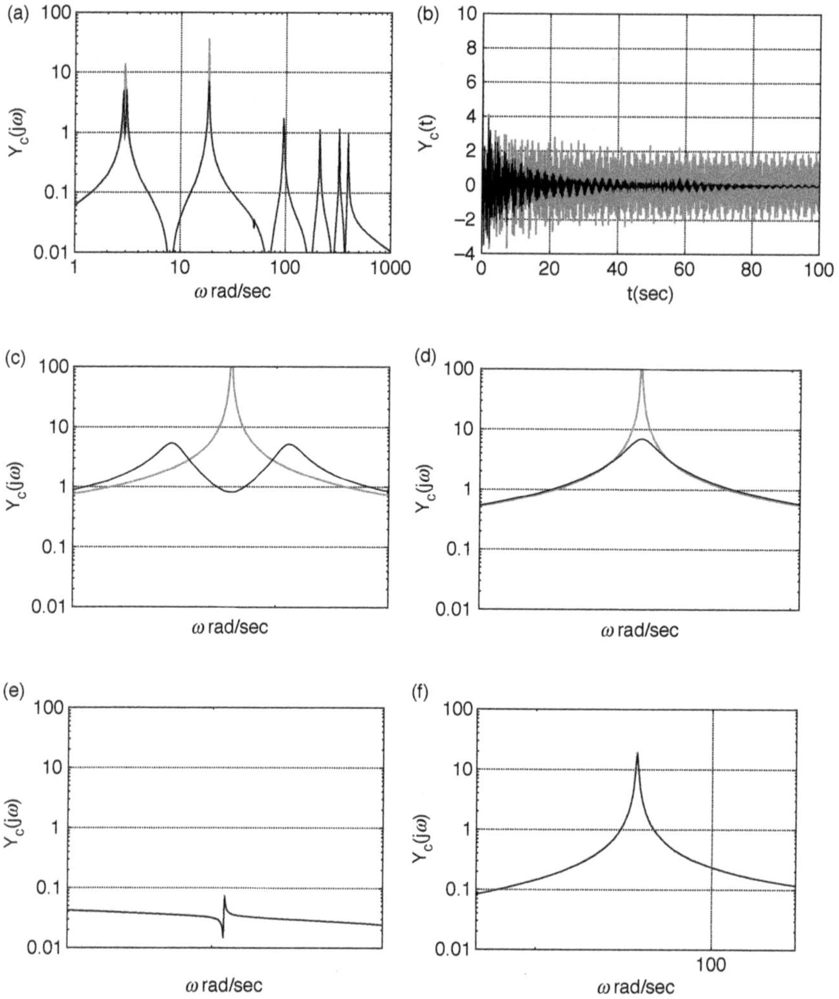


Figure 7. Cantilever beam with dual feedback. (a) Frequency responses $H(j\,\omega)$, (b) impulse responses $h(t)$, (c) $H(j\,\omega)$ – first resonance, (d) $H(j\,\omega)$ – second resonance, (e) $H(j\,\omega)$ – third resonance and (f) $H(j\,\omega)$ – fourth resonance.

### *5.1. Flexible joint*

The Quanser cantilever beam experiment (Apkarian, 1995) has two parts that may be considered using the substructuring configuration shown in Figure 2(c). The physical component, **Phy**, consists of a rigid arm which is mounted to a platform via a pivot. This pivot exhibits a stiffness due to two linear springs mounted between the platform and the arm. A position disturbance is provided to the system via the rotational servo motor on which the platform is mounted (the pivot is directly above the motor). The vibration absorber, **Num**, has a torque input $F$. Because of the springs in **Phy**, this torque is proportional to the joint deflection angle $\theta$ (the arm rotation relative to the platform rotation), and so is generated from this measurement. The output of **Num** is a rotational displacement $x$ which, along with the disturbance $x_0$, is imposed on **Phy** using the servo motor by setting the servo motor PD controller demand to $x_0 - x$ (Figure 10).

The open-loop properties of the system were investigated by applying a square-wave reference signal with a period of 10 s to the servo and measuring both the servo angle $x_a$ and the joint angle $\theta$ for 5 periods. Because all of the signals are periodic, the methods of Pintelon & Schoukens (2001) were used to generate the frequency response of the system at the discrete frequencies corresponding to the periodic input. Figure 11 gives two measured frequency responses:

(1) + indicates the response from servo angle $x_a$ to joint angle $\theta$.
(2) ○ indicates the response from servo reference to $\theta$.

These responses match at low frequencies, but the gain of the second transfer function falls at the higher frequencies due to the limited servo bandwidth of about 10 Hz.

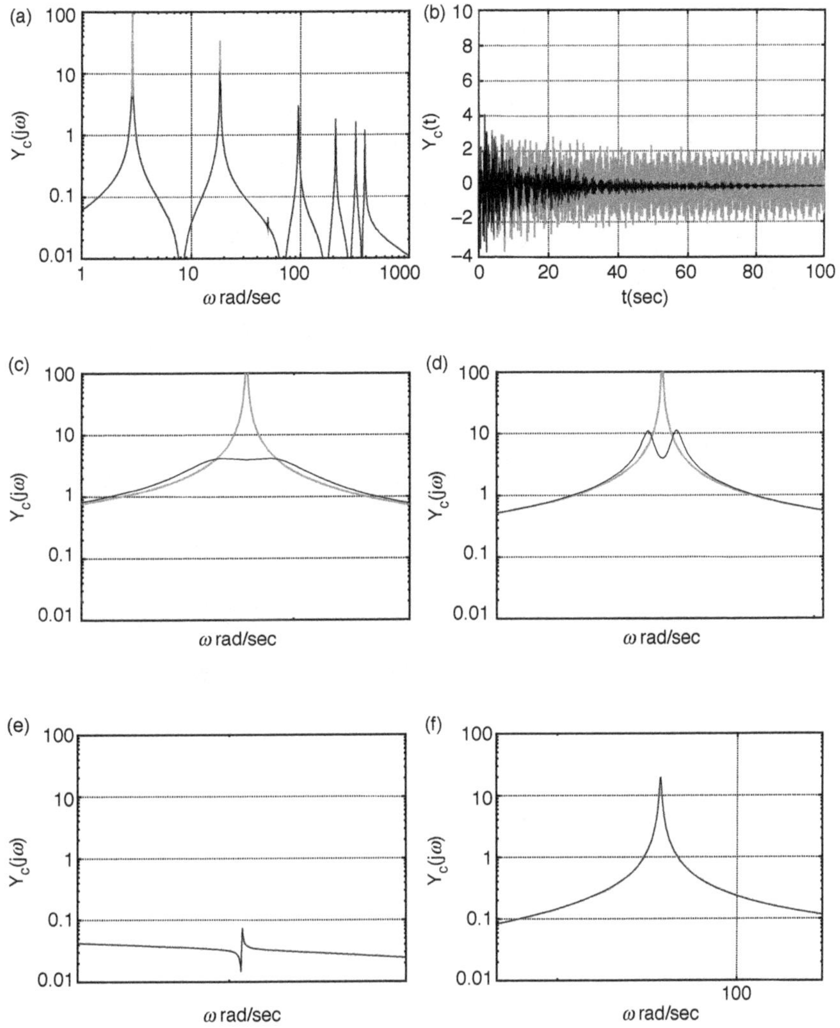


Figure 8. Cantilever with dual modal feedback. (a) Frequency responses $H(j\omega)$, (b) impulse responses $h(t)$, (c) $H(j\omega)$ – first resonance, (d) $H(j\omega)$ – second resonance, (e) $H(j\omega)$ – third resonance and (f) $H(j\omega)$ – fourth resonance.

With reference to Figures 2(c) and 10(a) the physical system **Phy** relating input displacement to output force is of the form

$$s\bar{p}(s,\theta_p) = \frac{g_0 s^2}{s^2 + 2\xi_0\omega_0 s + \omega_0^2}. \tag{30}$$

The parameters $\theta_p$ were fitted to the first measured frequency response with $\omega_0 = 15.1\,\mathrm{rad\,s^{-1}}$ and $\xi_o = 0.02$.

Following the methodology of Section 3.3 in the dual version of Figure 2(c), the feedback transfer function was chosen to be of the form:

$$\frac{\bar{n}(s,\theta_n)}{s} = \frac{g_c}{s^2 + 2\xi_c\omega_0 s + \omega_0^2}, \tag{31}$$

where $g_c = g^2 g_0$ is a variable positive gain factor and the damping ratio $\xi_c = 0.3$.

The periodic input experimental method described above was used. Figure 12 shows the experimental frequency results for three values of $g_c$: $g_c = 0$, $g_c = 20$, and $g_c = 40$. $g_c = 0$ corresponds to Figure 11. The height of the resonant peak is reduced in the two non-zero cases and the peak splitting of Figure 5(c) is evident in Figure 12 for the highest gain of $g_c = 40$. Figure 13 shows the periodic data corresponding to the joint angle $\theta$ for the three gain values. The five consecutive periods have been superimposed to form the figures; the variability between periods is essentially high-frequency noise. As indicated by the frequency responses, the time responses show damping increasing with gain.

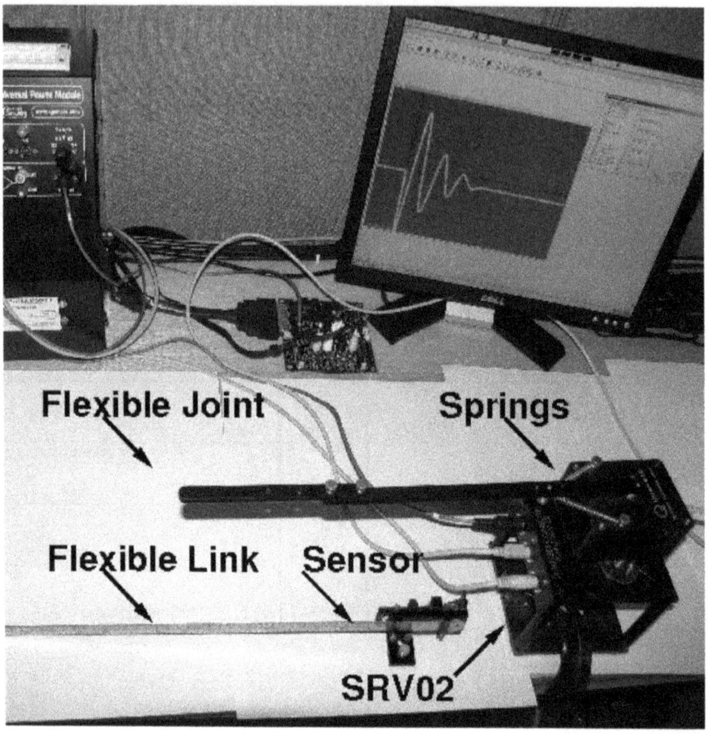


Figure 9. Experimental systems. Experimental systems. The SRV02 servomotor module is in the bottom right-hand corner and the associated power module in the top left-hand corner. The computer display is at the top right and the computer interface board near the centre. The flexible joint module is shown mounted on the SRV02 and rotates about a vertical axis driven though the two springs. The cantilever beam module is shown unmounted and replaces the flexible joint module in the second set of experiments.

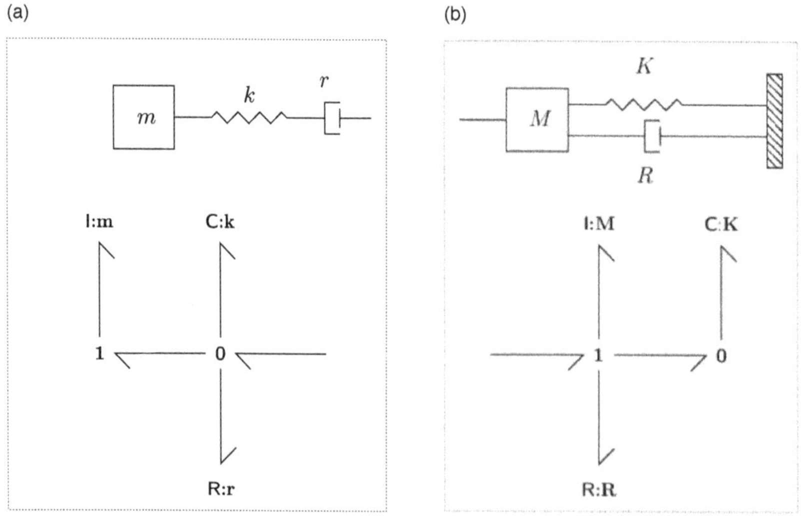


Figure 10. Rotational joint experiment. (a) With the components interpreted in a rotational sense and $r \to \infty$, **Phy** represents the rotating arm with the attached springs. (b) **Num** is the modified scaled dual of **Phy**. (a) **Phy**: the physical system and (b) **Num**: the dual physical system.

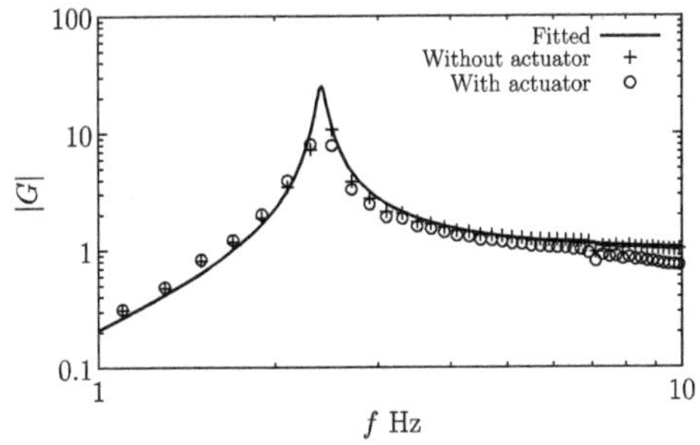


Figure 11. Flexible joint: open-loop frequency responses. Flexible joint: open-loop frequency responses. + indicates the response from servo angle $x$ to joint angle $\theta$; o indicates the response from servo reference to $\theta$. The solid line is the fitted frequency response.

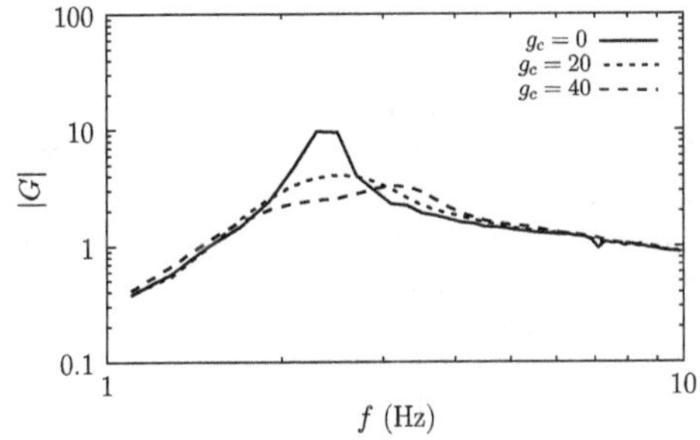


Figure 12. Flexible joint: closed-loop frequency responses.

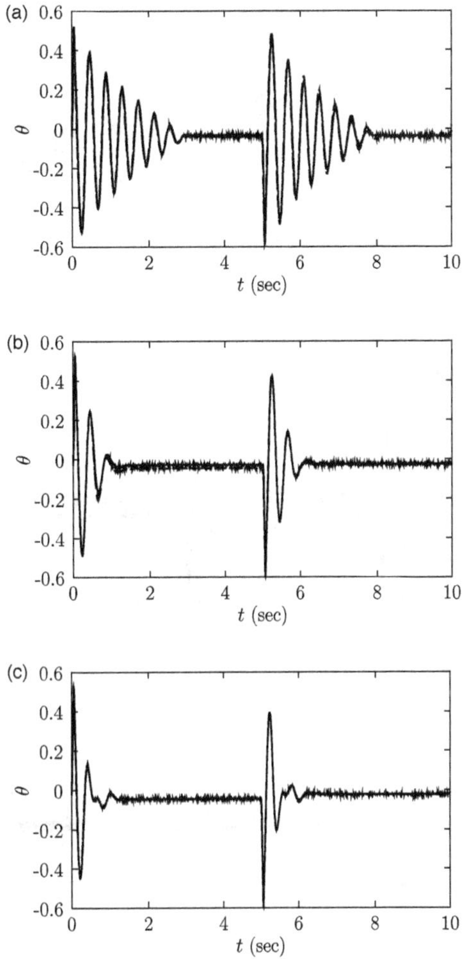


Figure 13. Flexible joint: time responses. (a) $g_c = 0$, (b) $g_c = 20$ and (c) $g_c = 40$.

### 5.2. *Cantilever beam*

The flexible joint module was replaced by the cantilever beam module in Figure 9. A strain gauge measures the curvature at the root of the cantilever beam. In the same way as the joint potentiometer of Section 5.1 provided a voltage proportional to torque $F$, the strain gauge sensor provides a voltage proportional to torque $F$. The open-loop response was measured using the same methods. Two resonances and one anti-resonance appear in the measured frequency response and similarly to Section 5.1, this was fitted by a transfer function of the form:

$$s\bar{p}(s, \theta_p) = g_0 \left[ \frac{\kappa_1 s^2}{s^2 + 2\xi_1 \omega_1 s + \omega_1^2} + \frac{\kappa_2 s^2}{s^2 + 2\xi_2 \omega_2 s + \omega_2^2} \right] \tag{32}$$

with $\omega_1 = 23.25\,\text{rad s}{-}1, \omega_2 = 159.00\,\text{rad s}{-}1$, $\xi_1 = \xi_2 = 0.04$, $\kappa_1 = 0.36$ and $\kappa_2 = 1 - \kappa_1 = 0.64$. Because of the 10 Hz servo bandwidth, the discrepancy between measured and fitted transfer function is large above 10 Hz.

Following the methodology of Section 4 a two-mode transfer function corresponding to Equation (33) was

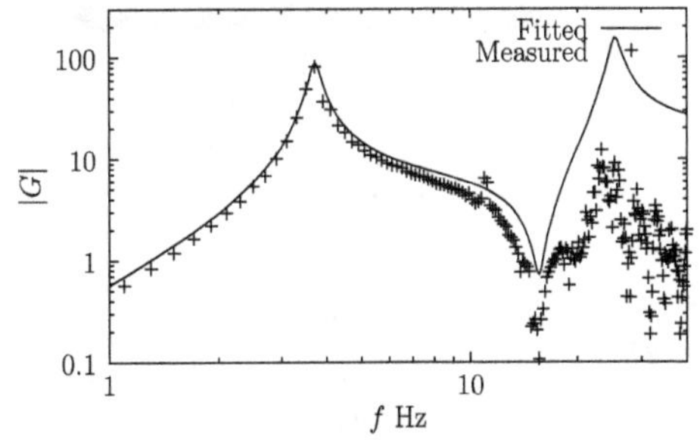


Figure 14. Cantilever beam: open-loop frequency responses. Cantilever beam: open-loop frequency responses. + indicates the response from servo angle $x$ to joint angle $\theta$. The firm line is the fitted frequency response.

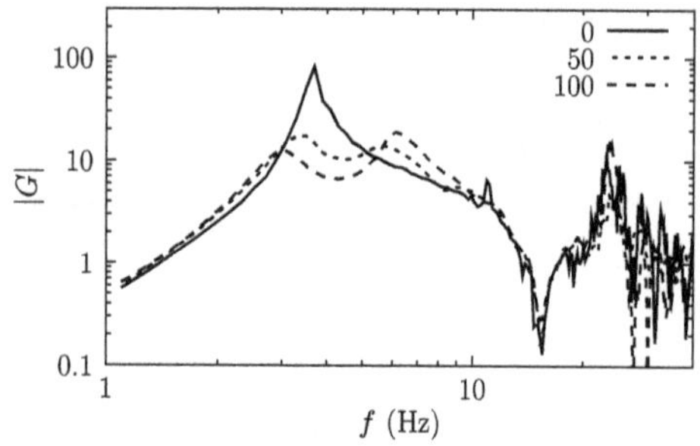


Figure 15. Cantilever beam: closed-loop frequency responses for the cases where $g_c = 0$, 50 and 100.

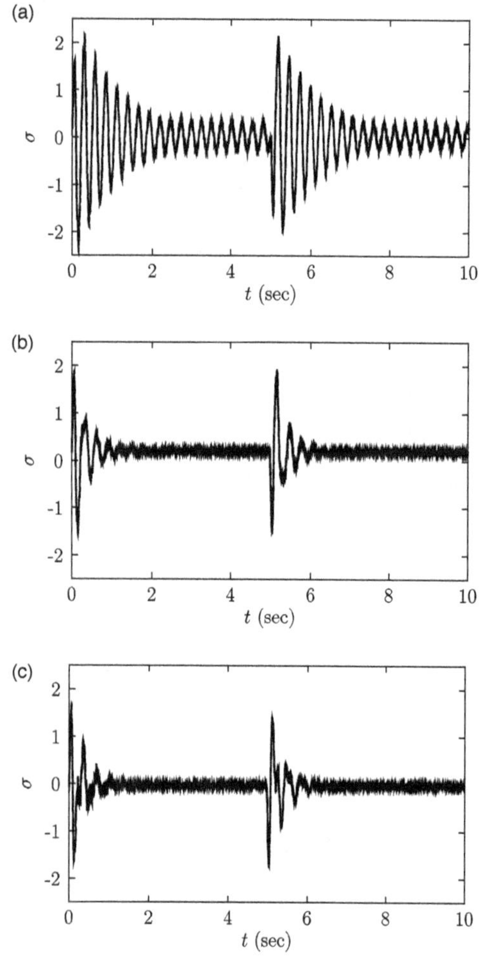


Figure 16. Cantilever beam: time responses. (a) $g_c = 0$, (b) $g_c = 50$ and (c) $g_c = 100$.

chosen as:

$$\frac{\bar{n}(s,\theta_n)}{s} = g_c \left[ \frac{\kappa_1 s^2}{s^2 + 2\xi_c \omega_1 s + \omega_1^2} + \frac{\kappa_2 s^2}{s^2 + 2\xi_c \omega_2 s + \omega_2^2} \right] \tag{33}$$

with $\xi_c = 0.3$.

The periodic input experimental method described above was used. Figure 15 shows the experimental frequency results for three values of $g_c$: $g_c = 0$, $g_c = 50$, and $g_c = 100$. $g_c = 0$ corresponds to Figure 14. The height of the first resonant peak is reduced in the two non-zero cases and the peak splitting of Figure 8(a) is evident in Figure 15 for both cases. The second resonance is largely unaffected; we attribute this to the limited actuator bandwidth. Figure 16 shows the periodic data corresponding to the measured strain voltage $\sigma$ for the three gain values. As with Figure 13, five consecutive periods have been superimposed to form the figures showing that the variability between periods is essentially high-frequency noise. Again, as indicated by the frequency responses, the time responses show damping increasing with gain.

## 6. Conclusion

The DDVA approach has been shown to provide a novel method to design vibration absorbers in the physical domain. In particular, the method is a natural generalisation of not only the classical single-degree of freedom vibration absorber of Den Hartog (1985, Section 3.3) but also of acceleration feedback (Preumont, 2002). Placing these two well-known design methods into the wider context of the DDVA of this paper has the following advantages: the methods immediately extend to multi degree of freedom systems and the implementation and theoretical results (including robustness to sensor/actuator dynamics) from substructuring can be directly applied.

The DDVA approach has been illustrated using numerical simulations of single mode and multi-mode systems and verified using two experimental systems: a rigid beam with an flexible joint and a flexible cantilever beam. Future work will apply the results to more complex dynamical systems including those with multiple sensor-actuator pairs.

The location of the sensor-actuator pairs has not been considered in this paper even though it certainly affects controllability and observability issues (Balas, 1978). Future work in this area will extend bond graph approaches (for example those of Marquis-Favre & Jardin (2011) and Gawthrop & Rizwi (2011)) to sensor/actuator placement in this context.

In principle, the method is equally applicable to the control nonlinear vibrations where dynamical dual of the nonlinear physical system provides the basis for a nonlinear controller. This is also an area for future work.


## Acknowledgements

Peter Gawthrop was a Visiting Research Fellow at The University of Bristol when this work was accomplished; he is now supported by a Professorial Fellowship at the University of Melbourne.

## Disclosure statement

No potential conflict of interest was reported by the authors.

## Funding

Simon Neild is supported by EPSRC fellowship EP/K005375/1. David Wagg is supported by EPSRC [grant EP/K003836/2].

## Notes

1. As the word 'dual' has many meanings, the term dynamically dual is used for the specific meaning of this paper.
2. The bond directions have been changed for this paper to correspond to the usual sign convention for feedback control block diagrams.
3. 'One-port' refers to the single *energy port* associated with force and velocity
4. In the sense that it consumes but does not produce energy.
5. Other models such as the Timoshenko model, as well as non-uniform beams, could similarly be handled using this approach

## References

Ali, S. F., & Padhi, R. (2009). Active vibration suppression of non-linear beams using optimal dynamic inversion. *Journal of Systems and Control Engineering*, *223*(5), 657–672.

Anderson, B. D. O., & Vongpanitlerd, S. (2006). *Network analysis and synthesis a modern systems theory approach*. Dover. First published 1973 by Prentice-Hall.

Apkarian, J. (1995). *A comprehensive and modular laboratory for control systems design and implementation*. Markham, Ontario: Quanser Consulting.

Balas, M. J. (1978). Feedback control of flexible systems. *IEEE Transactions on Automatic Control*, *23*(4), 673–679.

Blakeborough, A., Williams, M. S., Darby, A. P., & Williams, D. M. (2001). The development of real-time substructure testing. *Philosophical Transactions of the Royal Society Part A*, *359*, 1869–1891.

Borutzky, W. (2011). *Bond graph modelling of engineering systems: Theory, applications and software support*. New York, NY: Springer.

Bucher, R., & Balemi, S. (2006). Rapid controller prototyping with matlab/simulink and linux. *Control Engineering Practice*, *14*(2), 185–192.

Cellier, F. E. (1991). *Continuous system modelling*. Berlin: Springer-Verlag.

Den Hartog, J. P. (1985). *Mechanical vibrations*. Dover. Reprint of 4th ed. Published by McGraw-Hill 1956.

Desoer, C. A., & Vidyasagar, M. (1975). *Feedback systems: Input-output properties*. London: Academic Press.

Fleming, A. J., & Moheimani, S. O. R. (2005). Control orientated synthesis of high-performance piezoelectric shunt impedances for structural vibration control. *IEEE Transactions on Control Systems Technology*, *13*(1), 98–112.

Gawthrop, P. J. (1995). Physical model-based control: A bond graph approach. *Journal of the Franklin Institute*, *332B*(3), 285–305.

Gawthrop, P. J. (2004). Bond graph based control using virtual actuators. *Proceedings of the Institution of Mechanical Engineers Pt. I: Journal of Systems and Control Engineering*, *218*(4), 251–268.

Gawthrop, P. J., & Bevan, G. P. (2007). Bond-graph modeling: A tutorial introduction for control engineers. *IEEE Control Systems Magazine*, *27*(2), 24–45.

Gawthrop, P. J., & Rizwi, F. (2011). Coaxially coupled inverted pendula: Bond graph-based modelling, design and control. In W. Borutzky (Ed.), *Bond graph modelling of engineering systems* (pp. 179–194). New York, NY: Springer.

Gawthrop, P. J., & Smith, L. P. S. (1996). *Metamodelling: Bond graphs and dynamic systems*. Hemel Hempstead: Prentice Hall.

Gawthrop, P. J., Bhikkaji, B., & Moheimani, S. O. R. (2010). Physical-model-based control of a piezoelectric tube for nano-scale positioning applications. *Mechatronics*, *20*(1), 74–84. Available online 13 October 2009.

Gawthrop, P. J., Wagg, D. J., & Neild, S. A. (2009). Bond graph based control and substructuring. *Simulation Modelling Practice and Theory*, *17*(1), 211–227. Available online 19 November 2007.

Gawthrop, P. J., Wallace, M. I., Neild, S. A., & Wagg, D. J. (2007). Robust real-time substructuring techniques for under-damped systems. *Structural Control and Health Monitoring*, *14*(4), 591–608. Published on-line: 19 May 2006.

Gawthrop, P. J., Wallace, M. I., & Wagg, D. J. (2005). Bond-graph based substructuring of dynamical systems. *Earthquake Engineering & Structural Dynamics*, *34*(6), 687–703.

Hagood, N. W., & von Flotow, A. (1991). Damping of structural vibrations with piezoelectric materials and passive electrical networks. *Journal of Sound and Vibration*, *146*(2), 243–268.

Hogan, N. (1985). Impedance control: An approach to manipulation. part I – theory. *ASME Journal of Dynamic Systems, Measurement and Control*, *107*, 1–7.

Høgsberg, J. R., & Krenk, S. (2006). Linear control strategies for damping of flexible structures. *Journal of Sound and Vibration*, *293*(1–2), 59–77.

Hong, J. H., & Bernstein, D. S. (1998). Bode integral constraints, colocation, and spillover in active noise and vibration control. *IEEE Transactions on Control Systems Technology*, *6*(1), 111–120.

Karnopp, D. (1966). Coupled vibratory-system analysis, using the dual formulation. *The Journal of the Acoustical Society of America*, *40*(2), 380–384.

Karnopp, D., Margolis, D. L., & Rosenberg, R. C. (2000). *System dynamics: Modeling and simulation of mechatronic systems* (3rd ed.). New York, NY: Horizon Publishers and Distributors Inc.

Karnopp, D. C., Margolis, D. L., & Rosenberg, R. C. (2012). *System dynamics: Modeling, simulation, and control of mechatronic systems* (5th ed.). John Wiley & Sons.

Krenk, S., & Hogsberg, J. (2009). Optimal resonant control of flexible structures. *Journal of Sound and Vibration*. *323* (3–5), 530–554.

Lozano, R., Brogliato, B., Egelund, O., & Maschke, B. (2000). *Dissipative systems: Analysis and control*. New York, NY: Springer.

Margolis, D. L. (1985). A survey of bond graph modelling for interacting lumped and distributed systems. *Journal of the Franklin Institute*, *319*, 125–135.

Marquis-Favre, W., & Jardin, A. (2011). Bond graphs and inverse modeling for mechatronic system design. In W. Borutzky (Ed.), *Bond graph modelling of engineering systems* (pp. 195–226). New York, NY: Springer.

Moheimani, S. O. R, & Behrens, S. (2004). Multimode piezoelectric shunt damping with a highly resonant impedance. *IEEE Transactions on Control Systems Technology*, *12*(3), 484–491.

Moheimani, S. O. R., & Fleming, A. J. (2006). *Piezoelectric transducers for vibration control and damping*. Advances in Industrial Control. New York, NY: Springer.
Mukherjee, A., Karmaker, R., & Samantaray, A. K. (2006). *Bond graph in modeling, simulation and fault indentification*. New Delhi: I.K. International.
Ortega, R., Loria, A., Nicklasson, P. J., & Sira-Ramirez, H. (1998). *Passivity-based control of Euler-lagrange systems*. London: Springer.
Ortega, R., Praly, L., & Landau, I. D. (1985). Robustness of discrete-time direct adaptive controllers. *IEEE Transactions on Automatic Control, AC-30*(12), 1179–1187.
Ortega, R., van der Schaft, A. J., Mareels, I., & Maschke, B. (2001). Putting energy back in control. *IEEE Control Systems Magazine, 21*(2), 18–33.
Pintelon, R., & Schoukens, J. (2001). *System identification. A frequency domain approach*. New York, NY: IEEE Press.
Preumont, A. (2002). Vibration control of active structures: An introduction, volume 96 of *Solid mechanics and its applications*. Dordrecht: Kluwer.
Samanta, B., & Mukherjee, A. (1985). A bond graph based analysis of coupled vibratory systems taking advantage of the dual formulation. *Journal of the Franklin Institute, 320*, 111–131.
Samanta, B., & Mukherjee, A. (1990). Analysis of acoustoelastic systems using modal bond graphs. *Journal of Dynamic Systems, Measurement, and Control, 112*(1), 108–115.
Sharon, A., Hogan, N., & Hardt, D. E. (1991). Controller design in the physical domain. *Journal of the Franklin Institute, 328*(5–6), 697–721.
Shearer, J. L., Murphy, A. T., & Richardson, H. H. (1971). *Introduction to system dynamics*. Reading, MA: Addison-Wesley.
Slotine, J. E., & Li, W. (1991). *Applied nonlinear control*. Englewood Cliffs: Prentice-Hall.
Vink, D., Ballance, D., & Gawthrop, P. (2006). Bond graphs in model matching control. *Mathematical and Computer Modelling of Dynamical Systems, 12*(2–3), 249–261.
Wagg, D., Neild, S., & Gawthrop, P. (2008). Real-time testing with dynamic substructuring. In O. S. Bursi and D. Wagg (Eds.), *Modern testing techniques for structural systems*, volume 502 of *CISM courses and lectures*, Chapter 7 (pp. 293–342). Wien, NY: Springer.
Willems, J. C. (1972). Dissipative dynamical systems, part I: General theory, part II: Linear system with quadratic supply rates. *Arch. Rational Mechanics and Analysis, 45*(5), 321–351.

## Appendix. Derivation of Equation (14)

Equation (14) can be derived directly from the bond graph of Figure 4(a). Letting $F$ and $v$ be the force and velocity at the component interface, letting $F_{mr}$ be the force acting on the mass and damper and $F_c$ the spring force, it follows that the components represented by **I**:**m**, **C**:**k** and **R**:**r** have equations:

$$m\frac{\mathrm{d}v_m}{\mathrm{d}t} = F_{mr}, \tag{A1}$$

$$\frac{\mathrm{d}F_c}{\mathrm{d}t} = kv, \tag{A2}$$

$$v_r = \frac{1}{r}F_{mr}. \tag{A3}$$

Taking Laplace transforms (with zero initial conditions) it follows that:

$$\begin{aligned} v &= v_r + v_m \\ &= \left[\frac{1}{r} + \frac{1}{ms}\right] F_{mr} \\ &= \left[\frac{1}{r} + \frac{1}{ms}\right] (F - F_c) \\ &= \left[\frac{1}{r} + \frac{1}{ms}\right] \left[F - \frac{k}{s}v\right] \\ &= \frac{ms+r}{mrs}\left[F - \frac{k}{s}v\right]. \end{aligned} \tag{A4}$$

Collecting terms in Equation (A4) gives:

$$\frac{k(ms+r) + mrs^2}{s(ms+r)} v = F. \tag{A5}$$

Hence, rearranging Equation (A5):

$$\frac{v}{F} = \frac{s(ms+r)}{mrs^2 + kms + kr}. \tag{A6}$$

The right-hand side of Equation (A6) corresponds to the transfer function of Equation (14).

# Estimation of tire–road friction coefficient and its application in chassis control systems

Kanwar Bharat Singh[a]* and Saied Taheri[a,b]

[a]*Department of Mechanical Engineering, Virginia Tech, Randolph Hall (MC0238), Blacksburg, VA 24061, USA;* [b]*NSF I/UCRC Center for Tire Research (CenTiRe), Virginia Tech, Blacksburg, VA, USA*

Knowledge of tire–road friction conditions is indispensable for many vehicle control systems. In particular, friction information can be used to enhance the performance of wheel slip control systems, for example, knowledge of the current maximum coefficient of friction would allow an anti-lock brake system (ABS) controller to start braking with the optimal brake pressure, meaning the early cycles of operation are more efficient, resulting in shorter stopping distances. Also, from a passive safety perspective, it may be useful to present the driver with friction information so they can adjust their driving style to the road conditions. Hence, it is highly desirable to estimate friction using existing onboard vehicle sensor information. Many approaches for estimating tire–road friction estimation have been proposed in the literature with different sensor requirements and relative excitation levels. This paper aims at estimating the tire–road friction coefficient by using a well-defined model of the tire behavior. The model adopted for this purpose is the physically based brush tire model. In its simplest formulation, the brush model describes the relationship between the tire force and the slip as a function of two parameters, namely, tire stiffness and the tire–road friction coefficient. Knowledge of the shape of the force–slip characteristics of the tire, possibly obtained through the estimation of both friction and tire stiffness using the brush model, provides information about the slip values at which maximum friction is obtained. This information could be used to generate a target slip set point value for controllers, such as an ABS or a traction control system. It is also important to realize that a model-based approach is inherently limited to providing road surface friction information when the tire is exposed to an excitation with high utilization levels (i.e. under high-slip conditions). To be of greatest use to active safety control systems, an estimation method needs to offer earlier knowledge of the limits. In order to achieve the aforementioned objective, an integrated approach using an intelligent tire-based friction estimator and the brush tire model-based estimator is presented. An integrated approach gives us the capability to reliably estimate friction for a wider range of excitations (both low-slip and high-slip conditions).

**Keywords:** Brush model; friction estimation; Levenberg–Marquardt; nonlinear least squares; intelligent tire

## 1. Introduction

Tire friction forces, as the primary forces affecting planar vehicle motions, are physically limited by the road surface coefficient of friction ($\mu$) and the instantaneous tire normal forces (Figure 1). Therefore, the ability to reliably estimate the tire–road friction coefficient is important for maximizing the performance of vehicle control systems, which work well only when the tire force command computed by the safety systems is within the friction limit.

Instantaneous knowledge of the friction potential will result in improved performance by several of the active chassis control systems. Examples of vehicle control systems that can benefit from the knowledge of tire–road friction include anti-lock braking systems (ABS), electronic stability control (ESC), adaptive cruise control (ACC), and collision warning or collision avoidance systems (Braghin et al., 2009; Cheli, Leo, Melzi, & Sabbioni, 2010; Cheli et al., 2011a, 2011b; Erdogan, Hong, Borrelli, & Hedrick, 2011; Sabbioni, Kakalis, & Cheli, 2010; Singh, Arat, & Taheri, 2012, 2013). The quality of traffic management and road maintenance work (e.g. salt application and snow plowing) can also be improved if the estimated friction value is communicated to the traffic and highway authorities.

The importance of friction estimation is reflected by the considerable amount of work that has been done in this field (Google Scholar) (Table 1). In normal driving conditions, the frictional force is not fully utilized, and the developed tire forces will be somewhere in the interior of the friction circle. When inputs are imposed on a tire, a relative motion between the tire structure and the road surface will arise. This relative motion is referred to as tire slip. The relation between the resulting tire forces and slip depends on many factors, namely, tire inflation pressure, vertical load, tire wear state, temperature, etc., and contains information about the available friction. When the

*Corresponding author. Email: kbsingh@vt.edu

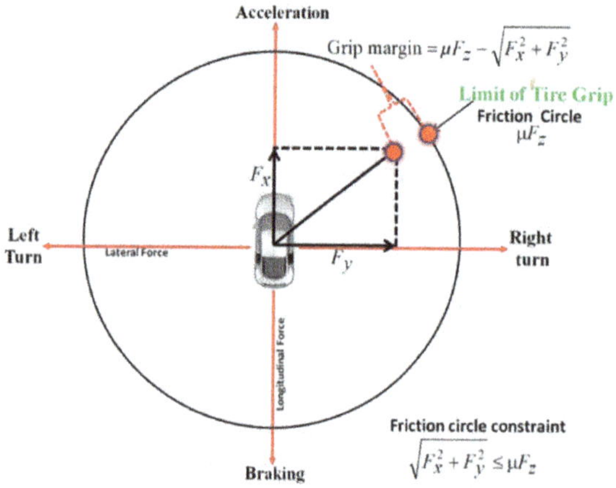


Figure 1. Friction circle of a tire.

Table 1. State-of-the-art literature review.

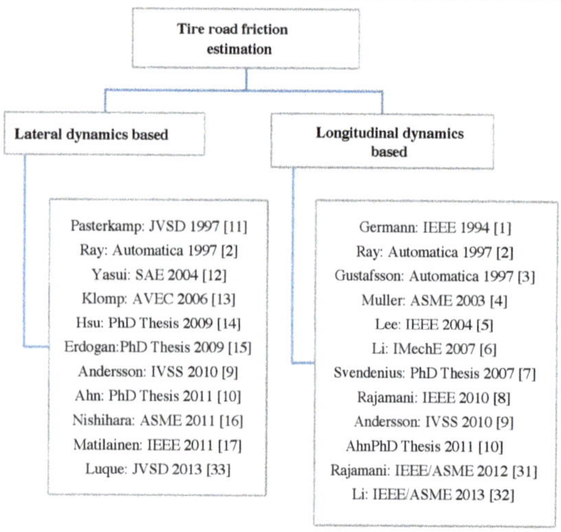


tire is exposed to excitation with high utilization, beyond the point corresponding to the maximum available friction force, the tire starts sliding and the resulting tire force directly corresponds to the friction coefficient.

Hence, determination of friction coefficient is straightforward in cases where tire forces are saturated, such as under hard braking conditions. The difficulty lies in obtaining a friction estimate under more normal driving circumstances, in which the tire slip is smaller (lower utilization conditions). In these cases, a model-based approach can be advantageous (Andersson et al., 2010). By fitting tire force and moment data to a model of the tire, the model parameters, including friction coefficient ($\mu$), may be estimated. This approach may allow the estimation of friction without requiring tire force saturation. This study investigates the use of a model-based approach to estimate the tire–road friction coefficient and, more importantly, determines the level of tire force excitation necessary to allow the estimate to converge within a specified range of accuracy.

## 2. Estimation of vehicle states and tire forces

A model-based friction estimation method is based on the assumption that the lateral force, traction/braking force, aligning torque, vertical load and the two tire kinematic variables, slip angle and wheel slip, can be estimated indirectly using observers developed based on vehicle dynamics measurements (acceleration, yaw and roll rates, suspension deflections, etc.), or measured directly using certain sensor-based advanced tire concepts. In this study, we propose using an integrated vehicle state estimator, comprising a series of model-based and kinematics-based observers and an effectively designed merging scheme that ensures robust estimation performance even during the vehicle maneuvers which show highly nonlinear tire characteristics and in the existence of road inclination or bank angle. It is assumed that measurements from a six-axis inertial measurement unit (three axes of rotation rate measurement and three axes of acceleration measurement), wheel speed sensors and steering wheel angle sensor are available.

The block diagram in Figure 2 explicitly shows the estimation process in its entirety.

The entire process is separated into five blocks: the first block serves to identify the road bank and grade angles (using a kinematics-based observer) and vehicle chassis roll (using a Kalman filter) and pitch angles (with vehicle mass adaptation); the second block contains a bias compensation algorithm (gravity compensation in accelerometer measurements), a vehicle longitudinal speed estimation algorithm (based on the measurements of the four wheel rotational speeds and the gravity-compensated longitudinal vehicle acceleration) and a tire load estimation algorithm (using gravity-compensated acceleration information and roll/pitch states); the third block contains a tire longitudinal/lateral force estimation observer (sliding-mode observer based), while the fourth block contains a nonlinear vehicle longitudinal and lateral velocity observer (based on an unscented Kalman filter), designed for the purpose of vehicle sideslip estimation. Finally, the fifth block makes use of the estimations provided by the third and fourth blocks to estimate the tire slip ratio and slip angle (Luenberger observer based).

It is worth stressing the fact that the present work is concerned with examining the feasibility of estimating the tire–road friction coefficient ($\mu$) in real time. Hence, a complete description of the integrated vehicle state estimator as shown in Figure 2 is beyond the scope of this work. A thorough coverage of this topic with specific details about the different schemes used for tire/vehicle state and parameter estimation is included in our previous work (Arat, Singh, & Taheri, 2013, 2014; Singh, 2012; Singh et al., 2012; Singh and Taheri, 2013).

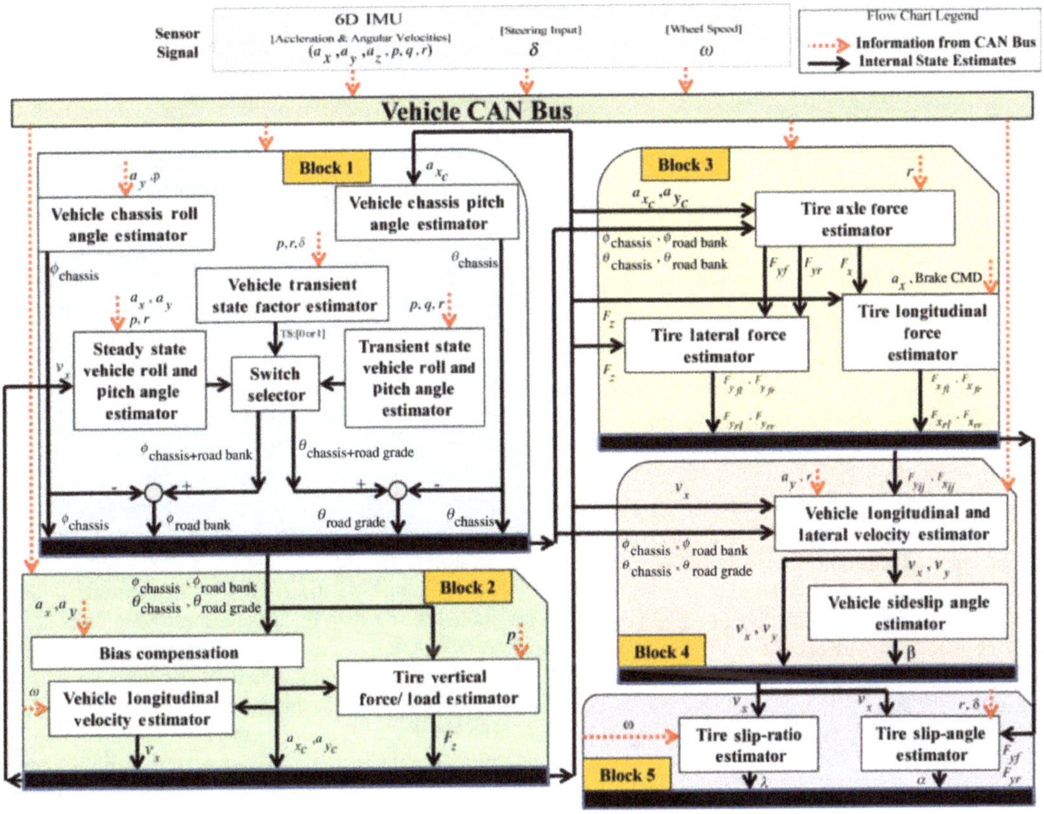


Figure 2. Functional diagram of the estimation process.

## 3. Tire model selection

Critical to the success of a model-based approach is the choice of model structure. Li, Fei-Yue, and Qunzhi (2006) provide a comprehensive summary of various models that have been developed to describe the complex nonlinear behavior of a tire. As this study focuses on parameter estimation, it is desirable to choose a model with a small number of parameters. The brush model (Pacejka, 2005) is well suited to these requirements, containing only stiffness and friction parameters. The basic concept of the brush model is to represent the tire as a row of elastic bristles which touch the road plane and can deflect in a direction that is parallel to the road surface (Figure 3).

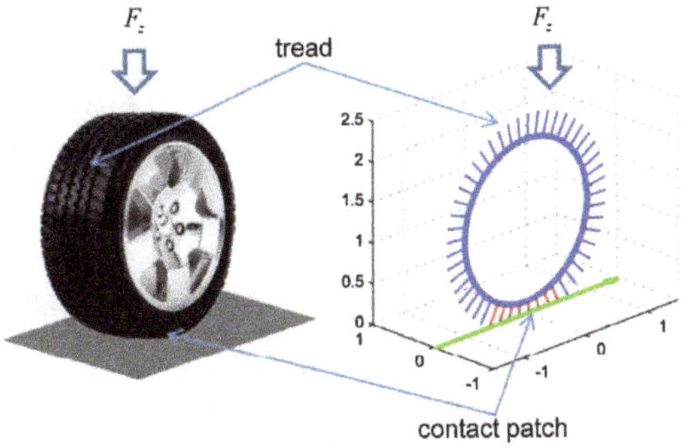


Figure 3. Tire and brush model (Ahn, 2011).

As a result, a tire can be modeled as a thin disk with brushes along the circumference that represent the tire treads. Treads in the contact patch are compressed and experience vertical stresses. The distribution of vertical stress is assumed to be parabolic. The generated forces or moment can be computed by integrating the stress of all brushes in the contact patch. A thorough coverage of the brush model is included in Pacejka (2005).

In a purely longitudinal slip case, the tire longitudinal force can be represented as follows:

$$F_x = \begin{cases} C_x\left(\frac{\lambda}{\lambda+1}\right) - \left(\frac{1}{3}\frac{C_x^2|\lambda/(\lambda+1)|(\lambda/(\lambda+1))}{\mu F_z}\right) \\ \qquad - \left(\frac{1}{27}\frac{C_x^3(\lambda/\lambda+1)^3}{(\mu F_z)^2}\right) \quad \text{for} \quad |\lambda| \le |\lambda_{sl}|, \\ \mu F_z \,\text{sign}(\lambda) \quad \text{for} \quad |\lambda| > |\lambda_{sl}|, \end{cases} \tag{1}$$

where $F_x$, $F_z$, $C_x$, $\mu$, $\lambda$ and $\lambda_{sl}$ stand for tire longitudinal force, tire normal force, tire longitudinal stiffness, tire–road friction coefficient, slip ratio and slip ration where transition from partial to full sliding occurs, respectively.

In a purely lateral slip case, the tire lateral force and tire aligning moment can be represented as follows:

$$F_y = \begin{cases} -C_y \tan(\alpha) + \left(\frac{1}{3}\frac{C_y^2|\tan(\alpha)|\tan(\alpha)}{\mu F_z}\right) \\ -\left(\frac{1}{27}\frac{C_y^3\tan^3(\alpha)}{(\mu F_z)^2}\right) \quad \text{for} \quad |\alpha| \leq |\alpha_{\text{sl}}|, \\ -\mu F_z \text{sign}(\alpha) \quad \text{for} \quad |\alpha| > |\alpha_{\text{sl}}|, \end{cases} \tag{2}$$

$$\tau_a = \begin{cases} \frac{C_y \tan(\alpha) a_{\text{cpl}}}{3}\left(1 - \left|\frac{C_y \tan(\alpha)}{3\mu F_z}\right|\right)^3 & \text{for} \quad |\alpha| \leq |\alpha_{sl}|, \\ 0 & \text{for} \quad |\alpha| > |\alpha_{sl}|, \end{cases} \tag{3}$$

where in addition to the above terms $F_y$, $\tau_a$, $C_y$, $\alpha$, $\alpha_{\text{sl}}$ and $a_{cpl}$ stand for tire lateral force, tire aligning moment, tire cornering stiffness, slip angle, slip angle where transition from partial to full sliding occur and half of tire contact patch length, respectively.

The force and moment equations in combined slip cases are similar to the equations for pure slip cases. If both lateral slip and longitudinal slip exist, the treads are deformed in the direction determined by the magnitudes of both slips. The brush model for the combined slip case can be represented by the following equation:

$$F_x = F\frac{\sigma_x}{\sigma}, \quad F_y = F\frac{\sigma_y}{\sigma}, \quad M_z = -t(\sigma) \times F_y, \tag{4}$$

where

$$F(\lambda, \alpha, \mu) = \begin{cases} \mu F_z(1 - \rho^3) & \text{for} \quad |\sigma| \leq |\sigma_{\text{sl}}|, \\ \mu F_z \text{ sgn}(\alpha) & \text{for} \quad |\sigma| > |\sigma_{\text{sl}}|, \end{cases}$$

$$\sigma_x = \frac{\lambda}{\lambda + 1}, \quad \sigma_y = \frac{\tan(\alpha)}{\lambda + 1}, \quad \sigma = \sqrt{\sigma_x^2 + \sigma_y^2},$$

$$a_{\text{cpl}} = a_{\text{cpl}_0}\sqrt{\frac{F_z}{F_{z0}}}, \quad C = 2c_p a_{\text{cpl}}^2$$

$$\theta = \frac{C}{3\mu F_z}, \quad \sigma_{\text{sl}} = \frac{1}{\theta}, \quad \rho = 1 - \theta\sigma,$$

$$t(\sigma) = \frac{l(1 - |\theta\sigma|)^3}{(3 - 3|\theta\sigma| + |\theta\sigma|^2)}.$$

## 4. Brush model adaptation

In order to use the brush model as a basis for friction estimation, it is desirable to validate the model. For validation purposes, tire force and moment data were created using "magic formula" tire model coefficients available in the literature (Pacejka, 2005). To account for the influence of road friction on the tire force characteristics, "magic formula" scaling factors previously published in the literature (Arosio, Braghin, Cheli, & Sabbioni, 2005; Braghin, Cheli, & Sabbioni, 2006) were used. To approximate the measurement/estimation uncertainty, the simulation data were corrupted with zero mean white noise. The model fitting algorithm is based on storing data points in the force–slip/moment–slip plane and using those points to compute the tire longitudinal/cornering stiffness and the tire–road friction coefficient by optimization. In other words, it is a method for the identification of parameters through determining the best fit between modeled and observed data. The optimization algorithm used is the method of Levenberg–Marquardt (LM) (Lourakis, 2005; Roweis, 1996). The algorithm of LM is an iterative technique to locate the minimum of a function with several variables, which is expressed as the sum of squares of real-valued nonlinear functions. It has become a standard technique in numerical solution of nonlinear least-squares problems and widely adopted in a broad spectrum of disciplines. The LM method can be thought of as a combination of gradient descent and Gauss–Newton methods. When the current solution is far from the correct one, the algorithm follows the gradient descent scheme, with a slower but guaranteed rate of convergence, whereas when the current solution is close to the correct solution, the algorithm reduces to the Gauss–Newton approximation.

In Figure 4, the results of adaptation of the brush model to tire data for different test conditions are shown. A description of the test conditions are given in Table 2.

From the results shown in Figure 4, it can be seen that, for the pure longitudinal slip, the coherence between the brush model and the reference curve is good. For the pure lateral slip, there are discrepancies in the lateral force and the self-aligning torque (SAT). As mentioned in previous research (Svendenius, 2003), the main reason for this discrepancy is the assumption of a stiff carcass.

## 5. Real-time implementation

The real-time parameter estimation algorithm used in this study is similar to the one presented in Hsu (2009). The complete real-time estimation algorithm is outlined below:

- Iteratively perform nonlinear least squares (NLLS) to the brush model on the batch of force–slip/moment–slip data, starting with initial estimates of braking/cornering stiffness and friction coefficient.
- To ensure that there is enough data for the NLLS fit to be meaningful, first initialize the process by placing a tire slip level threshold. The tire slip must exceed the threshold value before parameter estimation begins.
- The next step is to determine whether the tire force/moment has saturated sufficiently enough to estimate $\mu$. In parallel to the NLLS fit, apply the method of least squares to the data points to find the slope of the line through the origin. Calculate the incremental mean-squared error of both fits from the most recent vector of data points of length $N$. If

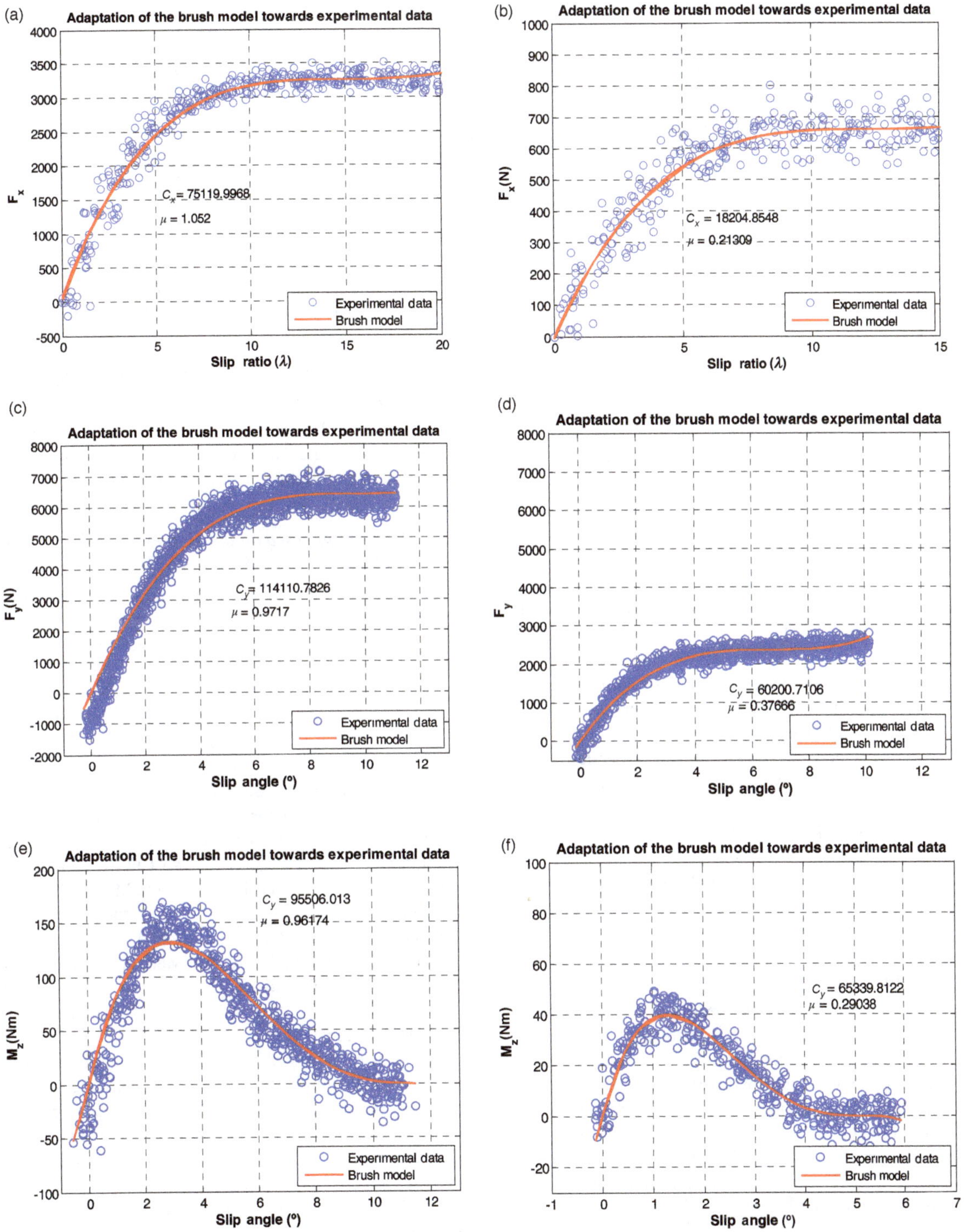


Figure 4. (a)–(f) Adaptation of the brush–tire model to tire measurement data (ref. Table 2 for a description of the test conditions). (a) Case 1, (b) case 2, (c) case 3, (d) case 4, (e) case 5 and (f) case 6.

Table 2. Tire test conditions.

| | Test description | Surface condition | Measured/estimated signals | Estimated parameters |
|---|---|---|---|---|
| Case 1 | Longitudinal force test | High $\mu$ | $F_x, F_z, \lambda$ | $C_x, \mu$ |
| Case 2 | Longitudinal force test | Low $\mu$ | $F_x, F_z, \lambda$ | $C_x, \mu$ |
| Case 3 | Lateral force test | High $\mu$ | $F_y, F_z, \alpha$ | $C_y, \mu$ |
| Case 4 | Lateral force test | Low $\mu$ | $F_y, F_z, \alpha$ | $C_y, \mu$ |
| Case 5 | Self-aligning moment test | High $\mu$ | $\tau_a, F_z, \alpha$ | $C_y, \mu$ |
| Case 6 | Self-aligning moment test | Low $\mu$ | $\tau_a, F_z, \alpha$ | $C_y, \mu$ |

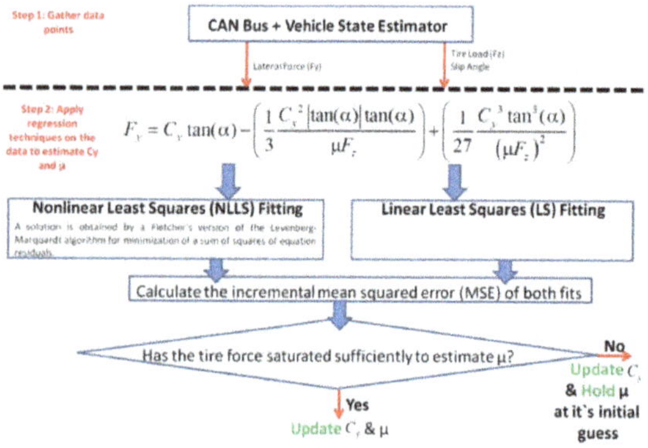


Figure 5. Tire parameter estimation algorithm – Lateral Force–Slip Regression Method.

the force–slip/moment–slip data are sufficiently nonlinear, update the value of $\mu$. Otherwise, hold the coefficient of friction estimate at its previous value.

- As new data are collected and appended to the existing batch, repeat the above algorithm.

A schematic representation of the "Lateral Force–Slip Regression Method" is shown in Figure 5.

Similar estimation schemes were used for the "Longitudinal Force–Slip Regression Method" and the "Moment–Slip Regression Method."

## 6. Parameter estimation results

Friction estimation using the real-time estimation technique presented in Section 4 has been tested on tire measurement data. This section investigates how the value of the highest available slip (tire force utilization level) affects the accuracy of the friction estimation. The following subsections present the results for each of the estimation methods, namely, "Longitudinal Force–Slip Regression Method," "Lateral Force–Slip Regression Method" and "Moment–Slip Regression Method."

### 6.1. *Estimation based on the Longitudinal Force–Slip Regression Method*

Using the estimation algorithm described previously with $\lambda_{\text{thres}} = 0.1$, $\mu_{\text{initial}} = 0.8$ and $C_{x_{\text{initial}}} = 110000$, force and slip ratio data are post-processed to yield longitudinal stiffness and friction coefficient estimates (Figure 6).

### 6.2. *Estimation based on the Lateral Force–Slip Regression Method*

With $\alpha_{\text{thres}} = 2°$, $\mu_{\text{initial}} = 0.8$ and $C_{y_{\text{initial}}} = 110000$, force and slip angle data is post-processed to yield cornering stiffness and friction coefficient estimates (Figure 7).

### 6.3. *Estimation based on the Moment–Slip Regression Method*

With $\alpha_{\text{thres}} = 0.5°$, $\mu_{\text{initial}} = 0.8$ and $C_{y_{\text{initial}}} = 110000$, moment and slip angle data are post-processed to yield cornering stiffness and friction coefficient estimates (Figure 8).

Based on the results shown in Figures 6–8, the required utilization of friction necessary to provide a friction estimate within the specified accuracy of $\pm 10\%$ is presented in Table 3. In the case of the "Force–Slip Regression Method," more than 75–80% of the available friction force must be generated before an accurate estimate can be derived. However, it is possible to estimate the tire road friction coefficient for lower levels of utilization ($\sim$30–40%) if SAT ("Moment–Slip Regression Method") is used as a basis for the estimator instead of the lateral force ("Lateral Force–Slip Regression Method").

The "Moment–Slip Regression Method" presents a better opportunity of estimating the friction coefficient for lower levels of utilization, since the SAT saturates before the lateral force is saturated (Figure 9).

It is also important to realize that the force-based approach is inherently limited to operate during either longitudinal or lateral excitation, that is, either during acceleration/braking or cornering. Since they are active during different instants, it is advisable to combine two or more methods (Figure 10) as also suggested in previous work (Ahn, 2011) and hence provide a continuous estimate of friction.

To be of greatest use to active safety control systems, an estimation method needs to offer earlier knowledge of the limits. The next section presents an implementation strategy for estimating the tire–road friction coefficient under low-slip conditions.

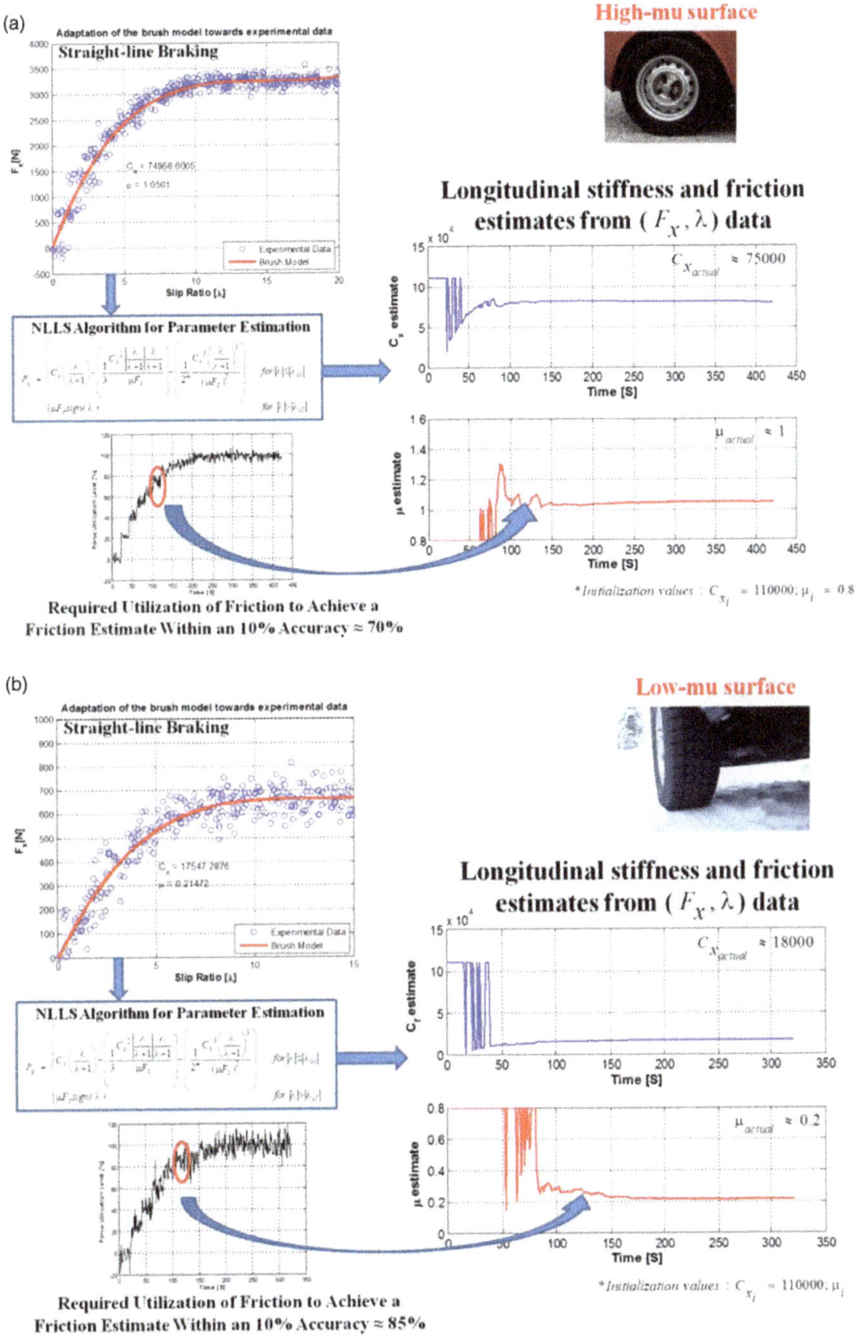


Figure 6. Longitudinal stiffness and friction coefficient estimates under (a) high $\mu$ conditions (case 1) and (b) low $\mu$ conditions (case 2).

## 7. Integrated tire–road friction estimation scheme

Availability of certain new technologies, popularly known as “intelligent tires” or “smart tires” (Morinaga; Yasushi Hanatsuka and Morinaga, 2013), hold the potential of providing real-time road surface condition information under low-slip rolling conditions. The implementation strategy for one such algorithm that utilizes sensor signals from an instrumented tire is presented in this section. The instrumented tire system was developed by placing accelerometers on the inner liner of a tire (Figure 11(a)). Figure 11(b) shows the final assembly of the instrumented tire with a high-speed slip ring attached to the wheel. Extensive

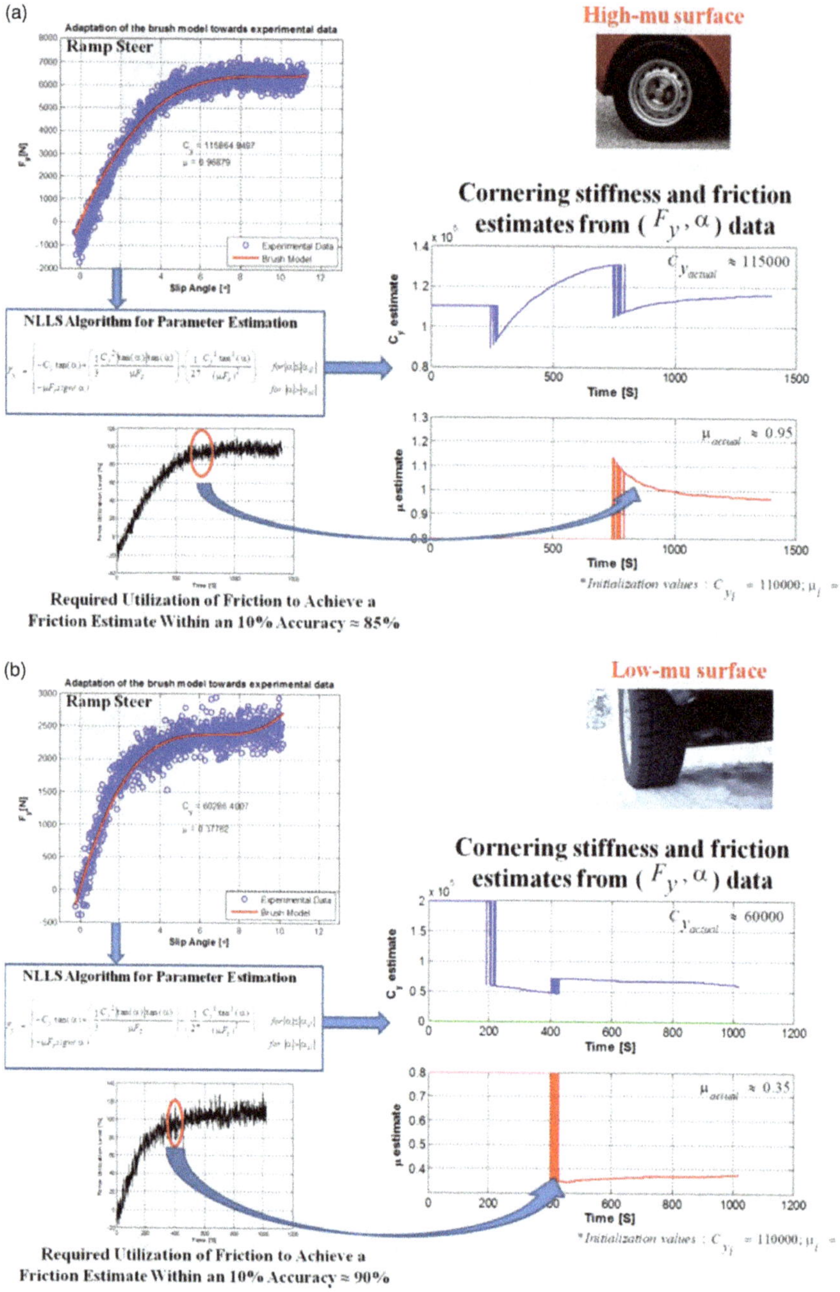


Figure 7. Lateral stiffness and friction coefficient estimates under (a) high $\mu$ conditions (case 1) and (b) low $\mu$ conditions (case 2).

dynamic tests of the tire were conducted using the in-house mobile tire test rig shown in Figure 11(c) and 11(d). An example of the acceleration signal is shown in Figure 12.

The effect of tire load, translational speed, varying pressure conditions and road surface roughness on the tire vibration spectra were studied by varying each of these parameters by carrying out extensive outdoor testing of the instrumented tire under free-rolling, traction/braking and steering conditions.

The power spectrum road of each accelerometer signal from these tests was computed using Welch's averaged modified periodogram method for spectral estimation (Figure 13). Analyzing the dynamic test results, it was

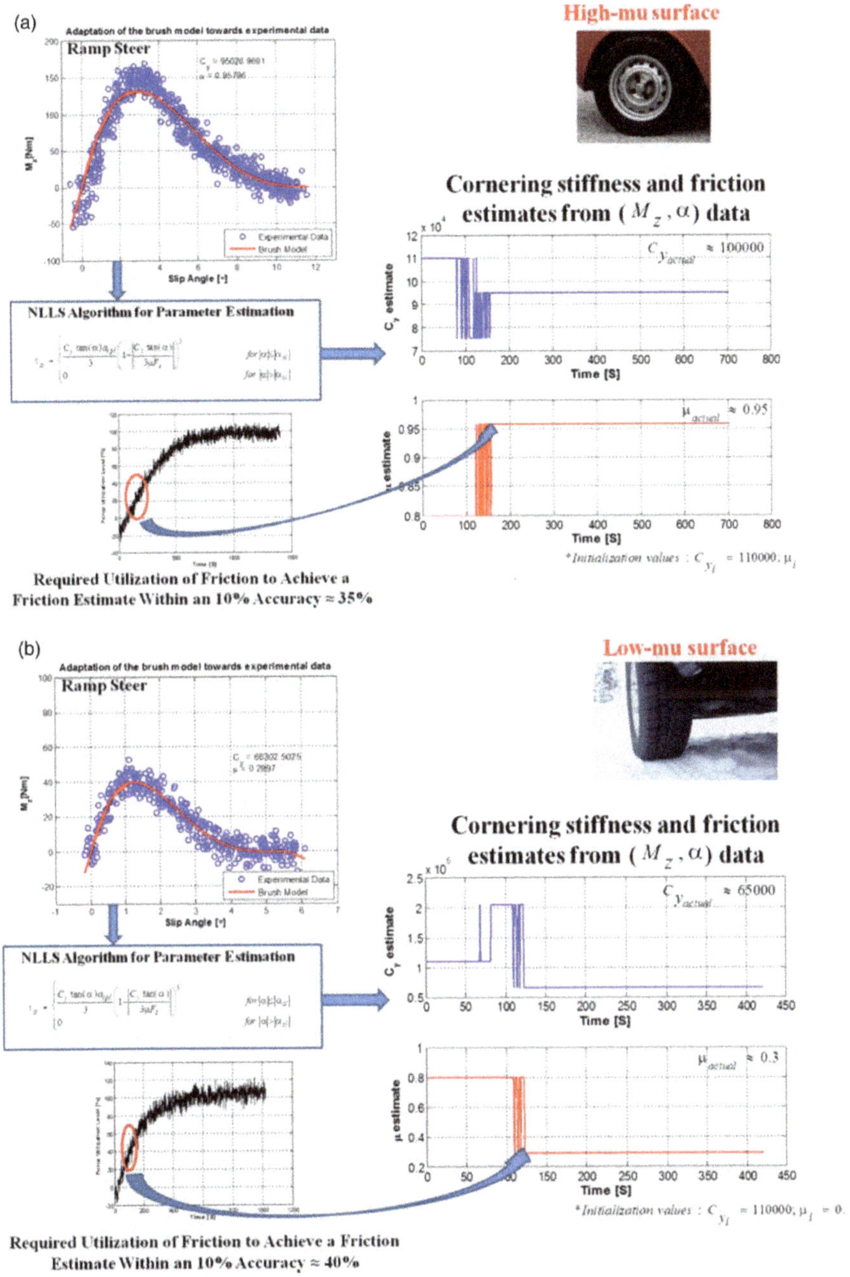


Figure 8. Lateral stiffness and friction coefficient estimates under (a) high $\mu$ conditions (case 1) and (b) low $\mu$ conditions (case 2).

concluded that, a marked difference was noticed in the concentration of the higher frequencies on the spectrum of the circumferential acceleration signal of the tire tested on different surface conditions (Figure 13(d)). This variation in the circumferential acceleration signal power spectral density (PSD) on different road surface conditions presented an opportunity to characterize the road condition using the tire vibration pattern information.

The proposed intelligent tire-based surface condition estimating algorithm consists of detecting the circumferential vibration of a tire of a running vehicle; dividing the detected tire vibration into vibration in a pre-trailing domain, the domain existing before a trailing edge position; and vibration in a post-trailing domain, the domain existing after a trailing edge position. Thereafter extracting signals of tire vibration only from the pre-trailing domain;

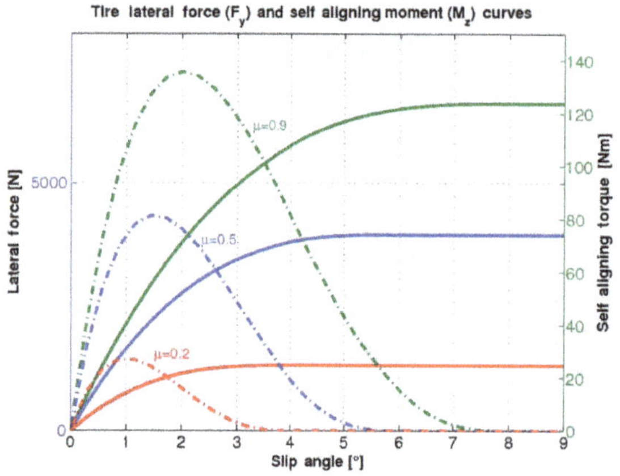


Figure 9. Lateral force and aligning torque versus slip angle. General behavior for a pneumatic tire.

Table 3. Required utilization of friction (in percent) to achieve a friction estimate within an accuracy of $\pm 10\%$.

| Estimation methodology | Friction coefficient ($\mu$) | Required utilization of friction (%) |
|---|---|---|
| Longitudinal Force–Slip Regression Method | High $\mu$ | 70 |
| | Low $\mu$ | 85 |
| Lateral Force–Slip Regression Method | High $\mu$ | 85 |
| | Low $\mu$ | 90 |
| Moment–Slip Regression Method | High $\mu$ | 35 |
| | Low $\mu$ | 40 |

obtaining a time-series waveform of tire vibration including only the frequencies in a predetermined frequency band by passing the extracted signals through a band-pass filter of the predetermined frequency band; calculating a vibration level in the predetermined frequency band and estimating a road surface condition based on the calculated vibration level.

The predefined frequency bands being a low-frequency band (e.g. 10–500 Hz band) and a high-frequency band (e.g. 600–2500 Hz). The motivation for only using the pre-trailing domain signal is the larger difference in the PSD of the pre-trailing domain signal, when compared with the PSDs obtained using the entire signal or using the signal from the post-trailing domain. To determine these differences, the instrumented tire was first driven on a dry surface and then on a wet road surface at different speeds and the change in the vibration level ratio ($R$) was measured, where $R$ is the ratio of the aforementioned

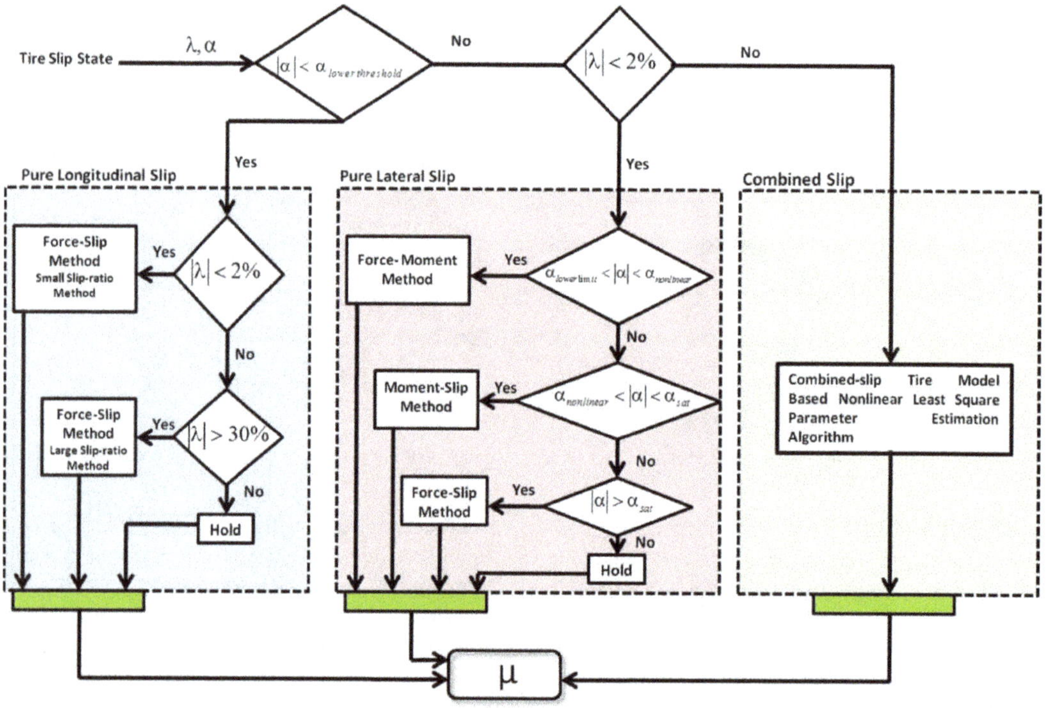


Figure 10. Integrated friction estimation algorithm – flow diagram.

Figure 11. Intelligent tire application: (a) sensor mounting location, (b) instrumented tire assembly, (c) mobile tire test rig and (d) test rig attached to the towing vehicle.

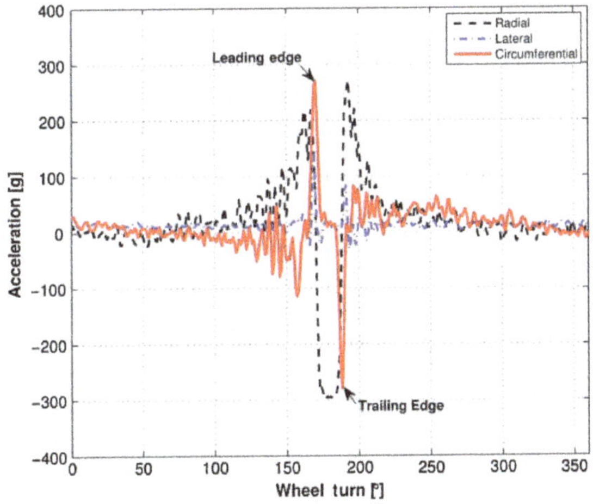


Figure 12. Measured acceleration signal for one rotation.

high-frequency band and the low-frequency band vibration level. It is evident from Figure 14 that the vibration level ratio $R$ increased as the tire was tested on the wet road surface. This change in the vibration level ratio can be attributed to the increased slippage of the tire, and thus, it has been confirmed that the slipperiness of a road surface can be decided by setting a proper threshold value.

For this purpose, a fuzzy logic classification approach was developed for the real-time implementation of the proposed algorithm. The application of fuzzy logic to solve the classification problem is motivated by its noise tolerance to the vibration data retrieved from sensors, and its ability for real-time implementation while ensuring robustness with respect to imprecise or uncertain signal interpretation. Figure 15 shows the fuzzy controller architecture.

Based on the interdependence of all the inputs for a given road surface condition and the way they effect the vibration spectra of a tire, a set of linguistic rules were developed to identify the road surface condition. The classifier performance was validated on smooth asphalt, regular asphalt, rough asphalt and wet asphalt (Figure 16).

Two different tests were performed to study the classifier performance. The first test involved testing the tire under free-rolling and low-slip conditions (low force utilization). The second test involved testing the tire under high-slip conditions (high force utilization). For the first test (free-rolling and low-slip conditions), the classifier was successfully able to distinguish between the different road surface conditions as shown in Figure 17. However, for the second test (high-slip conditions), classifier performance was unsatisfactory (Figure 18). Higher misclassification rates under high-slip conditions were attributed to the increased vibration levels in the circumferential

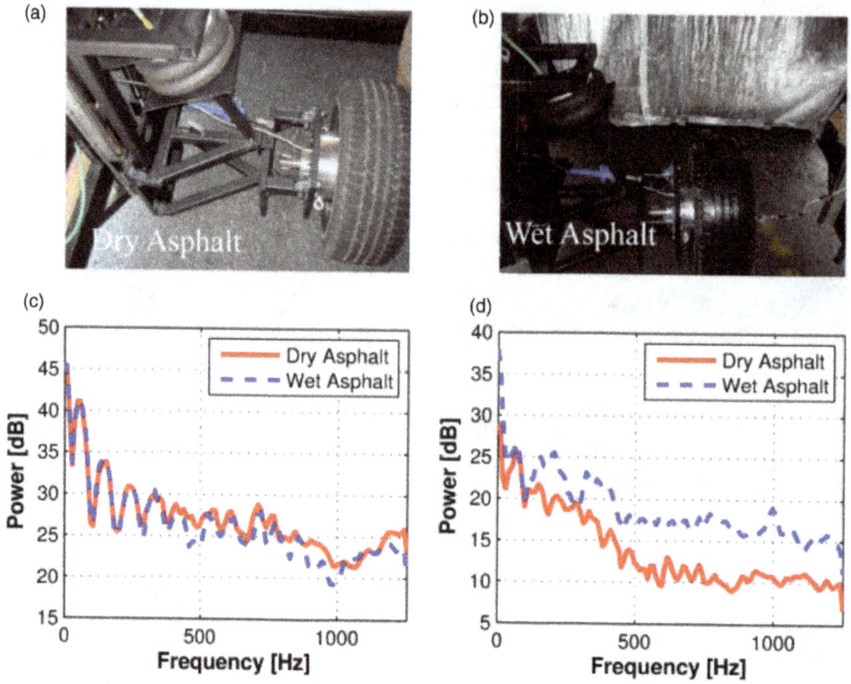


Figure 13. Tire tested on different road surface conditions: (a) dry surface testing and (b) wet surface testing; roughness dependence study: (c) radial signal PSD and (d) circumferential signal PSD.

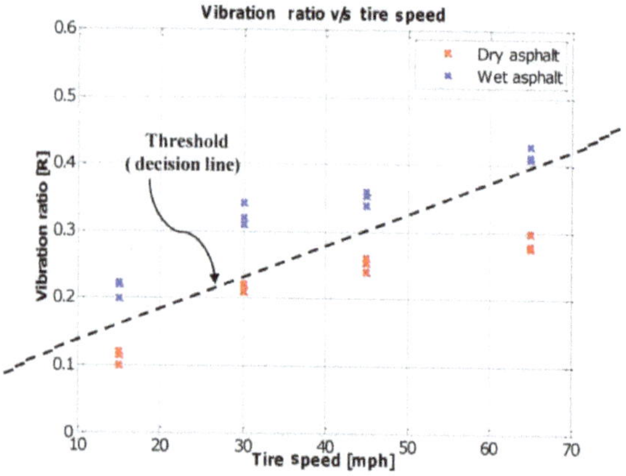


Figure 14. Vibration ratio on dry and wet surface conditions for a range of tire speeds.

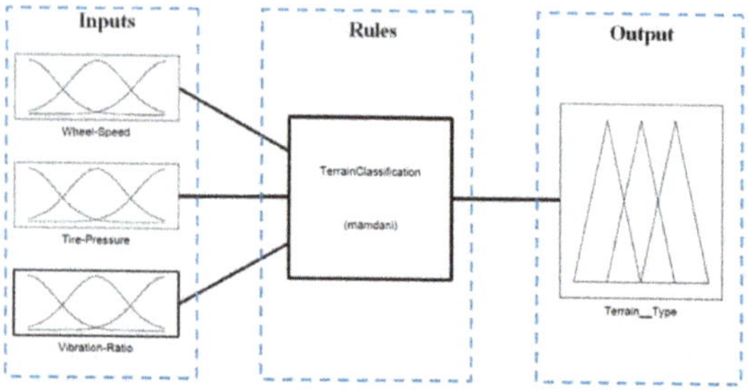


Figure 15. Fuzzy logic-based controller architecture.

acceleration signal due to the stick/slip phenomenon linked to the tread block vibration modes (Figure 19).

The requirement of a more complex event detection algorithm for distinguishing the high-frequency content of a signal due to the tread block mobility effects, under high-slip conditions, makes the proposed fuzzy logic classifier unsuitable for friction estimation under high-slip conditions.

Hence, it was proposed to use a model-based approach as presented in the previous section to estimate road surface friction under high-slip conditions. Finally, it was proposed to use an integrated approach (Figure 20) using the intelligent tire-based friction estimator and the model-based estimator which would reliably estimate friction for a wider range of excitations (both low-slip and high-slip conditions). Using an integrated approach, the road surface friction classifier was successfully able to distinguish between the different road surface conditions as shown in Figures 21 and 22.

## 8. Application of road friction information in vehicle control systems – development of new control strategies

A change in the peak grip potential of the tire (Figure 23(a)) not only affects the "limit" handling behavior of the vehicle, but is also known to affect the vehicle

Figure 16. Tire tested on different road surface conditions: (a) rough asphalt, (b) regular asphalt, (c) smooth asphalt and (d) wet asphalt.

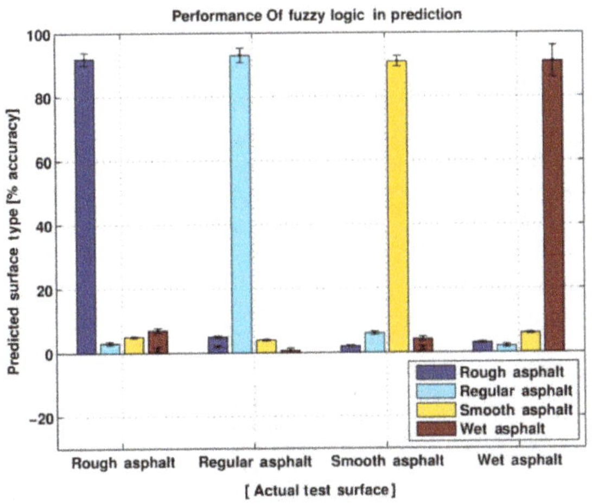


Figure 17. Performance of the fuzzy logic classifier – low-slip conditions.

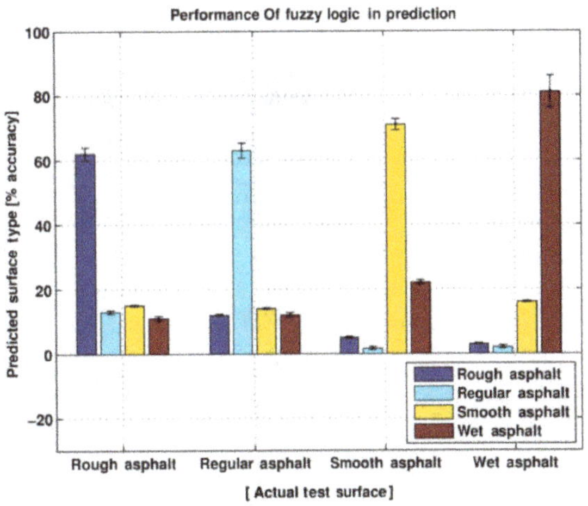


Figure 18. Performance of the fuzzy logic classifier – high-slip conditions.

"linear range" handling behavior. This change in the vehicle "linear range" behavior is due to the influence of road friction on the tire stiffness in the low-slip region (Figure 23(b)). Moreover, the peak slip ratio position of the maximum coefficient of friction varies for different road conditions (Figure 24). Hence, in the context of an anti-lock brake system (ABS) based on a fixed thresholding rule-based algorithm, it cannot be expected that an ABS controller that is optimized for dry asphalt performs as reliably and efficiently on wet or icy surfaces.

To quantify the performance benefits for an ABS controller using road friction condition information, a modified ABS algorithm has been developed, as shown in Figure 25. The modified controller leverages friction information

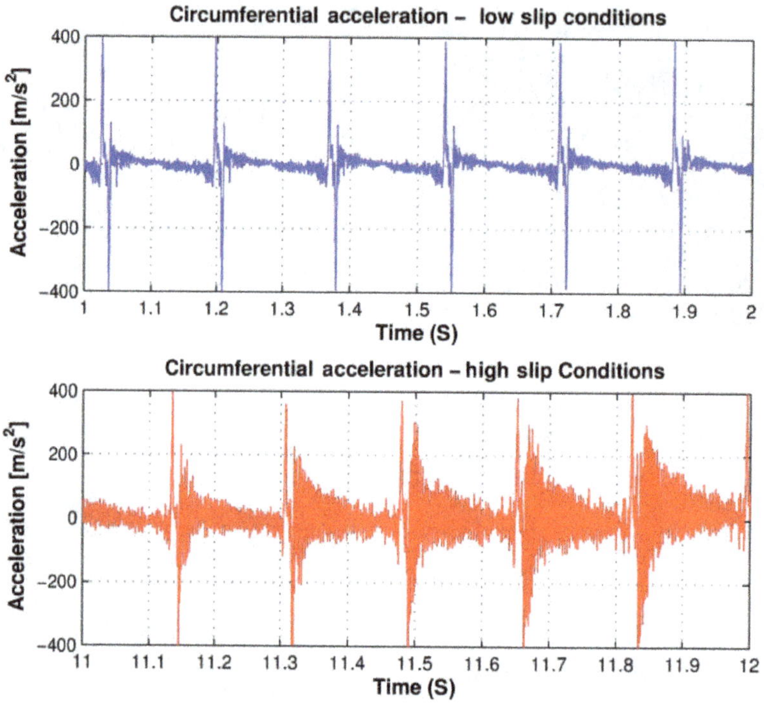


Figure 19. Circumferential acceleration signal under low-slip conditions (top), and increased vibration levels in the circumferential acceleration signal under high-slip conditions (bottom).

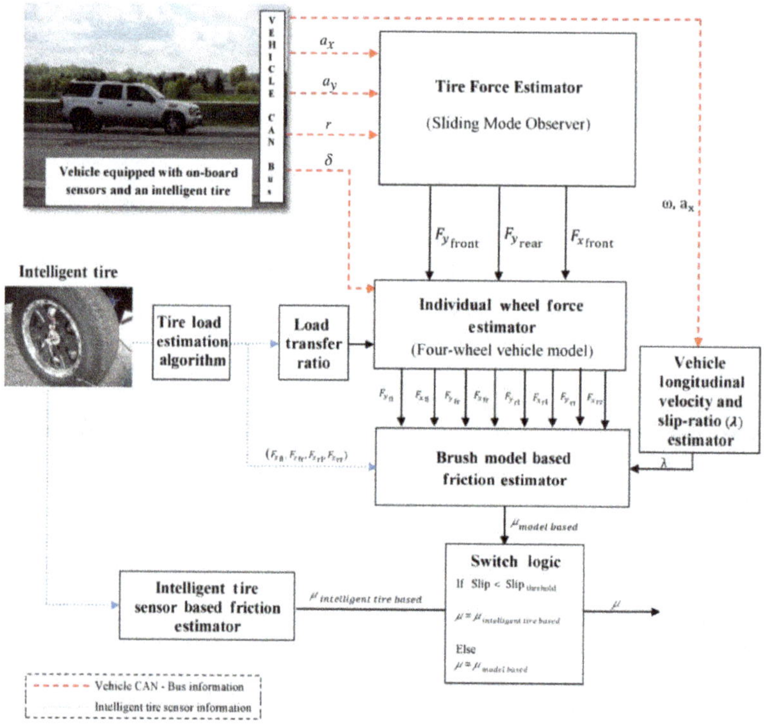


Figure 20. Architecture of the proposed integrated approach using an intelligent tire-based friction estimator and the model-based estimator.

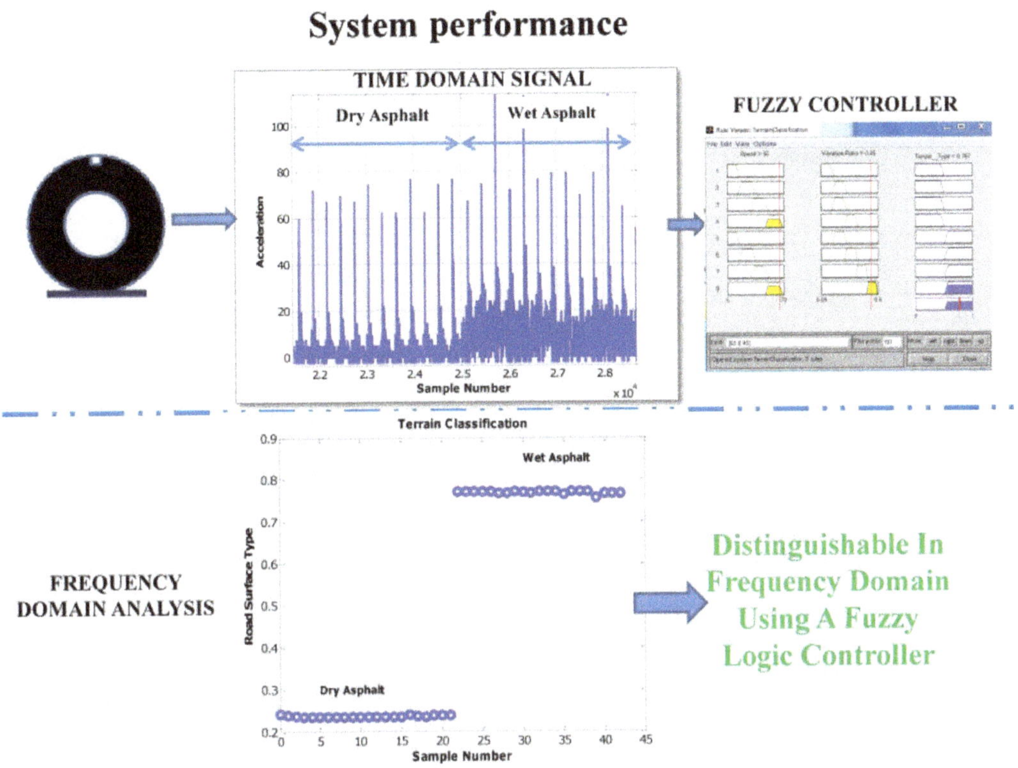


Figure 21. Classification performance on dry and wet asphalt.

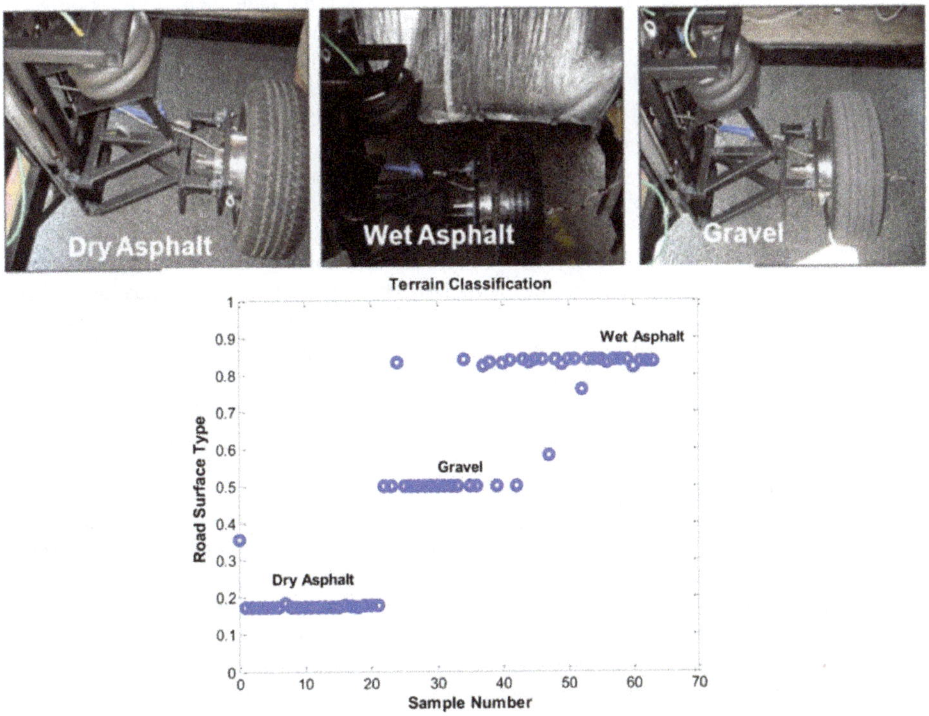


Figure 22. Classification performance on dry asphalt, gravel and wet asphalt.

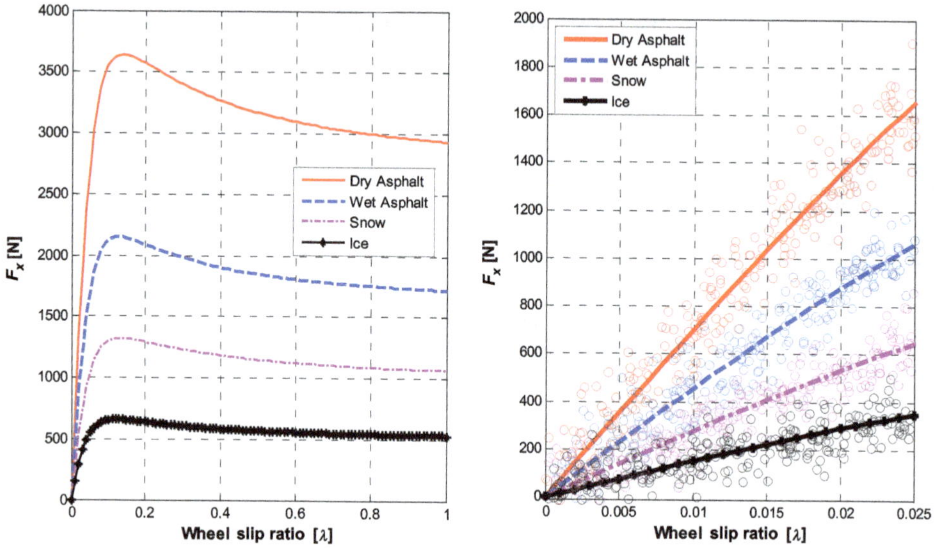


Figure 23. (a) Longitudinal tire force under different road surface conditions, and (b) longitudinal tire force in the small-slip region under different road conditions.

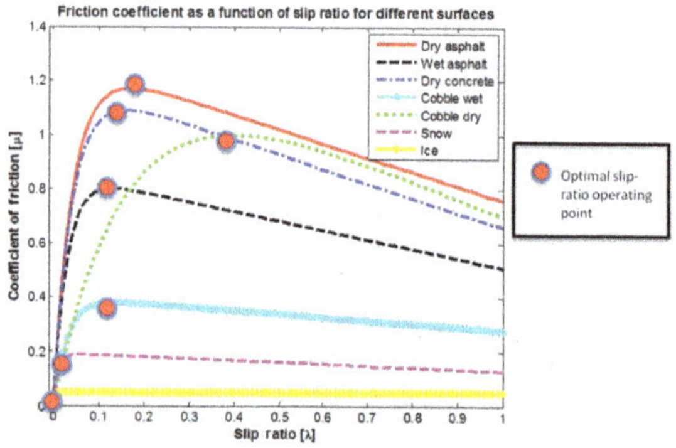


Figure 24. Position of slip values at which maximum friction (braking force) is obtained.

to change the optimal slip point thresholds, thus maximizing the braking force. Moreover, the controller uses a brake preconditioning algorithm with a friction adaptation strategy to start braking with the optimal brake pressure, and thus ensuring that the early cycles of braking are more efficient. Simulations were carried out on a series of braking maneuvers to examine the possible improvements in the ABS system performance. The results reveal that the presence of road condition information allows for a considerable decrease in the stopping distance (Singh et al., 2013). Most impressive improvements are obtained for the jump-$\mu$ tests (Figure 26). In light of these results, we conclude that the knowledge of road surface condition can be quite favorable for enhancing the current ABS algorithms.

In the context of a classical ESC system based on a model reference approach, the desired values of yaw rate and body sideslip angle are generated from a reference model, which takes into account the vehicle velocity, the driver input, tire/axle load and cornering stiffness (understeer/oversteer behavior). The weighting factor, which determines the balance between the yaw rate tracking and sideslip regulation, depends primarily on a term combining the estimated rear axle sideslip angle and its derivative. Therefore, an accurate online estimate of the vehicle sideslip angle is critical for the effective operation of an ESC system. In production cars, the vehicle sideslip angle is not measured because this measurement requires expensive equipment such as optical correlation sensors. Most production vehicles rely on observers based on vehicle dynamics models for indirectly estimating the vehicle sideslip angle. These observers perform reasonably well in normal driving situations, when the steering characteristics specify a tight connection between the steering wheel angle, yaw rate, lateral acceleration and vehicle sideslip angle. When a vehicle is near or at the limit of adhesion, tire forces and consequently the yaw dynamics strongly depend on the surface coefficient of friction. For example, limit tire forces on ice can be about 10 times smaller than on dry surface. The vehicle model used within the observer should therefore be adapted to the changing surface friction. The coefficient of friction, however, is unknown and has to be estimated. Thus, the estimation of sideslip angle depends on another estimate, which increases the potential for errors. Uncertainties in sideslip angle estimation lead the ESC system to be conservatively calibrated in order to balance robustness concerns with performance. This inherently sacrifices some level of performance.

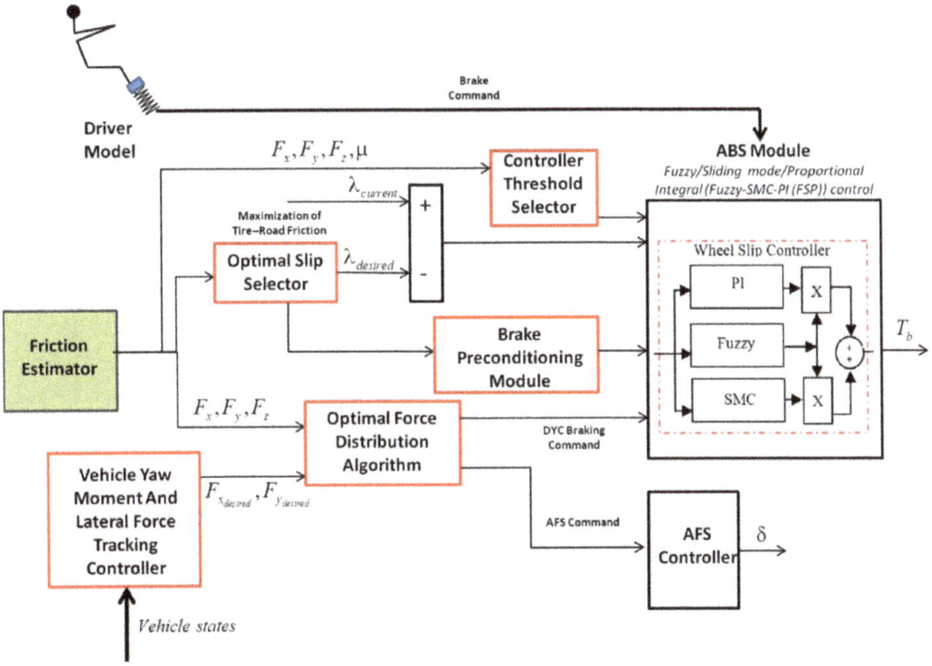


Figure 25. Modified ABS algorithm designed to use road friction information.

From the foregoing discussion, it is apparent that the knowledge of road surface condition would be beneficial in improving the accuracy of sideslip angle observers, which eventually would enhance the performance of ESC systems. To quantify the performance benefits, an enhanced ESC controller based on active front steering (AFS) and direct yaw moment control (DYC) has been developed (Figure 27). The proposed controller consists of a full vehicle state estimator with friction adaptation (Singh, 2012). The vehicle sideslip is estimated using an extended Kalman filter (EKF)-based observer. More details pertaining to the modified ESC algorithm are given in Table 4.

Simulation results show that the new control strategy aiming to use all of the information available from the vehicle state estimator can significantly enhance vehicle stability during emergency evasive maneuvers on various road conditions ranging from dry asphalt to very slippery packed snow road surfaces (Figure 28).

Another vehicle safety system that is becoming more prevalent in the vehicle industry is the advanced driver assistance system (ADAS). Typically, ADAS features three technologies: collision mitigation braking system (CMBS), lane keeping assist system and ACC. CMBS is an active safety system that helps the driver to avoid or mitigate rear-end collisions. It uses forward-looking sensors to detect obstacles ahead of the vehicle. The systems use relative distance, relative velocity and vehicle velocity information to warn the driver or control the vehicle. Specifically, a warning critical distance is defined as a function of vehicle velocity and relative velocity.

From Figure 29, we can see that if friction information would be available, the critical warning and critical braking distances could be calculated more precisely (since the deceleration rate for the vehicles ($\alpha_1$, $\alpha_2$) depends on the maximum tire–road friction available). Using a high default value for the friction coefficient causes the systems to lose some of their safety potential on low-friction surfaces. Using a low or medium default value would on the other hand cause the safety systems to activate too early in high-friction conditions, taking the driver "out of the loop" possibly unnecessarily.

The modified algorithm used in this study assumes to have full knowledge of the road conditions (Figure 30). Consequently, the collision mitigation algorithm adapts its critical distance (warning/braking distance) definitions when the road conditions change (Figure 31).

A parametric analysis aimed at evaluating the benefits induced by the introduction of friction information has been carried out. These simulations will be used to show the benefit of using friction estimation in conjunction with a collision mitigation brake system algorithm. In the test case, the host and the lead car are both traveling at 27.8 m/s with a separation of 50 m. The lead car suddenly applies the brakes and decelerates. The host vehicle maintains its velocity, which simulates a driver

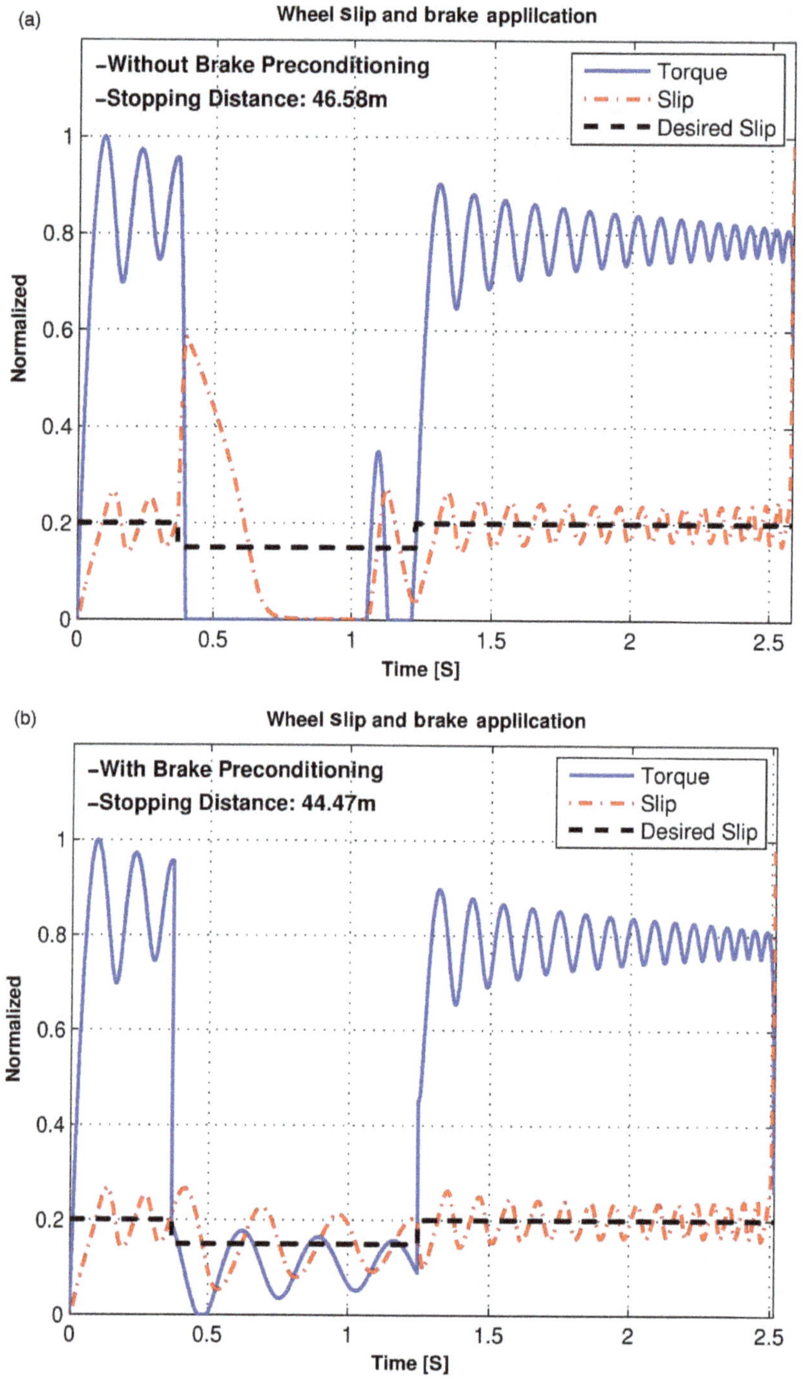


Figure 26. ABS performance – (a) without knowledge of road friction conditions and (b) with knowledge of road friction conditions.

who is unaware of the critical nature of the situation. Figure 32 shows the vehicle response when the collision mitigation brake system algorithm without the friction adaptation strategy is used (i.e. friction information is unavailable). The relative velocity at impact in this case is 14.5 m/s.

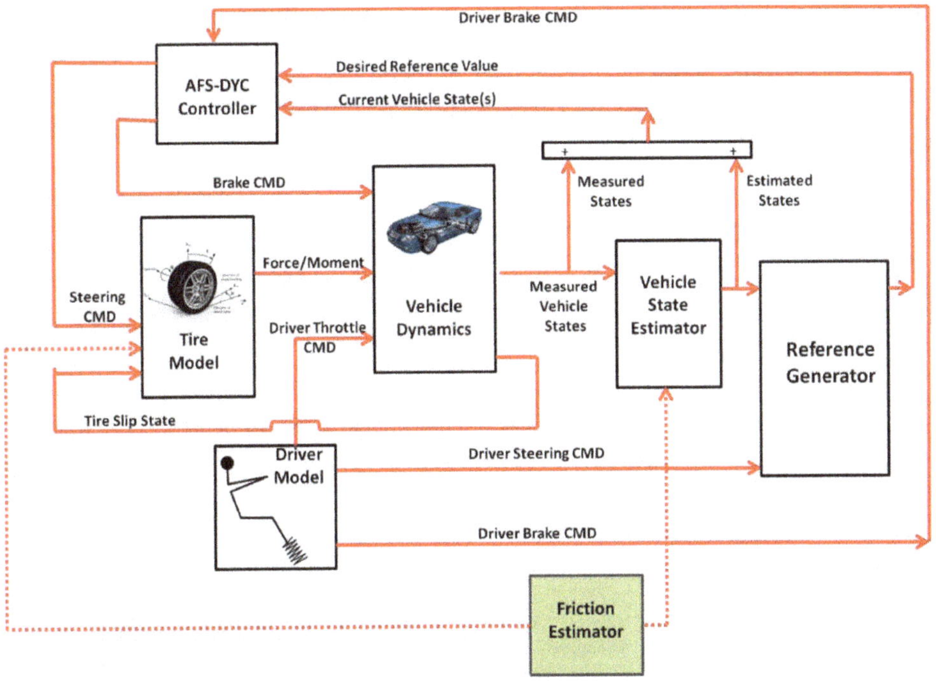


Figure 27. Modified ESC algorithm designed to use road friction information.

Table 4. Enhanced ESC system model details.

| | | |
|---|---|---|
| Reference generator | Sliding mode control (SMC) strategy | The model responses of vehicle sideslip angle and yaw rate are described. Total lateral force and the total yaw moment required for the controlled vehicle to follow the model responses are estimated using an SMC strategy |
| AFS controller | SMC strategy | Difference between vehicle measured yaw rate and desired yaw rate is considered as the sliding surface. Control law is based on tire force feedback and is obtained from a nonlinear eight-degree-of-freedom vehicle model |
| DYC controller | Rule based | Implemented through braking one of the four wheels based on detection of understeer or oversteer driving situations |
| Tire model | Combined slip model | Tire forces are modeled using magic formula tire model *with friction adaptation* |
| Vehicle state estimator (Singh, 2012) | Full state estimation using a nonlinear vehicle model | Vehicle sideslip is estimated using an EKF-based observer *with friction adaptation* |

Figure 33 shows the vehicle response when the collision mitigation brake system algorithm with the friction adaptation strategy is used (i.e. friction information is available). In this simulation, the driver is completely out of the loop, so the collision mitigation brake system brings the vehicle to a rest. Notice that the plot shows that the vehicles collide at $\sim$7.7 s. The relative velocity at impact is 9.5 m/s. Clearly, the modified algorithm with friction information (Figure 21) applies the brakes sooner during degraded road conditions, which gives the vehicle more time to slow down. As a result, the impact speed and the impact energy are reduced. These results can be improved even further by increasing the friction adaption scaling factors.

From the foregoing discussion, we can thus conclude that road friction condition information would enable slip control systems (ABS, traction control system, ESC, etc.) to be started with the optimal initial parameters for the friction situation at hand. Moreover, accurate friction information would also enable the CMBS to start intervention from a more optimal distance in every road condition.

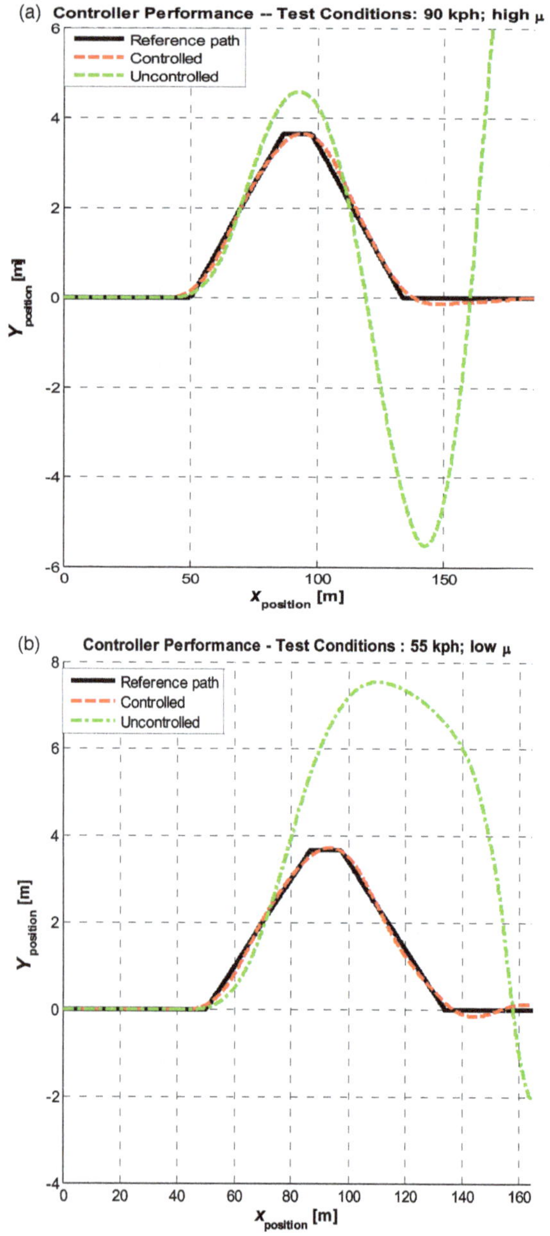


Figure 28. ESC performance: (a) high $\mu$ conditions and (b) low $\mu$ conditions.

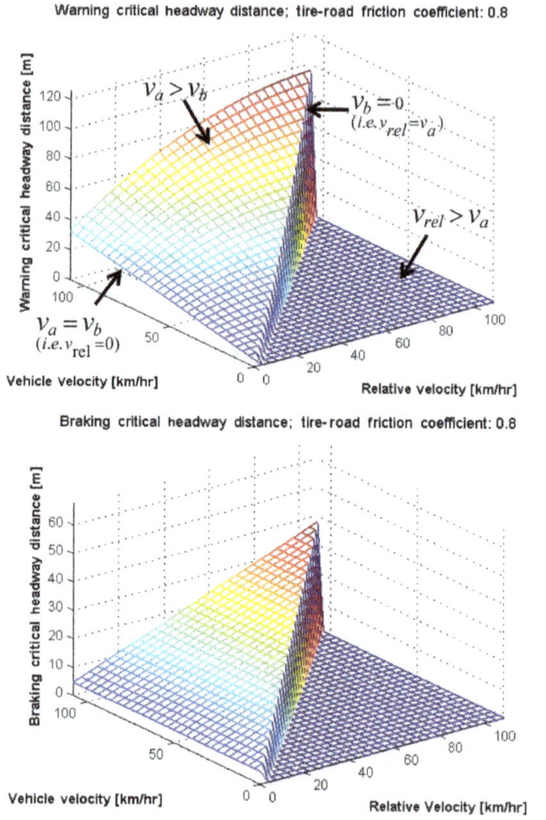


Figure 29. (a) Critical warning distance and (b) critical braking distance.

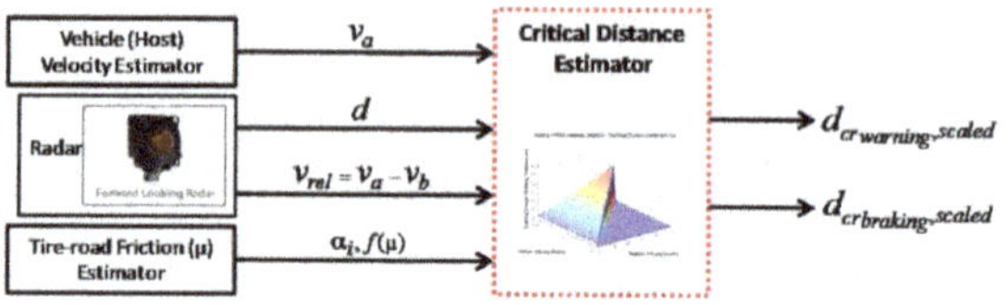


Figure 30. Modified algorithm designed use tire–road friction information to adapt its critical distance definitions.

## 9. Conclusion

Road friction is an important parameter for vehicle safety applications, but difficult to estimate accurately in all driving situations and weather conditions. Friction estimation methods based on the applied slip angle and slip ratio have been proposed earlier, but in the case of a freely rolling tire, the friction estimation is still an unsolved topic. This study uses a three-axis accelerometer on the inner liner of a tire to detect the tire road friction potential. More specifically, a two pronged approach was adopted to estimate tire–road friction coefficient. Firstly, a "force–slip" and "moment–slip" model-based approach is proposed. The primary shortcoming of the "Force–Slip Regression Method" identified is the requirement that the vehicle must enter the nonlinear region of handling before friction can be estimated. Thus, in the case of the "Force–Slip Regression Method," an estimation algorithm based on the brush model will work, but high friction utilization ($\sim$75–80%) is required for an accurate friction estimate. The "Moment–Slip Regression Method" based on total aligning moment has the benefit of knowing the coefficient of friction earlier, that is, under lower levels of utilization ($\sim$30–40%).

(a) **The scaled warning distance is defined as follows:**

$$d_{\text{cr warning, scaled}} = \left(\frac{v_a^2}{\mu g} - \frac{(v_a - v_{rel})^2}{\mu g}\right) + v_a \cdot \tau_1 + (v_a - v_b) \cdot \tau_2 + d_o$$

$\alpha_1 = \alpha_2 \approx \mu g$ } Maximum deceleration condition – depends on tire-road friction coefficient

$$^*ma = \sum F_x \Rightarrow ma = \mu(F_{z1} + F_{z2} + F_{z3} + F_{z4})$$
$$\text{or } ma = \mu mg \Rightarrow a = \mu g$$

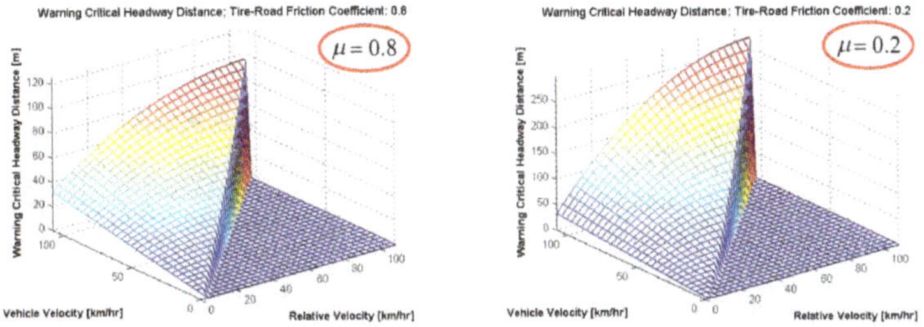


(b) **The scaled braking distance is defined as follows:**

$$d_{\text{cr braking, scaled}} = \left(v_{rel} \cdot (\tau_1 + \tau_2) + 0.5 \cdot \alpha_2 \cdot (\tau_1 + \tau_2)^2\right) \cdot f(\mu)$$

*where*:

$$f(\mu) = \begin{cases} f(\mu_{\min}) & \text{if } \mu \le \mu_{\min} \\ f(\mu_{\min}) + \dfrac{f(\mu_{norm}) - f(\mu_{\min})}{\mu_{norm} - \mu_{\min}} \cdot (\mu - \mu_{\min}) & \text{if } \mu_{\min} < \mu < \mu_{norm} \\ f(\mu_{norm}) & \text{if } \mu \ge \mu_{norm} \end{cases}$$

(piece-wise linear scaling function)

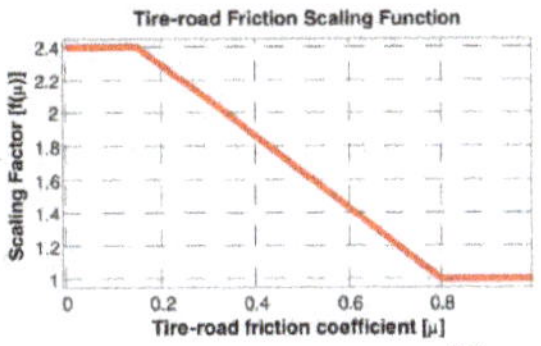


*Parameters:
$f(\mu_{norm}) = 1$
$f(\mu_{\min}) = 2.4$
$\mu_{norm} = 0.8$
$\mu_{\min} = 0.15$

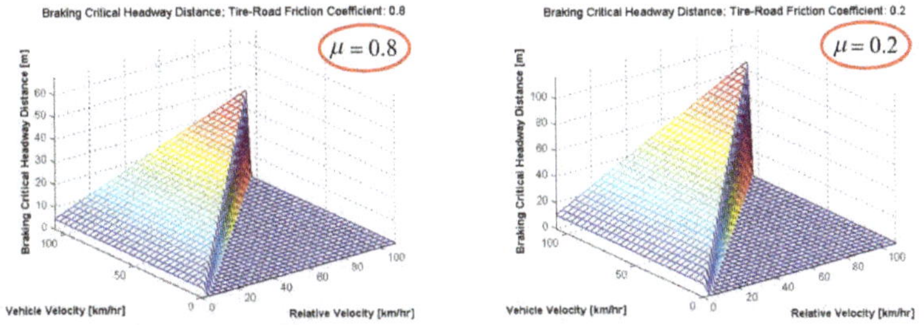


*Parameters:
$\tau_1 = 0.9s$
$\tau_2 = 0.2s$
$d_o = 5m$
$\alpha_2 = 7.5m/s^2$

Figure 31. (a) Scaled warning distance and (b) scaled braking distance.

To be of greatest use to active safety control systems, an estimation method needs to offer even earlier knowledge of the limits, that is, ideally offer knowledge on peak friction level under free-rolling conditions. To achieve this, an integrated approach using the intelligent tire-based friction estimator and the model-based estimator is proposed. This would give us the capability to reliably estimate friction for a wider range of excitations. The proposed intelligent tire-based method characterizes the road surface friction level using the measured frequency response of the tire vibrations and provides the capability to estimate the tire road friction coefficient under extremely lower levels of force utilization, that is, under free-rolling to low-slip excitation conditions. The ability to reliably estimate tire–road friction coefficient is important for maximizing the performance of vehicle control systems, which work well only when the tire force command computed by the safety systems is within the friction limit. The development of a sensorized smart/intelligent tire system is expected to eliminate some of the vehicle sensors and provide accurate, reliable and real-time information about magnitudes, directions and limits of force for each tire. Benefits of application of knowledge of friction potential have been demonstrated for an ABS, ESC system and CMBS.

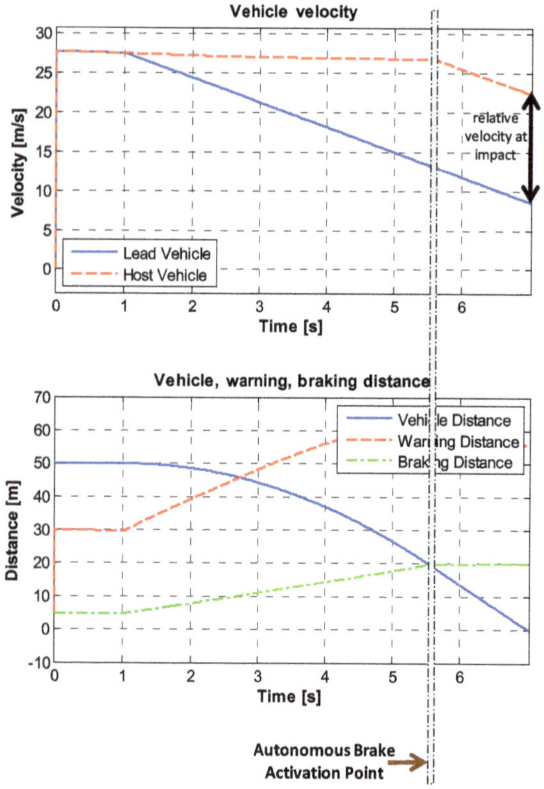


Figure 32. Friction condition unknown (assumed $\mu = 0.8$; actual $\mu = 0.25$).

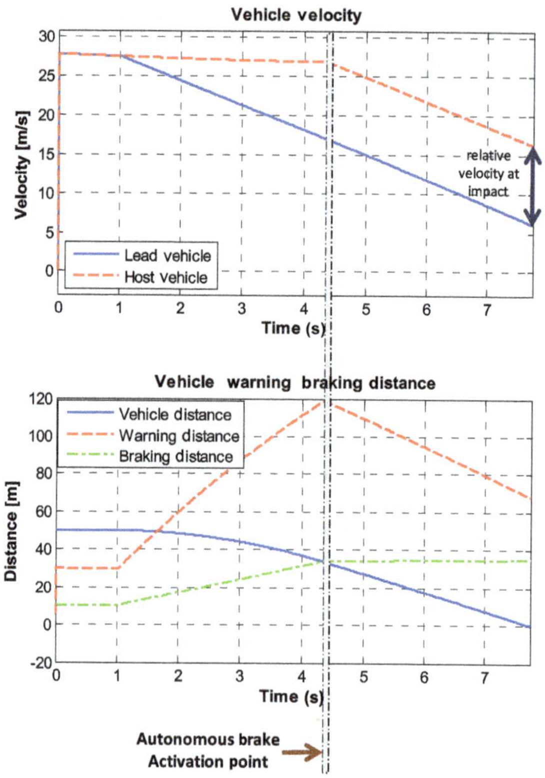


Figure 33. Friction condition known (assumed $\mu = 0.25$; actual $\mu = 0.25$).

## Disclosure statement

No potential conflict of interest was reported by the author(s).

## ORCID

Kanwar Bharat Singh http://orcid.org/0000-0001-7795-007X

## References

Ahn, C. S. (2011). *Robust estimation of road friction coefficient for vehicle active safety systems*. Ann Arbor: Department of Mechanical Engineering, The University of Michigan.

Andersson, M., Bruzelius, F., Casselgren, J., Hjort, M., Löfving, S., Olsson, G., … Yngve, S. (2010). *Road friction estimation, Part II – IVSS project report*. Retrieved from http://www.ivss.se/upload/ivss_refll_slutrapport.pdf

Arat, M. A., Singh, K. B., & Taheri, S. (2013). Optimal tire force allocation by means of smart tire technology. *SAE International Journal of Passenger Cars–Mechanical Systems*, *6*, 163–176.

Arat, M. A., Singh, K. B., & Taheri, S. (2014). An intelligent tire based adaptive vehicle stability controller. *International Journal of Vehicle Design*, *65*(2/3), 118–143.

Arosio, D., Braghin, F., Cheli, F., & Sabbioni, E. (2005). Identification of Pacejka's scaling factors from full-scale experimental tests. *Vehicle System Dynamics*, *43*, 457–474.

Braghin, F., Cheli, F., & Sabbioni, E. (2006). Environmental effects on Pacejka's scaling factors. *Vehicle System Dynamics*, *44*, 547–568.

Braghin, F., Cheli, F., Melzi, S., Sabbioni, E., Mancosu, F., & Brusarosco, M. (2009). *Development of a cyber tire to enhance performances of active control systems*. Presented at the 7th EUROMECH Solid Mechanics Conference, Lisbon, Portugal.

Chankyu, L., Hedrick, K., & Kyongsu, Y. (2004). Real-time slip-based estimation of maximum tire–road friction coefficient. *IEEE/ASME Transactions on Mechatronics*, *9*, 454–458.

Cheli, F., audisio, G., brusarosco, M., mancosu, F., Cavaglieri, D., & melzi, S. (2011a). *Cyber tyre: A novel sensor to improve vehicle's safety*. Presented at the SAE 2011 World Congress & Exhibition, Detroit, MI.

Cheli, F., Sabbioni, E., Sbrosi, M., Brusarosco, M., Melzi, S., & d'alessandro, V. (2011b). *Enhancement of ABS performance through on-board estimation of the tires response by means of smart tires*. Presented at the SAE 2011 World Congress & Exhibition, Detroit, MI.

Cheli, F., Leo, E., Melzi, S., & Sabbioni, E. (2010). On the impact of "smart tyres" on existing ABS/EBD control systems. *Vehicle System Dynamics*, *48*, 255–270.

Erdogan, G. (2009). *New sensors and estimation systems for the measurement of tire–road friction coefficient and tire slip variables*. Minneapolis, MN: Mechanical Engineering, University of Minnesota.

Erdogan, G., Hong, S., Borrelli, F., & Hedrick, K. (2011). *Tire sensors for the measurement of slip angle and friction coefficient and their use in stability control systems*. Presented at the SAE 2011 World Congress & Exhibition, Detroit, MI.

Germann, S., Wurtenberger, M., & Daiss, A. (1994). Monitoring of the friction coefficient between tyre and road surface. In *Control applications*. Proceedings of the Third IEEE Conference on, 1, 613–618.

Google Scholar search articles on tire road friction estimation. Retrieved from http://scholar.google.com/scholar?q = tire + road + friction + estimation&hl = en&as_sdt = 0&as_vis = 1&oi = scholart&sa = X&ei = eSU8VJbFLcf3yQTyy YLIAg&ved = 0CBsQgQMwAA

Gustafsson, F. (1997). Slip-based estimation of tire–road friction. *Automatica*, *33*, 1087–1099.

Hsu, Y.-H. J. (2009). *Estimation and control of lateral tire forces using steering torque*. Stanford, CA: Department of Mechanical Engineering, Stanford University.

Klomp, M., & Lidberg, M. (2006). *Safety margin estimation in steady state maneuvers*. Presented at the Proceedings of AVEC '06 – The 8th International Symposium on Advanced Vehicle Control, Taipei, Taiwan.

Li, B., Du, H., & Li, W. (2013). *A novel cost effective method for vehicle tire–road friction coefficient estimation*. Advanced Intelligent Mechatronics (AIM), 2013 IEEE/ASME International Conference on, 1528–1533.

Li, K., Misener, J. A., & Hedrick, K. (March 1, 2007). On-board road condition monitoring system using slip-based tyre–road friction estimation and wheel speed signal analysis. *Proceedings of the Institution of Mechanical Engineers, Part K: Journal of Multi-body Dynamics*, *221*, 129–146.

Li, L., W. Fei-Yue, & Qunzhi, Z. (2006). Integrated longitudinal and lateral tire/road friction modeling and monitoring for vehicle motion control. *IEEE Transactions on Intelligent Transportation Systems*, *7*, 1–19.

Lourakis, M. I. A. (2005). A brief description of the Levenberg-Marquardt algorithm implemented by levmar. *Foundation of Research and Technology, 4*, 1–6.

Luque, P., Mántaras, D. A., Fidalgo, E., Álvarez, J., Riva, P., Girón, P., … Ferran, J. (2013). Tyre–road grip coefficient assessment – Part II: online estimation using instrumented vehicle, extended Kalman filter, and neural network. *Vehicle System Dynamics*, *51*(12), 1872–1893.

Matilainen, M. J., & Tuononen, A. J. (2011). *Tire friction potential estimation from measured tie rod forces*. Intelligent Vehicles Symposium (IV), 2011 IEEE, 320–325.

Morinaga, H., *CAIS technology for detailed classification of road surface condition*. Presented at the Tire Technology Expo, Cologne, Germany.

Muller, S., Uchanski, M., & Hedrick, K. (2003). Estimation of the maximum tire–road friction coefficient. *Journal of Dynamic Systems, Measurement, and Control*, *125*, 607–617.

Nishihara, O., & Masahiko, K. (2011). Estimation of road friction coefficient based on the brush model. *Journal of Dynamic Systems, Measurement, and Control*, *133*, 041006–9.

Pacejka, H. B. (2005). Tyre brush model. In *Tyre and vehicle dynamics* (pp. 93–134). 2nd ed. Oxford: Elsevier.

Pasterkamp, W. R., & Pacejka, H. B. (1997). The tyre as a sensor to estimate friction. *Vehicle System Dynamics*, *27*, 409–422.

Rajamani, R., Phanomchoeng, G., Piyabongkarn, D., & Lew, J. Y. (2012). Algorithms for real-time estimation of individual wheel tire–road friction coefficients. *IEEE/ASME Transactions on Mechatronics*, *17*, 1183–1195.

Rajamani, R., Piyabongkarn, N., Lew, J., Yi, K., & Phanomchoeng, G. (2010). Tire–road friction-coefficient estimation. *IEEE Control Systems*, *30*, 54–69.

Ray, L. R. (1997). Nonlinear tire force estimation and road friction identification: Simulation and experiments. *Automatica*, *33*, 1819–1833.

Roweis, S. (1996). Levenberg–Marquardt optimization. Lecture notes, Department of Computer Science, University of Toronto. Retrieved from https://www.cs.nyu.edu/ ~ roweis/notes/lm.pdf

Sabbioni, E., Kakalis, L., & Cheli, F. (2010). *On the impact of the maximum available tire–road friction coefficient awareness in a brake-based torque vectoring system*. Presented at the SAE 2010 World Congress & Exhibition, Detroit, MI.

Singh, K. B. (2012). *Development of an intelligent tire based tire – vehicle state estimator for application to global chassis control* (Master's thesis). Dept of Mechanical Engineering, Virginia Tech.

Singh, K. B., Arat, M., & Taheri, S. (2012). Enhancement of collision mitigation braking system performance through real-time estimation of tire–road friction coefficient by means of smart tires. *SAE International Journal of Passenger Cars – Electronic and Electrical Systems*, *5*(2), 607–624.

Singh, K. B., Arat, M., & Taheri, S. (2013). An intelligent tire based tire–road friction estimation technique and adaptive wheel slip controller for antilock brake system. *Journal of Dynamic Systems, Measurement, and Control*, *135*, 031002–031002.

Singh, K. B., Bedekar, V., Taheri, S., & Priya, S. (2012). Piezoelectric vibration energy harvesting system with an adaptive frequency tuning mechanism for intelligent tires. *Mechatronics*, *22*, 970–988.

Singh, K. B., & Taheri, S. (October–December, 2013). Piezoelectric vibration-based energy harvesters for next-generation intelligent tires. *Tire Science and Technology*, *41*(4), 262–293.

Svendenius, J. (2003). *Tire models for use in braking applications*. Masters, Lund, Sweden: Department of Automatic Control, Lund Institute of Technology.

Svendenius, J. (2007). *Tire modeling and friction estimation*. Lund: Department of Automatic Control, Lund University.

Yasushi Hanatsuka, Y. W., & Morinaga, Hiroshi. (2013). *Method for estimating condition of road surface* (US 20130116972 A1). USA Patent.

Yasui, Y., Tanaka, W., Muragishi, Y., Ono, E., Momiyama, M., Katoh, H., … Imoto, Y. (2004). *Estimation of lateral grip margin based on self-aligning torque for vehicle dynamics enhancement*. Presented at the SAE 2004 World Congress & Exhibition, Detroit, MI.

# 19

# Fabrication and characterization of smart fabric using energy storage fibres

Ruirong Zhang[a], Yanmeng Xu[a]*, David Harrison[a], John Fyson[a], Darren Southee[b] and Anan Tanwilaisiri[a]

[a]Cleaner Electronics, College of Engineering, Design and Physical Sciences, Brunel University London, London, UK; [b]Loughborough Design School, Loughborough University, Leicestershire, UK

Fibre supercapacitors were designed and manufactured using a dip-coating method. Their electrochemical properties were characterized using a VersaSTAT 3 workstation. Chinese ink with a fine dispersion of carbon and binder was coated as the electrode material. The specific capacitance per unit length of a copper fibre supercapacitor with the length of 41 cm reached 34.5 mF/cm. When this fibre supercapacitor was bent on rods with a diameter of 10.5 cm, the specific capacitance per length was 93% of the original value (without bending). It showed that these fibre supercapacitors have good flexibility and energy storage capacity. Furthermore, the fibre supercapacitor in the fabric showed the same capacitance before and after weaving.

**Keywords:** supercapacitors; Chinese ink; energy storage fibre; gel electrolyte; flexible

## 1. Introduction

Recently, supercapacitors (also named electrochemical capacitors) have attracted much attention as energy storage devices. Because of their high power density, long cycle life and high reversibility (Kötz & Carlen, 2000), supercapacitors have a big potential to be used as high-power energy source in electrical vehicles, hybrid electric vehicles, portable electronic devices and emergency power supplies. Electrochemical energy can be stored in two ways in supercapacitors (Gómez et al., 2011; Kötz & Carlen, 2000; Seo, Yang, Kim, & Park, 2010; Shen et al., 2012). In electrical double-layer capacitors (EDLCs), there are no faradic reactions in the charge storage process; the capacitance arises from the charge separation at the electrode/electrolyte interface. Pseudocapacitors are another kind of supercapacitors, in which there is a faradic reaction in the energy storage process.

In recent years, there has been great interest in developing flexible, lightweight, low-cost and environmentally friendly energy storage devices. Supercapacitors can deliver higher power than batteries and store more energy than conventional capacitors (Seo et al., 2010). However, lots of existing supercapacitors are still too heavy and bulky for the intended applications. It is a challenge to develop highly efficient miniaturized flexible supercapacitors for future energy storage. Recently, some attempts have been made to manufacture flexible and weaveable supercapacitors (Bae et al., 2011; Cherenack, Zysset, Kinkeldei, Münzenrieder, & Tröster, 2010; Fu et al., 2012; Harrison et al., 2013; Jost et al., 2011; Le et al., 2013; Milczarek, Ciszewski, & Stepniak, 2011; Pushparaj et al., 2007; Qiu, Harrison, Fyson, & Southee, 2014; Wang et al., 2011; Zhang et al., 2014). Bae et al. (2011) developed a kind of novel fibre supercapacitor using ZnO nano-wires as electrodes, which showed a high specific capacitance of 2.4 mF/cm$^2$ and 0.2 mF/cm using polyvinyl alcohol (PVA)/$H_3PO_4$ as the gel electrolyte. Fu et al. (2012) developed a novel flexible fibre supercapacitor which consisted of two parallel fibre electrodes using Chinese ink as the active electrode materials, a helical spacer wire and an electrolyte that showed good capacitance. A low-cost and flexible mesh-based supercapacitor using commercial pen ink as active material has been developed by Shi, Zhao, Li, Liao, and Yu (2014). This device showed good electrochemical performance, with a specific capacitance of 47.4 mFcm$^{-2}$ at the scan rate of 2 mVs$^{-1}$. Pen ink (Chinese ink) as the active material of the electrodes showed good adhesive and porous morphology. A coaxial fibre supercapacitor was designed (Harrison et al., 2013; Qiu et al., 2014; Zhang et al., 2014) and manufactured in this study. This coaxial fibre structure contains five layers of materials: two electrode layers, two current collector layers and a gel electrolyte layer. The structure of this coaxial fibre supercapacitor developed was different from that of others (Fu et al., 2012; Shi et al., 2014). The coaxial fibre was all-solid and had the useful potential of being able to be woven to produce different-shaped energy storage fabrics. Herein, coaxial fibre supercapacitors, using Chinese ink as active materials, were designed, manufactured and characterized.

*Corresponding author. Email: yanmeng.xu@brunel.ac.uk

The results are discussed in this study. The particular novelty reported in this paper is the significant increase in specific capacitance (from 0.5 to 34.5 mF/cm), achieved by increasing the number of coatings on the carbon electrodes from 4 to 24 (Zhang et al., 2014).

## 2. Experiment

### *2.1. Materials*

Phosphoric acid ($H_3PO_4$, dry) and PVA (MW 146,000–186,000, >99% hydrolysed) were used without further purification. The gel electrolyte was made by dissolving 0.8 g $H_3PO_4$ and 1 g PVA in 10 mL deionized water. Copper wire (50 μm in diameter) was used as the core conductor material. Commercial carbon-based Chinese ink purchased from an art shop was used as the active coating material.

### *2.2. Design of the structure of the energy storage fibre*

Based on the working mechanism (Harrison et al., 2013; Kötz & Carlen, 2000), fibre supercapacitors have been designed (Zhang et al., 2014). As shown in Figure 1(a), the typical EDLCs consist of five layers which are two current collectors, two active layers with electrolyte and a separator with electrolyte. Following this typical structure of an EDLC, the coaxial single fibre supercapacitor with five layers was designed and manufactured (see Figure 1(b)). The central metal wire and the outer layer of silver paint are current collectors. Two active layers made of Chinese ink are separated by a gel electrolyte layer and serve as electrodes. The energy is stored by the accumulation of electrical charges at the boundary layers between the two electrodes and the electrolyte.

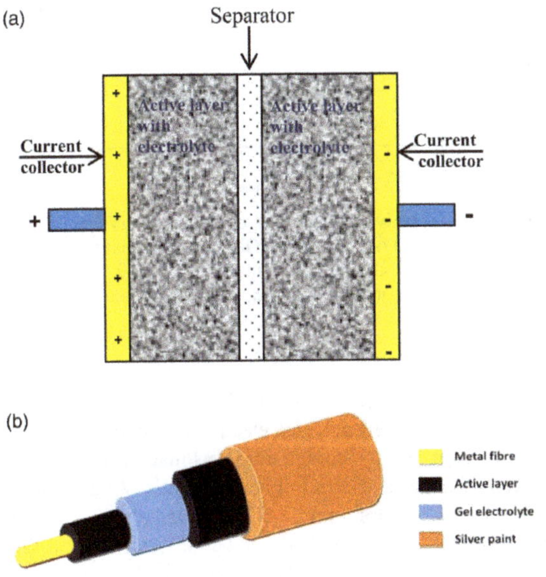


Figure 1. (a) Typical structure of an EDLC; (b) 3D schematic of four coating layers on the metal fibre.

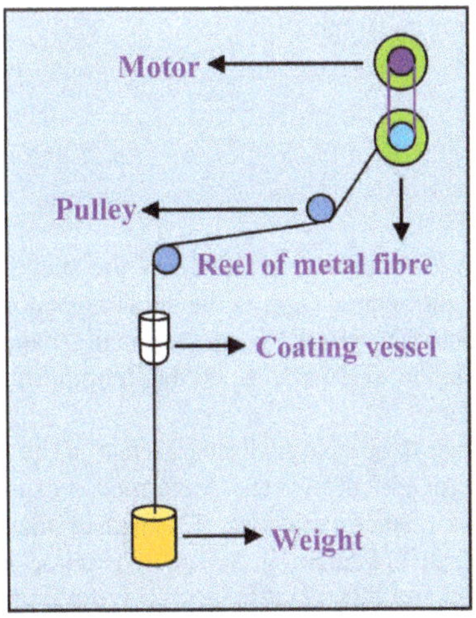


Figure 2. The schematic of coating method.

### *2.3. Manufacturing method*

Figure 2 schematically shows the experimental set-up for coating the energy storage fibre. A reel of metal wire and two pulleys are fixed horizontally onto a plate. A weight is clamped to the bottom end of the core wire to keep the wire straight and in alignment with the coating vessel. A motor with a two-direction controller was used to drive the reel axle to control the coating speed. When the coating process is carried out, a coating liquid is filled in the coating vessel. During the movement of the core metal wire through the hole, the wire drags the coating liquid with it, the solvent is evaporated and the coating materials deposit on the wire.

### *2.4. Characterization of the electrochemical properties*

The electrochemical performance of the energy storage fibre developed was studied by cyclic voltammetry (CV) and galvanostatic charge/discharge using a VersaSTAT 3 electrochemical workstation. The structure of a cross section of the supercapacitor was studied by optical microscopy, and the morphology of the active layer was characterized using a scanning electron microscope (SEM).

A CV test is carried out by applying a positive (charging) voltage sweeping at a rate of $dV/dt$ (scan rate) in a specific voltage range and then reversing (discharging) the voltage sweep polarity immediately after the maximum voltage range is achieved. The electrochemical behaviour of a supercapacitor could be evaluated based on the corresponding current response against the applied voltage by the following equations:

$$C = \frac{Q_{\text{total}}/2}{\Delta V}, \tag{1}$$

$$C_L = \frac{C}{L}, \tag{2}$$

where $C$ is the capacitance and $C_L$ is the specific capacitance per unit length. $Q_{\text{total}}$ is the supercapacitor's charge in coulombs. The charge is automatically calculated by the workstation software. $L$ is the length of the fibre supercapacitor.

A galvanostatic charge/discharge test is the most preferred DC (direct current) test performed on supercapacitors for performance evaluation. The measurement consists of two steps: (1) charging a supercapacitor at a constant current and then (2) discharging at a specific voltage range or charge/discharge time. The capacitance $C$ can be directly calculated by the following equation:

$$C = \frac{i \cdot \Delta t}{\Delta V}, \tag{3}$$

where $i$ is the discharge current in amperes (A), $\Delta t$ is the discharging time (s) and $\Delta V$ is the voltage of the discharge excluding $iR$ drop (V).

## 3. Results and discussion

Following the schematic of fibre supercapacitors (as shown in Figure 1), the energy storage fibres were made using the commercial Chinese ink as the active material.

A hand-made fibre supercapacitor with a length of 5 cm was made by a dip-coating method, and was tested using the electrochemical workstation. The typical galvanostatic charge/discharge curve record (first five cycles at a charge current of 0.5 mA) is shown in Figure 3(a). The charge/discharge curve is well defined. All the charge/discharge curves showed that the iR drop happened at the early stage, and the size of the iR drop is similar in each charge/discharge curve. This suggests that the performance of this short fibre supercapacitor is stable. Figure 3(b) shows the cyclic charge/discharge curves run on this fibre supercapacitor at different currents. It can be seen that the charge and discharge time decreases as the charge current increases, and the iR drop increases as the charging current increases as one would expect. The capacitance can be calculated using Equation (3). When the charging current increased from 0.5 to 1.5 mA, the capacitance decreased from 1.32 to 1.04 mF; the specific capacitance per length unit decreased from 0.26 to 0.21 mF/cm (Figure 3(c)). This may be caused by the slow diffusion of electrolyte ions through the double layer on the electrode surface or the slower permeation of ions through the pores in the carbon particles at higher currents. The capacitance at the charging current 1.5 mA was still 79% of that measured at the charging current of 0.5 mA. This

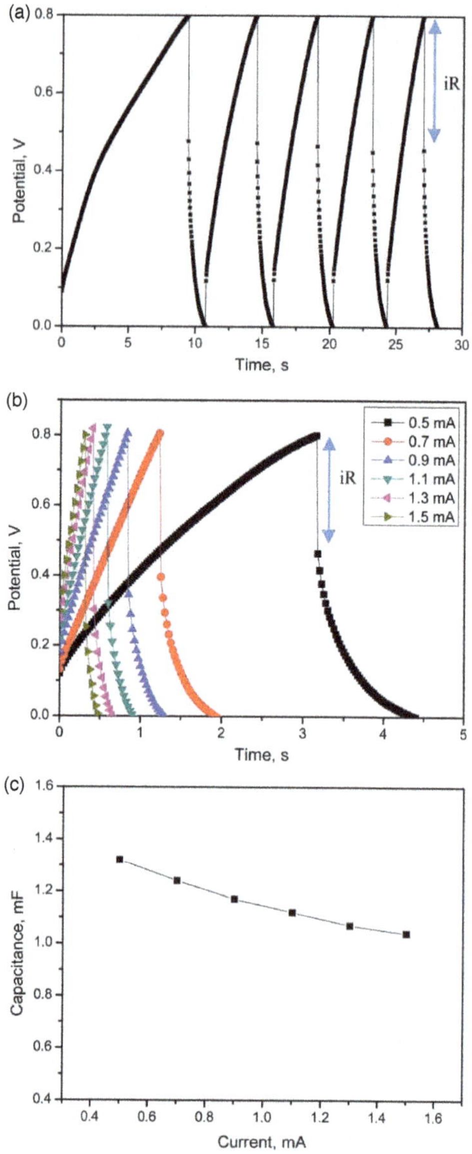


Figure 3. (a) Galvanostatic charge/discharge curve recorded at a charging current of 0.5 mA; (b) Galvanostatic charge/discharge curve at different charging currents (0.5, 0.7, 0.9, 1.1, 1.3 and 1.5 mA); (3) the corresponding capacitances calculated based on Figure 3(b).

suggests that the Chinese ink used as the activated material has reasonably well-behaved electrical properties.

A longer energy storage fibre was made using the dip-coating device as shown in Figure 2. The CV curve of the fibre supercapacitor sample with the length of 35 cm is illustrated in Figure 4. The CV curve is closed and regular; therefore, the capacitance and specific capacitance can be evaluated using Equations (1) and (2). The capacitance and specific capacitance per unit length were 18 mF

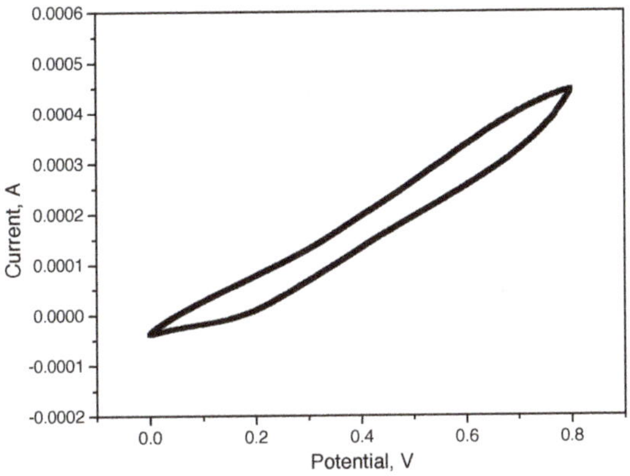


Figure 4. The cyclic voltammogram of the 35 cm long fibre supercapacitor sample.

and 0.5 mF/cm, respectively. This shows that this kind of energy storage fibre made by the dip-coating device is functional, although the capacitance was not as high as expected. This is probably because insufficient active material was coated on the fibre due to the manufacturing process.

A typical cross section of a copper fibre supercapacitor is shown in Figure 5(a). The five-layer structure can be seen clearly. The copper fibre core is easy to see. The second Chinese ink layer was not uniform. The gel electrolyte layer, the third layer, was complete but the thickness was not uniform. The fourth layer was also of Chinese ink and was also uneven. Figure 5(b) shows the typical porous structure of the Chinese ink layer surface. The diameter of the holes is about 200 nm. This micro structure will allow a good electrolyte accessibility to the inner surface of the active layer. This porous structure should also provide a large surface area for electrical charge storage.

A new fibre supercapacitor with the length of 41 cm was made based on the technical manufacturing skills gained from the experience of many laboratory trials (Figure 6). The first ink layer was deposited 24 times in this fibre supercapacitor, which was 6 times more than that of the initial 35 cm sample. The specific capacitance of the new sample was about 34.5 mF/cm, which is 69 times that of the trial 35 cm sample. Generally, when the number of coatings is increased substantially, the thickness of the ink layer increases dramatically as does the amount of the active materials in each of the electrodes. This should increase the electrical storage capacity of the fibre supercapacitor.

The flexibility of the fibre supercapacitors was also studied. The specific capacitance of the 41 cm long supercapacitor was examined when the fibre was bent on a glass rod with different curvatures. The diameters of the glass rods were 10.5, 3.0 and 1.5 cm. Figure 6 shows the CV curves of the fibre supercapacitor bent at different curvatures. It can be seen the CV curves were the same when the fibre supercapacitor was bent using the glass rods with the diameters of 1.5 and 3.0 cm. When the fibre supercapacitor was bent on the glass rod with the diameter of 10.5 cm, the specific capacitance decreased slightly to 32.2 mF/cm, which was about 93% of the original straight

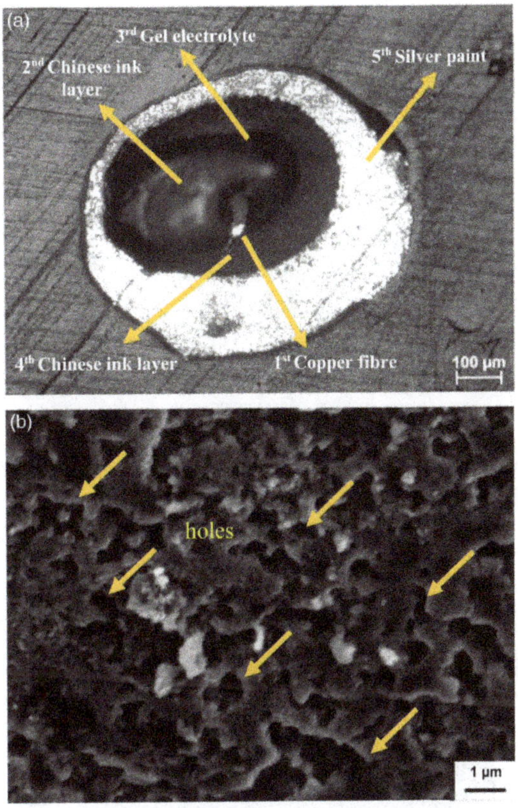


Figure 5. (a) Optical image of the cross section of a fibre supercapacitor; (b) SEM photo of the surface of the Chinese ink layer.

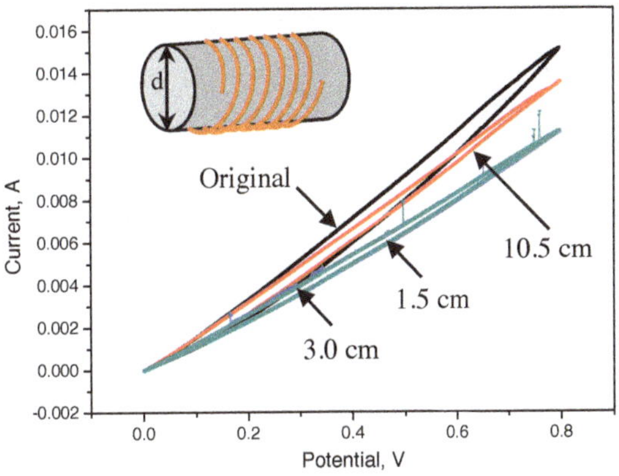


Figure 6. CV curves of the 41 cm long fibre supercapacitor at different testing conditions (straight and bent with different curvatures).

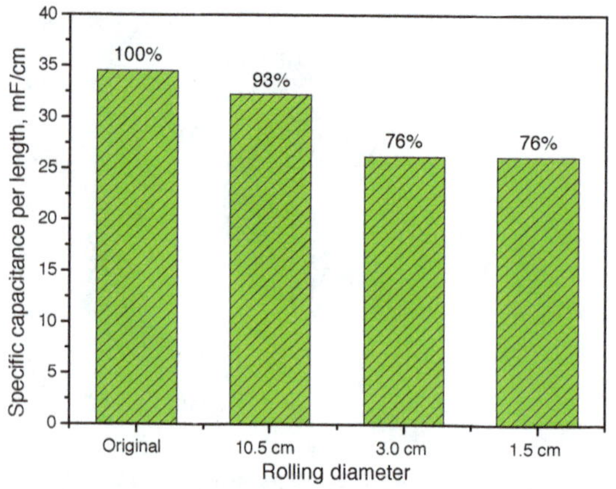


Figure 7. Comparison of the specific capacitance per length of the fibre supercapacitor being bent with different curvatures with the original straight sample.

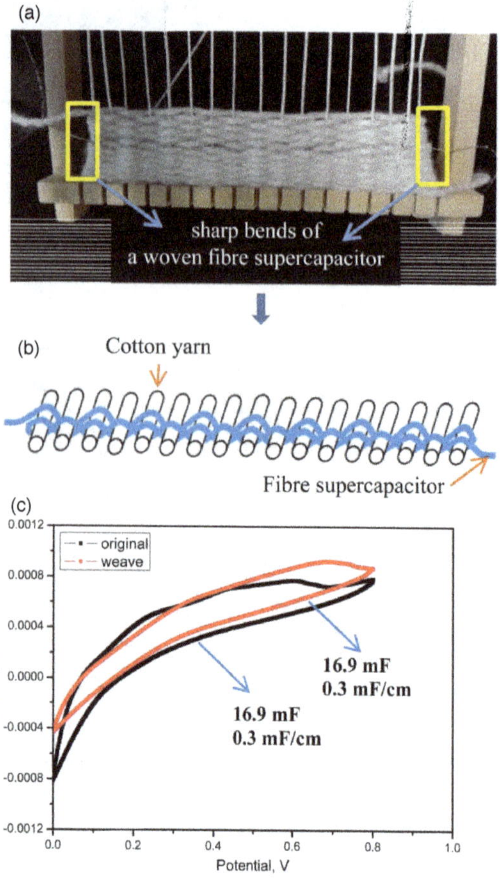


Figure 8. (a) Photo of the fibre supercapacitor woven into a fabric; (b) the schematic of the fibre supercapacitor in the fabric; (c) CV curves of the free fibre supercapacitor and that woven into fabric.

sample. When the diameter of rods decreased to 3.0 and 1.5 cm, the specific capacitance changed to 26.2 mF/cm for both cases, which was 76% of the straight sample.

The comparison of the specific capacitance of the fibre supercapacitor at different bending conditions with the original sample is shown in Figure 7. These results show that the fibre supercapacitors are able to work whilst being severely bent. As shown in Figure 8(a) and 8(b), a 50 cm long fire supercapacitor was woven with cotton yarn to produce a fabric using a simple loom. There are sharp bends along the two sides of the fabric. CV was employed to test the effect of weaving a fibre supercapacitor. Figure 8(c) shows the CV curves recorded at 30 mV/s of this fibre supercapacitor before and after being woven into the fabric. It demonstrates that even with a few sharp bends, the woven fibre supercapacitor can still work and its flexibility is more than sufficient for this use. There are a few differences between the features of these two CV curves. The CV curve of the fibre supercapacitor woven into the fabric showed a smaller slope, revealing its smaller resistance. The reason could be that the thickness of the PVA gel electrolyte layer decreases as it is being stretched in the weaving process. The capacitance of this woven supercapacitor calculated using Equation (1) is 16.9 mF, which was the same as the capacitance of the free fibre supercapacitor. This demonstrates that the charge storage capacity of this fibre supercapacitor is capable of being woven into fabric. It also shows that fibre supercapacitors are flexible enough to be woven into fabrics to make smart textiles.

## 4. Conclusions

Coaxial single fibre supercapacitors were successfully manufactured and characterized in this study. Chinese ink was used as the electrode material. A hand-made short sample proved that the fibre supercapacitor designed was functional. A dip-coating set-up was designed to produce longer fibre supercapacitors. When the coating number of carbon ink in each electrode layer increased substantially, the amount of active materials deposited on the electrodes increased dramatically. This improved the electrical storage of the fibre supercapacitor tens of times. The specific capacitance of the supercapacitors at various bending conditions was examined. The results showed that the supercapacitors manufactured using the present method kept a good electrochemical performance under severe bending conditions. The fibre supercapacitor kept the same capacitance as the original capacitance of the original free fibre when it was woven into a fabric. This indicated that the fibre supercapacitors designed and made in this study have very good flexibility. Based on these results, as the next step, the mechanical property will be tested and a new ink made with activated carbon, binder and an appropriate solvent will be used to optimize the energy storage performance. This could be used as a flexible energy storage and woven or perhaps embroidered into other fabrics to make smart textile materials.

## Disclosure statement

No potential conflict of interest was reported by the authors.

## Funding

We acknowledge funding support from the European Union Seventh Framework Programme (FP7/2007–2013) under grant agreement no. [281063].

## ORCID

*Yanmeng Xu* http://orcid.org/0000-0001-5549-1079

## References

Bae, J., Song, M. K., Park, Y. J., Kim, J. M., Liu, M., & Wang, Z. L. (2011). Fiber supercapacitors made of nanowire-fiber hybrid structures for wearable/flexible energy storage. *Angewandte Chemie International Edition*, *50*(7), 1683–1687.

Cherenack, K., Zysset, C., Kinkeldei, T., Münzenrieder, N., & Tröster, G. (2010). Wearable electronics: Woven electronic fibers with sensing and display functions for smart textiles. *Advanced Materials*, *22*(45), 5178–5182.

Fu, Y. P., Cai, X., Wu, H. W., Lv, Z. B., Hou, S. C., Peng, M., ... Zou, D. C. (2012). Fiber supercapacitors utilizing pen ink for flexible/wearable energy storage. *Advanced Materials*, *24*(42), 5713–5718.

Gómez, H., Ram, M. K., Alvi, F., Villalba, P., Stefanakos, E., & Kumar, A. (2011). Graphene-conducting polymer nanocomposite as novel electrode for supercapacitors. *Journal of Power Sources*, *196*(8), 4102–4108.

Harrison, D., Qiu, F., Fyson, J., Xu, Y., Evans, P., & Southee, D. (2013). A coaxial single fibre supercapacitor for energy storage. *Physical Chemistry Chemical Physics*, *15*, 12215–12219.

Jost, K., Perez, C. R., Mcdonough, J. K., Presser, V., Heon, M., Dion, G., & Gogotsi, Y. (2011). Carbon coated textiles for flexible energy storage. *Energy & Environmental Science*, *4*(12), 5060–5067.

Kötz, R., & Carlen, M. (2000). Principles and applications of electrochemical capacitors. *Electrochimica Acta*, *45*(15–16), 2483–2498.

Le, V., Kim, H., Ghosh, A., Kim, J., Chang, J., Vu, Q., ... Lee, Y. H. (2013). Coaxial fiber supercapacitor using all-carbon material electrodes. *ACS Nano*, *7*(7), 5940–5947.

Milczarek, G., Ciszewski, A., & Stepniak, I. (2011). Oxygen-doped activated carbon fiber cloth as electrode material for electrochemical capacitor. *Journal of Power Sources*, *196*(18), 7882–7885.

Pushparaj, V. L., Shaijumon, M. M., Kumar, A., Murugesan, S., Ci, L., Vajtai, R., ... & Ajayan, P. M. (2007). Flexible energy storage devices based on nanocomposite paper. *Proceedings of National Academy of Sciences of the United States of America*, *104*(34), 13574–13577.

Qiu, F., Harrison, D., Fyson, J., & Southee, D. (2014). Fabrication and characterisation of flexible coaxial thin thread supercapacitors. *Smart Science*, *2*(3), 107–115.

Seo, M. K., Yang, S., Kim, I. J., & Park, S. J. (2010). Preparation and electrochemical characteristics of mesoporous carbon spheres for supercapacitors. *Materials Research Bulletin*, *45*(1), 10–14.

Shen, H., Liu, E., Xiang, X., Huang, Z., Tian, Y., Wu, Y., ... Xie, H. (2012). A novel activated carbon for supercapacitors. *Materials Research Bulletin*, *47*(3), 662–666.

Shi, C., Zhao, Q., Li, H., Liao, Z., & Yu, D. (2014). Low cost and flexible mesh-based supercapacitors for promising large-area flexible/wearable energy storage. *Nano Energy*, *6*, 82–91.

Wang, K., Zou, W., Quan, B., Yu, A., Wu, H., Jiang, P., & Wei, Z. (2011). An all-solid-state flexible micro-supercapacitor on a chip. *Advanced Energy Materials*, *1*(6), 1068–1072.

Zhang, R., Xu, Y., Harrison, D., Fyson, J., Southee, D., & Tanwilaisiri, A. (2014). *Fabrication and characterisation of energy storage fibres*. Paper presented at the 20th International Conference on Automation & Computing, IEEE, Cranfield, UK, pp. 228–230.

# 20

## Interval Type-2 fuzzy logic controller design for the speed control of DC motors

Hossein Hassani and Jafar Zarei*

*Department of Electrical and Electronics Engineering, Shiraz University of Technology, Shiraz, Iran*

In this paper, an optimal interval Type-2 Fuzzy controller is designed for the speed control of DC motors. In this way, first, the importance and position of Type-2 fuzzy systems are mentioned. In addition, some properties of Type-2 operators are investigated as well as the properties of membership degree of Type-2 fuzzy sets. A comparison between different parts of Type-1 and Type-2 fuzzy systems, such as fuzzifier, fuzzy inference engine, rule-base and defuzzifier is given. Finally, an Interval type-2 Fuzzy logic controller is implemented for the speed control of DC motor for the cases of series and shunt. The motor is considered under both the load disturbances and disturbance free conditions. The obtained results for different conditions are compared in tables and figures. The results show that in the disturbance free case, both controllers have acceptable performance, however, when the system is affected by disturbance interval Type-2 controller has better performance.

**Keywords:** interval type-2 fuzzy logic systems; interval type-2 fuzzy sets; interval type-2 fuzzy membership functions; theoretical operations on type-2 fuzzy sets

### 1. Introduction

Series and shunt connected *DC* motors are widely used in control applications. These motors have relatively high torque for their weight, especially when compared to a similar size permanent magnet motors. Permanent magnet motors are linear while shunt and series motors are nonlinear. The non-linearity of the series and shunt connected *DC* motors complicates their use in applications that require automatic speed control. However, classical control approaches are not able to handle these issues. Therefore, the major challenge in the control problem of *DC* motors is overcoming to the nonlinear behavior.

Fuzzy sets (Type-1 fuzzy sets) were first introduced by Zadeh in 1965 (Zadeh, 1965). Type-1 fuzzy sets (T1FSs) are exploited to design type-1 fuzzy logic controllers (T1FLCs) (Mamdani and Assilian, 1975). The successful applications of T1FLCs are reported in many researches. For example, in control and modeling (Wang, 1999; Yager and Filev, 1994), predictions of time series (Kasabov and Song, 2002; Liao, Tang, and Liu, 2004; Versaci and Morabito, 2003) and other applications (Azar, 2010, 2012; Wang and Mendel, 1992).

In Yousef and Khalil (1995), two T1FLCs are used for the speed and current control of series *DC* motors. Control of uncertain highly nonlinear biological processes based on type-1 Takagi-Sugeno fuzzy models is investigated in Bououden, Chadli, and Karimi (2015). An adaptive fuzzy control scheme is considered to estimate the concentration in substrate at the outlet bioreactor. Fuzzy controller for electric power steering system is introduced in Saifia, Chadli, Karimi, and Labiod (2014). In order to overcome the friction and disturbances of the road, which are the main sources of nonlinearity in the electric power steering systems, Takagi-Sugeno fuzzy model is used to represent the non-linearity of the system, and stabilization conditions are established based on linear matrix inequality. In Zhao, Pawlus, Karimi, and Robbersmyr (2014), an adaptive neural-fuzzy inference systems are used in data-based modeling of vehicle crash. A robust observer for unknown input Takagi-Sugeno models is designed in Chadli and Karimi (2013).

Despite the apparent advantages of T1FSs, it has been shown that it is not able to handle the effect of uncertainties completely (Hagras, 2004, 2007; Mendel, 2001). This is because a T1FS is certain in the sense that its membership grades are crisp values. Type-2 fuzzy sets (T2FSs) were also introduced by Zadeh as an extension of T1FSs (Zadeh, 1975). T1FSs have certain membership functions, while T2FSs have membership functions that are fuzzy themselves. In the other hand, the membership grade of type-1 membership functions are crisp numbers, whereas the membership degree of type-2 membership functions can be any subset in the interval [0, 1] that are called *primary membership function* (PMF). In addition, according

*Corresponding author. Email: h.hassani@sutech.ac.ir

to any PM, there is a value that is called *secondary membership function* (SMF) that defines the probability of PMFs. Since this improvement increases the computational burden, interval type-2 fuzzy logic controllers (IT2FLCs) in which SMFs are zero or one, are developed (Liang and Mendel, 2000).

IT2FLCs are used widely because of their reasonable computations. A type-2 fuzzy controller is designed for liquid-level control in Wu and Tan (2004). Using genetic algorithm, an optimal type-2 fuzzy controller is implemented for the velocity regulation of a *DC* motor in Maldonado and Castillo (2012). In Hsiao, Li, Lee, Chao, and Tsai (2008), an interval type-2 fuzzy sliding mode controller is proposed for linear and nonlinear systems. This controller is the combination of IT2FLC and sliding mode controller. In order to reduce the effect of uncertainty associated with the available information, a T2FLC is designed to control a buck *DC–DC* converter (Lin, Hsu, and Lee, 2005).

In this paper, $A$ denotes a T1FS, and $\mu_A(x)$ shows the membership degree of $x$ in the T1FS $A$; $\tilde{A}$ denotes a T2FS and $\mu_{\tilde{A}}(x)$ denotes the membership degree of $x$ in the T2FS $\tilde{A}$; i.e. $\mu_{\tilde{A}}(x) = \int_u f_x(u)/u$, where $u \in J \subseteq [0, 1]$; $\sqcap$ is used to show the *meet* operator; and, $\sqcup$ denotes *join* operation.

The remainder of this paper is organized as follows. In Section 2, type-1 and type-2 fuzzy sets are introduced. Two examples of T2FSs are given in Section 2.1. The theoretical operations of T2FSs are presented in Section 2.2. The structure of IT2FLSs is introduced in Section 3. The simulation results are given in Section 4.

## 2. Type-1 and Type-2 fuzzy sets

This section introduces T1FSs and T2FSs. An example of T1 fuzzy set $A$ is demonstrated in Figure 1, while only integer numbers are considered in the $x$ domain. This fuzzy set can be represented as $\{\frac{0}{1}, \frac{1}{3}, \frac{1}{4}, \frac{1}{5}, \frac{0.67}{6}, \frac{0.33}{7}, \frac{0}{8}\}$, where for example $\frac{0.67}{6}$ means that the number 6 in the domain of $A$, has the membership degree of 0.67.

Membership functions are in different shapes such as Trapezoidal, Triangular, Gaussian and etc. The parameters of membership functions can be designed by experts or tuned using optimization methods (Horikawa, Furuhashi, and Uchikawa, 1992; Jammeh, Fleury, Wagner, Hagras, and Ghanbari, 2009; Wang and Mendel, 1992).

It is shown that T1FSs have some limitations to model and reduce the effects of uncertainties (Hagras, 2004, 2007; Wu and Wan Tan, 2006). Therefore, T2FS was developed as a powerful alternative for addressing these issues (Zadeh, 1974). The main disadvantage of the T2FLSs is computational burden. This is because of the type-reducer computations. Using IT2FLSs computations reduces to reasonable volume that make the implementation of IT2FLC easy.

IT2FSs as a special case of T2FSs, are currently the most widely used for their reduced computational burden. An example of IT2FS is illustrated in Figure 1. According to this figure, in T1FSs the membership degree of each element $x$ in the domain of the T1FS is a crisp number, while the membership degree of each element of $x$ in the domain of IT2FS is an interval. For example, the membership degree of 0.5 is the interval [0.9, 1].

As it is illustrated in Figure 1, IT2FSs are bounded from up and down with two T1FMs that are called *upper membership function* (UMF) and *lower membership function* (LMF), respectively. The area between *UMF* and *LMF* is called *footprint of uncertainty*.

### 2.1. Two examples of T2 fuzzy sets

*Example 1* Consider a Gaussian membership function with the mean $m$ and standard deviation $\sigma$ that can take values in $[\sigma_1, \sigma_2]$, i.e.

$$\mu(x) = e^{-\frac{1}{2}((x-m)/\sigma)^2}; \quad \sigma \in [\sigma_1, \sigma_2]. \tag{1}$$

Corresponding to each $\sigma$, we will get a different membership curve. Therefore, membership degree of each element $x$ (except $x = m$) according to each values of $\sigma$ can change.

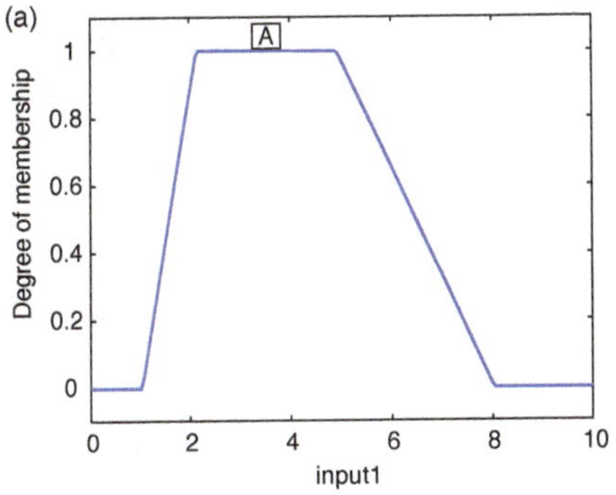


An example of T1FS.

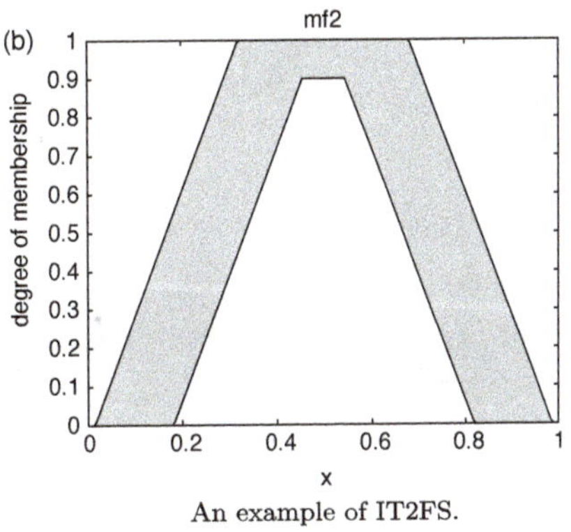


An example of IT2FS.

Figure 1. T1 and IT2 fuzzy sets. (a) An example of T1FS. (b) An example of IT2FS.

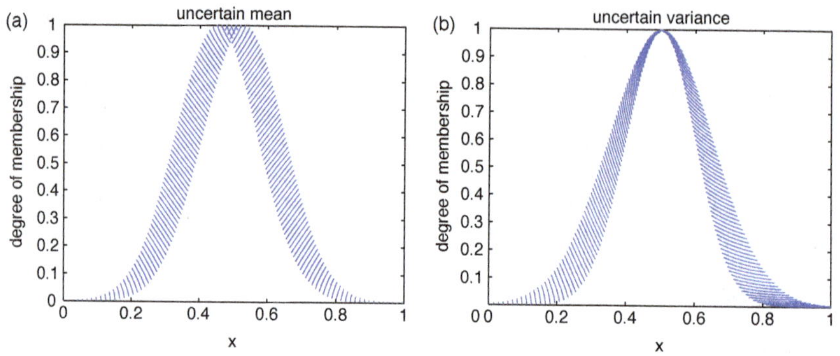


T2FSs that shows T1 Gaussian fuzz sets whit uncertainty in mean.

T2FS that shows T1 Gaussian fuzzy sets with uncertainty in standard deviation.

Figure 2. Two example of T2FSs. (a) T2FSs that shows T1 Gaussian fuzzy sets whit uncertainty in mean. (b) T2FS that shows T1 Gaussian fuzzy sets with uncertainty in standard deviation.

It is clear that the membership degree is not a crisp number, but is a fuzzy set. In this case, suppose that standard deviation can take values between $[0.1, 0.15]$ and mean is 0.5. This fuzzy set is shown in Figure 2(a).

*Example 2* As another case, consider a Gaussian membership function with standard deviation of $\sigma$ and uncertain mean $m$, that can take values in $[m_1, m_2]$, i.e.

$$\mu(x) = \mathrm{e}^{-\frac{1}{2}((x-m)/\sigma)^2}; \quad m \in [m_1, m_2]. \tag{2}$$

Consider a case that mean changes in the interval $[0.45, 0.55]$ and standard deviation is 0.12 as shown in Figure 2(a).

### 2.2. *Set theoretic operations on type-2 fuzzy sets*

Set theoretic operations are frequently used to implement an IT2FLS. Theoretical operations on type-2 fuzzy sets are defined using *Extension Principle* (Dubois, 1980; Zadeh, 1974). Consider two T2FSs $\tilde{A}$ and $\tilde{B}$ in the universe $X$. Suppose that they are characterized by their membership functions $\mu_{\tilde{A}}(x)$ and $\mu_{\tilde{B}}(x)$, respectively. These fuzzy sets can be described as

$$\mu_{\tilde{A}}(x) = \sum_i \frac{f_x(u_i)}{u_i}; \quad u_i \in J \subseteq [0, 1], \tag{3}$$

$$\mu_{\tilde{B}}(x) = \sum_i \frac{g_x(w_i)}{w_i}; \quad w_i \in J \subseteq [0, 1]. \tag{4}$$

Using *Extension Principle*, the membership degree of the union, intersection and negation of the T2FSs $\tilde{A}$ and $\tilde{B}$ can be written as (Mizumoto and Tanaka, 1976)

$$\text{Union}: \quad \tilde{A} \cup \tilde{B} \Longleftrightarrow \mu_{\tilde{A}\cup\tilde{B}}(x) = \mu_{\tilde{A}}(x) \bigsqcup \mu_{\tilde{B}}(x) = \sum_{i,j} \frac{(f_x(u_i) \star g_x(w_j))}{u_i \vee w_j}, \tag{5}$$

$$\text{Intersection}: \quad \tilde{A} \cap \tilde{B} \Longleftrightarrow \mu_{\tilde{A}\cap\tilde{B}}(x) = \mu_{\tilde{A}}(x) \prod \mu_{\tilde{B}}(x) = \sum_{i,j} \frac{(f_x(u_i) \star g_x(w_j))}{u_i \star w_j}, \tag{6}$$

$$\text{Complement}: \quad \bar{\tilde{A}} \Longleftrightarrow \mu_{\bar{\tilde{A}}}(x) = \neg\mu_{\tilde{A}}(x) = \sum_i \frac{f_x(u_i)}{(1 - u_i)}, \tag{7}$$

where $\vee$ represents maximum *t-conorm* and $\star$ is used to show the *t-norm* operator. Using Equations (5)–(7), the meet, join and negation operators can be obtained under minimum and product *t-norm*.

### 2.3. *Join and Meet operations under min t-norm*

THEOREM 1 *(Karnik and Mendel, 1998) Suppose that we have two convex, normal, type-1 real fuzzy sets F and G characterized by membership functions f and g, respectively. Let $v_0 \in R$ and $v_1 \in R$ is such that $v_0 \leq v_1$ and $f(v_0) = g(v_1) = 1$. Then, the membership functions of join and meet of F and G, using maximum t-conorm and minimum t-norm, can be expressed as*

$$\begin{aligned} \mu_{F\cup G}(\theta) &= f(\theta) \wedge g(\theta); \quad \theta < v_0, \\ &= g(\theta); \quad v_0 \leq \theta \leq v_1, \\ &= f(\theta) \vee g(\theta); \quad \theta > v_1 \end{aligned} \tag{8}$$

*and*

$$\begin{aligned} \mu_{F\cap G}(\theta) &= f(\theta) \vee g(\theta); \quad \theta < v_0, \\ &= f(\theta); \quad v_0 \leq \theta \leq v_1, \\ &= f(\theta) \wedge g(\theta); \quad \theta > v_1. \end{aligned} \tag{9}$$

Figure 3 shows Theorem 1 for Gaussian membership functions.

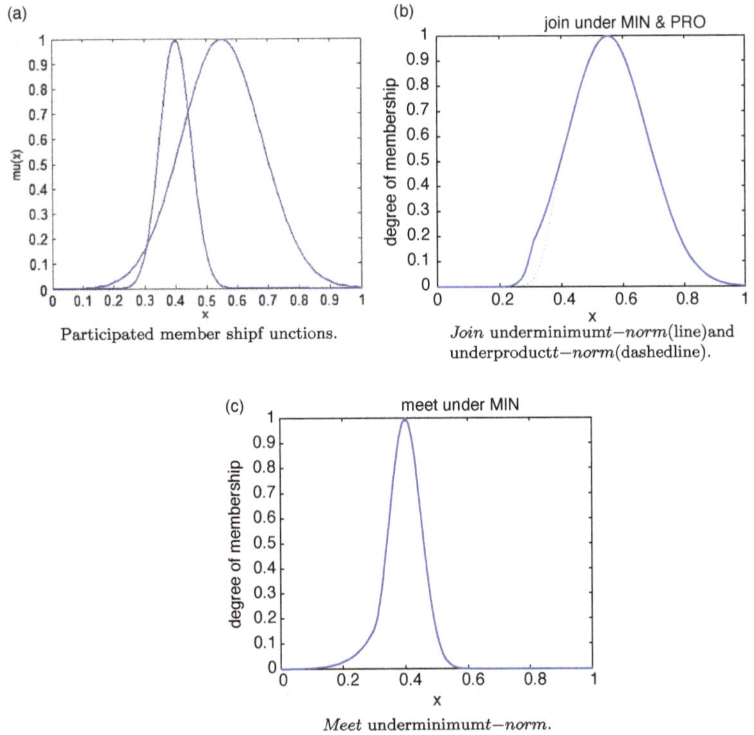


Figure 3. *Join* and *meet* operations. (a) Participated membership functions. (b) *Join* under minimum *t-norm* (line) and under product *t-norm* (dashed line). (c) *Meet* under minimum *t-norm*.

### 2.4. *Join under product t-norm*

THEOREM 2 (Karnik and Mendel, 1998) *Consider two convex, normal, type-1 real fuzzy sets F and G characterized by membership functions f and g, respectively. Let $v_0 \in R$ and $v_1 \in R$ be such that $v_0 \leq v_1$ and $f(v_0) = g(v_1) = 1$. Then, the membership functions of the join of F and G, using maximum t-conorm and product t-norm, can be expressed as*

$$\begin{aligned} \mu_{F \cup G}(\theta) &= f(\theta)g(\theta); \quad \theta < v_0, \\ &= g(\theta); \quad v_0 \leq \theta \leq v_1, \\ &= f(\theta) \vee g(\theta); \quad \theta > v_1. \end{aligned} \tag{10}$$

Figure 3 shows Theorem 2 for Gaussian membership functions.

### 2.5. *Meet under product t-norm*

There is no closed-form formula for meet operator under product *t-norm*. However, an approximation approach is suggested in Karnik and Mendel (1998). This method considers $n$ Gaussian membership functions with the mean $m_1, m_2, \ldots, m_n$ and standard deviations $\sigma_1, \sigma_2, \ldots, \sigma_n$, then,

$$\mu_{F_1 \cap F_2 \cap \cdots \cap F_n}(\theta) \approx e^{-\frac{1}{2}((\theta - m_1 m_2 \cdots m_n)/\bar{\sigma})}, \tag{11}$$

where for $i = 1, 2, \ldots, n$

$$\bar{\sigma} = \sqrt{\sigma_1^2 \prod_{i, i \neq 1} m_i^2 + \cdots + \sigma_n^2 \prod_{i, i \neq n} m_n^2} \tag{12}$$

Using Equation (12), an approximation of meet under product *t-norm* is obtained. After defining meet and join, *negation* operator is introduced as follows.

THEOREM 3 (Karnik and Mendel, 1998) *If a T1FS F has a membership function $f(v)(v \in R)$, then $\neg F$ has the membership function $f(1 - v)(v \in R)$.*

The *negation* operator for a typical MF is illustrated in Figure 4.

## 3. Interval type-2 fuzzy logic systems

General structure of IT2FLSs is illustrated in Figure 5. As it is clear, the structure is almost similar to the structure of T1FLSs. The main difference is that at least one of the FSs in the rule base is an IT2FS. Therefore, the outputs of the inference engine are IT2FSs and a type reducer is needed in order to convert them into a T1FS. Then, the T1FSs is defuzzified into a crisp number as the output of the IT2FLS. This process is shown in Figure 5.

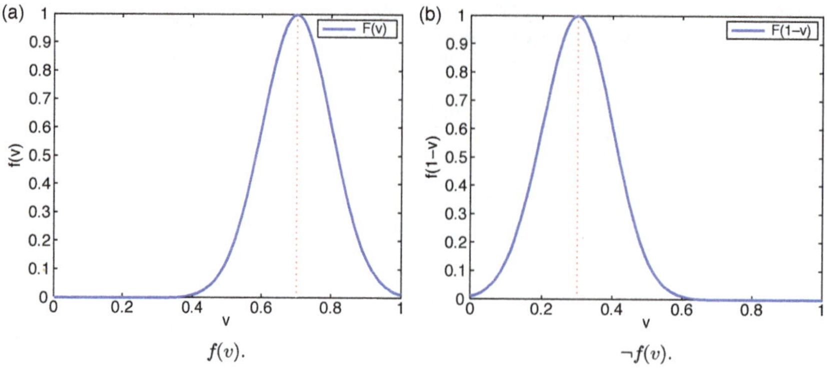


Figure 4. Negation operator. (a) $f(v)$. (b) $\neg f(v)$.

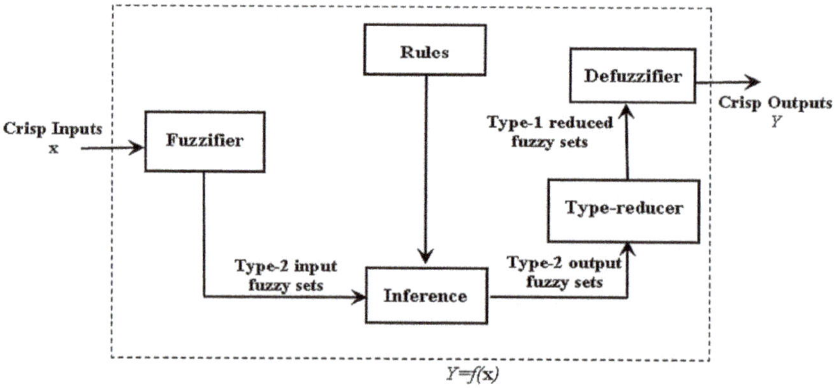


Figure 5. General structure of IT2FLSs.

A comparison between T1FLSs and IT2FLSs is given in the following. Fuzzifier is a mapping from crisp input to a fuzzy set. Singleton or non-singleton fuzzification can be considered. In the singleton case, in the input fuzzy set, only a single point has a non zero membership degree and is equal to 1. Non-singleton fuzzifier considers a fuzzy set (T1FS or T2FS) corresponding to each input point. The only difference is that in T2FLSs a type-2 non-singleton fuzzification can take place, i.e. a T2FS is consider for each input point.

In T1FLSs, the rules are in the general form of *if -then*. For example, the $l$th rule can be expressed as

$$R^l : IF\ x_1\ is\ F_1^l \text{ and } \cdots \text{ and } x_p\ is\ F_p^l,\ THEN\ y\ is\ G^l,$$

where $x_i$s are inputs, $F_i^l$s are antecedent sets, $y$ is output and $G^l$s are consequent sets.

The major difference between T1FLSs and IT2FLSs rule base refers to the nature of membership functions, and it does not have any effect on the general form of rules. Therefore, the structure of the rules will remain the same as T1FLSs, and the only difference is that in IT2FLSs membership functions are interval type-2.

Fuzzy inference engines are as a mapping from T1FSs into T1FSs. Antecedent sets in the rules connect to each other using *t-norm* (according to the intersection of sets). The input membership degree will be combined by the output membership degree according to the *sup-star* composition. These steps are also done in IT2FLSs. In addition, in T2FLSs, the computations of union and intersection of IT2FSs are needed as well as the compositions of T2 relations.

## 4. Simulation results

The design procedure of IT2FLSs with details is presented in Liang and Mendel (2000). These steps are ignored to present in this paper, however, the implementation of controllers are based on the results of Castro, Castillo, and Melin (2007) and Liang and Mendel (2000) and the structure of the proposed method has been shown in Figure 6.

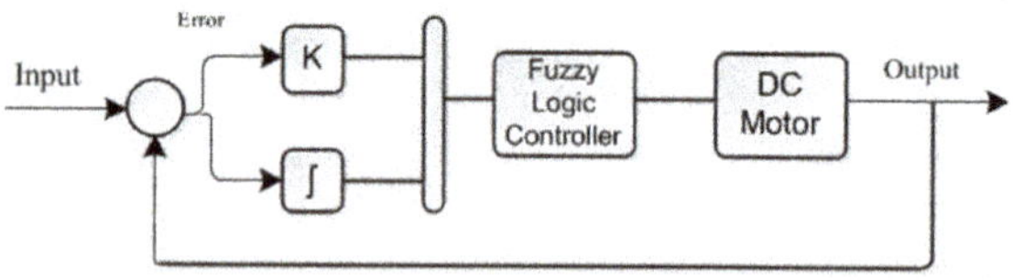


Figure 6. Type-reducer surface viewer.

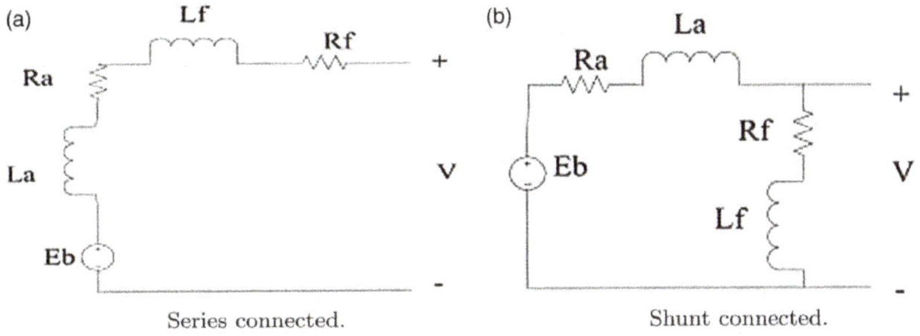


Figure 7. Equivalent circuits of *DC* motor.

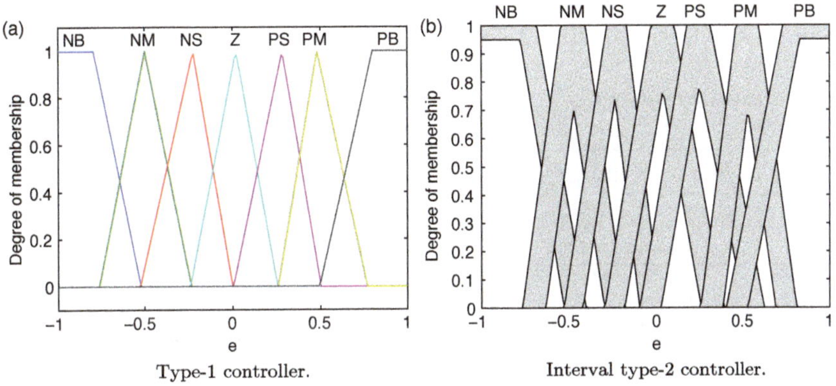


Figure 8. Membership functions of inputs and output of controllers. (a) Type-1 controller. (b) Interval type-2 controller.

In the current work, the developed toolbox for MATLAB in Castro et al. (2007) is used to implement the IT2FLC. This toolbox is considered specially for designing IT2FLCs. Like T1 toolbox of MATLAB, this toolbox has a graphical user interface, and it is also possible to design the system using the command line of MATLAB. Using this toolbox, both *Mamdani* and *Sugeno* types of fuzzy systems can be designed. This toolbox is constructed from IT2 fuzzy inference system editor, IT2MF editor, IT2 rule editor, surface and rule viewer. An interesting tool in this toolbox is the ability of type-reduced surface viewer that shows the type-reduced set.

The speed control of DC motors is under consideration. For this purpose, series and shunt connected DC motors are selected. The equivalent circuits of these motors are illustrated in Figure 7. The performance of T1FLC is considered to IT2FLC.

The controller inputs are error $e$ and its derivative $\dot{e}$. The only output of controller is the control signal $u$. Inputs and output membership functions of T1FLC and IT2FLC are demonstrated in Figure 8. As it can be seen from this figure,MFs are considered as a composition of trapezoidal

Table 1. Rule base of the proposed controller.

| $\dot{e}/e$ | NB | NM | NS | Z | PS | PM | PB |
|---|---|---|---|---|---|---|---|
| NB | NB | NB | NB | NM | NS | NS | Z |
| NM | NB | NM | NM | NM | NS | Z | PS |
| NS | NB | NM | NS | NS | Z | PS | PM |
| Z | NB | NM | NS | Z | PS | PM | PB |
| PS | NM | NS | Z | PS | PS | PM | PB |
| PM | NS | Z | PS | PM | PM | PM | PB |
| PB | Z | PS | PS | PM | PB | PB | PB |

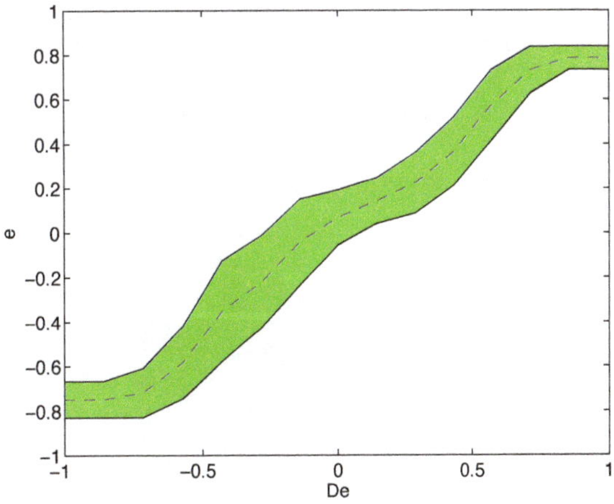


Figure 9. Type-reducer surface viewer.

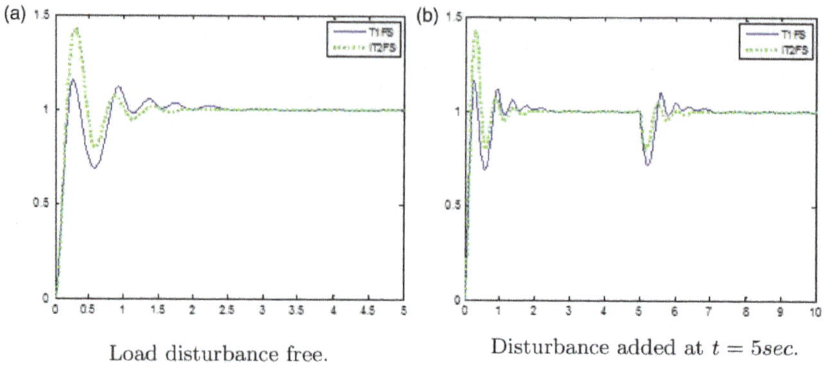


Figure 10. Performance of controllers in series connected *DC* motor. (a) Load disturbance free. (b) Disturbance added at $t = 5$ sec.

Table 2. Results of controllers for series disturbance free DC motor.

| Controller | %MP | Settling time | Rise time | IAE |
|---|---|---|---|---|
| T1C | 15.9289 | 1.4158 | 1.0644 | 0.1234 |
| IT2C | 43.3397 | 0.9563 | 1.0049 | 0.775 |

Table 3. Results of controllers for shunt disturbance free DC motor.

| Controller | %MP | Settling time | Rise time | IAE |
|---|---|---|---|---|
| T1C | 10.8405 | 1.3345 | 1.0643 | 0.1350 |
| IT2C | 13.3189 | 1.5882 | 1.0036 | 0.1161 |

and triangular shapes. Rule base for both controllers is in the same form as written in Table 1.

As mentioned in Section 3 about the type reducer, one can see in Figure 9 that the IT2MF converted to a T1MF, which is shown by dashed blue line, as the output of the type reducer. In Figure 10, the performance of both proposed controllers is compared in the speed control of series connected DC motors. The step information are collected in Table 2. From this table, IT2FLC has less *IntegralAbsoluteError* (*IAE*), *rise-time* and *settling-time* than T1FLC.

In addition, to evaluate the performance of the designed controllers, a load disturbance is inserted at $t = 5$ s. According to Figure 10, it is clear that IT2FLC has the better results than T1FLC. As it can be seen from Figure 10, IT2FLC has smaller *overshoot* and *settling-time* compared to T1FLC. In the case of shunt connected DC motor, in the load disturbance free case, T1 has smaller overshoot and settling time, but in general, using IAE index, it is clear that IT2C has better results. As shown in Figure 11, in the load disturbance case for shunt DC motor, better results from IT2C compared to T1C are obtained (Table 3). It is true that T1FLC has better performance in the disturbance free case. As the disturbance appears in the system, it is clear that IT2FLC has a better performance is sense of overshoot. In general, the system is always affected by some disturbances and as can be seen from simulation results, in such cases, IT2FLC has a better performance.

## 5. Conclusion

In this paper, T1FSs and IT2FSs were introduced and two examples of T2FSs were illustrated. Basic operations on T2FSs and T1FSs were described and formulated.

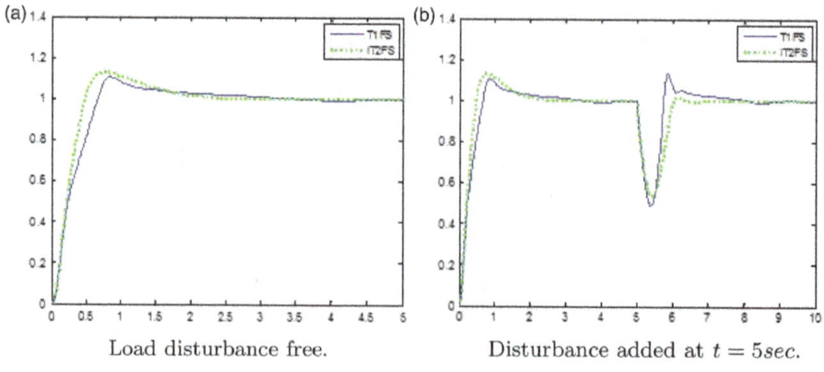


Figure 11. Performance of controllers in shunt connected *DC* motor. (a) Load disturbance free. (b) Disturbance added at $t = 5$ sec.

The main idea about IT2FLSs was presented. Differences between T1FLSs and IT2FLSs were highlighted. Next, the IT2FL toolbox introduced in order to implement an IT2FLC. Both T1FLC and IT2FLC were implemented to control series and shunt connected DC motors. In order to evaluate the performance of the controllers, the motor was also considered under load disturbance. Results showed that IT2FLC has better performance, specially in the case of appearance of disturbances. As the future work, the general T2FLC can be considered, or the controllers be optimized using evolutionary algorithms such as *PSO* or *Genetic*.

### Disclosure statement

No potential conflict of interest was reported by the author(s).

### References


Azar, A. T. (Ed.). (2010). Adaptive neuro-fuzzy systems. In *Fuzzy Systems*. (pp. 85–110). Crotia: INTECH.

Azar, A. T. (2012). Overview of type-2 fuzzy logic systems. *International Journal of Fuzzy System Applications*, *2*(4), 1–28.

Bououden, S., Chadli, M., & Karimi, H. R (2015). Control of uncertain highly nonlinear biological process based on Takagi–Sugeno fuzzy models. *Signal Processing*, *108*, 195–205.

Castro, J. R., Castillo, O., & Melin, P. (2007). An interval type-2 fuzzy logic toolbox for control applications. *Fuzzy systems conference, London, 2007. Fuzz-IEEE 2007. IEEE International*, London (pp. 1–6).

Chadli, M., & Karimi, H. R. (2013). Robust observer design for unknown inputs Takagi–Sugeno models. *IEEE Transactions on Fuzzy Systems*, *21*(1), 158–164.

Dubois, D. J. (1980). *Fuzzy sets and systems: Theory and applications*. New York: Academic Press.

Hagras, H. (2007). Type-2 flcs: A new generation of fuzzy controllers. *IEEE Computational Intelligence Magazine*, *2*(1), 30–43.

Hagras, H. A. (2004). A hierarchical type-2 fuzzy logic control architecture for autonomous mobile robots. *IEEE Transactions on Fuzzy Systems*, *12*(4), 524–539.

Horikawa, S.-i., Furuhashi, T., & Uchikawa, Y. (1992). On fuzzy modeling using fuzzy neural networks with the back-propagation algorithm. *IEEE transactions on Neural Networks*, *3*(5), 801–806.

Hsiao, M. Y., Li, T. H. S., Lee, J. Z., Chao, C. H., & Tsai, S. H. (2008). Design of interval type-2 fuzzy sliding-mode controller. *Information Sciences*, *178*(6), 1696–1716.

Jammeh, E. A., Fleury, M., Wagner, C., Hagras, H., & Ghanbari, M. (2009). Interval type-2 fuzzy logic congestion control for video streaming across IP networks. *IEEE Transactions on Fuzzy Systems*, *17*(5), 1123–1142.

Karnik, N. N., & Mendel, J. M. (1998). Introduction to type-2 fuzzy logic systems. *The 1998 IEEE international conference on fuzzy systems proceedings, 1998. IEEE world congress on computational intelligence* (Vol. 2, pp. 915–920).

Kasabov, N. K., & Song, Q. (2002). Denfis: dynamic evolving neural-fuzzy inference system and its application for time-series prediction. *IEEE Transactions on Fuzzy Systems*, *10*(2), 144–154.

Liang, Q., & Mendel, J. M. (2000). Interval type-2 fuzzy logic systems: theory and design. *IEEE Transactions on Fuzzy Systems*, *8*(5), 535–550.

Liao, S. S., Tang, T. H., & Liu, W. Y. (2004). Finding relevant sequences in time series containing crisp, interval, and fuzzy interval data. *IEEE Transactions on Systems, Man, and Cybernetics, Part B: Cybernetics*, *34*(5), 2071–2079.

Lin, P. Z., Hsu, C. F., & Lee, T. T. (2005). Type-2 fuzzy logic controller design for buck dc-dc converters. *The 14th IEEE international conference on Fuzzy systems, Reno, NV, 2005. fuzz'05* (pp. 365–370).

Maldonado, Y., & Castillo, O. (2012). Genetic design of an interval type-2 fuzzy controller for velocity regulation in a dc motor. *International Journal of Advanced Robotic Systems*, *9*(204), 204–212.

Mamdani, E. H., & Assilian, S. (1975). An experiment in linguistic synthesis with a fuzzy logic controller. *International Journal of Man–Machine Studies*, *7*(1), 1–13.

Mendel, J. M. (2001). *Uncertain rule-based fuzzy logic system: Introduction and new directions*. Upper Saddle River, NJ: Prentice Hall.

Mizumoto, M., & Tanaka, K. (1976). Some properties of fuzzy sets of type 2. *Information and control*, *31*(4), 312–340.

Saifia, D., Chadli, M., Karimi, H., & Labiod, S. (2014). Fuzzy control for electric power steering system with assist motor current input constraints. *Journal of the Franklin Institute*, in press. doi:0.1016/j.jfranklin.2014.05.007

Versaci, M., & Morabito, F. C. (2003). Fuzzy time series approach for disruption prediction in tokamak reactors. *IEEE Transactions on Magnetics*, *39*(3), 1503–1506.

Wang, L. X. (1999). *A course in fuzzy systems*. Hong Kong: Prentice-Hall Press.

Wang, L. X., & Mendel, J. M. (1992). Fuzzy basis functions, universal approximation, and orthogonal least-squares learning. *IEEE Transactions on Neural Networks*, *3*(5), 807–-814.

Wu, D., & Tan, W. W. (2004). A type-2 fuzzy logic controller for the liquid-level process. *2004 IEEE international conference on fuzzy systems, 2004. Proceedings*, Budapest, Hungary (Vol. 2, pp. 953–958).

Wu, D., & Wan Tan, W. (2006). Genetic learning and performance evaluation of interval type-2 fuzzy logic controllers. *Engineering Applications of Artificial Intelligence*, *19*(8), 829–841.

Yager, R. R., & Filev, D. P. (1994). *Essentials of fuzzy modeling and control*. New York: John Wiley & Sons.

Yousef, H. A., & Khalil, H. M. (1995). A fuzzy logic-based control of series dc motor drives. *Industrial electronics, 1995. ISIE'95, Proceedings of the IEEE international symposium, Athens* (Vol. 2, pp. 517–522).

Zadeh, L. A. (1965). Fuzzy sets. *Information and control*, *8*(3), 338–353.

Zadeh, L. A. (1974). *The concept of a linguistic variable and its application to approximate reasoning*. Springer.

Zadeh, L. A. (1975). The concept of a linguistic variable and its application to approximate reasoning – I. *Information Sciences*, *8*(3), 199–249.

Zhao, L., Pawlus, W., Karimi, H. R., & Robbersmyr, K. G. (2014). Data-based modeling of vehicle crash using adaptive neural-fuzzy inference system. *IEEE/ASME Transactions on Mechatronics*, *19*(2), 684–696.

# Permissions

All chapters in this book were first published in SSCE, by Taylor & Francis Online; hereby published with permission under the Creative Commons Attribution License or equivalent. Every chapter published in this book has been scrutinized by our experts. Their significance has been extensively debated. The topics covered herein carry significant findings which will fuel the growth of the discipline. They may even be implemented as practical applications or may be referred to as a beginning point for another development.

The contributors of this book come from diverse backgrounds, making this book a truly international effort. This book will bring forth new frontiers with its revolutionizing research information and detailed analysis of the nascent developments around the world.

We would like to thank all the contributing authors for lending their expertise to make the book truly unique. They have played a crucial role in the development of this book. Without their invaluable contributions this book wouldn't have been possible. They have made vital efforts to compile up to date information on the varied aspects of this subject to make this book a valuable addition to the collection of many professionals and students.

This book was conceptualized with the vision of imparting up-to-date information and advanced data in this field. To ensure the same, a matchless editorial board was set up. Every individual on the board went through rigorous rounds of assessment to prove their worth. After which they invested a large part of their time researching and compiling the most relevant data for our readers.

The editorial board has been involved in producing this book since its inception. They have spent rigorous hours researching and exploring the diverse topics which have resulted in the successful publishing of this book. They have passed on their knowledge of decades through this book. To expedite this challenging task, the publisher supported the team at every step. A small team of assistant editors was also appointed to further simplify the editing procedure and attain best results for the readers.

Apart from the editorial board, the designing team has also invested a significant amount of their time in understanding the subject and creating the most relevant covers. They scrutinized every image to scout for the most suitable representation of the subject and create an appropriate cover for the book.

The publishing team has been an ardent support to the editorial, designing and production team. Their endless efforts to recruit the best for this project, has resulted in the accomplishment of this book. They are a veteran in the field of academics and their pool of knowledge is as vast as their experience in printing. Their expertise and guidance has proved useful at every step. Their uncompromising quality standards have made this book an exceptional effort. Their encouragement from time to time has been an inspiration for everyone.

The publisher and the editorial board hope that this book will prove to be a valuable piece of knowledge for researchers, students, practitioners and scholars across the globe.

# List of Contributors


**B. Subathra**
ICE Department, Kalasalingam University, Madurai, Tamil Nadu, India

**S. Seshadhri**
International Research Centre, Kalasalingam University, Madurai, Tamil Nadu, India

**T.K. Radhakrishnan**
National Institute of Technology, Tiruchirappalli, Tamil Nadu, India

**Jinya Su and Wen-Hua Chen**
Department of Aeronautical and Automotive Engineering, Loughborough University, Loughborough LE11 3TU, UK

**Baibing Li**
School of Business and Economics, Loughborough University, Loughborough LE11 3TU, UK

**Rajeev Kumar Dohare and Kailash Singha**
Department of Chemical Engineering, Malaviya National Institute of Technology, Jaipur 302017, Rajasthan, India

**Rajesh Kumar**
Department of Electrical Engineering, Malaviya National Institute of Technology, Jaipur 302017, Rajasthan, India

**Ivailo Pandiev**
Department of Electronics, Faculty of Electronic Engineering and Technologies, Technical University - Sofia, Sofia, 1797, Bulgaria

**Mark Dooner and Jihong Wang**
School of Engineering, The University of Warwick, Coventry, UK

**Alexandros Mouzakitis**
Jaguar Land Rover Product Development Centre, Gaydon, Warwickshire, UK

**Jalal Javadi Moghaddam and Ahmad Bagheri**
Department of Mechanical Engineering, Faculty of Engineering, University of Guilan, PO Box 3756, Rasht, Iran

**Jacob Hostettler and Xin Wang**
Department of Electrical and Computer Engineering, Sourthern Illinois University Edwardsville, Edwardsville, IL, USA

**Tadashi Ishihara**
Faculty of Science and Technology, Fukushima University, Fukushima, Japan

**Hai-Jiao Guo**
Department of Electrical and Information Engineering, Tohoku Gakuin University, Tagajo, Japan

**Behrouz Safarinejadian and Mojtaba Yousefi**
Control Engineering Department, Shiraz University of Technology, Modarres Blvd., P.O. Box 71555-313, Shiraz, Iran

**Zhugang Ding, Guoliang Wei, Xueming Ding and Haidong Lv**
Shanghai Key Lab of Modern Optical System, Department of Control Science and Engineering, University of Shanghai for Science and Technology, Shanghai 200093, People's Republic of China

**Ajitha Thankappan**
Department of Civil Engineering, Government College of Engineering, Kannur, Kerala 670 563, India

**Amritha Sunny, Lelitha Vanajakshi and Shankar C. Subramanian**
Department of Civil Engineering, Indian Institute of Technology Madras, Chennai 600 036, India

**A. Nahvi, S. Azadi, R. Kazemi, A.R. Hatamian Haghighi and M.R. Ashouri**
Faculty of Mechanical Engineering, K.N. Toosi University of Technology, Tehran, Iran

**S. Samiee**
Faculty of Mechanical Engineering, K.N. Toosi University of Technology, Tehran, Iran

Institute of Automotive Engineering, Graz University of Technology, Graz, Austria

**Haidong Lv, Guoliang Wei, Zhugang Ding and Xueming Ding**
Department of Control Science and Engineering, University of Shanghai for Science and Technology, Shanghai 200093, People's Republic of China

**Abdo Abou Jaoude**
Department of Mathematics and Statistics, Faculty of Natural and Applied Sciences, Notre Dame University-Louaize, Lebanon

**Yue Wang, Jihong Wang, Xing Luo and Shen Guo**
School of Engineering, The University of Warwick, Coventry CV4 7AL, UK

**Junfu Lv and Qirui Gao**
Department of Thermal Engineering, Tsinghua University, Beijing 100084, People's Republic of China

**Emran Tohidi and Omid Reza Navid Samadi**
Department of Mathematics, Islamic Azad University, Zahedan Branch, Zahedan, Iran

**Peter Gawthrop**
aSystems Biology Laboratory, Melbourne School of Engineering, University of Melbourne, Victoria 3010, Australia

**S.A. Neild**
Department of Mechanical Engineering, Queens Building, University of Bristol, Bristol BS8 1TR., UK

**D.J. Wagg**
Department of Mechanical Engineering, Sir Frederick Mappin Building, University of Sheffield, Mappin Street Sheffield S1 3JD, UK

**Kanwar Bharat Singh**
Department of Mechanical Engineering, Virginia Tech, Randolph Hall (MC0238), Blacksburg, VA 24061, USA

**Saied Taheri**
Department of Mechanical Engineering, Virginia Tech, Randolph Hall (MC0238), Blacksburg, VA 24061, USA

NSF I/UCRC Center for Tire Research (CenTiRe), Virginia Tech, Blacksburg, VA, USA

**Ruirong Zhang, Yanmeng Xu, David Harrison, John Fyson and Anan Tanwilaisiri**
Cleaner Electronics, College of Engineering, Design and Physical Sciences, Brunel University London, London, UK

**Darren Southee**
Loughborough Design School, Loughborough University, Leicestershire, UK

**Hossein Hassani and Jafar Zarei**
Department of Electrical and Electronics Engineering, Shiraz University of Technology, Shiraz, Iran

# Index

www.ingramcontent.com/pod-product-compliance
Lightning Source LLC
Chambersburg PA
CBHW061946190326
41458CB00009B/2797
* 9 7 8 1 6 8 2 8 5 3 8 4 9 *